A TEXTBOOK OF ANIMAL BEHAVIOUR

A TEXTBOOK OF ANIMAL BEHAVIOUR

[For B.Sc., B.Sc. (Hons.) & M.Sc. Students]

HARJINDRA SINGH
Department of Zoology
M.M.H. College, Ghaziabad (U.P.)

Edited by
C.K. ARORA

ANMOL PUBLICATIONS PVT. LTD.
NEW DELHI - 110 002 (INDIA)

ANMOL PUBLICATIONS PVT. LTD.
H.O.: 4374/4B, Ansari Road, Darya Ganj,
New Delhi-110 002 (India)
Ph.: 23278000, 23261597
B.O.: No. 1015, Ist Main Road, BSK IIIrd Stage
IIIrd Phase, IIIrd Block
Bangalore - 560 085 (India)
Visit us at: www.anmolpublications.com

A Textbook of Animal Behaviour

First Edition, 1990
Second Edition, 1995
Reprint 1996, 1997, 1998, 1999, 2001
Third Revised & Enlarged Edition, 2003
Reprint, 2007
ISBN 81-261-1406-1

PRINTED IN INDIA

Printed at Mehra Offset Press, Delhi.

CONTENTS

1

INTRODUCTION

For many thousands of years, humans and their ancestors have been hunters and eaters of meat. The early hominids and the first *Homo erectus* practiced a crude variety of hunting. Peking man, a form of *Homo-erectus* of 4,00,000 years ago, was an accomplished hunters and users of fire and made tools from animal bones.

L.S.B. Leakey (1903-1972), an anthropologist known best for his discoveries of early hominid remains in Tanzania, proposed and tested a hunting strategy that was based on knowledge of animal behaviour—a strategy that early hunters may have used to capture a rabbit or other small animal. *Leakey* suggested that, upon sighting the prey at about fifteen metres distance, the hunter should dash directly toward the animal; a small animal initially freezes in such a situation. Within two or three metres of the prey, the hunter should turn sharply, either left or right, because the typical escape behaviour of the prey is to make a sudden dash in one direction or the other. If both prey and hunter go to the left, the hunter is upon the animal and can grab it bare-handed (as *Leakey* demonstrated), or the hunter may use a club or stone to strike the prey. If the hunter guesses incorrectly, he should stop, turn, and wait for the animal to stop so the process can be repeated, perhaps resulting in a successful capture.

Early *Homo spaiens* must have been close observers of animal habits and characteristics. They needed to be familiar with the behaviour of animals not only to know where and how to hunt their prey, but also to protect themselves from potential predators. Hunters of the Upper Paleolithic (35,000 to 10,000 years ago) probably used fire to drive animals over cliffs or into culs-de-sac where they could be slaughtered with rock or clubs. A ravine with at least 100 mammoth carcasses has been located in Czechoslovakia, and the remains of thousands of horses that had been stampeded over a cliff have been discovered in France.

Prehistoric cave paintings in France and Spain have revealed other aspects of humankind's relationship to animals. These paintings realistically depict game animals of all sorts in a way that suggests close observation of the animals at various times in their life cycles; in addition, some of the drawings are symbolic representations of actual hunting scenes. However, while early people were of necessity aware of the animals in their environment, their knowledge of animal behaviour was limited to strictly practical concerns.

What is Behaviour ?

The simplest definition of behaviour is movement, whether it is the movement of legs in walking, wings in flying, or heads in feeding. But some actions of animals, such as the honking of peacocks, which we should wish to count as behaviour, are not movements of the whole animal in the ordinary sense. The honking sound is produced as air is forced, by the contraction of muscles, out of the peacock's lungs, which causes a region of the throat to vibrate. There is movement here, of the pulmonary musculature, just as their is muscular movement when an animal feeds or walks in a more accurate sense, therefore, animal behaviour consists of a series of muscular contractions.

Naturalists had recorded incidental observations of behaviour for many centuries, but no real attempt at the scientific study of behaviour was made earlier than about a century ago. A crucial insight of the earliest workers—*Charles Darwin, Oskar Heinroth. Konrad Lorenz*—was that behaviour is orderly enough to allow that necessary criterion of all science—repeated observation. Behaviour, or muscular contractions (they noticed), comes in orderly sequences, recognizable patterns of behaviour which can be called behavioural 'units.' The same animal will produce the same pattern of movement again and again; different members of the same species will also behave in recognizable similar ways. Behaviour can only be studied because of this fact. It makes it possible for an observer to check his own evidence, and for different observers to check each others' evidence. Without recognized units of behaviour, anecdotes might accumulate, but each would be closed to criticism, and rigorous testing of theories would be impossible.

Is it correct to assert that behaviour is so regular ? Let us consider an example. We can illustrate the principle by that classic example of a behavioural unit, the 'egg retrieval response' of the greylag goose. The greylag goose, which was *Konrad Lorenz's* favourite study animal, breeds in monogamous pairs. It nests on the ground; the nest being little more than an area of grass shaped into a bowl with the edge built up, thought not enough to prevent an egg from occasionally rolling out. This is the occasion for the egg retrieval response. When a goose sees an egg just outside its nest, it enacts the following sequence of muscular movements. Standing in the nest, it first extends its neck outwards until its head is above the egg. It then puts the underside of its bill against the further side of the egg, and starts to roll it back. While rolling the egg, the goose moves its bill from side to side, to prevent the egg from slipping away to the side. The behaviour is not always effective, the egg may slip away. When it does, the goose does not immediately stop moving its bill backwards and re-establish contact with the egg. Instead it moves its bill all the way back to the nest and only then, when it again sees an egg (in fact the same one) outside the nest, does it place its bill against the egg, and try again. In other words, once started, the behavioural unit is continued until it is finished. Moreover, when *Lorenz* removed the egg from a goose while she was in the middle or rolling it back, the goose still continued and completed the sequence of movements. The two observations prove that sensory feedback, of the feel of the egg against the bill, is not needed to stimulate the continuing movement of the neck muscles.

Fig. 1.1. Egg retrieval by greylag goose.

Behaviour patterns can be recognized as units if they are performed often enough, and in similar enough form. Of course, the egg-retrieval response of different geese, and of the same goose on different occasions, will not be exactly identical. But identical repetition is not necessary to define a unit of behaviour. The behaviour pattern on different occasions only needs to be sufficiently similar to be recognizable, and then the behaviour units can be defined statistically.

Behaviour can be described inconsequentially, as a series of movements, as we have just done for the egg-retrieval response of the greylag goose, but it can also be described in terms of its consequences. 'Retrieve egg', for instance, is a consequential description; it does not mention the exact movements used, but does specify what results from them. For the general point being made here, it does not matter much which method we use to describe behavioural units; all that matters is that animals perform behaviour patterns which can be recognized by different observers. Then scientific study becomes possible.

The point can be made another way. *Darwin Heinroth, Lorenz,* and other early students of behaviour would have been educated to think biologically about the anatomical parts of animals, rather than their behaviour. They would have learned how to study parts such as limbs, urino-genital systems and circulatory systems. When they came to think about behaviour they naturally conceived and described behaviour in the form of units rather analogous to the limbs, kidneys and hearts that can be seen in the anatomy of an animal. Feeding, or courtship, in animals can be studied in the same way as their anatomy and physiology. We could ask the same kind of questions about them. Indeed, *Heinroth* and then *Lorenz* both started their work on animal behaviour by applying biological methods, such as tracing the course of development of units of behaviour in the life of an individual, and looking at different species and trying to see the equivalent units of behaviour in all of them. For it is not the greylag goose that uses the recognizable egg retrieval response. A similar sequence of muscular contractions is used by all other ground-nesting bird species to retrieve their eggs. For example, other species of geese and all species of gulls show an egg-retrieval response which is recognizably the same unit of behaviour. But modern ethology is different from Konrad Lorenz's early work, and it has moved on to more fascinating units of behaviour than the egg-retrieval response. However, this response illustrates the basis of the possibility of a scientific approach: recognizable sequences of muscular contractions and recognizable units of behaviour.

We have now said enough about the nature of the science and can move on to consider its content. Tinbergen's distinction of the four 'whys', the four different kinds of question that people ask about animal behaviour, is a useful structure for the material. But evolution and natural selection are of such all-pervasive importance that we must consider them, in relation to animal behaviour, immediately. Then, equipped with these fundamental concepts, we can move on to apply them to the behaviour of animals, as they maintain themselves against their environments, and against the society of members of their own species.

Foundations of Animal Behaviour Studies

The study of animal behaviour did not begin to emerge as a scientific discipline until the latter part of the nineteenth century. Three major developments contributed significantly to the study of behaviour: (1) publication of the theory of evolution by natural selection, (2) development of a systematic comparative method, and (3) studies in genetics and inheritance.

Theory of Evolution by Natural Selection

For several centuries European ships made voyages of exploration and discovery to all parts of the globe. Often scientists were officially attached to the voyages, as *Charles Darwin* (1809-1882) himself was. These scientists and other crew members made observations of exotic fauna and flora and brought back live and preserved specimens to zoos and laboratories in Europe, where scholars could observe, record, and speculate about the anatomy, behaviour, and interrelationships of these newly discovered species.

Like all major scientific paradigms, the theory of evolution drew on contributions by and suggestions from the work of other scientists. In 1798 *Thomas Malthus* (1766-1834), in his *Essay on the Principle of Population,* hypothesized that humans have the reproductive potential to rapidly overpopulate the world and outstrip the available food supply; that is, population increases geometrically, while the food supply increases arithmetically. The inevitable result is disease, famine, and war. Malthus's theory was an important influence on Darwin's thinking about the competition for survival among members of species. Geologist *Sir Charles Lyell* (1797-1875), a friend of Darwin's made observations of rock strata and successions of fossils which gave evidence of a process of continuous change in living material through time, an idea that was at odds with the biblical suggestion of simultaneous creation of all living things. This evidence of geological change led others to the idea that species themselves are not fixed entities. The artificial selective breeding of domesticated stocks by English farmers provided additional support for the thinking of both *Darwin* and *A.R. Wallace* (1823-1913).

Wallace's voyage to the Malay archipelago and Darwin's travels on the *Beagle* to South America and the South Pacific, combined with their other studies and the intellectual influences of the time, led each man independently to formulate the theory of evolution by natural selection. The theory states that while each animal species has a high reproductive potential, the number of animals of a species remains relatively constant over time. Thus, there is competition for survival. Variation in traits exists within animals of one species. Because some traits are more advantageous than others in the competition for survival, the operational process of natural selection occurs. Those species members that are able to survive to produce offspring contribute their characteristics to subsequent generations through their young.

The following passage from *The Origin of Species* illustrates that *Darwin* clearly recognized the central role of animal behaviour in determining the outcome of competition.

Amongst birds, the contest is often of a more peaceful character. All those who have attended to the subject, believe that there is the severest rivalry between the males of many species to attract, by singing, the females. The rock-thrush of Guiana, Birds of Paradise, and some others, congregate; and successive males display with the most elaborate care and show off in the best manner, their gorgeous plumage; their likewise perform strange antics before the females, which, standing as spectators at last choose the most attractive partner.

Darwin concluded that species were not fixed entities. The theory of evolution by natural selection was able to account for changes, through time, within a species and also for the gradual appearance of new species. Recent developments in other biological fields genetics in particular, have modified the theory of evolution by natural selection proposed by *Darwin* and *Wallace*. Today, some evolutionary biologists believe that evidence from the fossil record and genetic mechanisms supports the claim that rates of evolution vary through time (*Stanley* 1981). Change through evolution and, in particular, the appearance of new species may occur more rapidly during some time periods than at other times.

Comparative Method

George John Romanes (1848-1894) is generally credited with formalizing the use of the *comparative method* in studying animal behaviour. For Romanes the comparative method involved studying nonhuman animals to gain insights into the behaviour of humans. Romanes sought to support Darwin's theory of natural selection through his proposal that mental processes evolve from lower to higher forms and that there is a continuity of mental processes from one species to another. He argued that while people could really known only their own thoughts, they could infer the mental processes of animals, including other humans, from knowledge of their own. For Romanes the similarities between the behaviour of humans and that of other animals implied similar mental states and reasoning processes in humans and in nonhuman species. He suggested that a sequence could be constructed for the evolution of various emotional states in animals. Worms, which exhibit only surprise and fear, were placed lowest on this scale; insects were said to be capable of various social feelings and curiosity; fish showed play, jealousy, and anger; reptiles displayed affection; birds exhibited pride and terror; and finally, various mammals were credited with hate, cruelty, and shame.

Romanes's theory relied largely on inferences rather than on recorded facts; he made substantial use of anecdotes. A movement led by another Englishman, *C. Lloyd Morgan* (1852-1939), sought to counteract these faults by sing the *observational method.* Morgan's basic tenet was that only data gathered by direct experiment and observation could be used make generalizations and to develop theories.

Morgan is probably best known for his *law of parsimony,* which is now axiomatic in animal behaviour studies, "In no case may we interpret an action as the outcome of the exercise of a higher psychical faculty if it can be interpreted as the outcome of the exercise of one which stands lower in the psychological scale (*Morgan* 1896). This statement, also called *Morgan's canon,* has been interpreted to mean that in the analysis of behaviour we must seek out the simplest explanations for observed facts. Where possible, complex hypotheses should be reduced to their simplest terms to facilitate the clearest understanding of the mechanisms that control behaviour.

Theories of Genetics and Inheritance

The third development that greatly influenced research in animal behaviour was the birth of the science of genetics and the development of modern theories of inheritance. In the 1860s *Gregor Mendel* (1822-1884) reported on his findings from a series of breeding experiments on garden peas. These studies established key principles of the laws of inheritance of biological characteristics. Present day behavioural biology is based on the combination of evolutionary theory, which provides an explanation of the processes by which traits can change through time, and genetics, which explains how traits are passed from one generation to another.

Like the morphological and physiological traits, an animal's behaviour also has a genetic component. Thus, behaviour may change as a species evolves. This means that, as scientists, we can explore the genetic variation underlying various behaviours, just as other have investigated the role of genetic inheritance affecting morphology and physiology. Behaviour-genetic analysis had its beginnings in these early studies of inheritance and was then greatly expanded in the 1930s by the work of *R.A. Fisher* and others.

Behavioural Ecology

In the past five decades, a third approach to the study of animal behaviour has emerged. Behavioural ecology, with origins in zoology, examines the way in which animals interact with their living and non-living environments. Environment as used here includes *conspecifics*—animals of the same species—other living animals within the same ecological community, plants and inorganic features of the habitat.

Behavioural ecologists are concerned with both ultimate and proximate questions about behaviour. Consider, for example, the behaviour and habitat selection of a rabbit living on the edge of a large field. Ultimate constraints operating in this instance include the physiological tolerances of the rabbit. Variables like temperature and moisture level restrict the rabbit to certain environments, and its digestive system can break down only certain types of foods, primarily vegetation. Proximate factors include the rabbit's past experience with types of habitat and foods, which may lead to a particular preference for home site and diet. Ultimate factors establish the limits, and proximate factors affect the behaviour of an animal within those limits.

Behavioural ecologists, trained primarily in zoology, ecology, and related fields, are also greatly influenced by the thinking and methods of comparative animal psychology. Behavioural ecologists begin investigating a research problem in the field and define various questions, regarding, for example, population regulation or predator-prey relations. (For instance, does the predator maximize its net rate of energy intake by utilizing some of optimal foraging strategy ?) Certain aspects of the overall problem under investigation—for instance, determination of what features of the prey are most important for predator recognition and detection—may require more systernatic experiments than can be done in the field setting. The behavioural ecologist often brings specific, testable hypotheses into the laboratory, where controlled experiments are conducted, and then attempts to relate laboratory findings back to the natural field setting.

Among the investigations in behavioural ecology, those of *Emlen* (1952a, b) on bird behaviour and energy budgets, *Davis* (1951) on population biology, and *King* (1955) on social behaviour in relation to habitat were notable for the way in which they established topical areas for research work within the developing discipline. In more recent years, topics that have received particular attention by investigators include foraging strategy (*Krebs* 1978), the ecology of sex and strategies of reproduction (*Askenmo* 1984; *Clutton-Brock* et. al. 1979) and social systems in relation to ecology (*Christenson*, 1984).

Sociobiology

The most recent approach to animal behaviour reached maturity in 1975 with the publication of *Sociology : The New Synthesis* by *E.O. Wilson*. *Sociobiology* applies the principles of evolutionary biology to the study of social behaviour in animals. It is, in effect, a hybrid between behavioural biology from the ethological perspective, with an emphasis on ultimate questions, and population biology, with an ecological perspective. As with the other approaches to animal behaviour, sociobiology relies heavily on the comparative method. Similarities and differences in social systems are examined for diverse groups of animals living in a wide variety of habitats and situations to ascertain what, if any, general patterns or rules emerge to explain the social behaviour of a species.

The various theories and corollaries that constitute sociobiology have their rotos in many earlier works. Among the most significant are the writings of *Williams* (1966) on natural selection and the concept of adaptation, *Trivers* (1971, 1972) on evolutionary aspects of altruism and parental behaviour, and *Hamilton* (1964, 1971) on the genetic theory underlying the evolution of social behaviour. Studies conducted under the general heading of sociobiology range from those on altruism (*Sherman*, 1977) in ground squirrels, to research on strategies

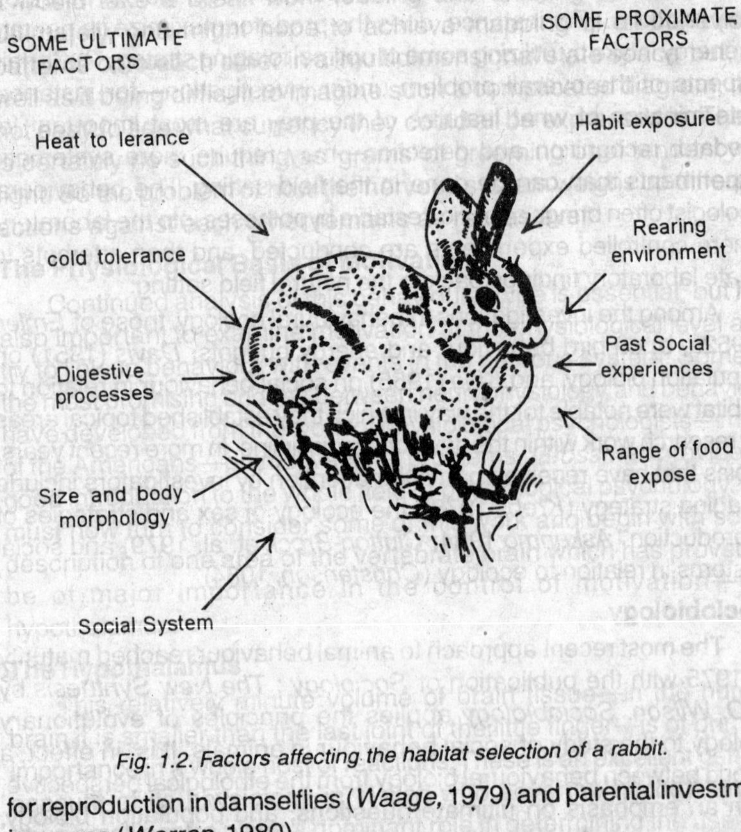

Fig. 1.2. Factors affecting the habitat selection of a rabbit.

for reproduction in damselflies (*Waage*, 1979) and parental investment in wasps (*Werren*, 1980).

Since the mid-1970s sociobiology has pervasively influenced research on animal behaviour. Faced with the challenge of devising new research questions and methods to test aspects of sociobiological theory, investigators have re-examined older data in light of new predictions. One area of prediction and hypothesis, and the source of considerable controversy, is the application of sociobiology to *Homo sapiens*. Sociobiologists argue that the same principles used to investigate the social behavior of animals can be applied to investigate the social behaviour of animals can be applied to investigate the social behaviour of humans. Other individuals argue that sociobiology is merely a form of biological determinism. A complete resolution of this controversy is probably not possible.

2

FEEDING AND ANTIPREDATOR BEHAVIOUR

The abstract principle of feeding is very simple: It is to find and catch food for yourself, while not being caught as food by another. The exact techniques used by different species are determined by the nature of their diet and predators, or, in other words by their ecology. (Ecology is the science that deals with the relations of species, and communities of species, with their environment. To understand the diversity of behavioural adaptations for feeding, therefore, we must first understand the ecological division of feeding types. For, although the methods by which different animal species obtain their food are about as diverse as the number of animal species, there are certain broad patterns to be seen. A first distinction is between herbivores and carnivores. The energy input to the world comes from the sun—plants grow on the sun's energy by photosynthesis—herbivores feed directly on the plants, carnivores feed on the herbivores. The chain has further steps. Carnivores feed on other carnivores (which make up 10 percent of the diet of the leopard, for example), scavengers feed on the dead carcases of all kinds of animal, and most important, the decomposing fungi and bacteria return the nutrients from dead bodies to the soil, from whence they can be re-used by plants.

We have some understanding of the relative proportions of the different ecological types. It was first observed early in the century that there is usually only about 10% as much energy (in the form of animal matter) at one level of the food chain as the next level down; at any one place, about 10% as much energy will be contained in the carnivores as in the herbivores. The reason was first realized by such early ecologists as the Englishman. *Charless Elton* and the American *Raymond Lindemann;* it is the energetic inefficiency of transferring food from one level to the next. For energy to be converted from zebra into lion, the lion has to chase and kill the zebra, and then digest the zebra meat. All of these processes burn up fuel and energy is, therefore lost in the transfer.

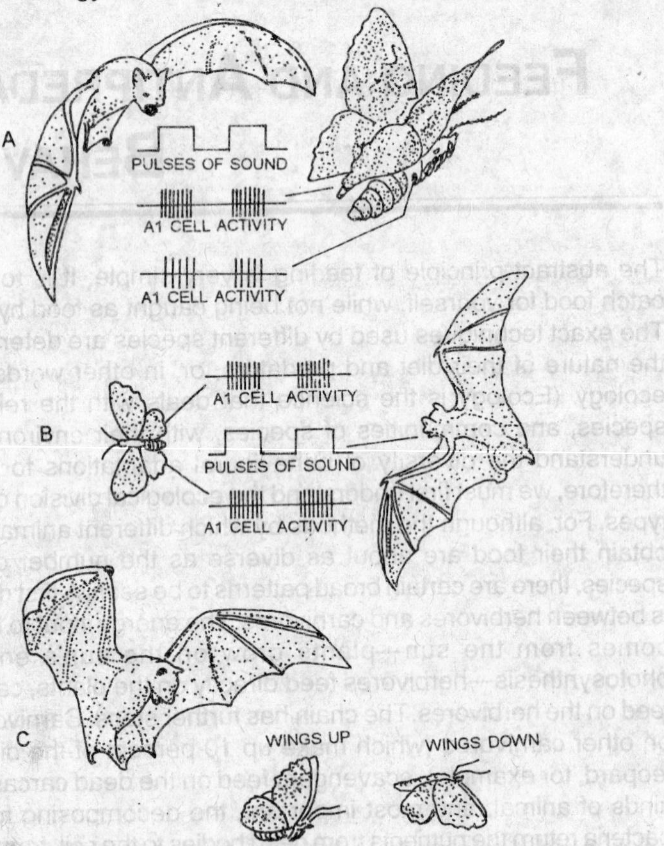

Fig. 2.1. Activity in the A1 Receptors of a moth's ears on detecting the cries of a bats located in different points in space. A—A bat is to one side of the moth; the receptor on the side closest to the predator fires sooner and more often than the shielded receptor. B—A bat is directly behind the moth; both A1 fibers fire at the same rate and time.

The behavioural adaptations of herbivore and carnivore will obviously differ. The herbivore has no difficulty in catching its food, although it may be difficult to select the most edible parts of the plants which are frequently tough, indigestible, short of nutritive value, and stuffed with poisons. The carnivore will at least have to be mobile to catch its moving prey. The distinctions at this high level are relatively unilluminating; we can better see how feeding ecology and feeding behaviour interact if we examine more closely a smaller group of species. Let us consider, for example, the feeding habits of the five main species of large herbivorous mammals the inhibit the Serengeti Plains of East Africa.

Food Preferences

All animals are selective in choosing food in their natural habitat, some more so than others. No two species living together at the same time and place eat exactly the same staple food. This is the competitive principle (*Lack* 1954, *Mayr* 1963). So feeding electivity is a subject of major biological importance, not only because is has a bearing upon studies of nutrition, but also because of its relationship to problems of interspecific competition. When animals are given a choice they show preferences. "Cattle exhibit preferences not only for certain plant species but for the same species at different stages of growth, and even for the various parts of an individual plant and for individual plants within a species" (*Hafez* and *Schein* 1962). The preferences of seed-eating birds (*Morris* 1955, *Wood-Gush* 1955, *Kear* 1962), rats (*Barnett* 1963) and primates (*Harlow* and *Meyer* 1952, *Berkson* 1962) are equally clear.

The mechanisms underlying such preferences are not well understood. In seed-eating birds ability to husk a seed quickly and efficiently may determine what seeds are selected. The choice of particular seeds for food correlates well with the size and structure of the bill as in, for example, the Galapagos finches (*Lack* 1947, *Bowman* 1961). Experience with different kinds of seeds leads each species to concentrate upon the seeds with the largest and most nutritious kernel which it can efficiently handle (*Kear* 1962). Body size and locomotor ability may play just a important a role as structure of the actual feeding equipment (*Davis* 1957, *Hinde* 1959). But within a species a strong element of individuality has been noted by many workers, indicating that a wide latitude in selection of food items is possible.

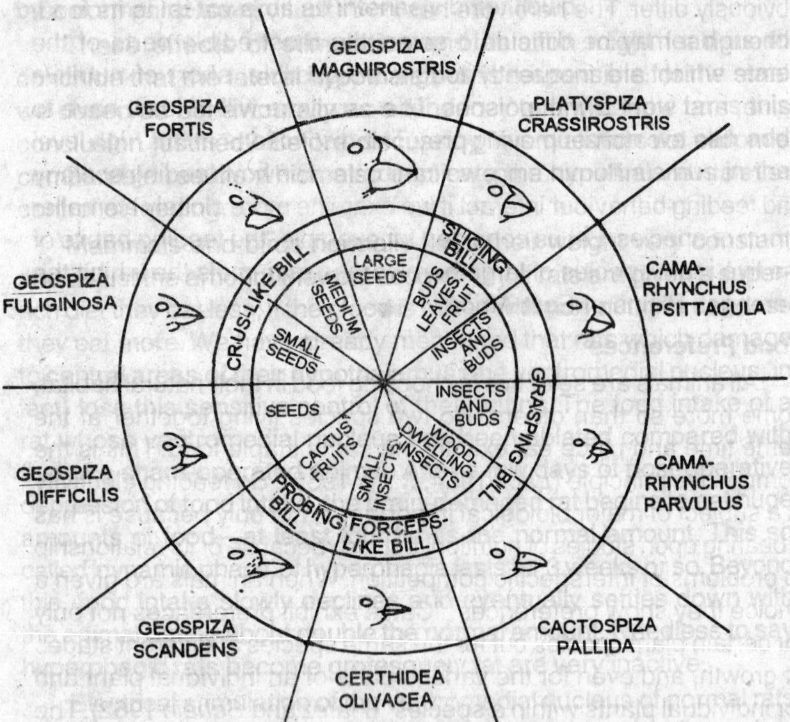

Fig. 2.2. Adaptive radiation in bill shape and diet that has taken place during the evolution of Darwin's finches.

Feeding Techniques

Natural selection has resulted in a variety of techniques for maximizing the net rate of energy intake. Some of these techniques are considered below.

Siphonophores of the phylum Coelenterata capture prey with tentacles armed with stinging cells called nematocysts. Some species have tentacles with branches and clustered nematocysts that look like zooplankton such as copepods or fish larvae. When the siphonophore moves its tentacles, it lures predators of zooplankton into the web of nematocysts and the predators become food for the siphonophore instead (*Purcell* 1980).

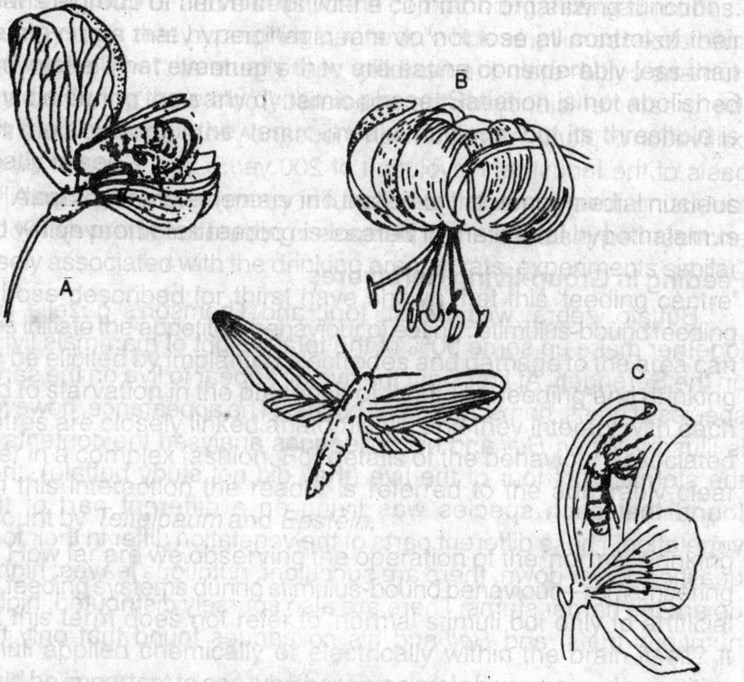

Fig. 2.3. Coadaptation of plants and their pollinators. A—A bumblebee weighs enough to trip open the locked flower of Scotch broom. B—The hawkmoth does not have to land to extract nectar from the delicately suspended Turk's cap lily flowers. Pollen adheres to its proboscis as it feeds.

Using Tools

Although the use of tools was once considered an exclusively human trait, it has evolved independently in several different species of animals. In most cases the tool is no unmodified inanimate object. The sea otter (*Enhydra lutris*), for example, holds a rock on its chest and cracks shellfish, such as mussels, against it. Similarly, the Egyptian vulture (*Neophron percnopterus*) picks up rocks and drops them on ostrich eggs. The chimpanzee (*Pan troglodytes*), however, sometimes makes a modified tool by striping leaves from a twig, which it then inserts into an ant or termite nest the insects cling to the stick, and the chimp eats those that hang on after it removes the sick. The woodpecker finch (*Cactospiza pallida*) of the Galapagos Islands uses sticks in a similar way to extract larvae from dead wood and may modify the stick by shortening it (*Lawick-Goodall* 1970; *Beck* 1980).

Tool use was important in enabling humans to compete with specialized carnivores and scavengers millions of years ago in Africa. Humans could retain the flexibility of the generalist while enjoying the greater efficiency of the specialist. Tools also permitted the cultivation of plants more than ten thousand years ago and were the basis of the Industrial Revolution of 200 years ago. Each of these tool-using developments increased our energy-gathering capability and resulted in a substantial increase in population (*Deevey* 1960).

Feeding in Group-living Herbivores

Buffalo, zebra, wildebeest, topi, and Thomson's gazelle live together make up some 90% of the total weight of mammals living on the Serengeti. At first sight the five all appear to live on the same species of grass, herbs, and small bushes. The appearance, however, is illusory. When *Bell* and his colleagues analysed the contents of the stomachs of four of the five (they did not study buffalo), they found that each species was living on a different part of the vegetation. These different parts of the vegetation differ in their food qualities: lower down, there are succulent, nutritious leaves; higher up are the harder stems. There are also sparsely distributed, highly nutritious fruits, and *Bell* and his colleagues found that only the

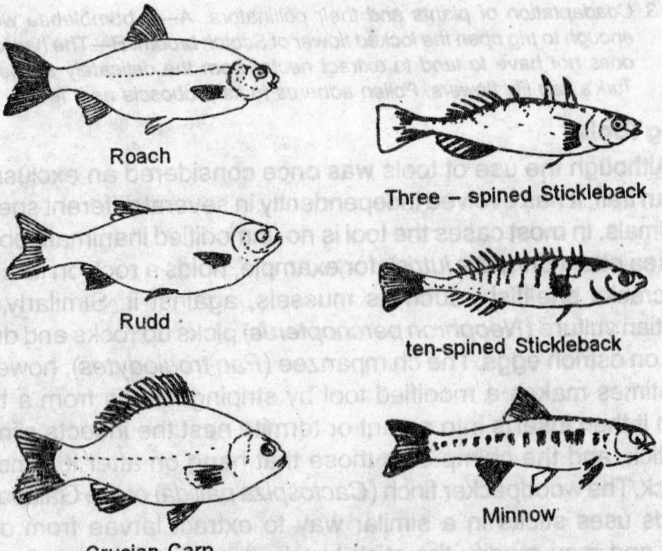

Fig. 2.4. Fish used in experiments on prey capture by pike.

Thomson's gazelle eats much of these. The other three species differ in the proportions of lower leaver and higher stems that they eat; the zebra eats the most stem material; the wildebeest the most of the leave; the topi is intermediate.

How are we to understand their different feeding preferences? The answer seems to lie in two associated differences among the species, one in their digestive systems and the other in their body sizes. The digestive systems can be divided into the non-ruminants (the zebra, which is like the horse) and the ruminants (wildebeest, topi, and gazelle, which are like the cow). Non-ruminants cannot extract much energy from the hard parts of the plant; however, this is more than made up for by the fact that food passes much more quickly through their guts. Thus, when there is only a short supply of poor quality food, the wildebeest, topi and gazelle enjoy an advantage. They are ruminants and have special structures in their stomaches (the rumen), containing special micro-organisms which can break down the hard parts of the plants. Food passes only slowly through the ruminant's guts because ruminating, digesting the hard parts, takes time. The ruminant continually regurgitates food from its stomach to its mouth to chew it up further (that is what a cow is doing when 'chewing cud'). Only when it has been chewed up almost to a liquid can the food pass through the rumen, and on through the gut. Larger particles cannot pass through until they have been chewed down to size. Therefore, when food is in short supply, a ruminant can last longer than a non-ruminant because it can extract more energy out of the same food. The differences can partially explain the eating habits of the Serengeti herbivores. The zebra chooses area where there is more low quality food. It migrates first to unexploited areas and chomps the abundant low quality stems before moving on. It is a fast-in/fast-out feeder, relying on a high throughput of incompletely digested food. By the time the wildbeest (and other ruminants) arrive, the grazing and trampling of the zebras will have worm the vegetation down. As the ruminants then set to work they eat down to the lower, leafier parts of the vegetation. All of which fits in with the differences of stomach contents with which we began.

The other part of the explanation is body size. Larger animals require more food than smaller animals, but smaller animals have a higher metabolic rate. Smaller animals can, therefore, live where there is less food, provided that it is of high energy content. That is why the smallest of the herbivores, Thomson's gazelle, lives on fruit, which is very nutritious but too thin on the ground to support a larger animal. By contrast, the large zebra lives on the masses of low quality stem material.

The differences in feeding preferences lead, in turn, to differences in migratory habits. We have seen that wildebeest follow, in their migration, the capricious pattern of local rainfall. The other species do likewise. But when a new area is fuelled, by rain, for exploitation, the mammals migrate towards it in an orderly pattern. The larger, less fastidious feeders, the zebras, move in first; the choosier, smaller wildebeest come later; and the smallest species of all, Thomson's gazelle, arrive last. The later species depend on the preparations of the earlier, for the action of the zebra fits the vegetation for the stomaches of the wildebeest and gazelle.

If we are to understand the feeding habits of the species, therefore, we must consider it in relation to the whole ecology of the species, and its relations to other species. Behaviour is an inseparable part of a whole system, made up (in this example) of body size, gut morphology, and the habits of associated species.

Social Carnivores

Among the mammalian carnivores, one member of the family Felidae and several members of the family Canidae have evolved complex social behaviour that is related to cooperative capture of prey. All of these species have been the subjects of intensive behavioural and ecological study in their natural habitats.

Lions

The lion (*Panthera leo*) lives in closed social unit and is most abundant in the grasslands and open woodlands of Africa. Schaller (1971) has studied the behaviour and ecology of this species, its competitors, and its prey.

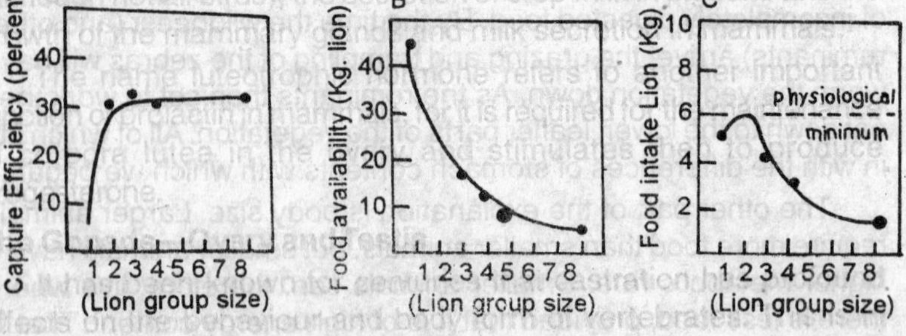

Fig. 2.5. Capture efficiency food availability, and estimated food intake as functions of lion group size for Thomson's gazelle prey.

Lions gain from cooperation in several ways. First, by hunting in groups, they increase the resource spectrum; specifically, they ad to their diet two species—buffalo (*Syncerus caffer*) and giraffe (*Giraffa camelopardalis*)—that an individual lion could never attack alone. Second, cooperative hunts are at least twice as successful as solitary ones. Third, a group can consume a captured prey more fully than can an individual. Fourth, the group can drive other predators and scavengers from the food. Schaller found that plains-dwelling prides kill less than half their food, relying on other predators, such as hyenas (*Crocuta crocuta*), to make kills for them.

Although lions hunt in groups, the amount of cooperation is not extensive. When the lions encounter a herd of prey, such as zebra (*Equus burchelli*), the pride female fan out, sometimes encircling the herd. Lions are not endurance runners and rely on a stalk-and rush tactic. Although lions are more effective when hunting upwind, Schaller found no evidence that they do so more often than would be expected by chance.

Once a kill has been made, the males, which are larger than the females but which do little hunting, may drive the females from the kill. The cubs are the last to get food. *Schaller* believes that lion populations are regulated directly by food supply through starvation of cubs. The reproductive rate remains constant, but the mortality rate of the young can be very high, exceeding 80 percent when food is scarce. Not all cub mortality is the result of starvation; strange males that move into a group may kill cubs.

Caraco and *Wolf* (1975) used Schaller's data to relate ecological factors to lion foraging group size. When hunting small prey, such as Thomson's gazelles (*Gazella thomsoni*), two lions are more than twice as efficient as one, but three or more do not better than two. Since the available food for each lion per kill decreases with increasing group size, we would expect foraging groups of about two. *Caraco* and *Wolf* went onto show that only pairs of lions can take in the minimum daily amount of food if they feed exclusively on Thomson's gazelles; Schaller reported that the mean foraging-group size ranged from 1.5 to 2, close to the expected. In hunting larger prey, such as zebras and wildebeests (*Connochaetes taurinus*), a foraging-group size of 2 was again the most efficient, but groups of 1 to 4 could still attain their minimum daily requirement; however, *Schaller* found larger than expected foraging groups of about 4 to 7. *Caraco* and *Wolf* were forced to suggest that factors other than prey size, such as the higher reproductive success documented for larger prides (lionesses share in the feeding of cubs), could influence group size.

African wild dogs

Weighing only about 18 kg (40 lbs), the African wild dog (*Lycaon pictus*), is unlikely big game predator but typically catches prey weighing as much as 250 kg (*Schaller* 1972). The dogs live in mixed-sex packs that average ten adults and, when there are young in a den, range out from it to hunt. The pack is tightly organized. Of particular interest here is the prehunt ceremony, in which dogs draw back their lips to expose their teeth, nibble and lick each other's

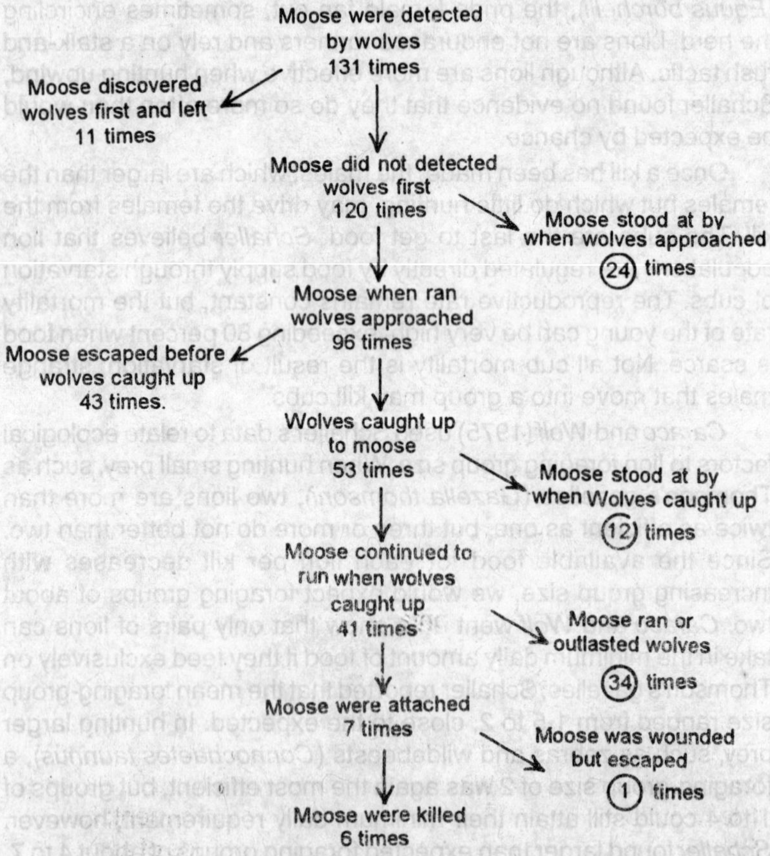

Fig. 2.6. Moose-wolf interactions.

mouths, and run whining from pack member to pack member. The greeting seems to represent ritualized food begging. Setting out on hunt, the dogs travel at a trot and fan loosely over the terrain; certain adults consistently lead some packs, but in other packs there seems to be no leader. *Schaller* (1972) describes the hunt in his field notes:

The hunting success rate for African wild dogs is extremely high; about 90 percent of all hunts result in the capture of at least one prey item.

Wolves

The social carnivore of northern temperature and arctic regions is the wolf (*Canis lupus*), which typically preys or deer (*Odocoileus* spp.), moose (*Alces americana*), buffalo (*Bison bison*), sheep (*Ovis* spp.), caribou (*Rangifer tarandus*), and elk (*Cerous canadensis*). Pack size varies greatly (*Mech* 1970), but most have fewer than eight members. Wolves hunt by traveling over their territory after consuming their previous kill. Most often they use scent; the lead animals stop when they detect the odor of prey. The group may stand nose to nose with tails wagging for a few seconds, and then all follow the leaders directly toward the prey. They may also use chance encounter and tracking. If they detect the prey at some distance, they proceed with restraint and stalk sometimes to within 10 meters of the prey. When the prey detects the wolves, it may stand its ground or flee. Once they prey is in flight, the wolves rush. Wolves must get close to their prey during the stalk and rush or they will quickly give up; if they cannot make on attack, they may chase the prey, usually for less than half a mile. Some earlier studies described wolves as coursers, but Mech has concluded that they are mainly stalkers and rushers.

One prey of the wolf, the musk ox (*Ovibos moschatus*), defends itself in an unusual way. Adults form a circle, facing outward and protecting the young inside. There is little evidence that wolves can break this defense; in areas where musk oxen are present, wolves eat mainly other prey or capture and occasional postreproductive individual.

When hunting moose, which are solitary, wolves rarely attack a prime individual (between one and six years old). Most studies of wolf predation have indicated that they are highly selective of young, old, and sick prey. Although earlier reports described the main killing tactic as hamstringing, wolves actually avoid the hooves and direct bites at the rump, flanks, shoulders, neck, and nose.

ANTI PREDATOR BEHAVIOUR

As most species are preyed upon by animals of at least some other species, avoidance of predation plays an important role in survival and reproduction.

Some defensive mechanisms operate throughout much of the life of an individual whereas others are used only when a predator is detected or when a predator attacks the individual. *Edmunds* (1947) refers to two types of mechanism, primary and secondary. Following *Kruuk* (1972) Edmunds define *primary defence mechanisms* as those which operate regardless of whether or not there is a predator in the vicinity. The behavioural components of such mechanisms may occur before any predator is detected. The mechanisms include (1) hiding in holes; (2) the use of crypsis; (3) mimicking inedible objects; (4) exhibiting a warning of danger to predators; (5) mimicking individuals in category (4); (6) timing activities so as to minimise the chance of detection by a predator; (7) remaining in a situation where any predator attack is likely to be unsuccessful because of possibilities for secondary defence; (8) maintaining vigilance so as to maximise the chance of detecting the advent of a predator. All of these mechanisms decrease the chance that a predator attack will occur. Many of them have a considerable influence upon habitat selection. For some species which are very vulnerable to predation, the avoidance of predators limits the distribution of the animals even more than do factors related to body maintenance and feeding. The section of a suitable habitat may be advantageous but which is an indicator of the presence of advantageous features, such as those which make possible feeding or predator avoidance. For example, a bird, such as a nightjar, which is camouflaged when amongst dead wood might fly over open downland towards a dark green mass of woodland because suitable hiding places are likely to be found around the edge of the wood.

Secondary defence mechanisms are defined by *Edmunds* as those which operate during an encounter with a predator. The encounter may just involve the supposed or actual detection of a predator by the prey or it may involved an attack by a predator. Some secondary defences are entirely passive. The position-spined sea urchin *Diadema* or the thick-skinned rhinoceros are adequately protected from attack. No behavioural specialisation is needed. Active forms of secondary defence include (1) exaggerating primary defence, e.g., a camouflaged animal remaining motionless; (2) withdrawal to a safe retreat; (3) flight; (4) use of a display which deters attack; (5) feigning death; (6) behaviour which deflects the attack to the least vulnerable part of the body, to an inanimate object, or to another individual; and (7) retaliation.

Both primary and secondary defence mechanisms are used by most animals and different mechanisms are often used for different predators. In an encounter with a predator, the prey animal may be able to use several alternative anti-predator behaviour. Decisions must be taken about, for example, whether to remain motionless, to flee, to display, or to fight. If one method is tried, without successful avoidance of attack, others may be tried. There are many possible strategies which involve alternative courses of action depending upon the characteristics and behaviour of the predator.

Concealment

Many prey species concerned themselves from predators by remaining in protected locations—such as burrows, crevices, or huts— when predators are present. Alternatively, concealment can be provided by the actual appearance of the animal itself. Cryptic colouration, which functions to enable animals to blend in with the background, can be found in virtually all animal taxa. Many extreme examples are found in insects, which may resembles leaves, twigs, or even bird droppings. Often, a particular behavioural pattern is associated with protective colouration such that the animal orients itself in a particular relationship to its environment, usually remaining motionless.

Camouflage

The nochid moth's defence is to seek escape in active flight. The opposite defence is to sit dead still and try to be invisible. Such is the method of camouflage in which a species evolves to resemble its background. Camouflage is of course an adaptation of appearance and colouration. But the most exquisite artistry will be wasted if the animal's behaviour is not suited to the camouflage. The world is a patch-work of different colours: the animal is only camouflaged if it settles in the right place, consider the European grass-hooper *Acrida turrita*. It comes in a green form and a yellow form. In nature the green form lives in green places and the yellow form in yellow and brown places, with rate exceptions. In a simple experiment, the German ethologist *S. Ergene* gave yellow and green grasshoopers a choice between yellow and green backgrounds. The green

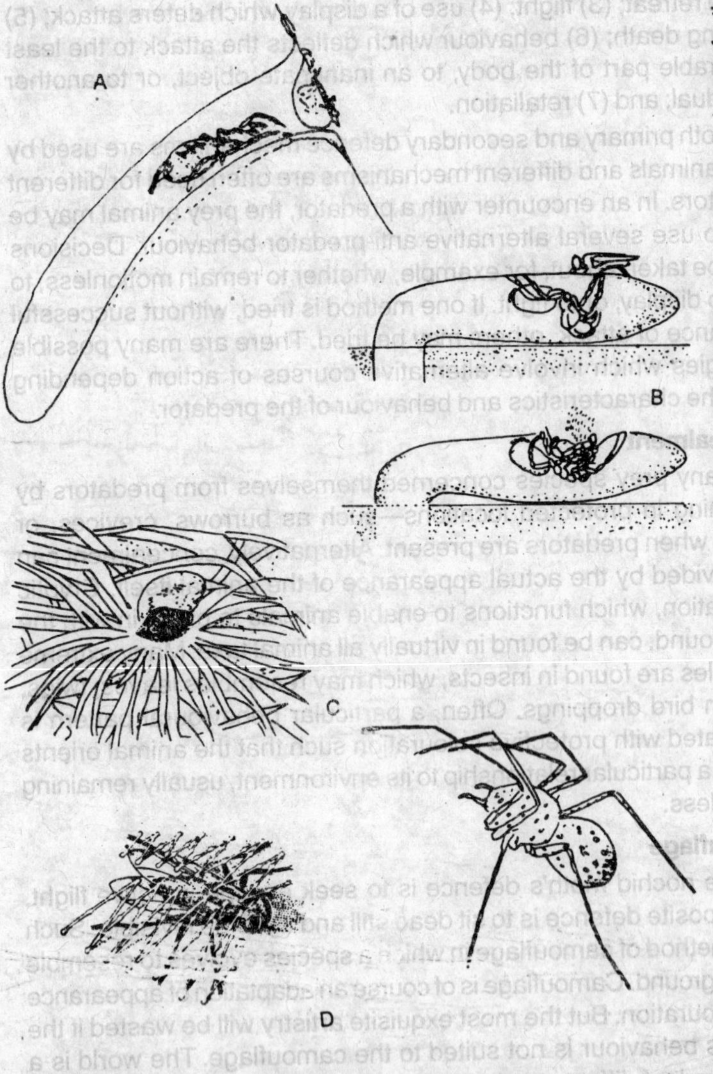

Fig. 2.7. Spiders that ambush their prey in various ways. A—A typical trapdoor spider waiting in its burrow for the approach of prey. B—*Atypus* in its silken tube in the act of capturint a fly.

grasshoopers fittingly tended to go and settle on the green backgrounds, and the yellow grasshoopers on the yellow.

The North American moth *Melanolophia canadaria* faces a more difficult problem in tining up with its background. It has striped wings and lives on the bark of trees. It much line its stripes up with the lines of the bark if it is to be camouflaged. In an experiment, *T.D. Sargent* allowed the moths to sit on cylinders that had regions of

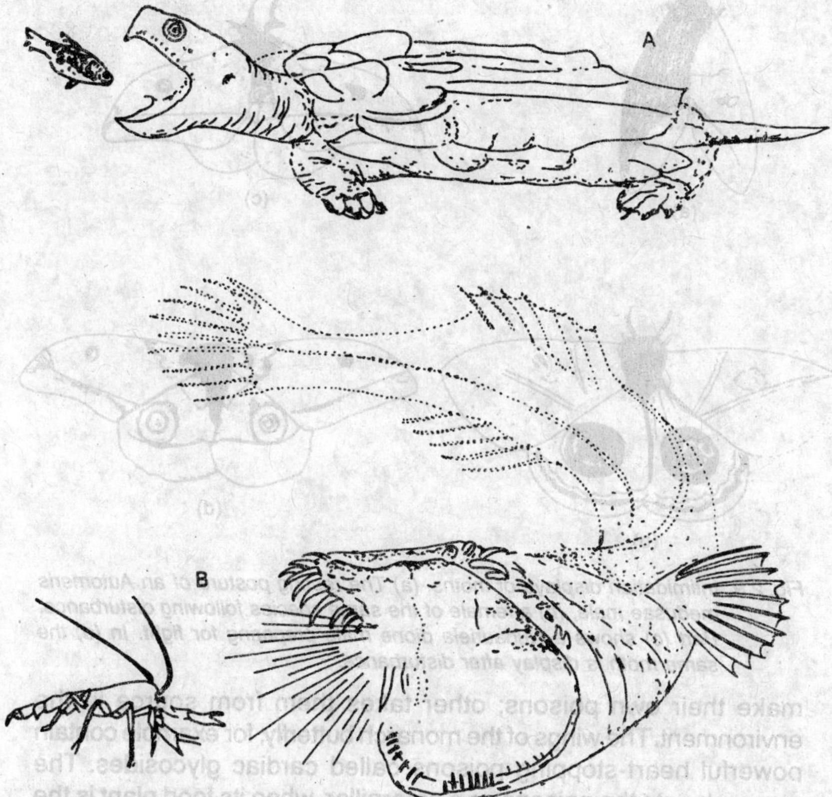

Fig. 2.8. Animals that employ lures to attract their prey. A—An alligator turtle using its tongue as a lure to draw a minnow to it. B—A deep sea anglerfish with a lighted bait in its mouth.

vertical stripes and regions of horizontal stripes. If the stripes (which were made a black tape, struck on a white surface) could be felt by the moths, then the moths usually iined up correctly. When *Sargent* covered up the stripes and surface with a transparent film, the moths no longer lined up correctly. The moths must be relying on the feel of the surface that they have to line up on. In nature they will be able to feel the stripes of then background, and ensure that they settle in a camouflaged posture.

Warning Colouration and Mimicry

Some animals protect themselves against being eaten by containing poisonous or sickening substances. Some such animals

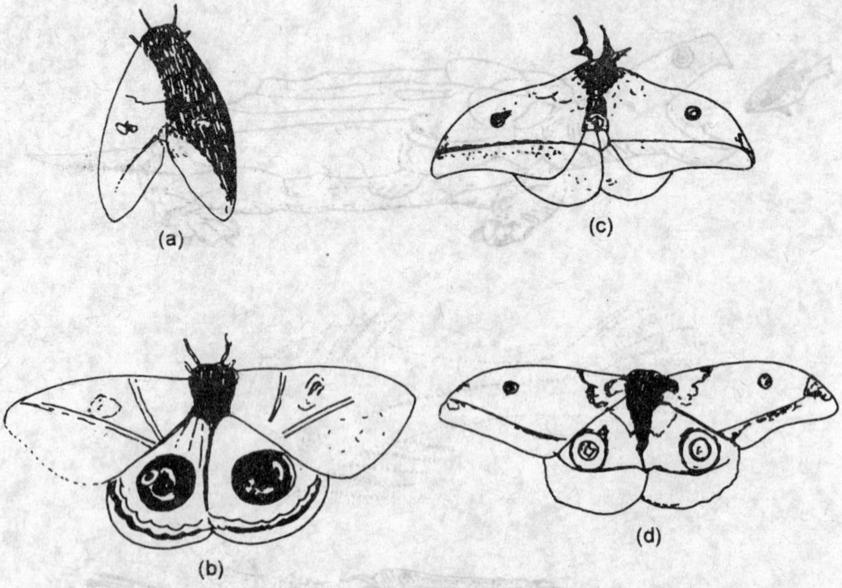

Fig. 2.9. Intimidation displays of moths. (a) The resting posture of an Automeris medusae male, (b) a female of the same species following disturbance. Part (c) shows a Nudauriela dione male preparing for fight. In (d) the same moth is display after disturbance.

make their own poisons; other takes them from source in the environment. The wings of the monarch butterfly, for example contain powerful heart-stopping poisons called cardiac glycosides. The monarch eats the poisons as a caterpillar, when its food plant is the asclepiad, or 'milkweed,' which contains cardiac glycosides. The caterpillar is not harmed by the poisons; it just stores them, and they are then retained by the adult.

For the behavioural problem of defence by poison, we must turn from the prey to the predator. If the defence is the work, the predator must learn not to eat poisonous animals, because natural selection will not favour a trait, by which an animal, after it is dead, makes its attacker sick. If the trait is to evolve, it must ensure survival. The tactic used is to enable predators to learn to recognize sick-making prey, in which case they will avoid them. It is a skill predators will readily learn. When a bird eats a monarch, as an adult or caterpillar, it will be violently sick within minutes, an experience it would learn not to repeat. The bird's problem then is to distinguish sickening from edible prey. Now, the colours of the monarch are bright gold,

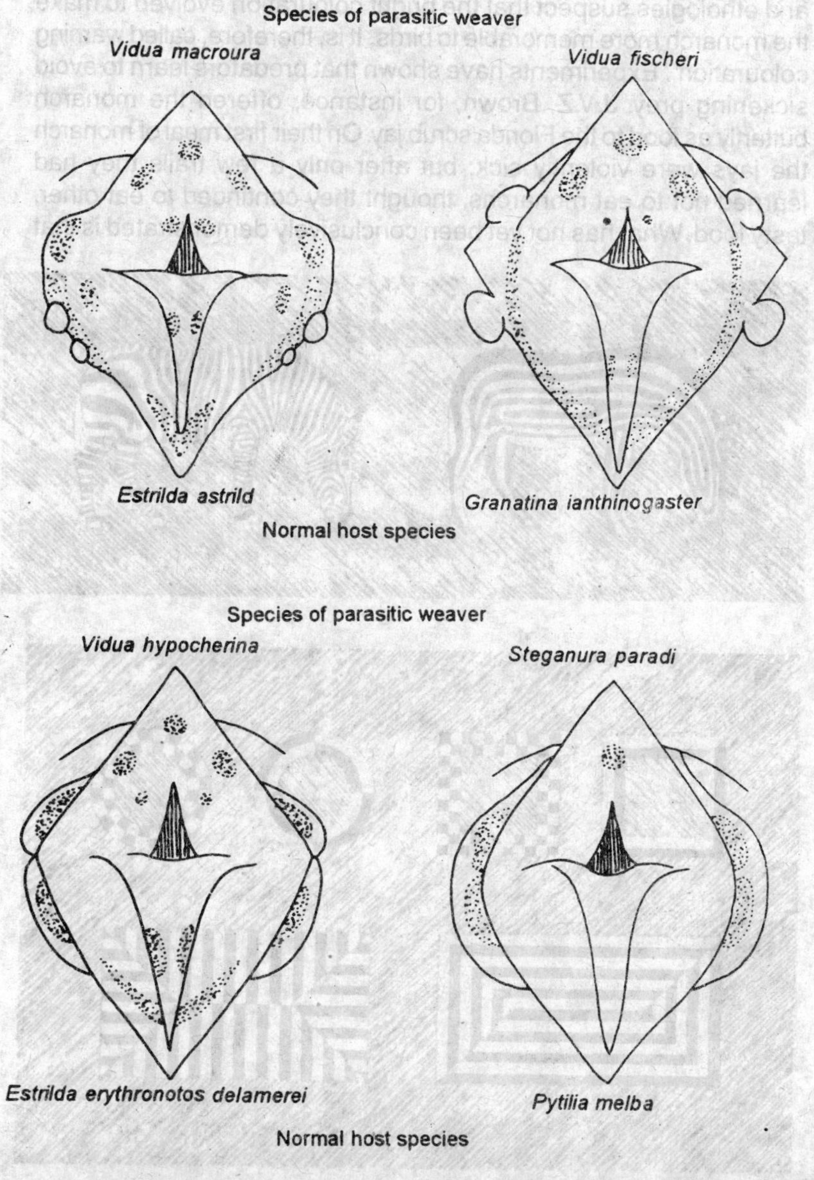

Fig. 2.10. Patterns of spots, colours, and reflecting surfaces on the palate of nestling finches are assumed to guide the parent in placing food in the semidarkness of a closed nest. Species of parasitic weavers, Vidua and Steganura, have similar nesting mouth markings to their normal hosts—here species of Estrilda, Granatina, and Pytilia.

and ethologies suspect that the bright colouration evolved to make the monarch more memorable to birds. It is, therefore, called warning colouration'. Experiments have shown that predators learn to avoid sickening prey. J.V.Z. Brown, for instance, offered the monarch butterfly as food to the Florida scrub jay. On their first meal of monarch the jays were violently sick; but after only a few trails they had learned not to eat monarchs, thought they continued to eat other, tasty food. What has not yet been conclusively demonstrated is that

Fig. 2.11. The principle of disruptive margins. Place the book in an upright position and back across the room until the zebra or "pseudozebra" disappears. The figures with disruptive edges also disappear first, illustrating the principle of the zebra pattern.

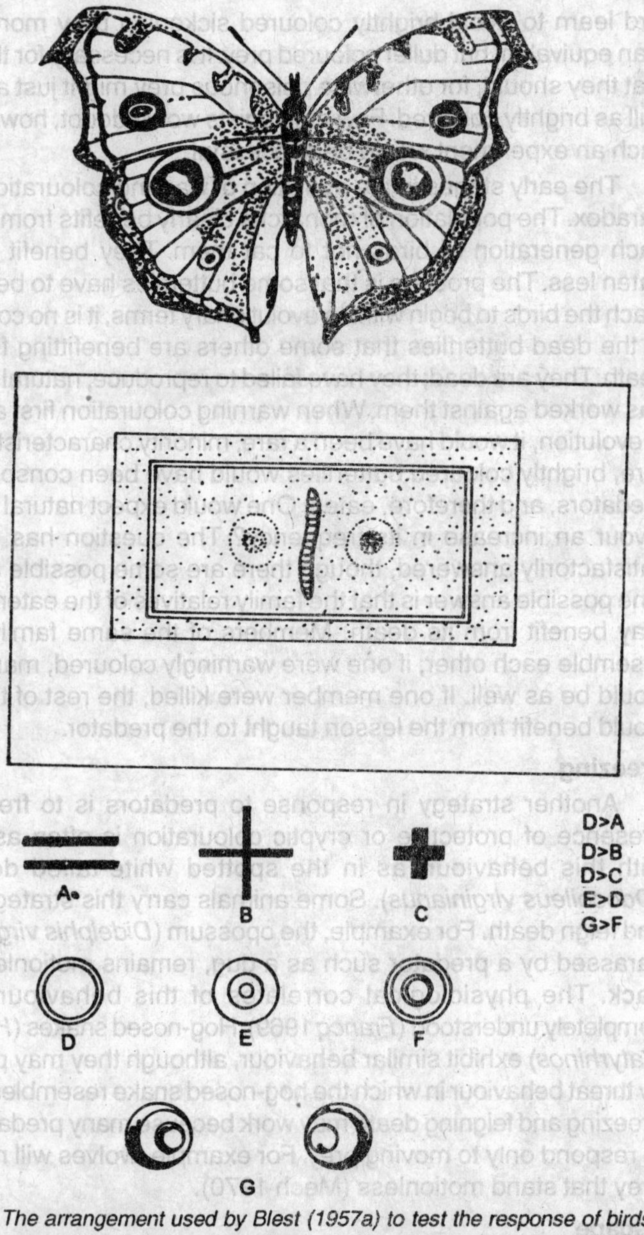

Fig. 2.12. The arrangement used by Blest (1957a) to test the response of birds to eyespot patterns such as those on the wings of the butterfly Precis almana given here. A model is projected onto a screen behind a mealworm placed on a slide where the bird has been trained to take food.

bird learn to avoid brightly coloured sickening prey more quickly than equivalent but duller coloured prey. It is necessary for the theory that they should, for otherwise poisonous prey might just as well be dull as brightly coloured. Few ethologists would doubt, however, that such an experiment would be successful.

The early stages in the evolution of warning colouration pose a paradox. The population of monarchs clearly benefits from teaching each generation of birds not to cat them. They benefit by being eaten less. The problem is that some butterflies have to be eaten to teach the birds to begin with. In evolutionary terms, it is no consolation to the dead butterflies that some others are benefitting from their death. They are dead; they have failed to reproduce; natural selection has worked against them. When warning colouration first appeared in evolution, it would have been a rare, minority characteristic. Those rare, brightly coloured butterflies would have been conspicuous to predators, and therefore, eaten. One would expect natural selection favour an increase in its frequency? The question has not been satisfactorily answered, though there are some possible answers. One possible answer is that the family relatives of the eaten butterfly may benefit from its death. Members of the same family tend to resemble each other; if one were warningly coloured, many others would be as well. If one member were killed, the rest of the family would benefit from the lesson taught to the predator.

Freezing

Another strategy in response to predators is to freeze. The presence of protective or cryptic colouration is often associated with this behaviour, as in the spotted white-tailed deer fawn (*Odocoileus virginianus*). Some animals carry this strategy further and feign death. For example, the opossum (*Didelphis virginiana*), if harassed by a predator such as a dug, remains motionless on its back. The physiological correlates of this behaviour are not completely understood (*Francq* 1969). Hog-nosed snakes (*Heterodon platyrhinos*) exhibit similar behaviour, although they may precede it by threat behaviour in which the hog-nosed snake resembles a cobra. Freezing and feigning death may work because many predators seen to respond only to moving prey. For example, wolves will not attach prey that stand motionless (Mech 1970).

Escape

Rapid and agile locomotion provides the best and probably the most common means of escaping from predators. Various species supplement their locomotor escape patterns with displays that function to distract or startle a potential predator. Still others may adopt a state of tonic immobility ("play possum") as a means of reducing the likelihood of attack (*Gallup,* 1974).

Social Antipredator Behaviour

Confusion Effect

One of the main benefits of sociality is though to be the protection gained, as stated in the edge, "safety in numbers". Living in flocks herds, or schools increases the chance of spotting predators. When approached or attacked by predators, groups usually become more compact. Fish schools, however, usually form a vacuole around a predator. Tightly packed prey make a difficult target for the predator which must select an individual toward which it can aim its attack. The predator tries to isolate an individual from the group; any individual that strays from the group or is in any way different from the rest is likely to be attacked. For example, when experiments presented hawks with set of ten mice, the hawks preferred the oddly coloured mouse (*Mueller* (1971).

A primary means of protection is the confusion effect, described by *Miller* (1922). A variety of predators that attack swarming prey exhibit a lower prey capture rate, a longer period of hesitation, and more irrelevant behaviour than when they attack solitary prey. Factors that enhance the confusion effect are swarm number, swarm density, and the uniformity in appearance of swarm members (*Milinski* 1979). The mechanism producing the confusion effect is not understood, but a prenator seems to be more easily frightened when it feeds in a dense swarm of prey; perhaps the predator finds it harder to detect other animals that might prey on it. In laboratory experiments with three-spined sticklebacks (*Gasterosteus aculeatus*) feeding on water fleas (*Daphina*), fish with experience in feeding on dense swarm fed more efficiently when tested with dense swarms that did those experienced only in feeding on low-density populations (*Milinski* 1979).

Detection

We might expect to find groups of mixed species if the combination of different sensory capabilities is advantageous. Baboons and ungulates are frequently together on the African Plains. The aboreal habits and keen vision of baboons and the well-developed olfactory systems of ungulates probably increase the distance at which prey can be detected. It is clear that these different species recognize and respond to each other's alarm calls.

Social living increases the likelihood that predators will be detected and allows group members to spend mor time in other activities. Individuals in large, dense colonies prairie drops (*Cynomys* spp.) spend less time in alert postures than do those in small colonies (*Hoogland* 1979). Once predators are detected, alarm calls alert neighbours to danger. *Marler* (1955) has argued that in some birds the alarm call is difficult to locate because it is a high-pitched note in a narrow-frequency range. The alarm-calling ground squirrels studied by *Sherman* (1977) emitted calls that were easily located, and calling squirrels were more likely to be caught than non-callers. Close relatives of the caller were likely to be close by and thus could benefit from this seemly altruistic behaviour.

The Development of Anti Predator Behaviour

A baboon which is attacked by a leopard but which survives, learns something about the speed and attack methods of leopards. Future encounters with leopards will be different as a consequence of the experience. Field studies of the improvement in techniques of antipredator behaviour are difficult, but many laboratory studies of avoidance conditioning minic certain aspects of this type of behavioural change. Suppose that a rat in a cage learns to avoid shock by pressing a set of levers in a particular sequence when a certain combination of stimuli is presented. That rat may be using abilities which would also be of use were if faced with successive exposures to a predator. The effects of experience are often easiest to study in developing animals since some aspects of previous experience can then be controlled.

In species in which parental care is prolonged, the most effective anti-predator behaviour which young animals can show is often to recognise and maintain contact with the parents and the place where they are put by the parents. The overriding importance of finding the parent leads to some apparently anomalous responses to predators. *Lorenz* (1935) describes how newly hatched precocial birds, such as goslings, show no avoidance behaviour to a variety of moving objects including, for example, man, a goose predator. Schaller and *Emlen* (1962) pushed objects towards the young of ten species of precocial birds and found very little avoidance reaction at 10 h of age. Avoidance in their experiments, and in those of *Bateson* (1964) on chicks exposed to novel moving objects in a runway, increased considerably during the first week of life but changed little after this time. The startle responses shown by young domestic chicks in their home pen, When a light-bulb on the wall was illuminated for the first time, increased in magnitude and duration with increasing age from 1 to 10 days (*Broom* 1969). The change with age of these studies was greatest during the first three days. As the animal gets older, it learns the characteristics of its immediate environment. If the degree of novelty of any environmental change depends upon a comparison with a 'model' of the familiar surroundings (*Salzen* 1962), novelty will increase whilst the model is established and hence responsiveness increases with age (*Broom* 1969).

A system which results in the young animal showing anti-predator behaviour to any novel change in its surroundings will operate when a real predator appears. Responsiveness to potential predators by young animals does not depend solely upon degree of novelty. It is also affected by the nature of the sensory input which results from the presence of the predator. Bodily damage, a looming object, or a loud noise will elicit more extreme anti-predator behaviour than the sight of small objects, the sound of quieter noises, etc. (*Hebb*, 1946). The existence of perceptual analysers in the brain which enable an individual to recognise dangerous events, for example a looming object (*Ewen* 1971). Some of these analysers are known to develop in a different way according to specific visual experience by others probably depend on general environmental factors such as light and metabolic stall at key times during embryological development. The average individual, therefore, would have some predator detection mechanism whether or not there had been previous visual experience of the predator. An elaborate and specific possibility is a detector for flying bird of prey. *Lorenz* (1939) and *Goethe* (1940) reported that young precoeial birds did not respond to flying geese but showed a flight response when a hawk or falcon passed overhead. Experiments by *Schleidt* (1961) showed that the result could have been due to habituation to the geese but not to hawks. Further work on nai mallard duckling by *Green, Oreen* and *Ca* (1966) and by *Mueller* and *Parker*, however, indicates that hawk and goose silhouet can be distinguished and do elicit different reactions. The reactions were not clear anti-predator reactions in either case. The linking of the perceptnal recognition to the command which initiates such reactions, therefore, may require experience which the experimental subjects of these experiments did not have.

The general effects of experience on anti-predator behaviour have been examined in many experiments with young animals. If the occurrence of such behaviour depends, in part, on the gradual formation of a model of familiar surroundings, with which future events can be compared, individuals reared in different surroundings will form different models which may be formed at various rates. Experiments with rhesus monkeys, chimpanzees and domestic chicks indicate that the magnitude and duration of reşponses to novel objects or novel environmental changes is less for individuals with more complex rearing conditions (*Harlow* and *Zimmermann* 1959, *Menzel, Davenport* and *Rogers* (1963, *Broom* 1969). The large literature on the effects on later behaviour in a strange pen (open field) of handling rodent pups during infancy, leads to the same general conclusion. It seems that the major effect of handling a rodent pup is to change the amount of time that the mother spends licking and sniffing that pup (*Richards* 1966, *Barnett* and *Barn* 1967). Perhaps as a consequence of this, handled pups show a less marked response in novel situations (*Deneberg* 1964). The extent and type of maternal care also has effects on the responses of young primates confronted with possible danger. For example *Ainsworth, Bell* and *Stayton* (1971) found that one-year old human infants, whose mothers were inadequate in their responses to them, explored less and fled to mother more readily in strange and hence potentially dangerous situations.

Defensive response by farm animals to man

To the ancestors of present day farm animals, man must have been recognised as a dangerous predator. Despite many generations in domestication, the anti-predator functional system still has considerable effects on the behaviour of farm animals. In some situations, predation by wild animals or dogs is still possible but anti-predator behaviour now is directed against environmental changes who origin is largely unknown to the animals, or against man. The occurrence of unexpected sounds or the arrival of olfactory or visual portents of danger may elicit adrenal and behavioural responses which are the same as those shown by their wild ancestors. The detection of a person who remains at a considerable distance from the animal may elicit no such responses but the close approach which is often necessary on farms must often elicit a defensive response. The fact that people are often treated as predators by farm animals is ignored by many of those who work on farms or who attempt to understand the behaviour of the animals but the competent stockman has learned observation about such behaviour. Good stockmen also known how to minimise the anti-predator behaviour which is directed at them, for if they can do this they can often improve production (*Seabrook* 1972, 1977) and enjoy their work more.

The importance of the abilities of the cowman in obtaining the best possible production from a dairy herd was stressed by *Albright* (1971). Milk let-down occurred faster if the cows saw the familiar milker (*Baryshnikoy* and *Kokorina* 1959). Yield was higher if the cowman was delioerate in movements, talked quietly to the cows and followed a regular routine with no panic activity (*Seabrook* 1977). Other effects of a 'good' cowman were that cows entered the milking parlour, and were more approachable in the field. It seems likely that a 'poor' cowman elicits a greater adrenal response and corresponding defensive behaviour so that management of the dairy herd is more difficult and milk production is impaired. Studies of the defensive responses of cows to cowmen and to disturbing aspects of yard and milking parlour design (*Albright* 1979), are helping to improve management and milk production.

The detection of a person in an animal-house may be a traumatic experience for comparatively small animals such as chickens or turkeys. If a person enters a house rapidly and noisly the nearest birds may show a violent escape response. Whether the birds are in separate cages, or in one large aggregation on the floor of the house, a wave of violent escape behaviour, sometimes called 'hysteria', may be transmitted down the house. Such behaviour can be minimised by a quite knock before entry and by slow, quiet movements while entering and when within the house. The economic importance of quiet slow movements in poultry and other livestock houses is obvious to the turkey farmer who has found the corpses of dead birds under a heap of turkeys disturbed by the entry of a noisy person. The effects of defensive behaviour and adrenal activity on larger animal may be obvious, for example, if a fleeing cabreaks a leg, but many effects are less obvious. Intermittent high adrenal activity may affect adversely milk or egg production, growth rate an disease resistance as well as ease of handling (*Kiley-Worthinberg* 1971).

3

Aggression

Aggression is a complex behaviour with many functions and many causes and it may include predatory behaviour. *The term with a more precise definition is agonistic behaviour which is a system of behaviour patterns that has the common function of adjustment to situations of conflict among conspecifics.* The term agonistic includes all forms (aspects) of conflict, such as *threat, submissions, chases* and *physical combat.*

Forms of Aggressive Behaviour

Aggression is expressed in two ways :

A. Social use of space.

B. Dominance hierarchy.

But we can list the forms of aggression in the following way :

1. *Terretorial.* Exclusion of others from some physical space.

2. *Dominance.* Central as a result of previous encounter, or the behaviour of a conspecific.

3. *Sexual.* Use of threats and physical punishment, usually by males to obtain and retain mates.

4. *Parent off-spring.* Disciplinary action of parent against off-spring.

5. *Weaning.* Restriction of access of offspring to milk.

6. *Moralistic.* Enforced conformity to group standards *in humans,* often based on religion and ideology.

7. *Predatory.* Act of predation, possibly including cannibalism.

8. *Antipredatory.* Defensive attack by prey on predator such as moblsing.

Aggression and Competition

Most agonistic behaviour *involves competition* for some resource. The competition is of two types :

(1) Scamble, and (2) Contest.

1. *Scamble.* To *win* something. For example, if we throw a piece of food between a couple of squirrel monkey they will both go after it, with the first one to reach the food and take it away.

2. *Contest.* To *compromise.* This will be well understood by the following examples :

If a piece of food is thrown between two *Rhesus* monkeys one would probably grivace and face away. The first would then calmly work and take the food.

TYPES OF AGGRESSIVE BEHAVIOUR

Now lets consider the two main types of aggressive behaviours in details.

A. Social Use of Space (Territoriality)

The area used habitually by an animal or group, and in which the animal spends most of its time, is the *home range.* The area of

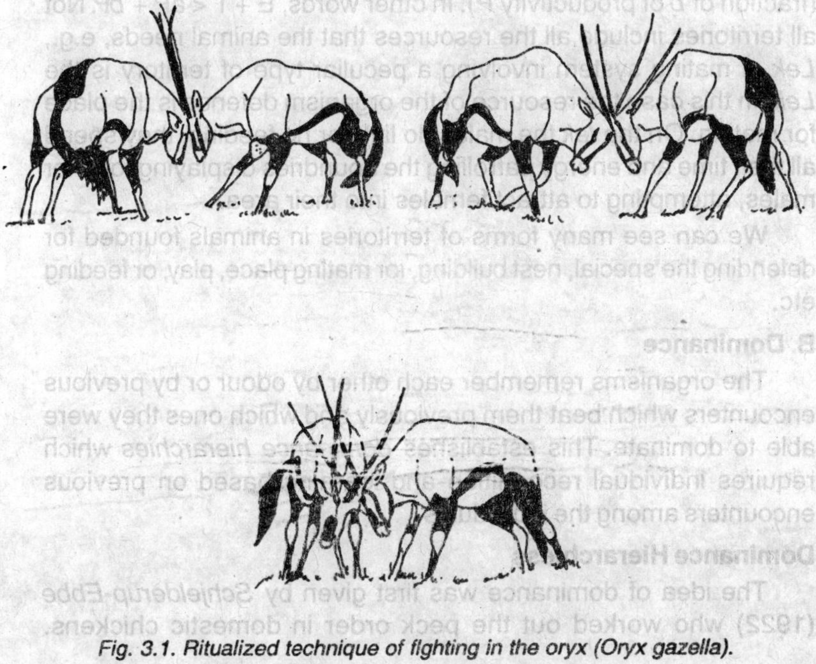

Fig. 3.1. Ritualized technique of fighting in the oryx (Oryx gazella).

heaviest use within the home range is the *core area.* This may include a nest, sleeping trees, water sources or feeding trees. The minimum distance that an animal keeps between itself and other members of the same species is its *individual distance.*

An area occupied more or less exclusively by an animal or group and defended by overt aggression or advertisement is a *territory.*

Size and Boundaries of Territory

Size depends on the size and diet of the animal as well as on the particular resource the animal is defending. Most vertebrates defend the food source.

Territorial animals spend much time patrolling the *boundaries* of their space, singing, visiting scent posts and making other displays.

Territorial Model

Carpenter and *Macmillian* (1976) demonstrated in their studies that the possibility of constructing a model to predict territorial behaviour. They argue that in order for territorial behaviour to occur E, the basic cost of living, plus T the added cost of the defending a territory must be less than the yield to the individual if not territorial plus the extra yield gained by the reduced competition of territorial (fraction of *b* of productivity P). In other words, $E + T < aP + bP$. Not all territories include all the resources that the animal needs, e.g., *Lek*, a mating system involving a peculiar type of territory is the *Lek.* In this case the resource of the organism defends is the place for mating. On the *lek* the males do little or no feeding, they spend all their time and energy patrolling the boundries displaying to other males, attempting to attract females into their area.

We can see many forms of territories in animals founded for defending the special, nest building, for mating place, play, or feeding etc.

B. Dominance

The organisms remember each other by odour or by previous encounters which beat them previously and which ones they were able to dominate. This establishes *dominance hierarchies* which requires individual recognition and learning based on previous encounters among the individuals.

Dominance Hierarchiess

The idea of dominance was first given by *Schjelderup-Ebbe* (1922) who worked out the peck order in domestic chickens.

Fig. 3.2. Darwin's illustrations of the expressive movements of dogs. The threat postures (a and b) include both 'serviceable associated habits' and 'movements due to direct action of the nervous system'. The submissive posture (c) is the reverse of threat behaviour in many respective illustrating Darwin's principle of antithesis. A similar contrast is seen between aggressive and submissive posture of kangaroo rats (d and e).

Dominance hierarchies from most often in arthopods and vertebrates that live in permanent or semipermanent social groups.

Hirarchic means the *ranking of individuals* for different objects such as assessment of water, favoured breeding site etc.

An animal is dominant if it controls the behaviour of another (*Scott* 1966). In another sense, a prediction is being made about the outcome of future competitive interactions (*Rowetl* 1974).

For example, if four or five mice that are strangers to each other are put together, several outcomes are possible. Most likely a despot will take over and all the subordinates will be more or less equal. Another possibility is a *linear hierarchy* where A dominates B, B dominates C and so on. (i.e., A → B → C → D). Sometimes triangular relationship is formed such as,

```
   A
 ↗   ↘
B ← C
```

In other species coalitions may affect dominance, such that

```
        A
      ↙   ↘
     B     C
but  C  +  B
        ↓
        A
```

Dominance hierarchies seem to be a common result when socially living organisms engage in context competition.

For example, Dominance hierarchy in *Rhesus monkeys.*

Dominance in Females

Position in the hierarchy is determined by the others rank (*Sade* 1967; *Maxdew* 1968). Adult females are linearly ranked with sons and daughters generally ranked just below their mothers.

Dominance in these families manifests itself sublity but constantly at a food or water source or increases to resting spots with the dominants supplanting subordinates.

Dominance in Males

Dominance among males is more complex because males have the natal group shortly after puberty while in the natal group they rank below their mothers. As males move in new groups dominance relationships are quickly formed. Size and previous fighting experiences seen not to influence rank directly as long as they stay in the new group, their rank is positively correlated with seniority those in the group the longest rank the highest (*Duckenar* and *Nessey* 1973).

Advantages of Dominance

1. Studies of group living birds and mammals indicate that the dominant animals are well fed and healthy.

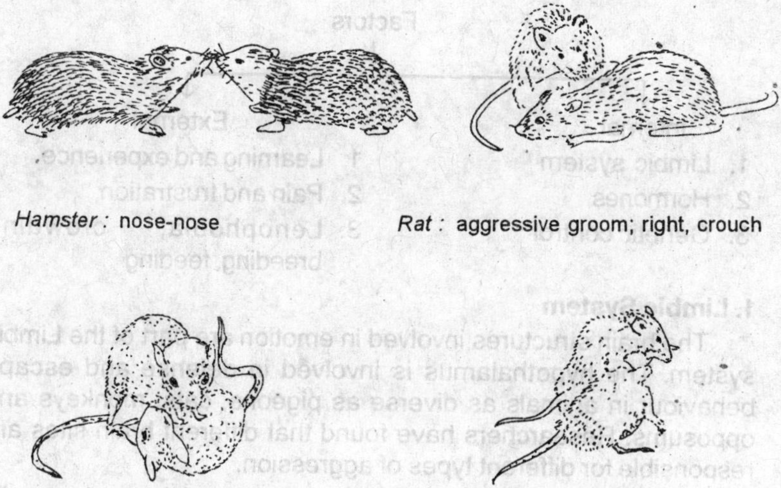

Hamster : nose-nose *Rat* : aggressive groom; right, crouch

Gerbil : left, aggressive; right submissive *Mouse* : left attacking right

Fig. 3.3. Aggressive behaviour patterns in small rodents.

2. *Christian* and *Davis* (1964) have reviewed data for mammals showing that low ranking individuals, frequent losers in fights have higher levels of adrenal cortical hormones than do dominants. These hormones elevate blood sugar and prepare the animal for fight or flight. The cost is a reduction in antigen—anti-body and inflaminatory responses—the bodies defense mechanisms—and a reduction in levels of reproductive hormones.

3. Reproductive success—Higher rank is often directly correlated with reproductive success, as *Le Boeuf* (1974) noted in his study on elephant seals.

4. Many popular treatments have emphasized the importance of dominance and aggression in keeping the species fit, since the survivors of fierice battles are the most vigorous, and therefore, "improve" the species by passing those traits on.

Factors Affecting Aggression

There are two factors which affect aggression :

A. Internal factors in aggression.

B. External factors in aggression.

Factors

↓

Internal	External
1. Limbic system	1. Learning and experience.
2. Hormones	2. Pain and frustration
3. Genetic control	3. Lenophobia, crowding breeding, feeding

1. Limbic System

The brain structures involved in emotion are part of the Limbic system. The hypothalamus is involved in defense and escape behaviour in animals as diverse as pigeons, cats, monkeys and opposums. Researchers have found that different brain sites are responsible for different types of aggression.

For example, electrical stimulation, in cats, of the ventromedial nucleons of the hypothalamus produces growling, hissing and attacking with claws.

2. Hormones

Interacting with the limbic system are neurosecretions and hormones. Epinephrine (adrenation and norepinephrine or noradrinaline) are related to physiological arousal. These are other substances, such as dopamine and serotonin, may act as neurotransmitters and may affect aggressiveness. Early work on chicken demonstrated that testosterone made them more aggressive and increased their dominance rank.

Reproductive hormone other than testosterone also affect aggression. L.H. increases aggression in some species of birds. Estrogen lowers aggression in some species and raises it in other (*Davist*, 1963).

3. Genetic Control

The synthesis of nervous structures, neurosecretions and other compounds in under genetic control, and agonistic behaviour is shaped by natural selection, as in any other behaviour. Artificial selection can lead to significant changes in levels of aggression within just a few generations.

External Factors in Aggression

1. Learning and Experience

Researchers generally accept the fact that some factor outside the animal triggers the response. Previous experience can produce semipermanent changes in the expression of agonistic behaviour and animals can be conditioned to win or lose.

Scott (1975-1976) has emphasized the role of experience and learning in the development of aggression. Although he acknowledges the genetic and physiological bases of aggression. He emphasizes the importance of early experience and learning in the development of aggression and suggests that one of the most important controls of agonistic behaviour is passive inhibition.

2. Pain and Frustration

More direct causes of aggression are pain and frustration.

Pain. Noxious stimuli such as wind noises, foot shocks, tail pinches and intense heat, cause different lab species to attack a wide variety of objects from conspecific cage mates to tennins balls (*Ulrich, Hutchinson* and *Azrin,* 1965).

Frustration. Researchers have explored the frustration-aggression convection in humans. In some early experiments researchers allowed children to view a room full to toys but denied them access to it, when they finally led them into the room, the children frequently smashed toys and fought more than did a control group that was allowed immediate access to the room.

Frustration is now considered to be only of several causes of aggression.

3. Xenophobia, Crowding, Breeding, Feeding

Xenophobia. This is another strong stimulus for aggression. Xenophobia means presence of a stranger. In most species, introducing a strange animal into an established social group produces the violent reaction of xenophobia, usually among members of the same see as stranger (Southwich et. al. 1974).

Crowding per setends to increase interaction rates but does not always increase aggression. The overall group size, presence of strangers, or restricted access to resources has a much more powerful effect on aggression than does reducing the space available to an already established social unit.

Breeding. Courtship and cooperation themselves seem to have aggressive components as evidenced by the appaling rackel mating cats make which makes the listener wonder whether it is a fight to death or courtship. Some of the aggression at the time of mating stems from the fact that males tend to court females rather discriminately, often making advances when the females are not sexually receptive.

Feeding. Some evidence has shown that feeding in groups of fishes intragroup agonistic behaviour. Albrecht (1966) proposed that predatory and aggressive behaviours, although functionally distinct, are motivationally linked such that motivational summation occurs.

Restraint of Aggression

1. Displays

Displays that communicate an animal's intentions and mood may inhibit attack be another Lorenz (1966), Scott (1966) and others have argued that such behaviours are necessary to avoid large scale killing and wounding.

2. Evolutionary Model

The species preservation argument has become unpopular recently because it implies that natural selection operates at the group or species level. Tinbergin (1951) had previously pointed out that retiralized contests would reduce the risk of injury to the aggression and that threats, which take less energy than physical combat are advantage to individuals as well as groups.

Whatever the evolutionary origins, the ritualization of agonistic displays into contests rather than struggles has occured in practically all species with social behaviour.

3. Relevance for Humans

Control of aggression in animals may have special relevance for humans.

Humans also become violent according to *Lorenz,* because we have few harmless outlets for aggression. *Lorenz* says that if aggression is inevitable then we must find harmless ways to vent it. This cathantic approach suggests the importance of ritualized tournaments e.g., sports events in which participants and spectators alike can work out their hostilities.

4. Social Control and Disorganization

Scott (1975) has developed the idea of the social control of aggression. Observation of agonistic behaviour in cichlid fish has shown that when strange fish of the same species are continually introduced into the group, fighting remains at a high level, if group membership is allowed to stabilize, fighting declines to a relatively low level.

One cause of *aggression* seems to be *social disorganization* when a high ranking male or female *Rhesus monkey* dies or leaves the group, aggression within the group may increase.

Conclusions

We can conclude by saying that maintenance of a territory or a position in a dominance hierarchy requires constant reinforcement by sings, threats, or other displays, individual recognition and memory of the results of the previous experiences are always involved. It has been suggested that the modification of the social environment through early experience and avoidance of social disorganization can help to control aggression.

4

LEARNING

Learning—the modification of behaviour in response to experiences—is one of the most characteristic attributes of animals. The most important feature of learning is that it is adaptive. The animal, having learned, responds in ways that improve its survival and reproductive success. Different animals, even of the same species, learn different things if they are exposed to different environments. Learning is demonstrated when animals in two groups, one given an experience denied to the other, develop different behaviour patterns. Sometimes these experiments reveal that experiences do modify behaviour, but at other times they show that experience is unnecessary. For example, most young birds begin "practice-flying" before they leave their nest. They stand tall and flap their wings vigorously as if they were testing them for the first take off. But experiments have shown that birds can develop the ability to fly purely by maturation. In one experiment, pigeons were reared in narrow tubes that prevented them from moving their wings. They could not undertake any form of practice flights. A control group of birds were allowed to practice each day. When the controls had "learned" to fly, the experimental birds were released from their tubes. Surprisingly, they flew just as well as the control birds.

According to *Hilgard* (1956) learning occurs when the probability of certain behaviour patterns in specific stimulus situation has been changed as a result of previous encontors or other similar stimulus situations.

Hinde (1970) has defined the learning as changes which can not be understood in terms of maturational growth process in the nervous system fatigue or sensory adaptations.

According to *Wallace* (1973) learning is the adaptive changes in behaviours that result from the individual experience and probably associated with physical changes in the central nervous system. It may function with innate pattern or but may ultimately be separable from these patterns.

FORMS OF LEARNING

Animal behaviourists have used many different procedures in studying learning. We shall next consider some of these. A description of the various forms of learning is analogous to the observation and description that provides the starting point for the study of all animal behaviour (*Tinbergen,* 1963). Many of the distinctions to be made among different "forms" of learning are based on differences in the procedures used in the study of learning. The fact that the procedures used differ need not necessarily imply that the processes underlying these different forms of learning are either similar or different. There may be one, two, or many kinds of processes underlying different forms of learning.

Habituation

Habituation may be defined as the decrease in probability or amplitude of a response, occurring when a stimulus which elicits that response is presented repeatedly. In an excellent discussion of the phenomenon of habituation, *Thompson* and *Spencer* (1966) listed nine characteristics of habituation that are manifested in most systems. We shall consider a few of these. If, after repeated stimulus presentation and habituation of the response, the stimulus is no longer presented, the response tends once again to be elicited by the stimulus. Within limits, the longer one withholds the stimulus the greater is the likelihood that the response originally elicited by the stimulus will again be elicited by it. This phenomenon is called *sportaneous recovery.* It may require a few minutes to several days, depending on the conditions. If a repeated series of habituation training trails and spontaneous recovery trails is given, the habituation occurs progressively more rapidly (*potentiation of habituation*). Habituation of a response to a given stimulus shows generalization to other similar stimuli (*stimulus generalization*). Thus, if an animal habituates a response to a tone of a given frequency, it is likely that responses to tones of frequencies close to the training stimulus will also show habituation. The more dissimilar the test tone from the habituation tone, the less would stimulus generalization be apparent.

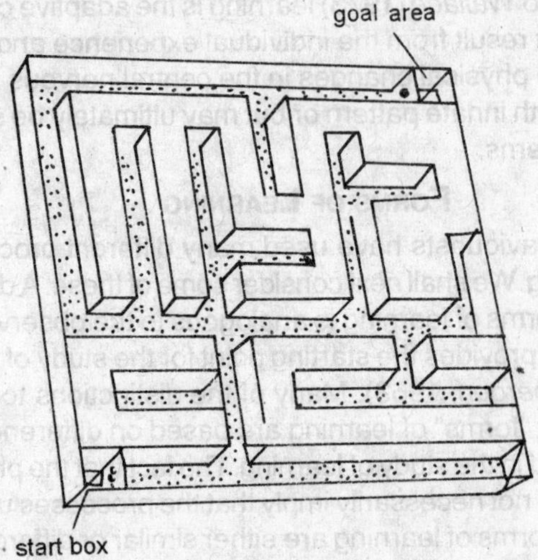

Fig. 4.1. A rat may learn a maze by trial-and-error processes, initially making many wrong turns and entering numerous culs-de-sac. Gradually the rat makes fewer errors, until, after a number of trials, it completes the maze without any errors. Also, a rat may perform with fewer errors when it has been given prior exposure to the maze with no rewards present, a process called latent learning.

Finally, presentation of another (usually very intense) stimulus results in a sudden recovery of the habituated response (*dishabituation*). In order to be considered true habituation, the response decrement must clearly be the result of changes in the central nervous system rather than of *adaptation* in the sensory system or fatigue in the effectors.

When a puff of air is applied to the gills of a horseshoe crab, it responds with a movement of its telson (tail). *Lahue, Kokkinidis,* and *Corning* (1975) studied the characteristics of habituation of the telson reflex, monitoring the muscle movements electro-physiologically. They found a clear habituation of the telson reflex to repeated presentations of the air-puff stimulus. A brief squirt of saline from a syringe aimed at the site of habituation resulted in dishabituation. Initial levels of responsivity were normal after an interval of 1 hour without stimulus presentation (*spontaneous recovery*). Potentiation of habituation was observed after about 2 hours. No generalization of the habituation was observed when stimuli were applied to other regions (no stimulus generalization). Habituation

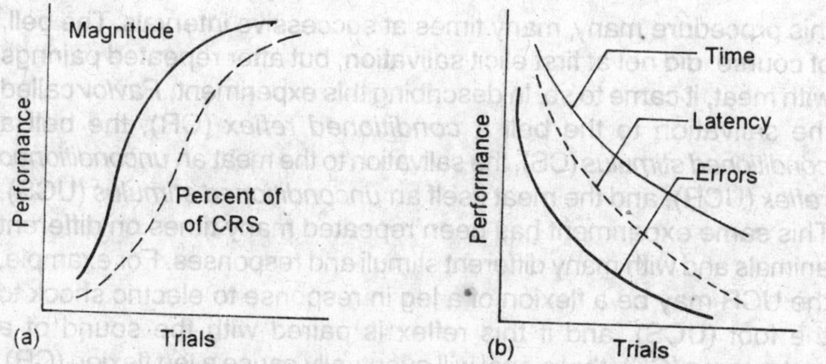

Fig. 4.2. Characteristic forms of learning curves. (a) The percentage of CRs and response magnitude increase with practice. (b) Latency measures and other time measures decrease as do error.

of contraction responses of sea anemones to water-stream stimulation was observed by *Logan* (1975). *Ratner* and *Gilpin* (1974) studied the habituation and retention of habituation of the response to air puffs in both normal and decerebrate earthworms.

Classical Conditioning

One very simple form of learning of the latter type is *classical conditioning,* so called because it was discovered by *Pavlov*, the father of conditioning. In his classical experiment, *Pavlov* lightly restrained a dog in a harness and repeatedly blew meat powder into its mouth and recorded accurately the amount it salivated. Then he associated the sound of a bell with the meat powder and repeated

Fig. 4.3. Pavlov's testing apparatus. Ivan Pavlov discovered classical conditioning through his work on the salivary reflex in dogs. The dog in the restraining apparatus is ready to be tested using the classical conditioning paradigm.

this procedure many, many times at successive intervals. The bell, of course, did not at first elicit salivation, but after repeated pairings with meat, it came to so. In describing this experiment, *Pavlov* called the salivation to the bell a *conditioned reflex* (CR), the bell a *conditioned stimulus* (CS), the salivation to the meat an *unconditioned reflex* (UCR), and the meat itself an *unconditioned stimulus* (UCS). This same experiment has been repeated many times on different animals and with many different stimuli and responses. For example, the UCR may be a flexion of a leg in response to electric shock to the foot (UCS), and if this reflex is paired with the sound of a metronome (CS), that signal will eventually cause a leg flexion (CR). Typically, the CR is very similar to the UCR, but it is never completely identical to it. Thus, the best way to describe classical conditioning is as *a process in which a previously neutral stimulus (CS=bell) is enabled to elicit a response (CR=salivation) that it never elicited before training.*

In his experiments, *Pavlov* found that the time relation between the CS and the UCS was critical. If the UCS preceded the CS. There was very little, if any, conditioning, and if the UCS by more than about a second, conditioning became more and more difficult to establish instrumental conditioning in which the sensory stimuli involved become more numerous and more complex and in which the animal is given a greater and greater measure of choice in the stimuli it uses. In general, as we go from the simple to the complex learning tasks, learning becomes more difficult and more easily disrupted by brain injury. With these points in mind, it will be interesting to compare the learning ability of different animals that represent different levels of the phylogenetic scale and thus different levels of development of the nervous system.

Operant Conditioning or Instrumental Learning

Operant conditioning or as it is sometimes called, *instrumental learning,* involves a wide variety of procedures. In each instance the animal learns to associate its behaviour with the consequences of that behaviour—that is, the sequence of events is dependent upon the behaviour of the animal. Usually some type of reward or punishment is involved. The task performed by the animal may be relatively simple, as in rat trained to press a lever in a skinner box. Pressing in this example is rewarded with food or water. Another example is when the animal has learned to run a maze to a goal box to receive the reinforcement. In nature a weasel may learn to associate the odor of mice with locating and catching a meal.

Mechanisms of Learning

It is easier to describe the learning abilities of animals than it is to understand the mechanisms are known. Completing the picture represents one of the greatest challenges to contemporary biology.

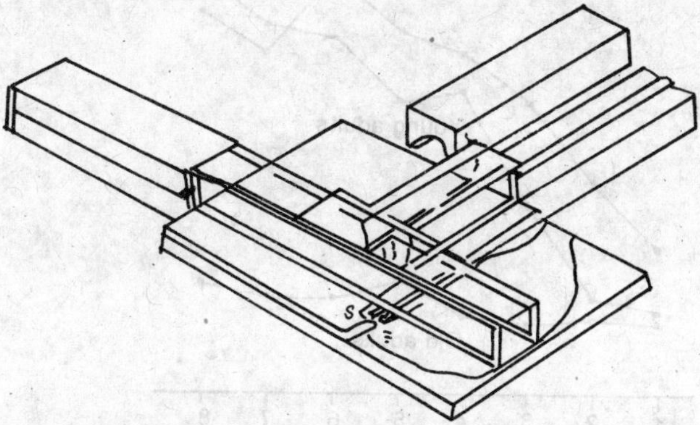

Fig. 4.4. A T-maze used by Yerkes for training earthworm to turn to the right to enter a dark, moist chamber (D) and to avoid electric shock at E on the left.

There is good evidence that the mechanisms by which new information is first stored in the brain are different from mechanisms of longer-term storage. Brain concussions produce *amnesia* (forgetting) for events immediately prior to the injury, which is why people usually cannot recall what happened to them just before accidents. Short term memory may be based on continuously circulating nervous impulses. A rat or a hamster, just having learned a simple maze, such as a T-shaped runway in which the correct solution is to make a right turn at the junction does not remember having learned the maze if it is given an electric shock to its brain within 5 minute of its training. Slight amnesia effects are still detectable if the shock is given up to an hour after the learning experience. In contrast, long-term memory is undisturbed by convulsions, electric shock, concussion, or deep anaesthesia.

For many decades investigators have been trying to determine whether specific memories are stored a particular cells or groups of cells in the brain. Experiments have involved stimulating specific parts of the brain to determine what behaviour is evoked and maling lesions (cuts) in the brain to determine which memories are destroyed by particular lesions. Extensive work with rats reveals that limited regions of the brain are essential for retention of some memories but that particular cells or small groups of cells are not responsible for specific memories.

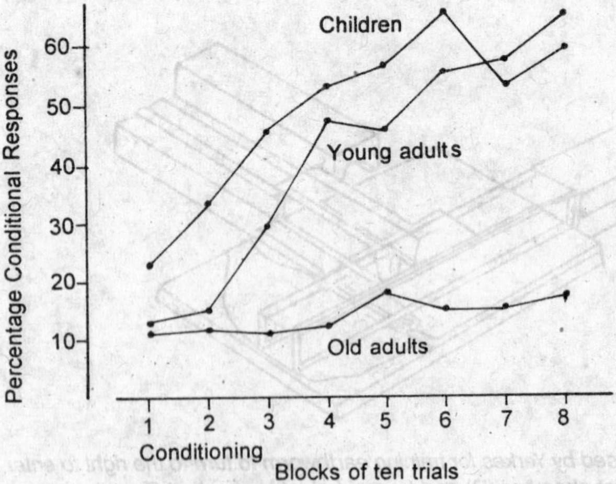

Fig. 4.5. The effect of age upon eyelid conditioning.

The search for the sites of specific memories a more easily carried out with simpler animals that have fewer cells in their nervous systems. An especially favourable animal has been the marine gastropod mollusc *Aplysia*, commonly known as the sea hare. As in most molluses, the nervous systems of *Aplysia* consists of a number of distinct ganglia that innervate different body structures. The cerebral ganglia innervate the anterior tentacles, mouth, eye and rhinophores. The buccal ganglia innervate the pharynx, salivary glands, esophagus, crop, gizzard and muscles that control the protraction and retraction of the odontophore. The pedal ganglia supply the foot, parapodia, head, caudal part of the penis, penis sheath, and penis retractor muscles. The abdominal ganglion controls a variety of respiratory, reproductive, circulatory and excretory functions and also controls movement of the external organs of the mantle. *Aplysia* is not a spectacular learner, but it does habituate to a number of stimuli. Habituation of the defensive withdrawal of the siphon and gill under control of the abdominal ganglion.

Electric shocks normally cause withdrawal of both organs, but if 10 stimuli are given at short (30-second intervals, the animal stops withdrawing its siphon and gills and does not resume doing so far several hours. Four such training sessions, each separated by approximately 1½ hours, produce habituation that lasts·up to 3 weeks. Several neural changes are involved in this learning. One is a reduction in transmission of messages from the receptors in the siphon skin and the motor neurons that cause withdrawal. *Eric Kandel* and his associates at Columbia University have shown that just a few neurons decrease the amount of neurotransmitters they release. Conversely, when withdrawal is enhanced by following shocks with other negative reinforcement, a few neutrons release an amine that acts on the terminals of sensory neurons to increase the release of transmitters. Although withdrawal of both organs involves a number of muscles, changes in just a few neurons (the memory traces) are needed for the learning (the modified behaviour) to occur.

Phylogeny of Learning

We cannot retrace the steps in the evolution of learning, but a brief survey of the learning capacities and capabilities of organisms from various animal phyla should provide some insight into the possible generality of any laws of learning and a sampling of the differences in learning ability across the various phyla. Four cautions should be noted. First, there has been a disproportionate examination of learning across the various phyla. For more studies have been conducted on learning processes in invertebrates than in invertebrates, and even within the vertebrates much of the attention has focused on mammals. Second, there are valid ongoing disagreements among scientists regarding what constitute proper criteria for demonstrating various types of learning and concerning the evaluation and interpretation of learning and concerning the evolution and interpretation of research results. Third exploring the physiological and behavioural aspects of learning in many animals will require more refined techniques and objective, bias-free methods. Fourth, we measure performance and not a genetically determined ability or mechanisms. Lots of factors (notably, test design) can influence performance. Additional problems regarding comparative learning studies will be discussed in two subsequent sections of this chapter. With these difficulties in mind we can now survey the known information regarding learning in a number of animal phyla.

Protozoa and Coelenterata

Whether protozoa are capable of learning is still a debated question. It has been reported, for example, that *Amoeba* and

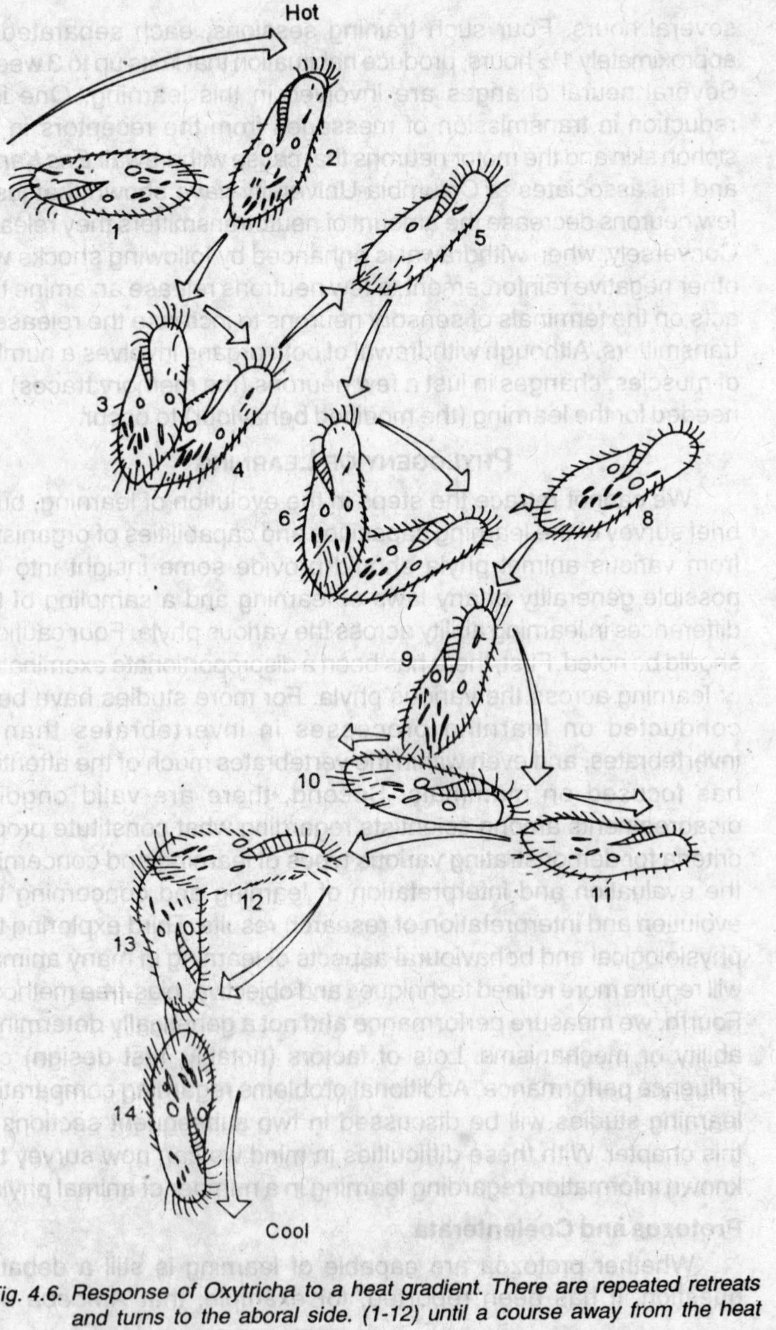

Fig. 4.6. Response of Oxytricha to a heat gradient. There are repeated retreats and turns to the aboral side. (1-12) until a course away from the heat source is achieved (13-14).

Paramecia are capable of habituation to noxious sensory stimulus such as strong light or mechanical shock, for upon repeated stimulations their responses grow weaker and weaker until in some cases they become totally unresponsive. However, two criticisms of this work have been offered. One is that none of these experiments has satisfactorily demonstrated that such diminished responses

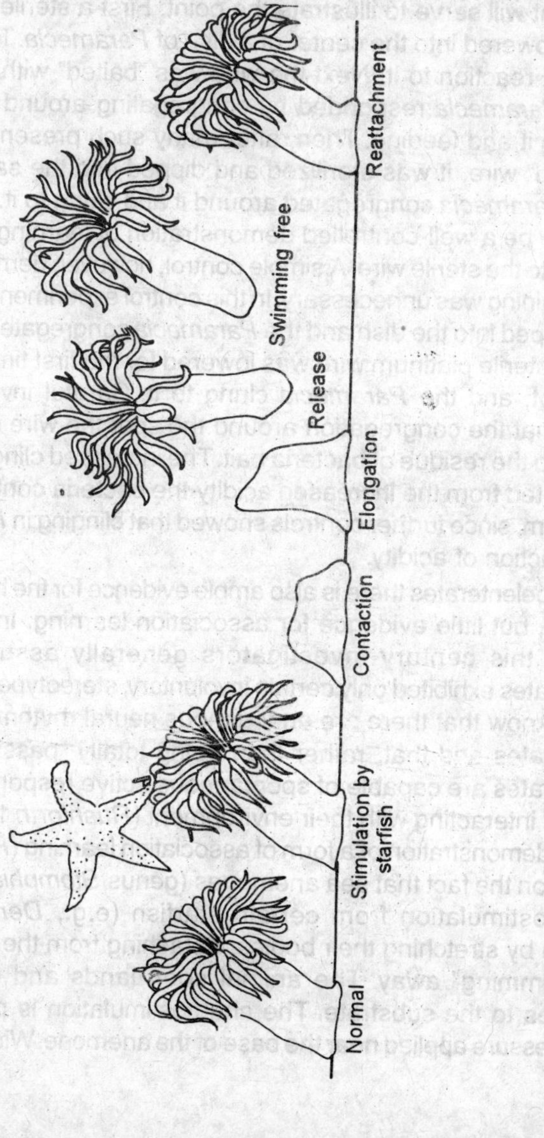

Fig. 4.7. Interaction between starfish and sea anemone. When a starfish (Dermasterias) makes contact with the sea anemone Stomphia cocinea, the anemone will release from its attachment, "swim" free for a period, and eventually reattach itself to the substrate.

lasted long enough to be anything more than adaptations to sensory stimuli. The other is simply that the noxious stimulation may have temporarily injured the organisms so that they were made capable of responding with each successive stimulation.

To get around these objections, attempts have been made to demonstrate some form of associative conditioning or learning. But each claim has again been met with cogent criticisms. One experiment will serve to illustrate the point. First a sterile platinum wire was lowered into the center of a dish of *Paramecia.* There was no special reaction to it. Next the wire was "baited" with bacteria, and the *Paramecia* responded by congregating around the wire, clinging to it and feeding. Then, after many such presentations of the "baited" wire, it was sterilized and dipped into the same spot, and the *Paramecia* congregated around it and clung to it. This was claimed to be a well-controlled demonstration of learning of a new response to the sterile wire. A simple control, however, demonstrated that the training was unnecessary. In this control experiment, bacteria were dropped into the dish and the *Paramecia* congregated and fed. Then the sterile platinum wire was lowered for the first time into the same spot, and the *Paramecia* clung to it. Careful investigation showed that the congregation around the spot the wire contacted was due to the residue of bacteria bait. The increased clinging to the wire resulted from the increased acidity the bacteria contributed to the medium, since further controls showed that clinging in *Paramecia* was a function of acidity.

For coelenterates there is also ample evidence for the habituation response, but little evidence for association learning. In the early years of this century investigators generally assumed that coelenterates exhibited only certain involuntary, stereotyped reflexes. We now know that there are endogenous neural rhythms in many coelenterates and that, rather than being totally "passive," many coelenterates are capable of spontaneous active responses in the course of interacting with their environment (*Rushforth* 1973). One possible demonstration of a form of association learning (*Ross* 1965) is based on the fact that sea anemones (genus *Stomphia*) respond to chemostimulation from certain starfish (e.g., *Dermasterias imbicata*) by stretching their bodies, detaching from the substrate, and "swimming" away. The animal soon lands and eventually reattaches to the substrate. The chemostimulation is paired with gentle pressure applied near the base of the anemone. With repeated trials the application of the pressure stimulus alone leads to some reduction in the "swimming" response; *Ross* labelled this as conditioned inhibition. However, the stimulus may have induced some related effect resulting in the anemone movement.

Platyhelminthes

Considerable publicity and controversy have surrounded the research conducted on learning to planarians. These flatworms are interesting because it is the platyhelminthes that a nervous system with bilateral symmetry and a primitive brain appears.

In this case, a classical conditioning technique was used. When the worms were gliding along in a trough of water, a light was turned on, followed two seconds later by electric shock which caused the worms to contract longitudinally. After 150 trails, the worms contracted to the light alone over 90 percent of the time. Following this, the worms were cut in half and allowed four week for regeneration. Both the regenerated head and tail sections showed a high degree of retention of what had been learned earlier. Even after these regenerated worms were cut and a second period of regeneration occurred, there was retention of the conditional response. Apparently, in this case, too, learning is not confined to the anterior portion of the nervous system, but has its effect throughout its extent.

Annelida

The first unequivocal evidence for learning is found at the level of the worms, where a bilaterally symmetrical, synaptic nervous system is already developed. Here we see clear instances of habituation and associative learning. In one experiment, earthworms were trained to go to one arm of a T-maze leading to a dark, moist chamber and to avoid the other arm which led to electric shock and irritating salt solution. In this simple learning of a position habit, it required about 200 trails on the average to reach a criterion of 90 percent correct responses. Interestingly enough, the worms were able to retain what they had learned after removal of the first five body segments containing the cephalic ganglion, and untrained worms were able to learn after removal of the head ganglion. Apparently, the neural changes involved in learning can take place in the ganglia of the lower body segments.

Mollusca

Among the molluscs, some evidence for the learning of a simple T-maze by snails has been demonstrated, but the experiment was not highly successful in that not all snails learned and those that did rarely reached high levels of consistency in their performance. Much better learning has been demonstrated in the octopus, which has a well developed eye and a rather elaborate brain. In these studies, for example, it was shown that the octopus could readily learn to discriminate the presence of a white card signifying electric shock. At first, the octopus was trained to come forward from its rocky nest to seize a crab that was lowered into the far end of the aquarium by a thread. Then on half the trails, a white card was lowered with the crab, and in these trials, the octopus was driven away from the crab by strong electric shock. After 12 trails, the octopus began to inhibit its approach to the card and by 24 trials consistently remained in its nest when the card was present and consistently came out to feed when the crab was presented alone. The octopus can also learn more difficult problems where two different cards were used, a small one associated with feeding and a large one associated with electric shock. Also, discriminations of various shapes have been

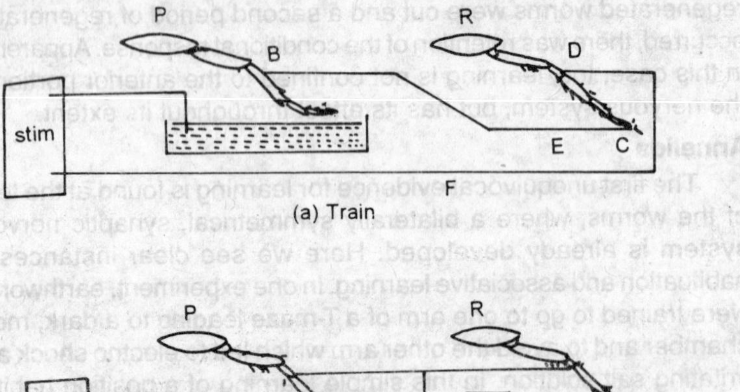

(a) Train

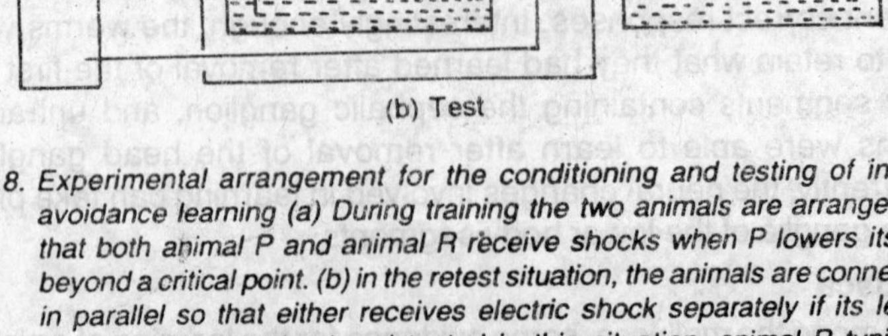

(b) Test

Fig. 4.8. Experimental arrangement for the conditioning and testing of insect avoidance learning (a) During training the two animals are arranged so that both animal P and animal R receive shocks when P lowers its leg beyond a critical point. (b) in the retest situation, the animals are connected in parallel so that either receives electric shock separately if its leg is lowered. When trained and tested in this situation, animal P typically demonstrates passive avoidance learning, as indicated by fewer instances, of the lowering of its than of animal R, the "yoked control."

demonstrated in a similar manner. It is interesting, however, that the octopus was rather poor in problems requiring it to "detour" around a barrier. Thus, when a glass partition was lowered between the crab and an octopus, the octopus persisted in swimming straight into the glass and failed to swim around it through an open space.

Arthropoda

Habituation has been demonstrated in several arthropod species including horseshoe crabs (*Lahue* et. al. 1975), crayfish (*Krasne,* 1969), and the pupae of grain beetles (*Hollis,* 1963). A well-controlled study of classical conditioning of proboscis extension in blowflies was reported by *Nelson* (1971). The UCS was an application of a sugar solution to the mouthparts of the fly, with the CS being the application of water or a saline solution to the feet.

Much attention has been devoted to the study of instrumental learning in arthropods. The studies of maze-learning in ants conducted by *Schneirla* (e.g., 1946) are classics in comparative psychology. Effects of maze-learning experiences in larvae have been shown to survive through metamorphosis in adult grain beetles (*Borsellino, Pierantoni,* and *Schieti-Cavazza,* 1970). Conditioning of responses related to food getting in honeybees has been demonstrated by *Bermant* and *Gary* (1966) and *Wenner* and *Johnson* (1966).

Horridge (1962, 1965) studied the acquisition of avoidance responses in insects, with the without their heads, by using a rather clever procedure. In the training situation, the legs of two cockroaches are connected in series to an electrical stimulator. The electrical circuit is completed and both animals receive shock whenever animal P lowers its leg into a saline solution, thus completing the electrical circuit. With training, animal P comes to hold up its leg and thus avoid shock. Animal R is a "yoked control" that receives the identical number and temporal patterning of shocks, but they are delivered in such a way as to not be contingent on its behaviour. Learning is demonstrated as a difference between animals P and R in a test situation when the animals are shocked independently for lowering their legs. Animal P experiences fewer shocks in the test situation than animal R. Both intact and headless locusts and cockroaches have displayed avoidance learning in this situation (*Horridge,* 1962, 1965; *Disterhoft,* 1972).

Considerable difficulty has been encountered in demonstrating learning in fruits flies, a species that, in other respects, provides

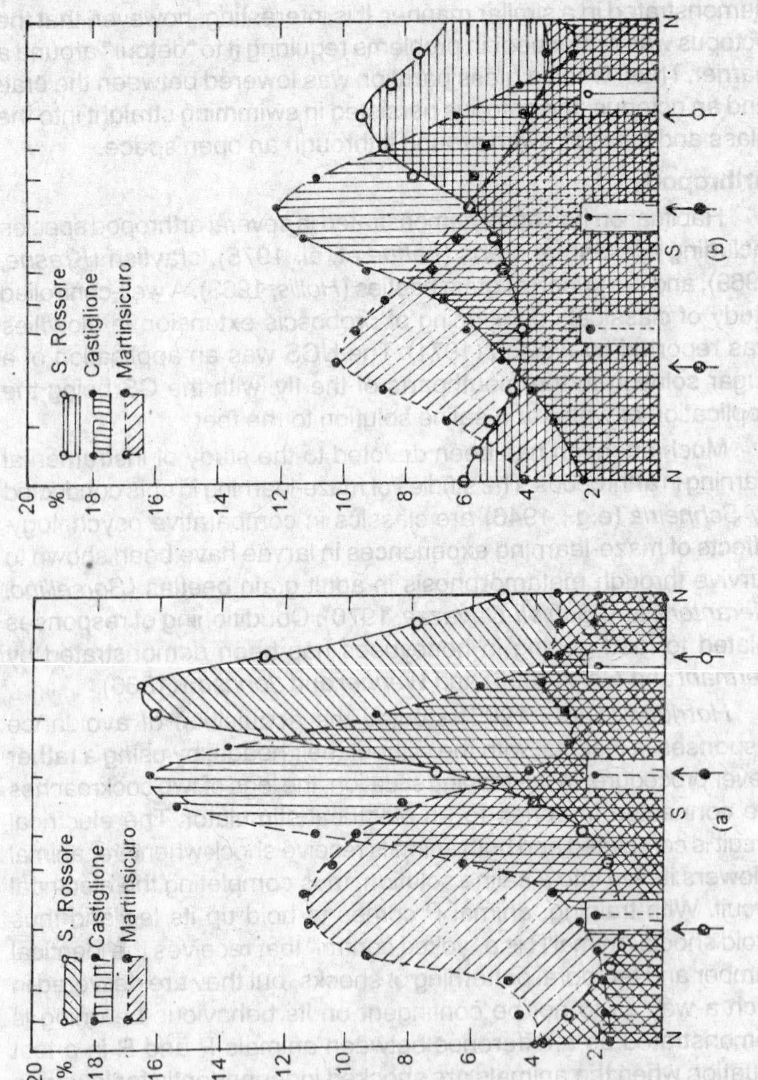

Fig. 4.9. The orientation direction of beach amphipods, Talitrus saltator, is characteristic of each population.

ideal subjects for behavioural studies. Although some learning studies have been quite controversial *Murphey*, 1967; *Yeatman* and *Hirsch*, 1971; *Hay*, 1971; *Bicker* and *Spatz*, (1976), it now appears that fruit flies can learn a discriminated avoidance task based on odor caes (*Quinn*, *Harris* and *Benzen*, 1974).

Learning in Vertebrates

A number of learning studies have been conducted on species from most vertebrate classes. We know considerably more about the complex processes which underlie these phenomena for this phylum (*Masterton* et. al. 1967); this is particularly true for the mammals. Since the rat and Rhesus monkey have been used in a large number of these studies, an example for each species will provide some flavor of the research and how it is conducted.

Rats can be trained in an operant conditioning apparatus to press a lever to receive food. They can also be trained in an apparatus with two levers (designated L for left and R for right) to press the levers alternately, LR or RL, for a food reward. Can rats be conditioned to press the levers in a LLRR or RRLL sequence, called double alternation? Investigators (*Travis-Niedeffer, Niedeffer* and *Davis* 1982) tested this question by conditioning rats first on the single lever task, either R or L, followed by the single alternation task, RL or LR. They then rewarded only double alternation performances. The rats learned this task at a level exceeding chance expectations, and their performance improved over days. It was necessary, however, in some instances, to permit the rats to give extra responses to the first level before pressing he second lever. Thus LLLRR was rewarded the same as LLRR. When the investigators attempted to condition the rats to a sequence LLRRLLRR or RRLLRRLL (double alternation with a fixed ratio schedule of two repetitions), no rats could successfully perform at better than a chance level. The results are significant because of the new information provided regarding the capacity of the rate to associate a series of responses required to obtain the reward and because previous attempts to condition double-alternation tasks in rats had failed.

One particular apparatus, the *Wisconsin General Test Apparatus* (WGTA) has been used in many of the studies of learning behaviour in macaques (*Meyer, Treichler,* and *Meyer* 1965 for the history of this and related techniques). One type of learning problem studied with this apparatus is delayed response problem (*Fletcher* 1965). The basic procedure starts with the test tray out of reach but in full vie of the test animal. Food is placed in one food well, and then two identical objects are placed over the wells. After a prescribed delay the test tray is moved closer to the monkey, which then responds by lifting one object or the other. The four stages in the procedure are the baiting phase, covering phase, delay phase, and response

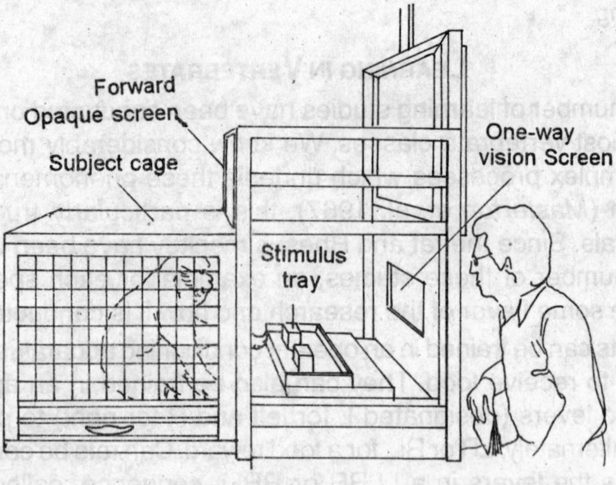

Fig. 4.10. Wisconsin General Test Apparatus. An early version of the Wisconsin General Test Apparatus (WGTA) used to test aspects of learning in primates and, with some modifications of the apparatus, cats and raccoons.

phase. A trail ends when the monkey picks up one of the objects, uncovering either the correct food well, containing a reward, or the empty well. A number of variables can be investigated with this procedure, and other animals, such as cats and raccoons, have been tested with slight modifications of the apparatus. Among the variables that have been manipulated are length of the delay phase, the nature and size of the reward, the nature and the similarity or dissimilarity of the objects used to cover the food wells, and whether the animal is permitted to watch the test tray during the delay phase or, instead, has an opaque screen lowered in front of the tray (*Meyer* and *Harlow* 1952). Among the conclusions are these : Rhesus monkeys are capable of learning basic discriminations in this procedure with delays of up to 30 seconds or more, more food reward leads to performance, and imposition of the opaque screen during the delay phase increases error rates by up to 50 percent. One interesting and striking finding in these studies is that the behaviours and performances of individual monkeys differ markedly. In general, monkeys that exhibit hyperactivity in the test situation and those that are more easily distracted during the delay phase exhibit lower levels of performance. Clearly, in studies of learning behaviour we must consider the significance of individual differences in performance, regardless of the species being tested, or the task being performed (*Warren* 1973).

NEURAL MECHANISMS OF LEARNING

If we were to attribute in improvement of learning ability to one thing in the phylogenetic series, it would be the evolution of the central nervous system. It is an article of faith that learning represents some change in the central nervous system, and that memory is the preservation of that change. From this starting point, many investigators have sought an answer to two major questions: (1) where does learning take place in the nervous system, and (2) what is the nature of the change? We shall take up each of these questions separately although it is obvious that they are interrelated.

The question of the *locus of learning* has been approached mainly by the technique of experimentally destroying parts of the nervous system. Most investigators have dealt with the cerebral cortex, since the earliest theories held that it was in this newly evolved part of the brain that mammalian learning occurs. *Pavlov* believed that the cortex is essential for conditioning, but studies have shown that simple conditioning is possible in the dog after its cortex has been removed. Such a decorticate dog often give emotional and generalized responses and is greatly deficient in sensory capacity, but it can be successfully trained in the classical conditioning technique using shock as the UCS.

Since total decortication grossly impairs the animal, many investigations of the cortex have involved the destruction of selected parts of the cortex. Thus, in his experiments on rats, *Lashley* just the visual area in the back of the cortex and tested the animals for visual learning and retention. He used a simultaneous discrimination situation in the jumping stand where the animal had to choose the correct one of two doors containing visual stimuli. When he used black and white doors, rats without the visual cortex could learn the discrimination almost normally. If they had learned to discriminate black from white before the lesion of the visual cortex, however, they lost the habit postoperatively and had to learn it all over again. When pattern discrimination was used involving a choice between a triangle and a circle, it turned out that the operated animals could never learn. Apparently in this case, they lost the capacity for form or detail vision, whereas in the brightness discrimination, capacity was unimpaired and only memory was affected.

Actually, however, further studies suggest that even in the brightness discrimination case, it was not memory that was affected, but rather it was a loss of sensory capacity needed to respond to

Fig. 4.11. Thorndike puzzle box.

the spatially separated black and white doors. To test this argument, dogs were confronted with a single, large, illuminated panel, shaped like a bowl so as to fill the entire visual field. Then they were conditioned to flex a leg every time the brightness of the field was changed. Here was a brightness discrimination not involving either spatial discrimination or the capacity to discriminate doors, and removal of the visual cortex had no effect on the animal's ability to retain it. So perhaps it was not a defect of memory that Lashley's brain lesions had caused.

Memory and learning ability also proved elusive in Lashely's maze experiments. Here he found that rats were affected in their ability to learn mazes, or retain them, in proportion to the size of the cortical lesions he made in their brains. In other words *Lashley* concluded that the cortex operates on a *mass action principle* in learning and memory so that the large the lesion, the poorer the ability. He also found that it did not mater where the lesion was in the cortex; a lesion of a given size had a given size had a given effect whether it was in the visual area in the back of the brain or the somatic sensory and motor areas in the front of the brain. From this finding, *Lashley* formulated his *principle of equiptentiality,* which says that all parts of the cortex are equal in their contribution to learning and memory. Again, it seems that these experiments may be as much matter of sensory deficit following cortical lesions as they are the result of defects in learning and memory capacities.

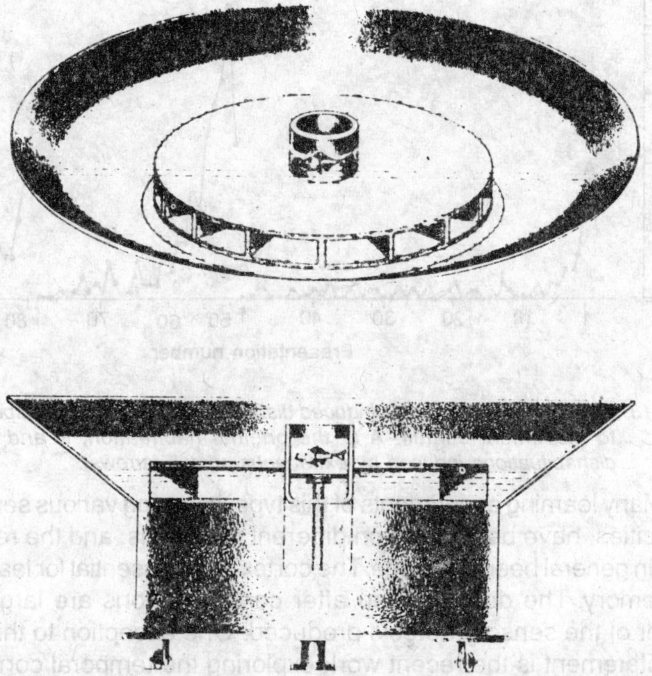

Fig. 4.12. *Orientation testing device for trained fish. The fish is restrained in the center of an arena of shallow water. When the shield is lowered the fish is free to select one of sixteen crevices at the end of the platform as a hiding place.*

Maze learning, we know, is a matter of the rat learning to use many different sensory cues throughout the maze (vision, sound, touch, proprioreception, smell). The more these cues are experimentally eliminated from the animal's sue by destruction of sense organs or by removing stimuli, the worse its performance, regardless of which particular sensory cues are eliminated. Since the rat's cortex is primarily a sensory cortex, it is reasonable to believe that the large the cortical lesion, the more it will impair the use of sensory cues and, therefore, the poorer will be maze performance. Arguing somewhat against this interpretation is an ingenious experiment that *Lashley* performed. He taught blind rats a maze and then removed the visual cortex. Because they showed defects in the retention of the maze habit, he concluded that the visual cortex had non-visual functions in learning and memory as well as visual functions.

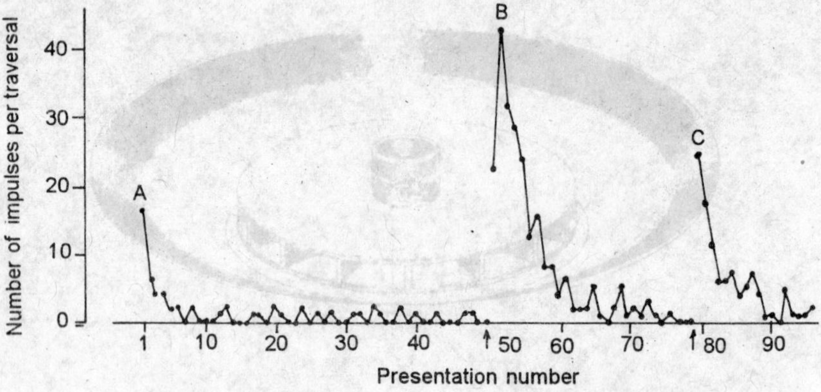

Fig. 4.13. Habituation and stimulus induced dishabituation in a neuron responding to movement stimuli. A is the original habituation, B and C are dishabituations induced by extraneous stimuli (arrows).

Many learning experiments of this type, involving various sensory capacities, have been done on different mammals, and the results have in general been the same. The cortex is not essential for learning or memory. The defects seen after cortical lesions are largely a matter of the sensory defects produced. One exception to this get real statement is the recent work exploring the temporal cortex of primates. Work with monkeys shows that there are defects in learning touch discriminations following lesions of the posterior borders of the temporal lobe and defects in learning visual discriminations after lesions somewhat more anterior in the temporal lobe. Also, it has been found that human patients with bilateral temporal cortex damage seem to have defects of memory, especially recent memory, as well will see at the end of this chapter. Finally, a most interesting related finding is the fact that electrical stimulation of the temporal cortex of fully awake epileptic patients evokes past memories in a vivid dreem-like sequence. However, this may be possibly as much through arousing subcortical structures as through the effect on the cortex itself.

Efforts to use the lesion method to explore the role of subcortical structures in learning have not been highly fruitful as yet. Not many such experiments have been done, and what information we have has been largely negative. Two recent studies, however, have provided hopeful leads. In one, the experiments tried to condition brain-wave responses after making irritative lesions on one side of the brain by

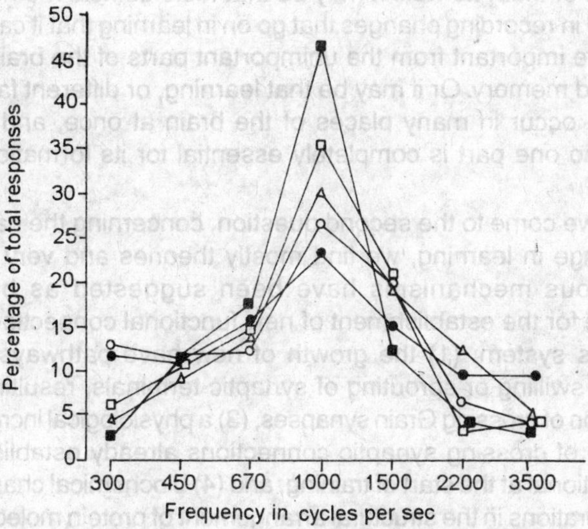

Fig. 4.14. Generalization gradients for individual pigeons responding to tones. The training tone had a frequency of 1000 cycles per second.

implanting aluminium cream. Of all the loci they investigated, placement of the irritative focus just below the temporal cortex in the amygdala and hippocampus was the most effective in impairing learning. In the second experiment monkeys were required to discriminate whether two patterns of tone or of light were the same or different, even though the two patterns might be separated in time by several seconds. In this case, surgical lesion of the region of the amygdala and hippocampus turned out to produce the most marked defects. Such operated monkeys could tell that the two patterns were the same only if one followed immediately after the other. It was as though the lesion made them unable to remember the first pattern over time, for they failed the test when the two patterns were separated by a few seconds.

Another approach to the understanding of brain mechanisms underlying learning is through the use of electrical recording methods in which it is possible to trace changes in the electrical activity of many parts of the brain during learning. In these experiments, the animal has electrodes chronically implanted in its brain, and changes in pattern of electrical activity are noted as the animal is trained. Striking thing here is that the changes take place in many parts of the brain, cortically and subcortically, within the sensory systems and outside of them as well. It may be that the electrical method is so sensitive in recording changes that go on in learning that it cannot separate the important from the unimportant parts of the brain for learning and memory. Or it may be that learning, or different facets of learning, occur in many places of the brain at once, and that therefore, no one part is completely essential for its formation or retention.

When we come to the second question, concerning the nature of the change in learning, we find mostly theories and very little facts. Various mechanisms have been suggested as being responsible for the establishment of new functional connections in the nervous system: (1) the growth of new have pathways, (2) anatomical swilling or sprouting of synaptic terminals, resulting in the facilitation of crossing Grain synapses, (3) a physiological increase in the ease of crossing synaptic connections already established but not functional at the start of training, and (4) biochemical changes such as alterations in the structural arrangement of protein molecules in nerve fibers. Of all these suggestions, one of the most intriguing has been the physiological concept of recurrent nerve circuit in which a loop or circle of connecting neurons is activated such that each neuron in the circle activates the next until the first one, having time to recover, is activated again. Such a loop could theoretically continue firing indefinitely and serve to add facilitation to any synapses in makes outside the loop and thus provides the basis for long-term memory. At the present time, however, neither this nor any of the other theoretical suggestions have any direct evidence bearing on them.

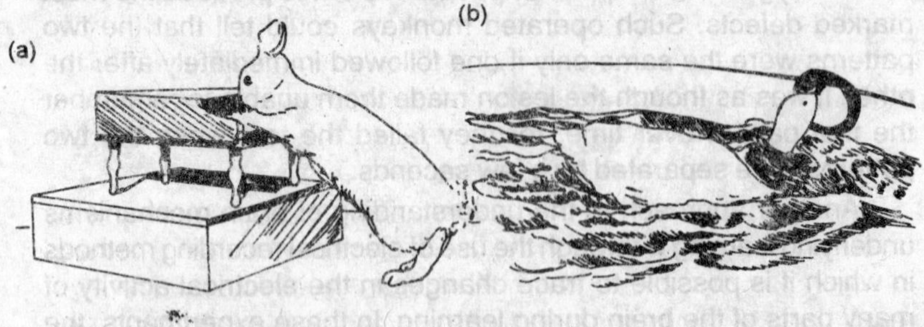

Fig. 4.15. Examples of tricks taught to animals by Keller Breland, using operant conditioning methods. (a) The rabbit plays up and down the keyboard several times to obtain a reward. (b) The duck must collect a number of rings before obtaining a reward.

Some insight into the nature of memory mechanisms has been gained by direct, experimental examination of temporal characteristics of the memory process. In one study, rats were trained to run from one compartment to another to avoid an electric shock. They were given one trial a day, and after each trial, they received an electro-convulsive shock through the head. Different groups of rats received the convulsive shock at different times after the learning trail 20 sec, 1 min, 4 min, 15 min, 1 hr, 4 hr. If the shock came within an hour, there was virtually no learning, but if it came after four, hours, learning was essentially normal. Apparently, memory takes time to "set" in the brain, and this consolidation requires at least an hour, which suggests that memory is a two-part process, consisting of an early phase when memory is vulnerable to convulsive shock and a later phase when it is not.

The same kind of conclusion turns up in two other rather different studies. In one, the octopus was trained, as we mentioned earlier, to discriminate between a crab it could eat and a crab, accompanied by a white card, that it could not approach under penalty of electric shock. If the octopus vertical lobe, an associational region of the brain, was removed, then a curious thing happened. If the trials were spaced more than an hour apart, the animal could not retain enough from trial to trial to improve its performance in normal fashion. If the trials were within fifteen minutes of each other, however, it was able

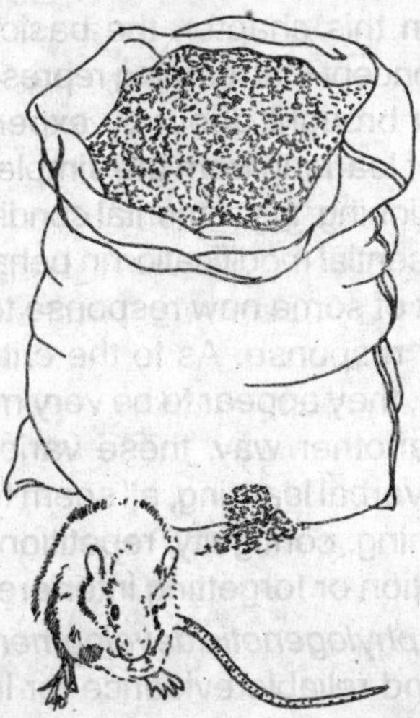

Fig. 4.16. A rat feeding in its natural environment.

to learn easily. Apparently, lesion of the vertical lobe affected the "permanent" laying down of memories, but did not disturb their "temporary" establishment.

In man, lesions of the temporal cortex on both sides of the brain may result in a similar defect. These patients can learn something simple and retain it for about fifteen minutes to an hour, but after the time, they forget completely and may not even remember having learned. Yet their life-long memories are left undisturbed. Taken together with the animal studies, this finding suggests that memory is a two-part process: (1) an initial, vulnerable, perhaps physiological process lasting fifteen minutes to an hour, and (2) a later, invulnerable, perhaps anatomical process, providing the permanent basis for memory.

Many mysteries remain in our quest for the physiological basic of simple learning. We know that learning is a property of is a property of at least all animals possessing a synaptic nervous system. The major question is whether the superior learning of mammals and especially primates is due to the development of superior neural mechanisms for learning and memory or to their greatly increased sensory and motor capacities or both. The development of superior neural mechanisms must be an important factor, for the capacities for complex learning, problem solution, and reasoning emerge with the evolution of the central nervous system.

We have reviewed, in this chapter, the basic facts of animal learning. We began with concept that learning represents an *enduring modification of behaviour* brought about by experience. Then we described various kinds of learning from the simple to the complex: habitation, classical conditioning, instrumental conditioning, and trial-and-error learning. The essential modification in behaviour in all these cases in the development of some new response to a stimulus that nerve before elicited that response. As to the critical elements of the experience in learning, they appear to be very much the same in all these cases. Or put another way, these various instances of learning including human verbal learning, all seem to obey the same fundamental laws of learning: contiguity, repetition, reinforcement, and for the case of extinction or forgetting interference.

When we took up the *phylogenetic development of learning,* we could not find clear-cut and reliable evidence for learning until the level of the worms. This is the point in phylogeny where the bilaterally symmetrical, synaptic nervous system first appears. With the cephalopods and arthropods, which have relatively large, concentrated ganglionic masses in the anterior regions of the nervous system, learning ability is much greater than in the worms. Finally, with the development of the vertebrate brain, learning capacity develops even further, gradually reaching an assymptote among the simpler mammals.

Despite this evidence that relates learning ability to the development of the central nervous system in phylogeny, it has not been easy to discover, with any degree of specificity, the neural basis of learning. The evidence from experimental brain lesions and, more particularly, from studies that record changes in the electrical activity of the brain during learning suggests that learning takes place in many places within the brain at once. The nature of the neural change in learning has proven elusive, however, despite the fact that many attractive and plausible theories have been proposed. At present, we know that plusible theories have been proposed. At present, we know that learning or, more particularly, the formation of memories is at least a two-part process. Initially, there is a temporary, perhaps physiological process lasting up to an hour in the mammalian nervous system. Following this and perhaps as a result of it, there a second, more permanent, perhaps anatomical change. Still a third mechanism may subserve the storage of long-standing memories, for it is possible to impair the permanent laying down of new memories by brain lesion without impairing either old, long-standing memories, or the temporary acquisition of new memories.

5

MOTIVATION

When we say that an animal, or a person, is motivated to do something, we generally imply that their behaviour is driven or directly by some internal force or urge.

Much of the causal analysis of animal behaviour has been concerned with identifying the relationship between responses and external stimuli. Many activities only occur in the presence of relevant stimuli an animal cannot eat in the absence of food, for example. It can, however, look for food when it is not immediately available, and such food searching is behaviour that can only be explained in terms of internal processes. A more general point is that in many of their activities, animals do not always respond in the same way to external stimuli. Whereas a rat may withdraw its foot every time it steps on a hot surface, it does not eat whenever it is presented with food. Such process, whatever they may be, are the stuff of motivation.

If asked to provide a causal explanation of why an animal does not eat when presented with food, our most likely answer is that the animal is not hungry. This appears to provide an answer by invoking an internal process but in fact it really only describes what we have seen in different words and serves only to arise more questions. What is hunger and what are the physiological events that produce it? The essence concepts such as hunger in terms either of observable physiological processes or of rules which may help us to predict when a particular behaviour patterns will occur.

Before going further, we should consider briefly other possible reason we might have used to explain an animal's failure to eat. One possibility is that the animal was engaged in another, perhaps more important activity, such as seeking a mate. The motivation of one activity is dependent on that of others. Another factor that may affect the way an animal responds to external stimuli is its stage of development. Young mammals being fed on their mother's milk do not respond to food they will eat as adults and, likewise, fully formed sexual responses generally do not appear until an animals is mature. Ontogenetic changes of this kind are not regarded as manifestations of motivational processes because they are irreversible; mature animals do not revert to immaturity. By contrast, being hungry, thirsty or sexually aroused are states that animals can experience frequently. Like maturational changes, learned responses to stimuli are usually also considered to be distinct from motivational process because of their typically long-lasting influence on behaviour. Particular noxious flavours may be experienced once and avoided thereafter, and an animal's mating preferences may be restricted by early experience of its parents or siblings. Such effects cannot be ignored, however, but must always be taken into account in studies of motivation; as with other aspects of behaviour, motivation is subject to maturation and is influenced by learning.

Many American psychologists emphasized the motivational aspects of instinct and start with the consumption that many patterns of instinctive behaviour can be analyzed upon drive directed ago. The attainment of which results in reduction of he drive. For example: The case of a 3 years old boy which an abnormal craving for salt. For early life he always preferred salty foods and would like the salt of crackers rather than eat them. When he was 18 months-old he discovered the salt sheckers and began eating salt by the spoonful. He learned to point to the cover cuppled to take the salt shekered. The first would he learned was salt. It turned-out that his craving for salt had kept him alived. When he was taken to the hospital for observation and placed on a standard hospital type with a limited salt, he died within 7 days. At autopcy it was learned that he had tumors of the adrenal glands and thus lacked constantly necessary to reabsorb salt at the kidney only by constantly replacing salt lost in his urine. He could maintain himself. Thus it can be seen that drive is a striking towards some goal to get salt.

Intervening Variables

Although hunger is a sensation of which we have direct experience, it is a hypothetical construct in the analysis of animal behaviour: it is something we cannot observe or measure directly, but can only infer from variations in overt behaviour. Since hunger is invoked to explain variations in response to food it is referred to as an intervening variable, because it has some modulating effect between a stimulus and its associated activity. Discussions of motivation are littered with intervening variables, such as thirst, sexual arousal, fear and, more generally, drives of one sort or another. The question we must ask if whether such intervening variables help in understanding motivation.

The basic problem with intervening variables is that any attempt to explain things in these term involves a circular argument. We may infer that because an animal is eating voraciously, it has a high feeding drive. However, our only evidence for that assertion is the nature of the animal's behaviour, which is the very thing we are seeking to explain. It may appear that we break this circularity if we identify factors that influence feeding drive. For example, animals generally eat more readily if they have been deprived of food, and so we could say that food deprivation increases feeding drive which stimulates feeding. We can express this effect more economically, however, by simply saying that food deprivation stimulates feeding.

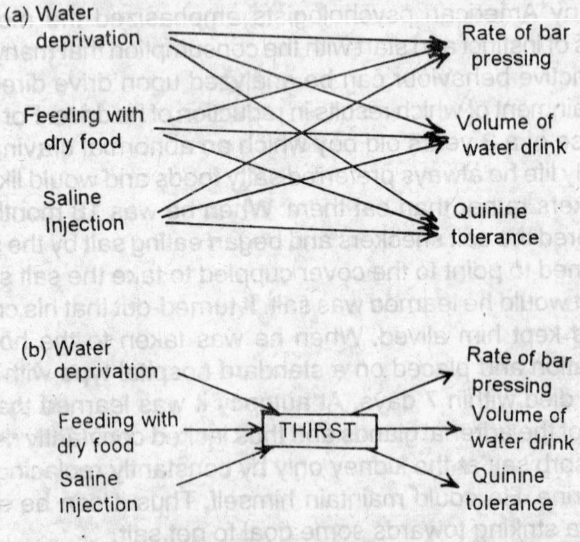

Fig. 5.1. The role of a hypothetical intervening variable in simplifying the relationships between casual factors and behavioural measures. (a) The tree treatments on the left all influence the three scores of drinking behaviour on the right, in rats, giving nine casual relationships. (b) By introducing 'thirst', the number of causal relationships is reduced to six.

A hypothetical intervening variable may have some explanatory value if it enables us to simplify observed relationships between several stimulus inputs and a number of different behavioural outputs (*Hinde,* 1982), three treatments that increase the tendency to drink in rats: water deprivation, feeding with dry food and injection with saline solution. The effect of all three treatments can be observed in three measures of drinking behaviour : the rate at which a rat presses a bar to obtain water, the volume of water it drinks, and the concentration of quinine (a distasteful substance) it will tolerate in the water it drinks. The number of relationships between these treatments and measures can be reduced from nine to six by introducing thirst as an intervening variable.

Invoking a single variable, such as thirst, could therefore be useful proved that the motivational process underlying the activity concerned constitute a single process. But there is no a *priori* basis for such as assumption and, indeed, there is abundant evidence that it is not justified in systems that have been studied in detail. There study of drinking in rats by *Miller* (1959) provides an example

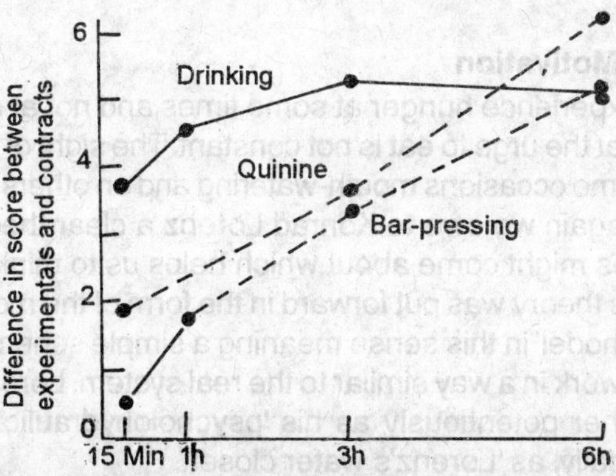

Fig. 5.2. Changes over time in three measures of drinking behaviour in rats following injection of saline solution into their stomachs.

of such evidence. He gave rats saline solution by means of a tube inserted into their stomachs and then recorded that three measures of drinking mentioned above the various time intervals after the treatment. One would assume that, as time passed, the rat's thirst would have increased and that, if their behaviour is indeed influenced by a single intervening variable, the three measures should be closely correlated. One measure, the concentration of quinine that rats will tolerate in their water, rises at a constant rate, as one would expect if thirst increased as a simple function of time elapsed since the treatment. By contrast, another measure, the volume of water drunk in 15 minutes, shows an increase only over the first three hours and then reaches an asymptote. On this scope rats seem to be no thirstier after six hours than after three. One explanation of this effect might be that there is a limit to the volume of water a rat can drink in 15 minutes. These results suggest that different measures correlate rather poorly with one another, a finding which is contrary to the idea that these measures of drinking are all influenced through a single variable that we could call thirst.

In summary, when an intervening variable is involved, we should not believe that anything has been explained; all that has been done is that some phenomenon that requires explanation has been identified. We cannot even assume that a unitary process is implicated. At best, intervening variables are labels for unknown physiological processes. The ultimate aim of much research into motivation is to identify and understood how such processes work, so that concepts such as hunger, thirst are drive become unnecessary.

A. Model of Motivation

We all experience hunger at some times and not at others and we realise that the urge to eat is not constant. The sight of a delicious meal is on some occasions mouth-watering and on others on interest at all. Once again we owe to Konrad Lorenz a clear theory of how such changes might come about which helps us to think about the problem. This theory was put forward in the form of the model shown in the word 'model' in this sense meaning a simple scheme which is proposed to work in a way similar to the real system. Lorenz's theory is known either potentiously as his 'psycholohydraulic' model or, more prosaically, as 'Lorenz's water closet.'

This model is not something one would look for inside an animal! It is what is called 'as if' model : the animal may be have as if it had

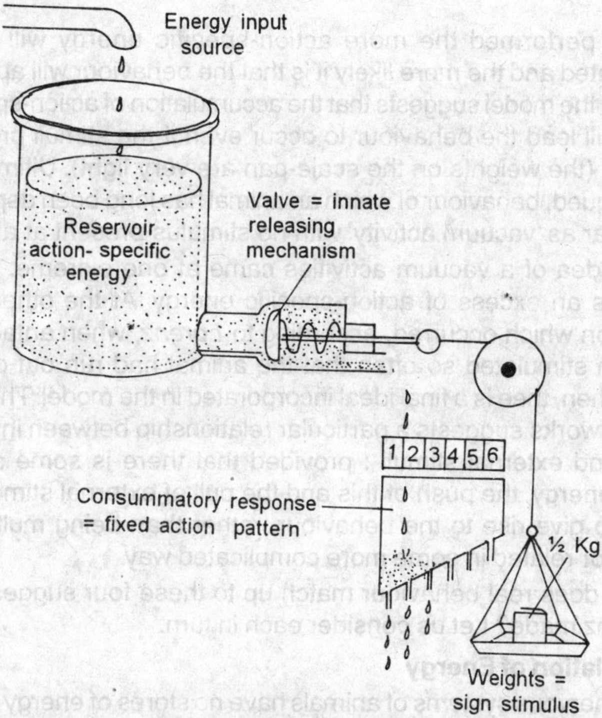

Fig. 5.3. One early scheme to explain motivation was developed by Lorenz, often called the psycho-hydraulic model.

such a system of organising its behaviour within it. The theory helps one to think about the problem and design experiments to see how the system really works rather than proposing specific mechanisms. Lorenz supposed that different actions depended for their appearance on a supply of 'action-specific energy' which accumulated with time since the animal last behaved that way and was used up as if performed the act. He visualised this as water accumulating in a tank out of which it could only escape through a valve at the bottom. The valve was, however, a rather strange one. It could be opened either by the water pushing within, or by a string attached to a scale-pan pulling from outside. To Lorenz, weights on this scale-pan were the equivalent of stimuli leading to the behaviour: the more appropriate the stimulation, the heavier the weights and the more likely the valve was to be opened.

Lorenz's model has some interesting properties which can be compared with real behaviour. First, the longer since the behaviour was last performed the more action-specific energy will have accumulated and the more likely it is that the behaviour will appear. Secondly, the model suggests that the accumulation of action-specific energy will lead the behaviour to occur even if the stimuli present are slight (the weights on the scale-pan are very light). Ultimately, lorenz argued, behaviour of which an animal has long been deprived will appear as 'vacuum activity' with no stimulus present at all.

The idea of a vacuum activities came at one extreme: when there was an excess of action-specific energy. At the other was exhaustion which occurred, according to *Lorenz,* when an activity had been stimulated so often that the animal had run out of this energy. Then, thee is a final ideal incorporated in the model. The way the valve works suggests a particular relationship between internal factors and external stimuli : provided that there is some action specific energy, the push of this and the pull of external stimuli will add up to give rise to the behaviour, rather than being multiplied together or related in some more complicated way.

How does real behaviour match up to these four suggestions the Lorenz made? Let us consider each in turn.

Accumulation of Energy

The nervous systems of animals have no stores of energy within them as Lorenz proposed, and this is obviously not a realistic aspect of the model. Nevertheless, animals might behave as if they did and, if so, we would expect them to become more likely to perform a particular action with the passage of time since they last did so. Do they do this?

There is no doubt that certain aspects of behaviour do become more likely the longer the gap since they last appeared. Feeding and drinking are the most obvious of examples, as we all know from our personal experience. There are very good reasons for this nutrients are used up by the body and water is lost constantly from it, so both must be replenished and the need to do so will rise with the interval since the last meal or the last drink. But many of the interval since the last meal or the last drink. But many of the other activities which animals perform are not concerned with regulating aspects of physiology, so there is no reason to think in advance that they might become more likely with time, and they are quite often found not to do so.

Many behaviour patterns, such as singing in birds, mating in fish or exploration in rodents, could be used to illustrate this point.

However a particularly appropriate one to take a aggression. Although he did not mention his psychohydraulic model in his book. *On Aggression,* Konard Lorenz obviously had it in mind. He suggested that aggression is an innate drive which rise with time and must somehow be expended. His view of innate behaviour is that it is inevitable and the only alternative is to sidetrack it into harmless channels: for human aggression the suggests that sport may play such a role, helping us to get rid otherwise destructive urges.

Leaving aside the issue of whether sportsmen are less aggressive than other people as a result of their activities, there are a great many objections to these views. Some of them concern whether the urge to behave aggressively does accumulate in an inevitable fashion. One of the few cases where it seems to do so is where an animal is isolated, for example a mouse placed in a cage on its own. The longer it has been in isolation the more it will fight another mouse which is put in with it. But this could simply be that, when in company, it habituates to the other animals with it and it needs some days on its won to recover its tendency to fight them. In fact, some fish fight less after isolation and need stimulation from others over a period

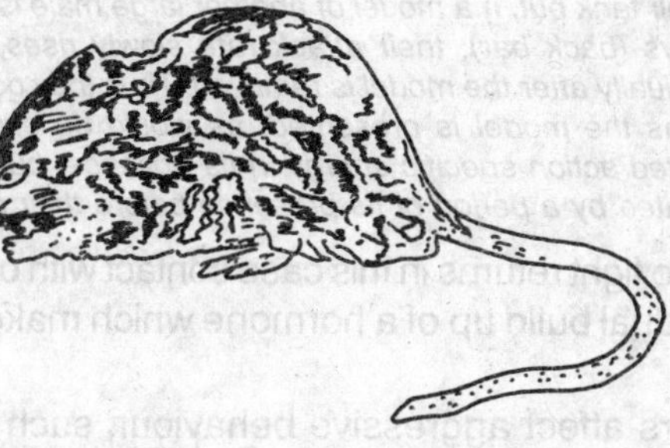

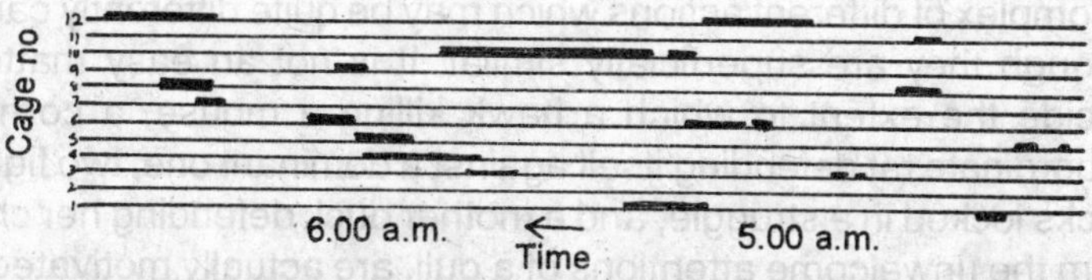

Fig. 5.4. Traces from a pen recorder to show the feeding behaviour of 12 rats over a period of about 2 hours. The animals tend not to nibble at food the whole time, but pattern their feeding into meals, taking one of these at fairly regular intervals. Thus the longer it is since its last meal, the more likely it is that a rat will eat.

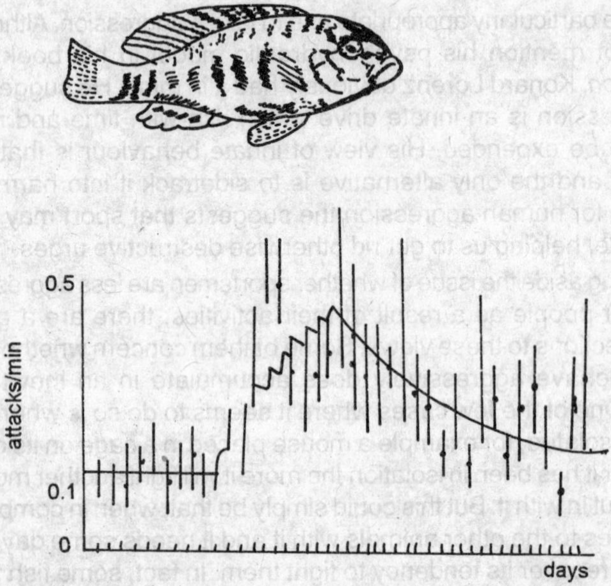

Fig. 5.5. Males of the cichlid fish Haplochromis burtoni seldom attack small fish kept in their tank but, if a model of another large male is put in with them for 10 days (back bar), their attack rate slowly rises, then falls back again gradually after the model is removed. Thus they do not attack most as soon as the model is presented, as would be expected had they accumulated action specific energy in its absence, but instead need to be stimulated by a period of its presence before their attack rate-rises.

before the urge to fight returns in this case contact with others probably leads to the gradual build up of a hormone which makes aggression more likely.

Many factors affect aggressive behaviour, such as hormones and shortage of food on the intake, and presence of rivals and contested resources on the outside. Furthermore, aggression is itself a complex of different actions which may be quite differently caused through they are superficially similar. It is not an easy matter to decide the extent to which a hawk killing a mouse, a cornered subordinate rat defending itself against a dominant one, two fighting cocks locked in a struggle, and a mother duck defending her chicks from the unwelcome attentions of a gull, are actually motivated in a similar way. Aggression serves many different functions within one species and its uses and the situations in which it appears may vary between species, so that the mechanisms underlying it are also likely to vary. Evolution matches behaviour marvellously well

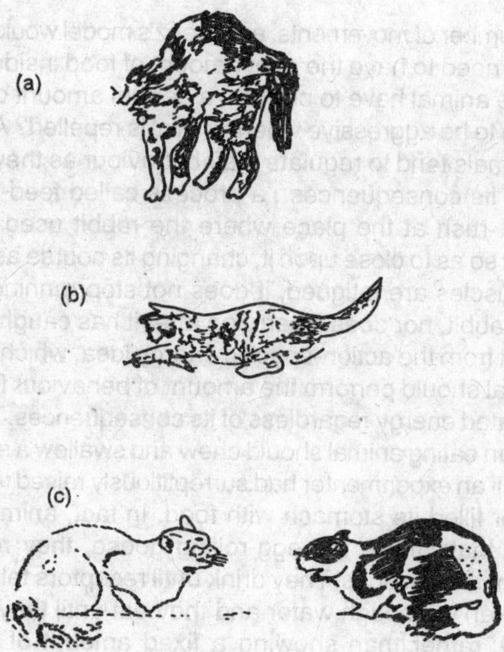

Fig. 5.6. A cat cornered by a dog shows defensive behaviour, fighting back as best it can, (a) Cats will also attack mice they encounter, here pouncing and showing other predatory actions as well as using their teeth and claws, (b) In territorial disputes cats may fight intruders of their own species, (c) Many of the actions involved in these different situations are similar. Although all of them involve attempting to inflict damage on other individual they probably have very different causes, so that it is not useful to think of them all simply as examples of aggression.

to an animal's particular environment and way of life. So it should be no surprise that the organisation of aggression varies a lot between species just as feeding differs between the lion that kills and eats every few days and the wildebeest that must crop it plant food for most of its waking hours. Likewise, there is no reason why the urge to behave aggressively should mount with time like hunger or thirst, and it is rather surprising that Konard Lorenz should have thought this, given that he is a biologist with great respect for the power this, given that he is a biologist with great respect for the power of evolution to adapt behaviour to an animal's exact requirements.

A final aspect of the accumulating energy idea concerns the extent to which an animal needs to perform the action as opposed to needling to achieve its results. Must a hungry animal make the appropriate number of movements, as Lorenz's model would propose, or does ti just need to have the right amount of food inside it? Does an aggressive animal have to perform a certain amount of fighting, or will it cease to be aggressive when its rival is repelled? As already seen that animals tend to regulate their behaviour as they go along in the light of its consequences : a process called feed-back. The dog does not rush at the place where the rabbit used to be but follows an arc so as to close upon it, changing its course as required. Unless its muscles are fatigued, it does not stop running before it reaches the rabbit, nor does it carry on after it has caught it. This is quite different from the action-specific energy idea, which proposed that the animal should perform the amount of behaviour for which it has accumulated energy regardless of its consequences. This would suggest that an eating animal should chew and swallow a set number of times even if an experimenter had surreptitiously raised the glucose in its blood or filled its stomach with food. In fact, animals do not generally do that; unlike the egg-rolling goose, they respond to feedback from their actions. They drink until receptors tell them that they have taken in enough water and they eat until they have had enough food, rather than showing a fixed amount of behaviour depending on the time since they last ate or drink.

Vacuum Activities

There are few examples of the idea that animals deprived of an opportunity to perform an action might eventually show it even in the absence of all stimuli. Lorenz himself described a pet starling that would flutter up to the ceiling to catch a non-existent fly, but it is hard to be certain that there was nothing there to which the bird was reaching. There is no doubt that behaviour can sometimes show stimulus generalisation, so that animals deprived of the usual stimulus will show the behaviour to a much less adequate one. In zoos, they will often mate with the wrong species when their own is not available : lions tigers, for example, will mate to produce 'tigers.' As we all know, the hungrier one is the more one is prepared to eat less tasty food (indeed, if very hungry, I suspect even I might eat rice pudding!). Thus, generalising to less adequate stimuli is a reality, but there is not so much evidence for behaviour begin shown in the total absence of any appropriate stimulus, as the vacuum activity idea suggests. Perhaps the best is an account by the ornithologist *David Lack* of the behaviour of a European robin he had just finished testing with a stuffed bird of its won species, a potent releaser of

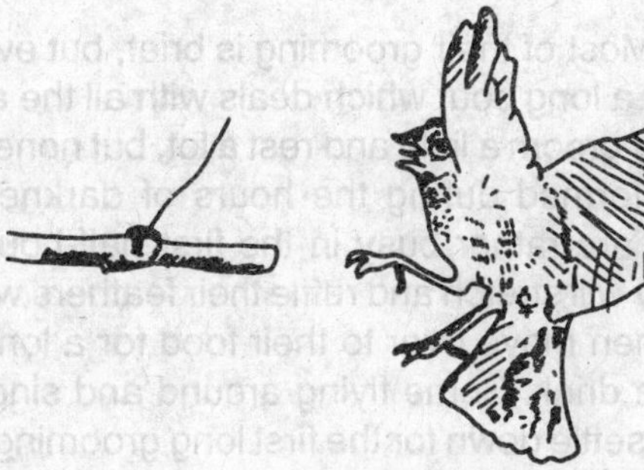

Fig. 5.7. A European robin attacks the spot in mid-air where previously a stuffed rival had been placed.

aggression for these birds. As he removed the specimen and walked off, he chanced to look back and saw the robin fluttering around, singing and delivering pecks to the position in mid-air where its apparent rival. Here the appropriate stimulus had clearly been removed, but the animal was certainly not suffering from deprivation of the opportunity behave aggressively.

Deciding what to do

Animals generally only do one thing at a time, yet they often have the need to perform several. For example, a caged zebra finch waking up after 12 hour night must have a long list of priorities. These birds normally eat and drink about every half-hour, sing and fly around their cage for a period after this, and groom and rest till

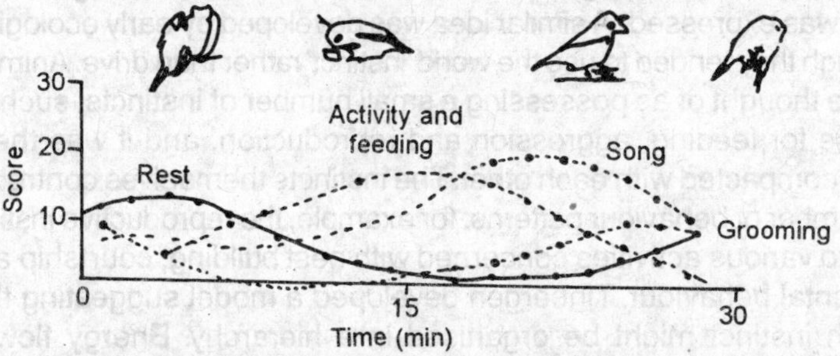

Fig. 5.8. The short-term cycle of behaviour shown by a male zebra finch. The illustration is based on observation of 12 cycles shown by a single bird synchronised at 30 minutes long by a cycle of changing light intensity. In each cycle the bird moves from resting to an active period during which it feels, through a peak of singing to grooming, and then back to resting once more.

the next meal is due. Most of their grooming is brief, but every two hours or so they show a long bout which deals with all the areas of their body. At night they groom a little and rest a lot, but none of their other activities is performed during the hours of darkness. Not surprisingly, the, they are rather busy in the first half-hour of the morning. Typically, they will stretch and ruffle their feathers when the lights come on, and then move over to their food for a long meal, perhaps followed by a drink. Some flying around and singing will follow this before they settle down for the first long grooming bout of the day.

Despite their need to do several things, these birds show a well-organised sequence of behaviour. They do not rush around doing a little bit of one thing and a little bit of another, nor do they try to sing and groom at the same time. The sequence shown by different birds is similar and suggests that they all have the same priorities. The question posed in this section is how they decide between them, and it is a difficult question to which there is as yet no definite answer. What is clear is that the animal must make decisions and that to do so it must have some means of weighing against each other their internal and external factors relevant to different activities.

The realisation that some sort of weighing up of opinions must go on within animals was one reason why psychologists postulated the existence of internal 'drives' which, like Lorenz's action—specific energy, were thought of as powering behaviour and were also seen as being measured against each other, the strongest being the one that was expressed. A similar idea was developed by early ecologists though they tended to use the world 'instinct' rather than drive. Animals were thought of as possessing a small number of instincts, such as those for feeding, aggression and reproduction, and it was these that compacted with each other. The instincts themselves controlled a number of behaviour patterns; for example, the reproductive instinct led to various activities concerned with nest building, courtship and parental behaviour. Tinbergen developed a model suggesting that each instinct might be organised in a hierarchy. Energy flowed downwards within this from the controlling centre at the top of centres responsible for groups of related activities and then further down to those concerned with individual actions. At each level the appropriate releasers had to be present, these acting to open gates through which the energy could flow to the level beneath and so on to give behaviour.

This is ingenious idea, and unlike the 'as if' model *Lorenz* proposed, was based by Tinbergen on the ideas of psychologists; he hoped it might have some neurophysiological reality, with various centres in the brain devoted to different instincts. Unfortunately, it turns out not to be that simple. As mentioned before, the nervous system does not store and use up energy as these models suggest. Nor, unfortunately, it is nearly compartmentalised into centres and pathways with clear and distinct behavioural functions. Furthermore, thought to some extent behavioural systems, like those controlling

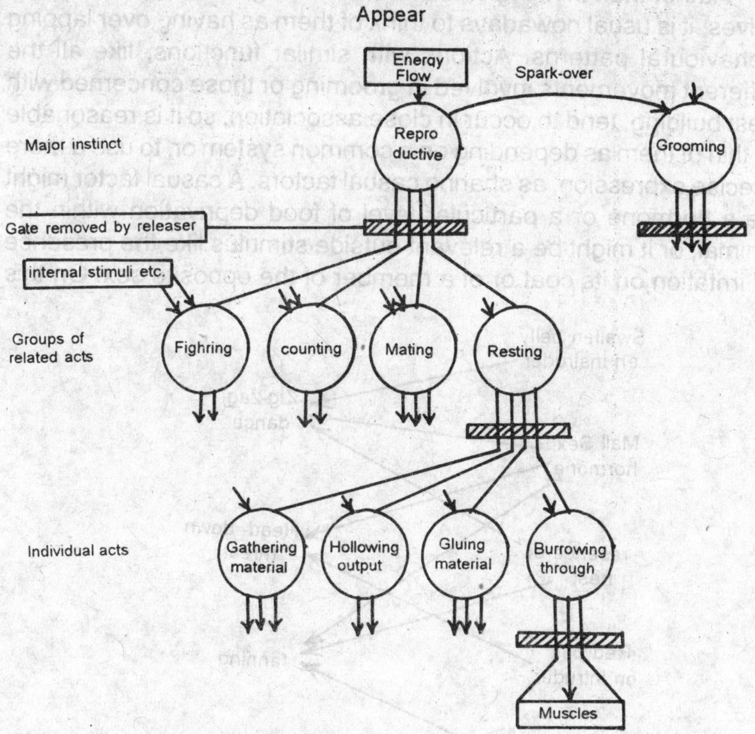

Fig. 5.9. The essence of Tinbergen's hierarchial model was that a centre in the brain controlled each 'major instinct.' Energy from this flowed down to a series of lower centres when a gate was opened by the presence of appropriate releasers. In his original scheme, Tinbergen saw the hierarchy being extended right down to the level of muscle units. Though the hierarchis of different instincts were separate, lack of releasers for one could lead energy from it to 'spark-over' to another and so cause a displacement activity to occur. Thus, as in the example shown here, if reproductive activities were thwarted, grooming might appear.

feeding, drinking or sexual behaviour, can be thought of as distinct, the factors affecting them overlap and they often influence each other. For example, the hormone oestrogen is an important internal factor making female rates receptive to the male, it also makes them very active so that they are more likely to come across a male, and it makes them less interested in food so that, when receptive, they spend less time eating and more looking for mates. This hormone, therefore, influences systems concerned with feeding, with activity and with sexual behaviour.

Rather than thinking of animals as having a number of discrete drives, it is usual nowadays to think of them as having over lapping behavioural patterns. Actions with similar functions, like all the different movements involved in grooming or those concerned with nest building, tend to occur in close association, so it is reasonable to thin of them as depending on a common system or, to use a more precise expression, as sharing casual factors. A casual factor might be a hormone or a particular level of food deprivation within the animal, or it might be a relevant outside stimulus like the presence of irritation on its coat or of a member of the opposite sex. On this

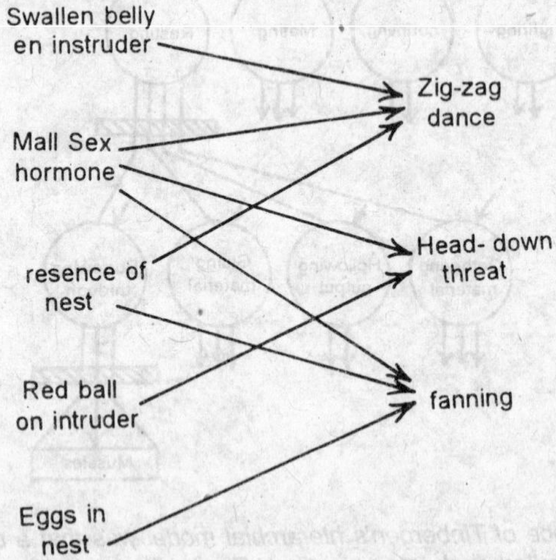

Fig. 5.10. Current views of motivation do not see animals as endowed with a series of separate 'drives' or 'instincts.' Instead different aspects of behaviour are thought of as influenced to varying degrees by numerous internal and external causal factors. Here some possible stimulating effects of five different causal factors on three-spined stickleback patterns are shown.

view of behaviour, whether or not a particular activity is shown depends on the combined level of its casual factors relative to those of all other possible actions.

For several reasons this is a more satisfying way of describing the motivation of behaviour than simply to consider it to be the result of a series of drives. One of these is that the drive idea is only useful if all aspect of the behaviour can be thought of as affected by the drive in the same way. Careful studies of motivation run counter to this. Most obviously, a causal factor may raise some aspects of behaviour and lower others, whereas if it was enhancing drive, all aspects of the behaviour should be increased. Thus a very hungry animal may chew less and swallow more, and a bird building a nest actively may fly to and from with material at a great rate but spend less time than usual carefully selecting and gathering material or in weaving it into the structure.

Another reason why viewing behaviour as dependent on a variety of internal and external causal factors is useful is that if helps to account for why behaviour patterns occur in association with one another to varying degrees. A grooming animal will wash its face, scratch its flank and shake its body, and these very different actions tend to occur close together rather than in association with feeding or mating movements. The most likely reason for this is that grooming actions share causal factors with each other, such as activity in a particular part of the brain or irritation of the coat. Another reason whey different behaviour patterns occur in association may be, quite simply, that one causes another. The fact that birds wipe their breaks after drinking almost certainly comes into this category: drinking makes the beak wet and so produces a stimulus which is only removed by wiping. Thus, as well as behaviour patterns sharing causal factors and thereby appearing close together, they can be linked because one acts as a causal factor for another.

The way in which behaviour patterns may complete for expression has been more extensively studied in the case of feeding and drinking, because these are two acts the causal factors for which can be manipulated by depriving animals of food and water for different periods of time. An animal deprived of water tends not to eat as much (doubtless dryness in the mouth makes it difficult to chew and swallow), so the two actions are not totally independent of each other. But just how hungry and thirsty a deprived animal can be established by seeing how much it cats and drinks when given food and water. We can then say it was 10 grams hungry and 1 gram thirsty if these are the amounts it consumes.

The most detailed experiments on the relationship between feeding and drinking have been those on Barbary doves (known as ring doves in North America) by *David McFarland.* His research is a good example of how the use of techniques developed by psychologists has helped to illuminate ethological problems. The doves he studied were placed in the 'Skinner boxes' that psychologists

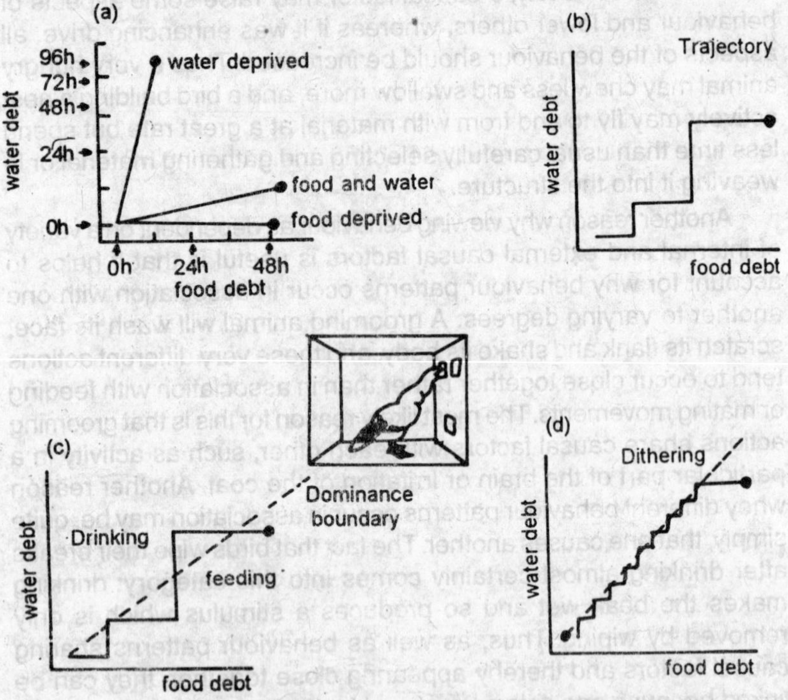

Fig. 5.11. The way in which a deprived Barbary dove replenishes its reserves when placed in a Skinner box where it can peck keys to earn food and water. (a) The food and water debt of the bird can be represented by a point in a 'state space.' Here three such points are shown for birds deprived of food alone, food and water, and water alone. Birds eat less when deprived of water, so the point for the last situation shows a small food debt as well. (b) A bird deprived of both food and water eats and drinks alternately, so tracing a trajectory in its state space which takes it down to the point where it has cancelled both debts. (c) The boundary drawn to separate areas where the animal starts to feed (food dominant) from those where it starts to drink (water dominant). The animal tends to feed until it is well into the water dominant area and vice versa, rather than dithering between the two, as shown in (d).

often use to study learning, in which they could peck one key to receive water and another to receive food. He found that a dove deprived of both food and water would alternate eating and drinking until it had satisfied both requirements. Its food and water needs, and how these changed as it ate and drank, could be neatly illustrated in the form of 'state space' plots. If the bird was more hungry than thirsty it would start by eating, and the amount that it ate could be traced by a line on the graph. After eating for a little it would then become more thirsty than hungry and would switch to drinking, the line on the graph moving so that the water deficit was shown as decreasing, while that for food remained the same. Finally, after several drinks and meals, the line would have zig-zagged down to zero on both axes, indicating that the animal was satiated.

On this graph we can assume that every time the animal switches from one action to the other it is moving to that which has greater causal factors. We can draw a diagonal line which gives a rough impression of the boundary between feeding having priority and drinking doing so. All switches in the drinking segment are from feeding to drinking, and those in the eating one are in the other direction. A feature which may seem curious here are in the other direction. A feature which may seem curious here is that the animal goes well over the line in each direction before it switches rather than doing so immediately.

If it did the latter, it would end up 'dithering' : after an initial large meal or drink to take it over the boundary line, it would oscillate rapidly from one activity to the other, taking tiny amounts of each in turn until eventually both needs were satisfied. This would obviously be a badly organised way of doing things, but what makes sure that it does not happen? Several factors are probably involved, but one is likely to be especially important. This is that an animal that has just taken an item of food is likely to be looking at a food dish rather than at stimuli appropriate to any other action. As the sight of food is a causal factor for feeding, and that of water for drinking, this will mean that the animal is likely to continue with what it is doing rather than switching to another behaviour for which internal factors may well be high, but external ones are absent.

In this simple example, the state space plot expresses the relationship between two activities, feeding and drinking, but a third dimension might be added so that the animal's tendency to groom was also mapped in the hope that one night be able to predict when it would take a break from feeding and drinking to clean itself. Ultimately, one might hope to achieve mapping of all behaviour patterns onto each other in a 'multidimensional state space' but, as well as it being difficult to imagine such a complicated diagram, it is not easy to see what currency they could all be expressed in. There is certainly no such thing as 'grams' of grooming or of tendency to fight. So the problem of how the nervous system weighs up different actions against each other remains as real as ever.

The Physiological Basis of Motivation

Continued analysis of this behavioural type is essential, but it is also important to examine motivation at the physiological level and try to link up behaviour with events in the nervous system. Some of the most promising bridges between neurophysiology and behaviour have developed from the work of physiological psychologists—most of the Americans—on motivational problems. *Grossman* provides a good introduction to the whole field of physiological psychology. We must now turn to consider some of this work and begin with some description of one area of the vertebrate brain which has proved to be of major importance in the control of motivation—the hypothalamus.

The Hypothalamus

This relatively minute volume of brain tissue—in the human brain it is smaller than the last joint of the little finger—is of primary importance in a whole host of reactions. These is an excellent general account of its physiology in *Walsh* who says the hypothalamus, 'thus small centre plays a dominant role in determining the use that is made of the resources of the body. It is difficult, indeed, to think of any function of the body that is not dependent, directly or indirectly, upon the hypothalamus'.

A little neuro-anatomy is needed at this point; the reader is referred to a clear, concise account in *Romer.*

The brain of all vertebrates is constructed on the same basic plan, and at an early stage in embryology consists of three swellings at the anterior end of the spinal cord. These are called the prosencephalon, mesencephalon and rhombencephalon or more simply the fore-, mid- and hind-brain. Primitively these swellings arose to cope with the increased amount of sensory information which flowed into the central nervous system from the sense organs of the head. The fore-brain originally dealt with olfaction, the mid-

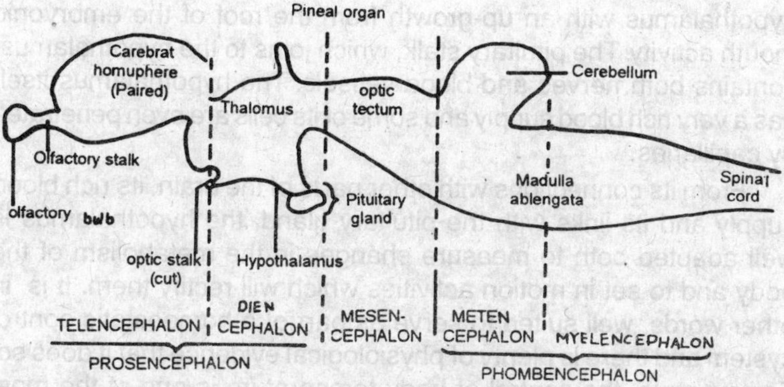

Fig. 5.12. The basic divisions of the vertebrate brain. The brains of all vertebrates pass through a stage rather like this during development, but in mammals and birds in particular, the adult brain is dominated by the enormous growth of the cerebral hemispheres and cerebellum.

brain vision and the hind-brain balance and hearing. In most living vertebrates their original functions have become greatly extended and complicated, but still the appropriate sensory data are led first to these regions, even if subsequently they are passed on elsewhere.

The fore-brain is easily sub-divided into two portions. The anterior portion has arising from its roof the cerebral hemispheres, which primitively were olfactory areas but have now come to dominate the whole nervous system in mammals. The posterior portion of the fore-brain—called the diencephalon—has on its dorsal surface the pineal organ. This was once associated with a light receptor or pineal eye which can still be seen in some living reptiles. The side walls of the diencephalon are thick and form the thalamus, and important 'staging place' in the brain where fibre tracts link up with one another in nuerous 'nuclei' or clusters of neutron cell bodies. On the floor of the diencephalon, below the thalamus as its name implies, is the hypothalamus.

There are nuclei in the hypothalamus, but they are not so well defined as in the thalamus above. However, there are several well-marked fibre tracts entering and leaving, and these put the hypothalamus into connection with the cerebral hemispheres and also with more posterior parts of the brain. From the behaviour point of view, one of the most significant structural features of the hypothalamus is its intimate connection with the pituitary gland. This endocrine gland, which controls the whole hormonal system of the body, develops from the fusion of a down-growth of the embryonic hypothalamus with an up-growth from the roof of the embryonic mouth activity. The pituitary stalk, which joins to the hypothalamus, contains both nerves and blood vessels. The hypothalamus itself has a very rich blood supply and some of its cells are even penetrated by capillaries.

From its connections with other parts of the brain, its rich blood supply and its links with the pituitary gland, the hypothalamus is well adapted both to measure changes in the metabolism of the body and to set in motion activities which will rectify them. It is, in other words, well suited to serve as part of a homeostatic control system and there is plenty of physiological evidence that it does so. For example, the control of body temperature is one of the most delicate homeostatic systems in a mammals or a bird. There are areas of the hypothalamus which are highly sensitive to changes in the temperature of the blood. If these areas are heated artificially by implanted wires, the animal starts sweating and painting. Sweating is controlled peripherally by the autonomic nervous system and the hypothalamus can initiate its activity. The reverse effect is produced when the temperature sensitive areas cooled; now the animal shivers—another autonomic response.

All this may appear to be pure physiology, having little to do with the study of behaviour. But homeostasis is one of those topics where the boundaries between traditional fields of study break down. It is not helpful to distinguish between physiology and behaviour in the matter of temperature control. If a rat is briefly cooled in the manner described. Shivering is started to generate some heat. We might classify this as a reflex activity and consign it to the realms of physiology. However, if the rat is cooled for a longer period shivering alone is inadequate and, if given the material, the rat begins to build a nest or to enlarge the one it already has, in order to insulate itself. The reflex response is now supported by a complex behavioural one. Both are initiated by the hypothalamus and both form part of the rat's homeostatic system, although nest building will involve a more elaborate neural mechanism than does shivering.

The Hypothalamus and Motivation

It is from studies of the role of the hypothalamus that some of the most important links between brain and behaviour have developed. Modern physiological techniques allow parts of the brain to be explored with electrodes, controlled injection of chemicals or by the destruction of very small selected areas.

As a typical example of the way such studies reveal the underlying physiology of motivation we may consider the role of the hypothalamus in thirst. There are cells in its lateral areas which respond to increased concentration of the circulating body fluids. They can set in action two compensating systems. The first, acting via the links with the pituitary gland, causes the secretion of antidiuretic hormone (ADH), which increases the resorption of water by the kidneys. The second system causes the animal to seek and drink water. If the link between the hypothalamus and the posterior lobe of the pituitary gland is damaged, ADH may never be secreted. In such a situation the kidneys continue to excrete copious quantities or urine and to compensate the animal drinks large quantities of water—a condition known as *diabetes insipidus.*

Normally, the amount of water taken in is exactly adjusted to the animal's needs, but if the lateral hypothalamic detector area is artificially stimulated either electrically or by injecting hypertonic saline, drinking is greatly increased. *Anderson* prepared a number of goats with fine hollow needles penetrating into the hypothalamus. He allowed them to drink as much water as they would take and then injected concentrated salt solution. There was little effect unless the needle tip was in the lateral hypothalamus. In this region the salt caused the goats to drink frantically within a minute or two of injection. These animals, previously satiated with water, would drink salt and bitter solutions which they would not normally touch even if extremely thirsty. This condition is not comparable to *diabetes insipidus*, because the goats are not compensating for water lost through the kidneys. They are drinking water which is surplus to their physiological needs.

Anderson could produce the same effect if he stimulated the lateral hypothalamus electrically and such drinking behaviour is often described as 'stimulus-bound,' because it continues only whilst the stimulus (saline or electrical) is actually being applied. Stimulus-bound drinking has also been produced in the rat and the importance of the lateral hypothalamus for drinking is further emphasized by the behaviour of rats with damage to this part of the brain. Such animals may stop drinking altogether. They will not drink even though in the last stages of dehydration and will eventually die in the presence of water, unless this is given artificially through a tube into the stomach. It is fascinating that a rat with a lateral hypothalamic lesion is not just disinterested in water, it becomes actively averse to it. If water is placed in its mouth it will not swallow but allows the water to run out, making tongue and lip movements identical to those made by a normal rat towards an intensely bitter liquid.

These effects of stimulation and ablation might lead us to conclude that the lateral hypothalamus is responsible for the state we observed behaviourally as a tendency to drink. How far is this conclusion justified? Before discussing this question we can add comparable results which also implicate the hypothalamus in the control of feeding.

Mammals and birds normally keep their weight very constant and adjust the amount they cat accordingly. If rats are given a super-rich diet they eat less; if their food is mixed with non-nutritive cellulose they eat more. We have already mentioned that rats which damage to central areas of their hypothalamus (the ventromedial nucleus, in fact) lose this sensitive control of their eating. The food intake of a rat whose ventromedial nucleus has been ablated compared with that of a sham-operated animal. After a few days of post-operative depression of food intake, the brain-damaged rat begins to eat huge amounts of food—at least four times the normal amount. This so called 'dynamic phase' of hyperphagia lasts for 3 weeks or so. Beyond this, food intake slowly declines and eventually settles down with the animal eating about double the normal amount. Needless to say hyperphagic rats become grotesquely fat are very inactive.

Electrical stimulation of the ventromedial nucleus of normal rats depresses their feeding, so that this region of the hypothalamus has often been called a 'satiety centre', it measures when the animal has eaten enough and inhibits further eating. (The term 'centre' here

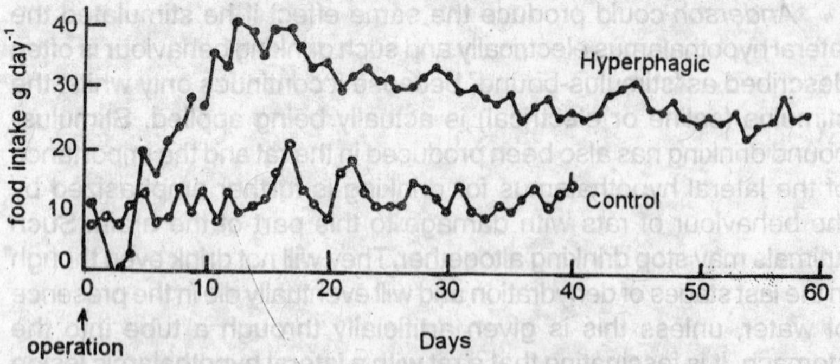

Fig. 5.13. The daily intake of normal rats and those with bi-lateral losions in the ventromedial nucleus of the hypothalamus.

means a group of nerve cells with a common organizing functions. We may note that hyperphagic rats do not lose all control of their food intake. That eventually they are eating considerably less than they did during the early dynamic phase. Satiation is not abolished with the ablation of the ventro-medial nucleus, but its threshold is greatly raised.

A centre complementary in function to the ventromedial nucleus and which promotes feeding is located in the lateral hypothalamus closely associated with the drinking area. In rats, experiments similar to those described for thirst have shown that this 'feeding centre' does initiate the appetitive behaviour of eating : stimulus-bound feeding can be elicited by implanted electrodes and damage to the area can lead to starvation in the presence of food. The feeding and drinking centres are closely linked anatomically and they interact with each other in a complex fashion. For details of the behaviour associated with this interaction the reader is referred to the admirably clear account by *Teitelbaum* and *Epstein.*

How far are we observing the operation of the 'normal' drinking and feeding systems during stimulus-bound behaviour—remembering that this term does not refer to 'normal stimuli but only to artificial stimuli applied chemically or electrically within the brain itself? It would be important to see whether an animal shows normal appetitive behaviour when being stimulated, because it might be argued that their feeding or drinking was merely a reflex response to the stimulation of neural pathways controlling the motor patterns concerned. It is possible to get fairly well co-ordinated lip and tongue movements by stimulating other areas of the brain—parts of the motor cortex of the cerebral hemispheres, for example.

Some of Anderson's observations appear to rule out this explanation because, when stimulated, his goats showed all the signs of normal thirst. They walked over to the corner of their pen and searched around for the water bowl, i.e., they showed normal appetitive behaviour and not just 'forced drinking'. Again with feeding in rats, *Coons* have shown that stimulus bound feeders will learn a new response (bar pressing) in order to acquire food, and that this response is transferred and used when subsequently the same animals are made normally hungry.

Such results are very convincing, but some doubts still remain and have been emphasized recently by *Valenstein* and others. There are a number of experiments which have shown that stimulus-bound feeders and drinkers are much more easily dissuaded from their goal than normally motivated animals. At least this is the case in rats, where slight adulteration of food or water with quinine is usually enough to stop them responding.

Further, the rate at which rats drink by lapping from a tube in normally very constant. Normal water deprivation simply leads to lengthening their bouts of drinking but does not affect their lapping speed. However, *White report* that stimulus-bound drinkers do change their rate of lapping with the intensity of electrical stimulation, which certainly suggests that the motor-organization centres for drinking are being affected.

We shall need more facts to be able to resolve this question and, even from the evidence we have at present, we must not expect the same details to apply to all vertebrates or even to all mammals. Nevertheless there can be little doubt that the basic function of the hypothalamus as a detector of physiological imbalance remains constant. Its cells will meter the temperature, food, water and hormonal concentrations of the blood-stream. As a result of any particular imbalance, the detector sites will initiate both physiological and behavioural action, as we outlined earlier in this section.

The behavioural problems then become more obscure, because whilst we may accept that the hypothalamus is essential for the initiation of motivational states, we cannot claim that it alone is essential for their control. Clearly this control involves many other areas of the brain also. *Grossman* reviews the extension evidence showing that other parts of the forebrain are involved in feeding, drinking, sexual and aggressive behaviour. Many experiments have involved making lesions and observing their behavioural effects. It has been found that damage to most areas of the cerebral hemispheres does not affect an animal's motivation in a specific way, but there are some significant exceptions. Many of these deprivation result in behaviour which is designed to restore such deficits and we result in behaviour which is designed to restore such deficits and we noted that together with behavioural responses there may also be hormonal ones; thirst leads to the release of antidiuretic hormone from the posterior pituitary. We can not consider the interaction between hormones and behaviour in more detail. This topic deserves closer attention because in many way the endocrine organs, which produce the hormones, and the nervous system share the common functions of communication and co-ordination both within the animal and between it and the outside world. The hormones from a chemical message system which is probably as old as the nervous system. Indeed the one may have developed from the other in part. Throughout the animal kingdom we find neurosecretory cells in the nervous system. These are modified neurons which can pass special chemicals down their axons and into the bloodstream. The vertebrate pituitary gland develops from the fusion of neural and epithelial tissue and remains closely connected to the hypothalamus. The pituitary regulates by its various secretions all the other endocrine glands and is in turn regulated by the nervous system. This system of control enables environmental changes picked up by the nervous system to be matched by an appropriate hormonal response. The most familiar example is the control of breeding seasons in mammals or birds. Changing day length, perceived by the eyes, results in changed activity in the hypothalamus which stimulates the pituitary. This in turn secretes hormones which start the various growth changes in the body associated with the onset of breeding condition.

Throughout all their interactions the functions of the two communication systems—endocrine and nervous—remain essentially complementary. The nervous system can only pass information by trains of nerve impulses. Its state can change very rapidly, but it is clearly less suited to transmit a steady unchanging message for a long period; it operates on a time scale from milliseconds to minutes. The endocrine system cannot respond so rapidly, but its cells can maintain a prolonged steady secretion into the bloodstream lasting for months if necessary. Moreover, hormones can reach every cell in the body via the bloodstream, whereas the nervous system generally controls only the muscles.

Circulating hormones are commonly regarded as prime motivating factors in animal behaviour and ethologists have often assumed that they act directly on brain control centres to increase drive. Certainly there are dramatic examples of hormones, apparently 'forcing' behaviour from an animal even in the most inappropriate circumstances. For instance, *Blum* and *Fiedler* describe how injections of the pituitary hormone prolactin will cause isolated males of the fish *Crenilabrus ocellalus* to perform the parental fanning movements. This is similar in form and function to the stickleback's fanning. Injected *Crenilabrus* males will fan in a bare tank devoid of all the normal external stimuli for fanning, CO_2 in the water, fertilized eggs, nest site etc., and the amount of time they spend fanning is dependent on the dose of prolactin. Examples of this type are convincing evidence for the central motivating role of hormones but they can also affect behaviour in other ways.

Before discussing examples of hormone action it is necessary to give a brief outline of those aspects of the vertebrate endocrine system which are most important for behaviour. *Fuller* accounts are given by *Gorbman* and *Bern* and by *Austin* and *Short.*

The Pituitary Gland

The pituitary gland secretes several hormones which affect the output of other endocrine organs and in this way the pituitary effectively controls the whole endocrine system. The hormones we shall be most concerned with are the *gonadotrophins* which act upon the gonads and promote both the growth of germ cells and the tissues of the gonads which secrete the sex hormones. There are two main gonadotrophins—*follicle stimulating hormone* (FSH) and *luteinizing hormone* (LH)—both were named after their action on the female gonads or ovaries but they are secreted by, and have similar functions in males. In females both are necessary for the growth of eggs and for their release into the oviduct ready for fertilization.

A third pituitary hormone important for behaviour is *prolactin*, also known as *lactogenic hormone* or *luteotrophic hormone* (LTH), which has a variety of physiological effects. We know that it is secreted by various classes of vertebrates but it often has completely different functions and 'target organs' (those parts of the body whose growth or functioning is affected by the hormone). Interestingly, most of its effects are on 'parental behaviour' in the broadest sense. As just mentioned, prolactin stimulates fanning of the eggs by male sticklebacks and some other fish; it promotes broodiness in chickens (although not all birds), the secretion of crop milk in pigeons and the growth of the mammary glands and milk secretion in mammals.

The name luteotrophic hormone refers to another important function of prolactin in mammals, for it is required for the maintenance of corpora lutea in the ovary and stimulates then to produce progesterone.

The Gonads—Ovary and Testis

It has been known for centuries that castration has profound effects on the behaviour and body form of vertebrates. This is in contrast to many invertebrate groups in which castration has little or no outward effect. Only in the vertebrates are the gonads important endocrine organs where, under stimulation from pituitary FSH and LH, they produce the sex hormones from special secretory cells. The female hormones are collectively called *oestrogens* and the male hormones, *androgens.* All these hormones are steroids which are closely related in chemical structure, and although different vertebrate groups secrete slightly different steroids, those from one group are usually quite effective in another. The commonest androgen secreted by mammals is called *testosterone.*

The sex hormones are responsible for the development of the secondary sexual characters and for the growth of the reproductive system in preparation for the shedding of eggs and sperm. Usually there are permanent differences between the body form of male and female, which are maintained throughout adult life. These are often augmented by the seasonal growth of secondary sexual characters under the influence of increased sex hormones. Stage can be distinguished from hinds all the year round but in addition they show seasonal growth of antlers.

Finally, we must mention another steroid hormone produced by the ovary. After a mammalian egg has been shed, its empty follicle enlarges and forms a prominent yellowish structure on the ovary surface—the corpus luteum. This begins to secrete *progesterone* under whose influence the lining of the uterus is prepared for receiving the egg after fertilization and development into a blastocyst. Progesterone also inhibits the contraction of the uterine muscles, which must be avoided if pregnancy is to continue. It is justly called 'the hormone of pregnancy' but it is not an exclusively mammalian hormone. Structures like the corpora lutea form when eggs are shed in fish, amphibians and reptiles but we known little about the presence or action of progesterone in these classes. Birds have progesterone which is almost certainly secreted by the ovary although their corpora lutea are not conspicuous : male birds of several species are also known to produce progesterone, probably in the testis.

Hormones and Behaviour

In the context of this chapter it is possible to give only the most superficial account of what is known of the role of hormones in motivation. Of all the internal factors that influence behaviour, hormones are the most fully understood. There are two reasons for this. First, hormone levels in the body can be precisely measured and related to behavioural changes. Secondly, and more significantly, the endocrine organs that secrete hormones can be removed with little damage to the animal. It is then possible to examine the role of a specific hormone by injecting known quantities of it, in effect mimicking the secretion of the removal organ. For example, the role of testosterone in the behaviour of male animals is commonly studied by injecting it into castrated individuals.

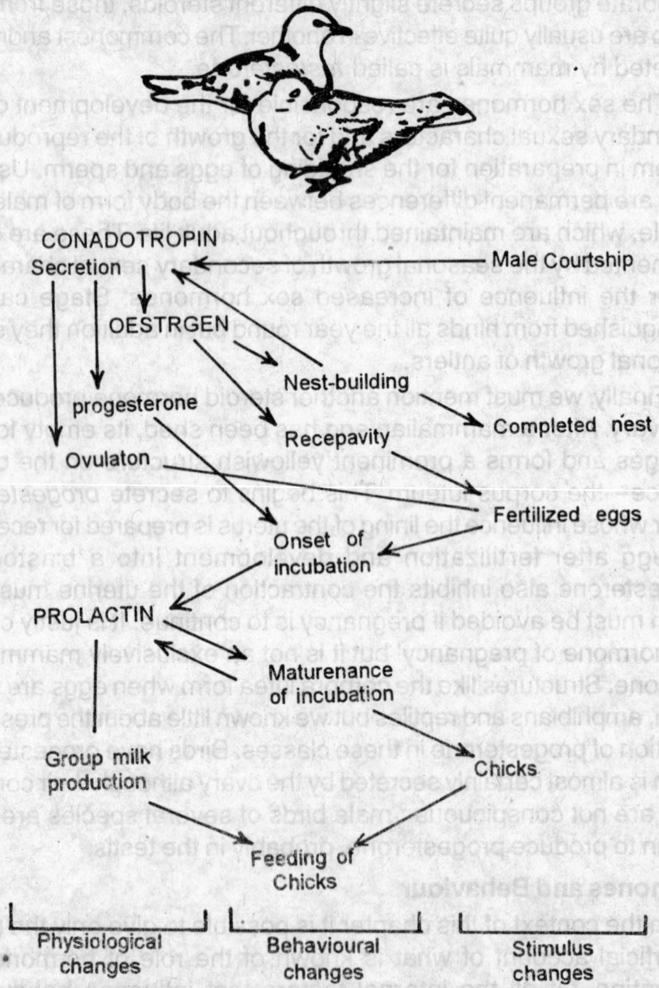

Fig. 5.14. The principal interactions between physiological changes, external stimuli and behaviour involved in the reproductive behaviour of female ring doves. Hormones are shown in capitals.

Hormones are general factors in motivation in two senses. First, they typically influence several aspects of behaviour and may be involved in the causation of a number of different activities. An obvious example is testosterone, which influences not only sexual behaviour but also aggression. It appears that testosterone may have other effects. In chicks and mice, it increases the persistence have other effects. In chicks and mice, it increases the persistence with which individuals search for a particular type of food (*Andrew* 1976). In courtsnip also, some of the effects of testosterone may be because it increases the male's persistence. The second sense in which hormones have general effects is that they influence behaviour over a relatively long time-scale. Reproductive hormones are generally necessary for the expression of sexual behaviour, but levels of these hormones do not change from moment to moment, so that changes in sexual behaviour cannot be attributed to concurrent changes in hormone levels but must be explained in terms of more immediate events, such as the behaviour of a male. The time-scale over which most hormones act on behaviour may be an animal's lifetime, as in sexual maturation, one or more years, as in animals that breed more sexual maturation, one or more years, as in animals that breed more or less annually, or a matter of days or weeks, as in the control of reproductive cycles in mammals. The 'fight' hormone adrenalin, which brings about the effects one feels when suddenly angered, is unusual in having effects on behaviour over a matter of minutes or even seconds, but future research may well reveal other such short-term influences of hormones on behaviour.

Although the effects of hormones on motivation are obviously internal, it is important to stress that they do not work in isolation but interact in complex ways with causal factors from the outside, Hormone secretion is commonly itself dependent on external stimuli; for instance, in male birds testosterone secretion starts in spring in response to increased daylength. Also, a hormone cannot usually elicit behaviour on its own without appropriate external stimuli, such as those provided by a mate. One of the most elegant features of the relationship between hormones and behaviour is the way in which the hormones appropriate to an aspect of behaviour have often been found to be secreted in advance of the situation in which that behaviour appears. This si a good example of a feed forward mechanism.

One of the most complete analyses of the interaction between hormones, external stimuli and behaviour is that which has been carried out on the reproductive behaviour of doves. Originally started by *D.S. Lehrman, this* work has been carried on by many others Silver (1978) and by *Cheng* (1979). If a pair of Barbary doves (*Streptopelia risoria*) are housed together with some nest material and a bowl in which to build a nest, they will go through a week-long courtship period, towards the end of which they mate and build a nest. Two eggs are then laid and, until these hatch about two weeks later, male and female share in incubating them. By the time hatching occurs, the crops of both parents have developed the capacity to secrete a milky fluid on which the nestling called squabs, are fed.

As the behaviour of a female dove is influenced by a number of physiological factors, including at least four different hormones, and by external stimuli from her mate, the nest, eggs and squabs. A feature of this system is that at each stage the female is stimulated to secrete the hormones that prepare her for the next stage. Male courtship stimulates the secretion of oestrogen and progesterone, which cause her to become sexually receptive and which are necessary for nest building and incubation. Incubation in both sexes stimulates prolactin secretion and this hormone is necessary both for incubation to be maintained and for the secretion of crop milk. That prolactin secretion is stimulated by and stimulates incubation indicates positive feedback between the behaviour and a causal factor that underlies it. This mechanism ensures not only that incubation is maintained but also that when it ends and the eggs hatch, the birds are secreting crop milk and are thus ready to feed the young immediately.

The reproductive behaviour of male doves appears to be controlled in a rather different way from that of females, although prolactin has similar effects on both sexes. Whereas it is clear that testosterone, which shows a marked rise after pair formation, is involved, many of the changes in a male's behaviour are primarily responses to the female's activities, not to his own hormonal state. For example, he begins nest building in response to the female starting to spend long periods at the nest site, not to any specific endocrine events (*Erickson* and *Martinez-Vargas* 1975). It appears that testosterone is involved in the causation of a wide range of male activities, including several different courtship displays, nest building and incubation, but that their precise coordination is determined by external cues from the female, the nest and its contents. Perhaps what this example illustrates most clearly is that it is a mistake to think of causal factors as either specific or general because some of them are more limited in their effects than others.

Testosterone does not affect all activities performed by male doves, nor are its effects limited to just one of them, but it does influence a variety of activities to varying degrees.

Displacement Activities

The idea that animals sometimes show displacement activities was a final proposal made by ethologists which deserves discussion. Watching a pair of courtship birds, the observer might be surprised to see one of them break off to preen briefly and hurriedly before carrying on with its courtship. Similarly, a fish in the middle of threatening a rival may suddenly swim down and start digging in the sand as if switching to nest-building, or a cock might interrupt fighting to take a few pecks of food. Such action seemed irrelevant to those who saw them, and they often shared other features such as incompleteness or a hurried and frantic appearance, so they came to be grouped together and labelled as displacement activities. It was suggested that they arose when the animal was unable to carry on which more relevant behaviour because it was in a conflict at to what to do or was thwarted from achieving its aims. Thus, the fighting gull might be in a conflict between attack and escape and so unable to do either; a courting male fish might have his sexual approaches thwarted by an unreceptive female and so, again, he unable to show the behaviour most relevant to the moment. In line with his hierarchical model of motivation. Tinbergen suggested that the thwarted energy from the reproductive instinct 'sparked over and motivated behaviour down a different channel, such as that concerned with grooming.

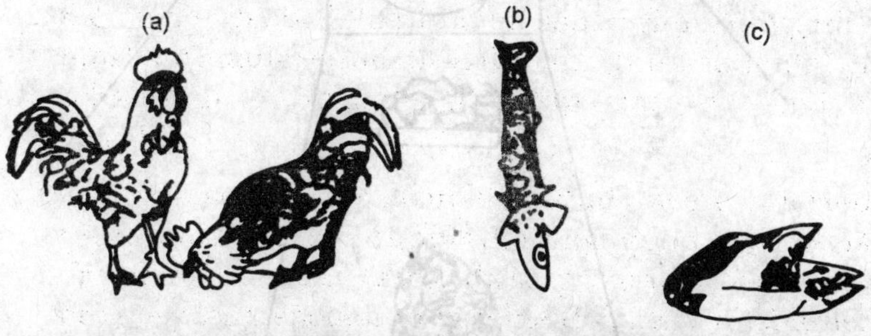

Fig. 5.15. Three classic examples of displacement activities: ground pecking in a fighting cock (a); sand digging in a three-spined stickleback confronted by an intruder (b); wing preening in a courting mallard drake (c).

Ideas on displacement activities have changed enormously since they were thought to be due to sparking over. An important finding was that the causal factors which affect the behaviour in its normal context also affected it when it appeared as a displacement activity. A courting male stickleback might break off to fan at the nest, a behaviour which serves to aerate his eggs but which is irrelevant when there are no eggs there yet. Nevertheless, in its courtship context as cell as when it when it appears later, the fanning is enhanced by carbon dioxide in the water. In the same way a bird which grooms when in conflict shows more of this grooming when its feathers are wet. Results such as these show that the behaviour patterns labelled as displacement activities are motivated in the usual way : their irrelevance lies only in their context.

The behaviour patterns labelled as displacement activities are undoubtedly a rag-bag of different actions, differently caused, the main thing they have in common being that a watching ethologist thought they were out of place. Conflict or thwarting may certainly be one reason why they arise : the animal, unable to carry out actions which are highest in priority, moves instead to ones which are next in line. This process is called disinhibition, and it is probably an important reason why displacement activities appear. It is perhaps

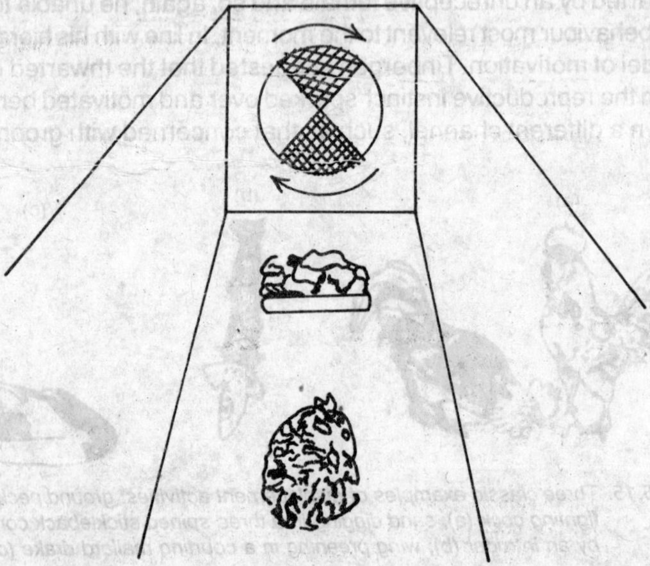

Fig. 5.16. A hungry guinea pig placed in a conflict by a rotating card near its food dish stops some distance from its goal and shows various behaviour patterns such as grooming.

not surprising that very often the displacement activities that have been described are grooming movements, because the animal carries around the stimuli for grooming with it so that, unlike food or water, they are always there on the outside of its body. Thus, if disinhibition occurs, grooming is very likely to follow. However, another reason why grooming often occurs out of context may be much simpler : the vigorous actions of fighting or courtship that the animal has been showing may lead its fur or feathers to be dishevelled and so actively enhance the stimuli for grooming till it breaks off for a quick preen to get rid of the itch.

In the days when ethologists thought of behaviour as being motivated by instincts or drives, it was these concepts that were thought to be in conflict with each other within the animal, or to be thwarted by a barrier to its behaviour on the outside. Thus conflict between the drives of aggression and a escape, or thwarting of the sex drive, were commonly referred to. However, these ideas were very vague; the drives themselves were hypothetical, so the conflict between them was doubly so. The theory was also virtually untestable: if everything was due to either to its own drive or to a conflict between other drives, it was hard to think of any observation or experiment that could disprove it. Thus, as drive theories have fallen from favour, so has the idea of conflict between drives. But this does not mean that conflict or thwarting is not a real phenomenon and an important cause of switches in behaviour. Thwarting can easily be arranged by placing a perspex screen between a hungry animal and its accustomed food dish. Similarly, an approach-avoidance conflict can be set up by placing a frightening object beside a food dish. In such circumstances a whole array of behaviour patterns may be disinhibited because the animal is unable to perform those that are its top priority.

6

COMMUNICATION

Many of the examples discussed throughout this book have involved communication between animals, for this has always been a subject of particular interest to ethologists. Indeed the theories they developed were in some ways biassed by this interest. Acts of communication do tend to be very fixed in form, to act as releasers for other actions, and to develop very similarly in all members of a species. Perhaps ideas such as that of the fixed action pattern or the innate releasing mechanism would have come less easily to mind if ethologists had been more interested in grooming or feeding behaviour than in aggressive or courtship displays. But then they had every reason to be enthusiastic about studying the latter, because the signals that animals produce in communication are often remarkably striking. Discovering just what they are saying to each other is one of the most fascinating fields of ethology.

Whenever we see an animal that is brightly coloured or boldly patterned, or one that is making a loud and obvious noise, we can be pretty certain that communication of some sort is taking place. Standing out from the background is dangerous and there must be some benefit to doing so which offsets the risks involved. Most often the communication is between members of the same species, and the risks come when predators cue in on this as a means of finding a meal. Sometimes, however, different species communicate with one another and here it is often prey animals that forsake their camouflage and communicate directly with predators that might eat them. These different sorts of relationship between predators and their prey make a useful starting point as they illustrate some basic principles of animal communication.

How Signals Convey Information

Discrete and Graded Signals

Some signals are discrete (digital) and others are graded (analog). The alarm calls of many species, which are typically given at the same intensity each time are relatively constant across species, permit communication among different species. These discrete signals are of a frequency and duration that make them difficult for predators to pinpoint (*Marler* 1957).

Graded signals may vary in intensity as a function of the strength of the stimulus. The waggle dance of the honeybee illustrates both the complexity and the graded properties of signals. The number of turns per minute is inversely proportional to the distance from the food source, and the duration and liveliness of the dance increase with the quality of the food. Signals that might at first seem discrete often, upon closer study, turn out to be graded. Although the bursts of light emitted by fireflies seem to be discrete and species-specific, they vary in intensity and duration under different conditions.

Distance and Duration

Although most species are limited to about twenty to forty different displays, signals can vary in several other ways that increase their information content. The distance a signal travels may vary. In some cases the detection threshold of the receiver is no more than several molecules of a substance, as with sex attractants of many insects. Thus, a small amount of material produced by a female can be detected by a male several kilometers downwind. Visual displays, at the other extreme, usually operate over much shorter distances. The duration of signal may also vary. Alarm signals, such as chemicals produced by many invertebrates, many have a localized, short-term effect and thus a rapid fade-out time (*Wilsom* 1975). Signals such as male secondary sexual characteristics in polygynous species can last for long periods of time. Information about reproductive status, like the antlers of male deer or the bright red perineal region of female rhesus monkeys in estrus, is conveyed during the breeding season. In contrast, the brightly coloured epaulets on the red-winged blackbird, although present during the entire breeding season, are conspicuous only when the male exposes and erects these feathers during the song spread.

Composite Signals, Syntax and Context

Two or more signals can be combined to form a composite signal with a new meaning. For example, equids such as zebras communicate hostile behaviour by flattening their ears and communicate friendliness by raising their ears (discrete signals). They indicate the intensity of either emotion by the degree to which the mother opens (graded signal). The mouth-opening pattern is the same for both hostile and friendly behaviour.

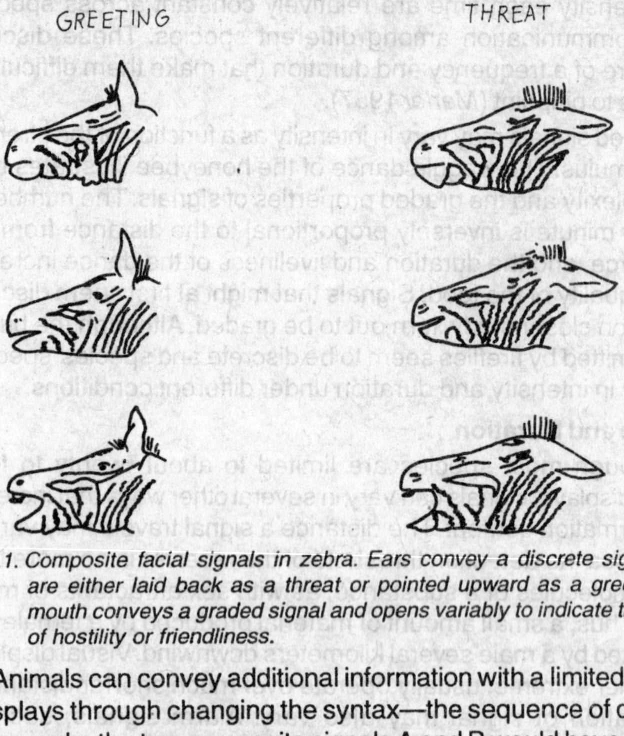

Fig. 6.1. Composite facial signals in zebra. Ears convey a discrete signal. They are either laid back as a threat or pointed upward as a greeting. The mouth conveys a graded signal and opens variably to indicate the degree of hostility or friendliness.

Animals can convey additional information with a limited number of displays through changing the syntax—the sequence of displays. For example, the two composite signals A and B would have different meanings depending on whether A or B came first. No evidence exists that non-human animals use syntax naturally. However, in the rather special case of language learning in chimpanzees, chimps assemble words in novel ways as they communicate with their human companions (*Rumbaugh* and *Gill* 1976).

The same signals can have different meanings depending on the context—that is, on what other stimuli are impinging on the receiver. For example, the lion's roar can function as a spacing device for neighbouring prides, as an aggressive display in fights between males, or as a means of maintaining contact among pride members. The "ruff-out" display of the male cowbird serves in courtship as well as in conflict with other males.

Metacommunication

Increasing the information content of display by *metacomm-unication*—communication about communication—is theoretically possible; one display changes the meaning of those that follow. We can see good examples in play behaviour: animals use aggressive, sexual, and other displays in play, but they precede such behaviour by an act that communicates the message that "what follows is play, join in" (*Bekoff* 1977). Canids such as dogs and wolves precede play with the play bow. Monkeys communicate play behaviour through a relaxed, open-mouthed face.

Disagreement exists about the importance of metacomm-unication (*Smith* 1984), and some observers have used the word a bit loosely. For example, male rhesus monkey (*Macaca Mulatta*) sometimes communicate dominance by carrying the tail elevated in an "S" shape over the back. While this posture conveys status and mood, it does not really change the meaning of behaviours that follow; rather, it communicates that aggressive behaviour is likely to follow.

Information and Manipulation

In the past, ethologists often used to refer to communication between animals of the same species as involving the sharing of information for their mutual benefit. Indeed, much communication may be just this. For example, when potential mates meet it is advantageous to both of them that they communicate to each other accurately information about the species to which they belong, for if they are not members of the same species both of them will waste their breeding effort. Strictly speaking, however, animals will produce signals when this is to their advantage, regardless of whether others benefit from receiving them. The examples in the last section showed this well, for prey animals send signals to predators when they gain by doing so and the gain they make is usually at the predator's expense.

Exactly the same rule applies within a species, as we might expect from the idea that animals are essentially selfish, behaving so as to maximise their own inclusive fitness. So, while the information that passes from one to another when a signal is transmitted may be useful to the one that receives it, this will not necessarily be the case, and there are some very good examples where it is better to think of animals as manipulating each other rather than sharing information. One of these is in the mating behaviour of the bluegill sunfish. Most males of this species do not

Fig. 6.2. Three bluegill sunfishes spawning. The large animal at the back is a full-grown adult male. Though they look similar, only the nearer of the two small individuals is a female. That in the middle is a young male which is a female mimic. By this deception these fish avoid being chased away by large males and are able to steal some fertilisations from them.

mature until they are seven or eight years old, when they set up territories to which they attract females or mating. The females are smaller and look quite different. Sometimes two of them may spawn simultaneously with one of these males, sperm and eggs being shed into the water as they swim around together on his territory. Deceit enters into this scheme of things because some males, around 20%, mature at only 2—3 years old, when much smaller than the norm. These look just like females and join in the mating of spawning pairs. The territorial male does not drive such a rival off as he cannot tell that he is not a second mate, but actually he is a male and he shares in the fertilisation of the eggs that the female produces. Of course all males cannot mature early, or there would be no territorial ones to attract females in the first place, so this is an example of a mixed evolutionarily stable strategy, with a balance between the two possible modes of behaviour. In this example males mimic females and the other males, unable to see through the deception, are manipulated by the mimics. The case of the pied flycatcher mentioned in the last chapter can be put in very similar terms. Here the female cannot tell if a male is mated or not and so cannot choose an unmated one who will help her rear her chicks rather than the one who will leave her to return to his first mate.

Deception is possible in these cases because there are no cues that signal receiver can use to discriminate between the classes of individuals involved, in other words to tell young male sunfish from females and mated flycatchers from unmated ones. In other cases, however, deceit is impossible because there are ways in which other animals can see through it. Some good examples here come from cases where it pays animals to appear to be larger than they really are. This is often so in aggressive displays for, the lager an animal appears to be, the more a rival is likely to be intimidated into retreating without a scrap. It is probably for this reason that Siamese fighting fish extend all their fins as far as possible when swimming alongside an opponent and raise their gill covers when facing one. Both actions make them as large and striking as they can be. If only some fish displayed like this, as was probably originally the case, they would be likely to have deceived others of the same size into retreat. Thus appearing bigger would have been to their advantage and natural selection would have made it spread through the population until, eventually, all fighting fish looked as large as they possibly could when displaying. At this stage, however, the trick breaks down for each just looks an exaggerated version of its own size. If every fish can increase its apparent size by 20%, all will look that much bigger, but their sizes relative to each other will remain the same. Their opponents will gain accurate information about their size rather than being misled in any way.

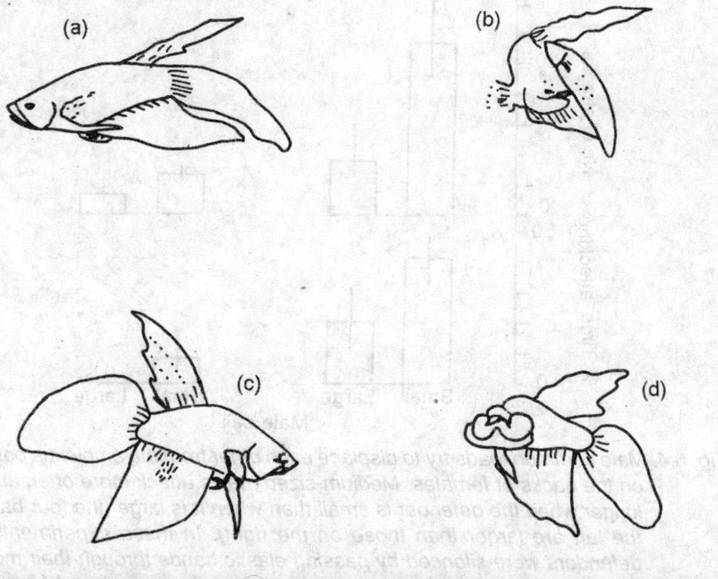

Fig. 6.3. The male siamese fighting fish is a fine example of an animal that looks larger when displaying (c and d) then it does normally (a and b). The displaying fish raises all its fins when broadside to its rival (c) and its gill covers when facing it (d), thus making itself appear as large as possible.

Another nice example is in the calling of male toads studied by *Nick Davies* and *Tim Holliday*. Male toads cling onto ripe females in a posture called amplexus which ensures that they fertilise the eggs as they emerge. Males often fight to take up this position and the male already in amplexus kicks and croaks to repeal boarders. Larger males generally win in such encounters, as one might expect from their greater strength. Being larger, however, they are also able to produce deeper sounds, and davies and Halliday showed that these too had an effect. They silenced males in amplexus by fitting elastic bands round their arms and through their mouths. They then saw whether other males attempted to displace them. As expected, medium sized males spent more time attempting to displace small males than large ones. However, if the deep croaks of a large male

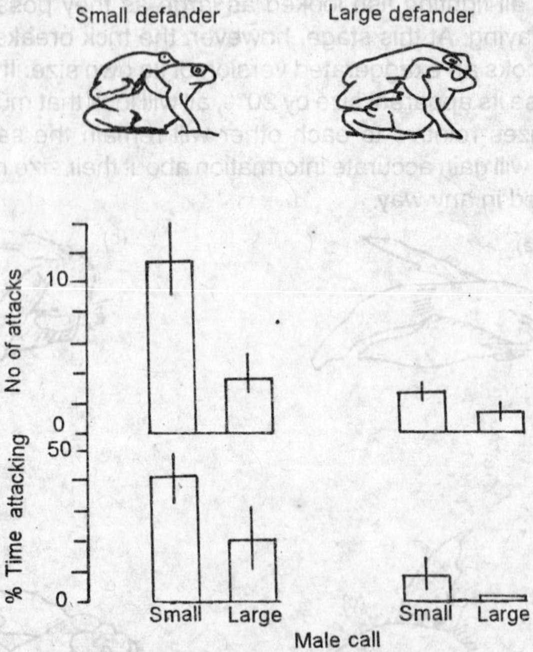

Fig. 6.4. Male common toads try to displace each other from the amplexus position on the backs of females. Medium-sized males attack more often and for longer when the defender is small than when it is large (the four bars on the left are larger than those on the right). In these experiments the defenders were silenced by passing elastic bands through their mouths and round their forelimbs, and other male calls were played from loud-speakers. No matter what the size of the defender, the attacker would be more persistent if the call played was that of a small toad than it it was that of a large one (the left bar of each pair is larger than the right one).

were played from a loudspeaker, a small male in amplexus was harassed much less by his larger rival. Clearly toads can use the calls to assess the size of others and will be less likely to fight with those whose calls are deeper. Here again then, the information transmitted is accurate and the animal receiving it acts on its appropriately. But why do small toads not cheat, developing deeper voices and so appearing larger than they really are ? The answer is probably that the larger an animal the deeper the sound it is capable of producing. So, just as with the size of fighting fish, if all toads calls as deeply as they can, the larger ones will still all more deeply than the smaller ones, and depth of croak will be an accurate reflection of size.

To summarise, communication involves the transmission of information, in the form of a signal, from one individual to another. As we have discovered, this 'information' may be true or false, the important point being that producing the signal should be of benefit to the animal that does so. Sometimes deception may be possible, and the signaller may gain from it, but this will not be possible if the animal receiving the signal can detect that the signal is misleading. Deception tends, therefore, to be rather rare, for the more animals practice it, the more it will be best for the recipient to assume that the information it is receiving is false. If almost every plover which appears to have a broken wing is really feigning injury, then a fox discovering one would gain most by not following the adult but searching the area for a nest or chicks.

Messages and their Meanings

A useful way of analysing communication was devised by the American ethologist W. John Smith. He pointed out that it could be looked at either from the point of view of the animal that sent the signal, in which case one would ask what the *message* encoded in it was, or from the viewpoint of recipients, to see what its *meaning* was for them.

Let us return to our singing bird. What is the message incorporated in the song and what meanings may it have for those that are listening? The most obvious point is that, as every bird-watcher knows, song is usually clearly distinct between species, so that it includes information as to he species to which the singer belongs. Furthermore, song tends only to be produced by male birds inbreeding condition, they usually only sing when they have a territory and in some cases they stop singing as soon as they obtain a mate. In such a species the basic message of song may be 'I am an unmated adult male sedge warbler on my territory and in breeding condition'. In addition, there may be many subtleties incorporated in the message. Signals often vary from place to place, forming dialects like those in human language. Individual animals may also have idiosyncrasies in the form f the signal that they employ. They may have a range of different signals which convey subtly different information. The singing bird may thus also be saying where he comes from, exactly who he is and perhaps even something about his motivation if the songs differ according to whether they are directed at rivals or at potential mates.

Many animal signals are, like song, primarily concerned with reproduction, attracting mates and repelling rivals being their main role in communication. However, especially in species that live in social groups, there are many other messages that it may be useful to convey. Animals that move around in groups often have obvious patterns on them which can be seen some distance off and enable them to maintain contact with their companions. Call notes can serve a similar function, especially where visibility is poor. These are simple sounds and patterns, and they tend not to vary much. Their messages may simply be 'I am here', though usually they also communicate the signaller's species, and there are cases where, despite their simplicity, they vary enough to indicate individual identity.

The alarm calls referred to earlier are an example of another type of message, individuals in this case warning their companions of danger. The 'seet' call of small birds is most reliably produced by the appearance of a hawk out, in the breeding season, birds with young may also call in this way when alarmed by a dog or a human near to their nest. Most animal signals are like this: the message they convey is a generalised one rather than being precise like a word in our language. A famous exception is the alarm calling of vervet monkeys. These animals have several different calls which they produce when alarmed and three of these are specific to particular sorts of predator: a snake call, a leopard call and an eagle call. Here the information in the call is very precise: the calling animal might as well be shouting 'snake', 'leopard' or 'eagle' provided that its listeners were able to recognise the meaning of those words as we can do.

What do its companions do when they hear one of these calls? In other words, what *meaning* does the call have for them. The answer

Fig. 6.5. The response of vervet monkeys to three different alarm calls. (a) That produced in the presence of a leopard leads them to run into trees, (b) the eagle call leads them to run down into the undergrowth and (c) the smake call causes them to look down and approach.

is that this too is very precisely defined. Animals hearing the leopard call rush up into trees, they respond to the eagle call by running down to the ground and hiding in thickets, and the snake call leads them to approach and look down. We cannot tell if they have a mental image of the predator in question, but they certainly behave as if they were aware of the particular threat it poses.

To understand messages, we have to study the particular situation in which an animal produces a signal, in other words try to discover what causes it to do so. In the case of meaning it is the recipient of the signal that we must study: how does it respond when the signal is received? This response may vary from one animal to another. In the case of a contact call, for example, other members of the flock may approach so that the group keeps together, animals from other flocks may keep their distance, predators may be attracted with the prospect of a meal, and animals of other species are likely to ignore it as irrelevant to their interests. The same signal can thus have many different meanings, depending on the listener and on the exact content in which it is received. To stress a point made earlier: the critical factor is that, for a signal to be produced, the advantages to the signaller must outweigh the disadvantages. Otherwise, there is no reason why evolution would favour signalling.

There are thus many different messages that animal signals may convey, though it is unusual for these to be very precise. Most often they provide rather general information about the motivational state of he signaller, though their features can also incorporate more specific information, such as the species or identity of the caller. The number of distinct signals that a species uses is not often large. What makes the signalling system flexible and varied is that the same signal may mean different things to different recipients and the same animal may respond to it in different ways depending on the context in which it is received. Thus, with a comparatively small repertoire of rather stereotyped signals, animals can convey a wealth of information from one to another.

Signals

The message is what a signal says about its sender, the meaning is what the receiver extract from it. Neither of these is tangible: they cannot be studied directly but must be inferred from behaviour. On the other hand, the signal is the physical form in which the message is embodied for its transmission through the environment.

For most animal signals, the form of the signal has no clear relationship to the message that it encodes, just as the word 'dog' does not look or sound anything like the animal to which it refers. However, singles are not absolutely arbitrary. The modality in which they are transmitted and the form of the signal within that modality may both be determined by considerations of how the signal is most effectively transmitted given its function and the need to avoid the attraction of predators. A well-known example of this is the 'seep' alarm call of passerine birds, found in very similar form in many species (*Marler* 1955) and which even has equivalents in some mammals (*Owings* and *Virginia* 1978). Thin, high-pitched whistles like these have probably evolved because the sources of signals of this form are especially hard to locate and this is important in a call produced when a predator is present. Conspecifics do not need to know where the signaller is in order to react appropriately, and it is essential to the caller that the chances of the predator locating it are

Fig. 6.6. Threatening and submissive postures in the dog.

minimised. Hence the calls of many species have converged on the same signal form, and these species will react to each other's signals, without it necessarily being advantageous for an animal of one species to tell those of another about the danger. There are other cases in which the form of a signal seems to be related to the message that it conveys. *Morton* (1977) argues that this is so with many bird and mammal sounds, harsh, low-pitched calls being used in hostile contexts and purer, higher pitched ones where the situation is friendly. He suggests that the former may be intimidating because harsh sounds of low frequency can only be produced by large individuals. Thus, animals might be expected to avoid those producing deeper sounds than themselves, as size is such an important factor is success in fights. Selection should then lead to aggressive signals being as harsh and low as possible. It may favour a contrast to this in appeasing and friendly sounds to make them easy to discriminate and so to minimise the chances of their eliciting a hostile response. That signals of opposite meaning are also often opposite in form, presumably to avoid ambiguity, was first noticed by *Darwin* (1872), who referred to it as the principle of antithesis; Morton's idea is smaller. Darwin himself pointed to the extreme difference in posture between a hostile dog and the same animal 'in a humble and affectionate frame of mind', a similar contrast is seen in gulls where the beak is shown off in aggressive displays but is hidden, by turning the head away, in appeasing ones (*Tinbergen* 1959).

These examples show that signals used to convey similar messages may have features in common. Often this will be because the best possible form of signal for a particular function is moulded by environmental constraints but, in some cases, as *Morton* suggests, the form of the signal may be related to the content of the message itself.

Measurement of Communication

Suppose that we wish to learn something about the way crayfish (*Oronectes rusticus*) communicate. If we place crayfish that are stranger together in a tank in the lab, they assume various postures with the large claws and a dominance hierarchy, which is positively related to the size of the animal, develops (*Bovjberg* 1956).

Observation

Our first step is to identify motor patterns that are relatively constant by observing the animals and describing the movements they make. We might divide the behaviour patterns into the following acts: *retreat*—a rapid, backward swimming motion; *cheliped presentation*—the movement of the large claw from a downward-facing position to one that is horizontal to the substrate; *cheliped extension*—the rapid movement of the opened claw toward the other crayfish; *forward locomotion*—movement toward the other crayfish; and *fighting*—striking and pinching the other crayfish.

Simple observation might suggest that communication is occurring since certain behaviour occurs only in the presence of other animals. Perhaps a behaviour performed by one individual—say, cheliped extension—is usually followed by a particular behaviour performed by the other animal—say, retreat. In that case we suspect that communication has occurred. But the occurrence of any behaviour is probabilistic (stochastic). He can we be sure that what one crayfish does affects the behaviour of the other?

Quantification

Observed and Expected Frequencies. For a more thorough analysis, we can compare the frequency of each behavioural response against the expected frequency if the response were random, and without regard to the behaviour of the other crayfish.

The results of the chisquare test are presented in the table 6.1. Note that some behaviours facilitate response and other inhibit them. Some, such as forward locomotion, seem to have little effect on other behaviours.

Channels of Communication

Odor

From an evolutionary standpoint, the earliest type of communication was chemical, for odor is used throughout the animal kingdom, with the exception of most bird species. Most pheromones are involved in mate identification and attraction, spacing mechanisms, or alarm. The greatest amount of research has been done on insects and mammals. There are several probable reasons why the wide-spread use of chemical signals has evolved: such signals can transmit information in the dark, can travel around solid object, can last for hours or days, and are efficient in terms of the cost to production (*Wilson* 1975). However, because these pheromones must diffuse through air or water, they are show to act and have a long "fade-out" time.

We have a relatively good understanding of insect pheromones. For example, the sex attractant bombykol, produced by the female

Table 6.1. Frequency distributions of agonistic behaviour patterns in crayfish. The frequencies expected, if random, are given in parenthesis

Signaler's Singla	Receiver's Response						
	No Change	Retreat	Cheliped Presentation	Cheliped Extension	Forward Locomotion	Fighting	Total Observed
no change	4 (0.9)	1 (5.6)	0 (0.4)	4 (1.2)	0 (0.1)	0 (0.8)	9
retreat	6 (4.0)	14 (24.3)	0 (1.8)	7 (5.2)	3 (0.3)	9 (2.7)	39
cheliped presentation	6 (2.7)	6 (16.2)	6 (1.2)	0 (3.5)	0 (0.2)	8 (2.3)	26
cheliped extension	8 (17.5)	126 (106.1)	3 (7.7)	24 (22.6)	0 (1.3)	9 (14.9)	170
forward locomotion	17 (8.5)	39 (51.8)	6 (3.7)	15 (11.0)	0 (0.6)	6 (7.3)	83
fighting	0 (7.4)	63 (44.9)	3 (3.2)	3 (9.6)	0 (0.5)	3 (6.3)	72
Total observed	41	249	18	53	3	35	399

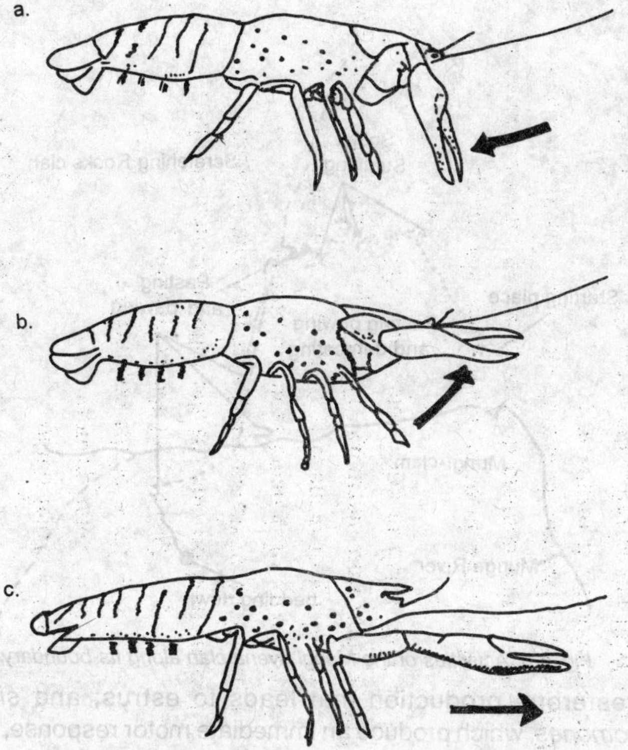

Fig. 6.7. Cheliped (large claw) presentation and extension in the crayfish. In the first phase of cheliped presentation (a), the cheliped tip is lowered and the claw is opened and in the second phase (b), the cheliped is raised. Pard (c) shows cheliped extension. Arrows denote direction of movement.

silk mouth, has been isolated, and we are reasonably familiar with the male's perceptual system. A single molecule of bombykol triggers a nerve impulse in a receptor cell in an antenna of a male. About 200 receptor-cell firings in one second lead to a behavioural response (*Schneider* 1974). The male responds by flying upwind, equalizing the pheromone concentration on both antennae, until he reaches the female. With the commercial synthesis of insect sex attractants, traps baited with pheromone to lure males are used to control such pests as the gypsy moth in the north-eastern United States.

Many of the recent studies of pheromones in mammals have been done on rodents. Two general classes of substances, which differ in effect, have been identified: *priming pheromones,* which produce a generalized response, such as triggering estrogen and

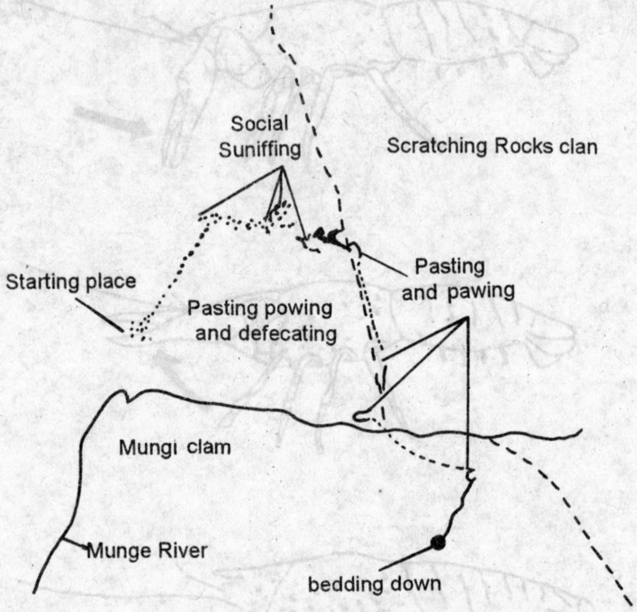

Fig. 6.8. Activities of the Mungi hyena clan along its boundary.

progesterone production that leads to estrus; and *signaling pheromones,* which produce an immediate motor response, such as the initiation of a mounting sequence. *Bronson* (1971) suggested hat pheromones in mice have the following functions:

Signaling pheromones

Fear substance
Male sex attractant
Female sex attractant
Aggression inducer
Aggression inhibitor.

Priming pheromones

Estrus inducer
Estrus inhibitor
Adrenocortical activator.

In addition, substances produced in male mouse urine speed up maturation in young female *Lombardi* and (*Vandenberg* 1977). Other urinary products inhibit aggression, increase aggression, stimulate the adrenal cortex, block implantation of embryos, and so on. At this point it is not clear how many different chemicals are involved; different functions maybe served by the same pheromone. The sources of these products include the sexual accessory glands and even the plantar tubercles on the mouse' feet.

Many of the substances produced by mammals function as a means of staking out territories or home ranges, much as does birdsong. The advantage of pheromones, as mentioned previously, is that the odor may last for many days and nights, a significant asset since many animals are nocturnal. Since these substances are often associated with the urinary and digestive systems, eliminative behaviour is often highly specialized. Hyena clans mark the boundaries of their territories by establishing latrine areas. Clan members defecate simultaneously in a area, then paw the ground. The feces turn white and become quite conspicuous. In the same or in another area, hyenas engage in "pasting." Both sexes possess two anal glands that open into the rectum, just inside the anal opening. When pasting, the hyena straddles long stalks of grass; as the stems pass underneath, the animal events its rectum and deposits a strong smelling whitish substance on the grass stems (*Kruuk* 1972).

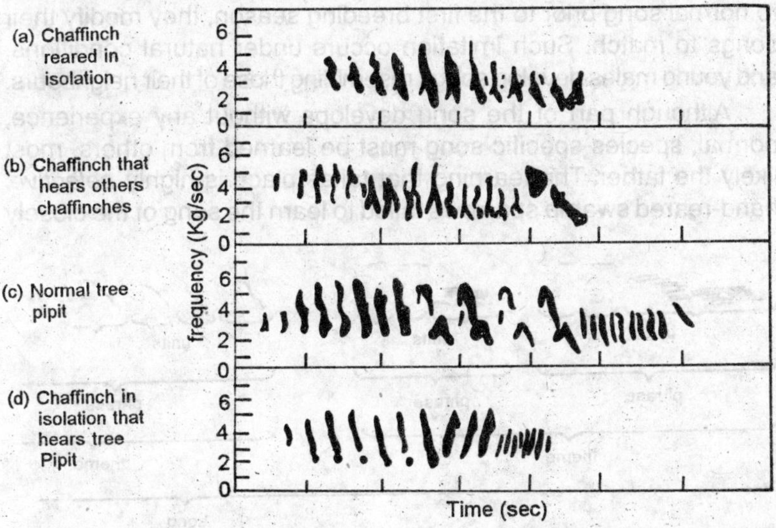

Fig. 6.9. Chaffinch song experiments. A male chaffinch (Fringilla coelebs) reared in isolation has a simpler song (a) than one that hears other male chaffinches singing. (b) If a tree pipit (Anthus trivialis) song (c) is played to a male chaffinch while it is reared in isolation, some elements of the tree pipit song are incorporated into the chaffinch song (d).

Sound

Much more information about immediate conditions can be transmitted faster by sound than by chemicals. Sound can be produced by a single organ, and it can also travel around objects, through dense vegetation, and can be used in the dark. Information can be conveyed by both frequency and amplitude modulation. As we can see in sound spectrographs, a unit of birdsong contains sounds of various frequencies and amplitudes. Low-frequency sounds travel great distances and are used by animals with large home ranges. For example, howler monkeys (*Alouatta*) in the Neotropical rain forest signal to other groups with low-frequency calls. Animals with smaller home ranges—for example, squirrel monkeys (*Saimiri sciureus*)—use higher-frequency sounds, which dissipate rapidly. Such calls serve to maintain contact among group members.

Aside from human speech, birdsong, whose development has been analyzed be *Thorpe* (1958), *Marler* and *Tamura* (1962), and others, is probably the most complex auditory communication. Birds raised in isolation develop abnormal songs; but if they are exposed to normal song prior to the first breeding season, they modify their songs to match. Such imitation occurs under natural conditions, and young males develop songs resembling those of their neighbours.

Although part of the song develops without any experience, normal, species-specific song must be learned from others, most likely the father. The learning that takes place is highly *selective*: hand-reared swamp sparrows failed to learn the song of the closely

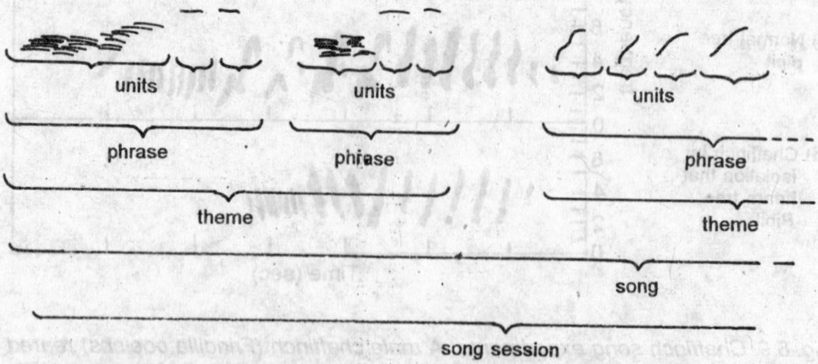

Fig. 6.10. Song of humpback whale. The song of a humpback whale can be broken up into units, phrases, themes, songs and song sessions. Each whale sings its own variation of the song, which may last up to a half hour.

related song sparrow (both in the genus *Melospiza*) but readily learned the song of other swamp sparrows (*Marler* and *Peters* 1977). Birds may modify their song to match the local dialect if they move into a new area at the start of a breeding season. Females may not mate as readily with strange-sounding males (*Baker* 1983).

High-frequency sound are used by a variety of animals, particularly mammals. The distress calls of young rodents and some of the vocalizations of dogs and wolves are well above the range of human hearing, as are the echolocation sounds of bats. Although bat sound are used mainly to locate food objects, communication also occurs between predator and prey. Noctuid moths, for example, do not produce sounds themselves, but they possess tympanic membranes on each side of the body that receive sonar pulses from bats (*Roeder* and *Treat* 1961). Depending on the location and intensity of sound stimulation, the mouth may fly away in the opposite direction, dive, or desynchronize its wingbeat to produce erratic flight.

Underwater sound has properties somewhat different from sound in air. Fish and invertebrates produce a variety of sounds, some of which have only recently been investigated. Marine mammals produce clicks, squeals, and longer, more-complex sounds in corporating many frequencies; most of these are short-duration sounds, which are thought to function in echolocation. Baleen whales (family Mysticeti) produce lower and longer sounds than do the toothed whales, such as dolphins. *Payne* and *McVay* (1971) analyzed the sounds of the humpback whale (*Megaptera navaneangliae*) which are varied and occur in sequences of seven to thirty minutes duration and are then repeated. The songs have a great deal of individuality, and each whale adheres to its own for many months before developing a new one. Researchers have as yet ascribed no clear function to these sounds, but they serve to maintain group cohesion across thousands of miles because of the low attenuation of low-frequency sounds in water.

Touch

Short-range communication in the form of physical contact is used by many invertebrates whose antennae, covered with receptors, are the first part of the body to contact other objects and organisms. Antennae are used by non-social insects, such as cockroaches, and by social insects, such as bees. The honeybee of the performs the waggle dance in a dark hive; therefore, much of he information about the type and location of food comes from tactile communication as the workers' antennae contact the dancer and pick up taste cues.

Perhaps the most widespread use of tactile displays occurs during copulation. In many rodents stimulation of the back end of an estrous female produces concave arching of the back and immobility (lordosis). In some mammals vaginal stimulation induces ovulation.

In most primates grooming is an important social activity and seems to have a function not only in the removal of ectoparasites but also as a "social cement" in the reaffirmation of social bonds. Most grooming takes place between close relatives, but long-term relationships, indicated by grooming patterns, may exist between non-relatives (*Sade* 1965). In the large, multimale groups characteristic of macaques (*Macaca* spp.) and baboons (*Papio* spp.) grooming between the sexes is confined largely to the mating season. South American titi monkeys (*Callicebus*), which live in monogamous groups, entwine their tails when resting. These signals may not be complex, but they are no less important than otner signals.

Surface Vibration

Information can be conveyed by patterns of surface vibrations. Males of one species of water strider (*Gerris remigis*) send out ripples of a certain frequency, and receptive females respond by moving toward the source. When a female gets within a certain distance, the male switches to courtship waves (*Wilcox* 1972). *Wilcox* (1979) also observed that males generate high-frequency (HF) waves when they are close to another water strider. If return HF waves are not picked up from the second strider, the first attempts copulation. In order to demonstrate the function of HF, *Wilcox* used an ingenious playback method or program *females* to send out HF. He glued a magnet to the female's foreleg and allowed her to move freely inside an electrical coil. When *Wilcox* played an electrical copy of the male HF signal through the coil, the magnet moved the female's leg and she involuntarily sent out the HF signal. When *Wilcox* placed a rubber mask over the male's eyes to eliminate possible visual cues, he found that when the male approached a female, it always attempted to copulate when the female was not sending an HF signal. We can apply this type of playback experiment to studies of other types of substrate-transmitted signals, such as web communication among spiders.

Electric Field

Some sharks and electric fish have electroreceptors that they use both passively and actively in detecting objects and in communicating socially. Sharks (*Scyliorhinus caniculus*) detect the electric field produced by prey flatfish that are buried in the sand (*Kalmijn* 1971). In addition to electrolocating objects, electric fish of the African family Mormyridae communicate information about species identity (*Hopkins* and *Bass* 1981), individual identity, and

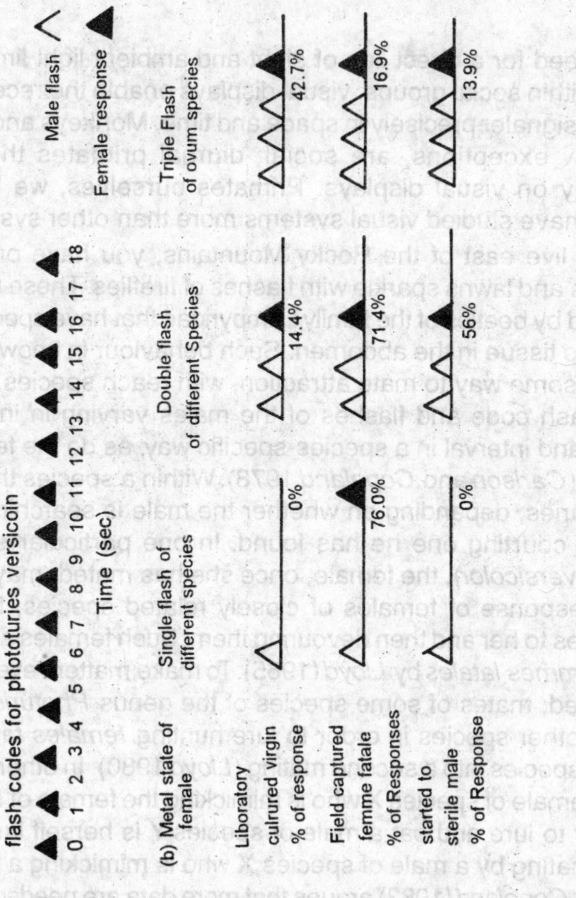

Fig. 6.11. Responses of female Photuris vereicolor to flashes of males. Unmated virtin fireflies of the same species (top row) usually answer only the triple flash of males of their own species. Mated females become femmes fatales, answering flashes of different species (middle row). Females mated to sterile males do not become femmes fatales (bottom row) and respond significantly less.

sex by modulating the shape of the electric organ discharge *Moller* (1976) demonstrated that members of this family also use electric organ discharges to maintain group coordination in schools. By altering either wavelength or pulse duration, they can communicate threat, warning, submission, and so on (*Bullock* 1973). The advantages of this sensory mode are that it is useful in dark, murky. Water, it can travel around and even through certain objects, and it provides precise information on location.

Vision

The need for a direct lien of sight and ambient light limit their use, but within social groups, visual displays enable the receiver to locate the signaler precisely in space and time. Monkeys and apes, with a few exceptions, are social, diurnal primates that rely extensively on visual displays. Primates ourselves, we human observes have studied visual systems more than other systems.

If you live east of the Rocky Mountains, you have probably seen fields and lawns sparkle with flashes of fireflies. These flashes are emitted by beetles of the family Lampyridae that have specialized photogenic tissue in the abdomen. Such behaviour in known to be related in some way to mate attraction, with reach species having its own flash code and flashes of the males varying in intensity, duration, and interval in a species-specific way, as do the females' response (*Carlson* and *Copeland* 1978). Within a species the flash interval varies, depending on whether the male is searching for a female or courting one he has found. In one particular species (*Photuris versicolor*), the female, once she has mated, may mimic the fish response of females of closely related species, thereby luring males to her and then devouring them. Such females are aptly termed *femmes fatales* by *Lloyd* (1965). To make matters even more complicated, mates of some species of the genus *Photuris* mimic males of other species in order to lure hunting *females fatales* of their *own* species into a second mating (*Lloyd* 1980). In other words, a mated female of species X who is mimicking the female of species Y in order to lure and eat a male of species Y is herself lured into another mating by a male of species X who is mimicking a male of species Y! *Copeland* (1983) argues that more data are needed before male mimicry can be assumed.

Fish and some invertebrates are able to change colour within seconds by expanding and contracting chromatophores beneath the skin. The most spectacular in this regard is the octopus; waves of

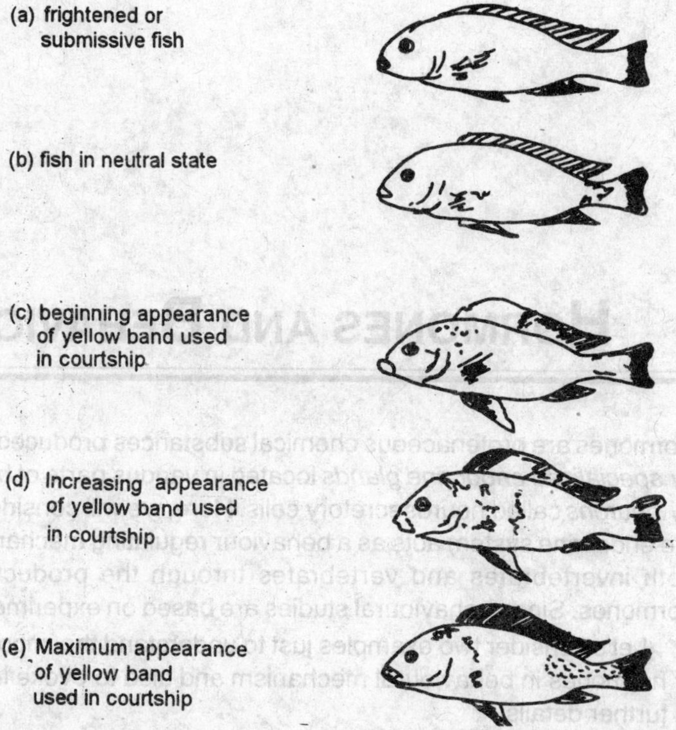

Fig. 6.12. Graded visual signals in the mouthbrooder cichlid fish. An increasing expression of yellow band used in courtship shows in parts (c) through (e).

colour advance and recede according to the animal's mood. Because of the limitations of visual displays, these displays are usually coupled with other modes of communication, such as audition. For instance, in the song spread, the redwing spreads its tail, lowers its wings, and raises its epaulets at the same time as it renders the song.

7

HORMONES AND BEHAVIOUR

Hormones are protenaceous chemical substances produced either by *specialized endocrine glands* located in various parts of body or by *neurons* called neurosecretory cells. Here, we will consider how the endocrine system acts as a behaviour regulating mechanism in both invertebrates and vertebrates through the production of hormones. Since behavioural studies are based on experiments.

Let's consider two examples just to understand the importance of hormones in behavioural mechanism and also to evoke interest in further details.

Eg. I. If a male rat is placed with a sexually mature female rat it will mount the female in a matter of seconds and computate which a castrated male rat takes longer to initiate mounting.

Since the testes are a source of androgens, hormones that affect reproductive behaviour, injections of synthetic androgens can replace the hormone normally and restore mount latencies to normal levels.

Eg. II. If we remove the *prothoracic gland* which secretes *ecdysone*—the hormone which controls moulting from the body of grasshopper, we will see that in an early nymphal stage, the normal developmental sequence stops. If, however, a prothoracic gland from another grasshopper is transplanted or if ecdysone is supplied artificially the sequence of nymphal stages resumes.

Let's examine various studies to gain appreciation of the effects of hormones on behaviour and look closely at the techniques that have been used to explore these relationships.

Dual System of Relationship between Nervous and Endocrine System

Before going deep in the details, we should have a clear concept that both the nervous and endocrine system are feedback systems that form key points of the body's mechanism. Both in vertebrates and invertebrates these two systems show a *dual system of interrelationship,* i.e., glands are interconnected through the circulatory system or they are closely tied to the nervous system through names connecting with brain.

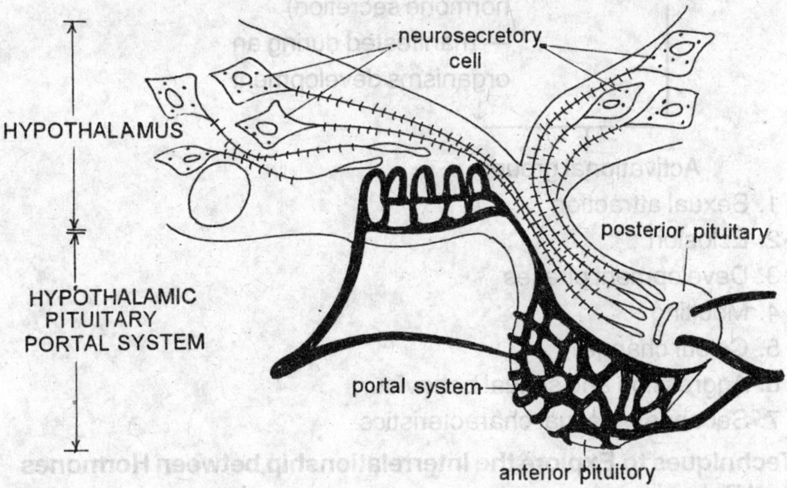

Fig. 7.1. Hypothalamus and pituitary of a generalized vertebrate neuroendocrine system.

Experimental Methods and their Effects

Hormonal influences on behaviour can be divided roughly in two categories: educational and organizational effects.

In *educational effects* hormones act as triggering influences on the expression an informance of behavioural patterns. The *organizational* effects of hormones are manifested during an organism's development.

The following classification will give a clear understanding of different types of effects (in behavioural aspects).

Hormonal influences on behaviour

Activational effects | Organizational effects

Activational effects:

- Direct (hormone secretion leads to a rapid response)
- Indirect (require more complete sequences of stimulation and hormone secretion) — manifested during an organisms development

↓

Activational groups

1. Sexual attraction
2. Eclosion
3. Development phases
4. Moulting
5. Colour change
6. Aggression and sexual behaviour
7. Secondary sexual characteristics

Techniques to Explore the Interrelationship between Hormones and Behaviour

Investigations have used several techniques to explore the interrelationship between hormones and behaviour.

These include —

1. *Exterpation or removal, of a particular endocrine gland* to assess the absence of a specific hormone on behaviour.
2. *Hormone replacement therapy* injection of specific hormones or transplantation of a gland.
3. *Blood transfusion* to transfer the "hormonal state" of one animal to another in order to observe the behavioural effect.
4. *Bioassays* to indirectly assess circulating hormones levels by measuring a secondary characteristic such as a skin gland, that is dependant on a particular hormone.

5. *Radioimmunoassay* to directly measure circulating levels of a hormone though the use of immunological methods.
6. *Autoradiography* to localize the sites at which hormone uptake occurs.

ACTIVATIONAL EFFECTS

(a) Sexual Attraction

In certain species of *cockroaches* and *moths* the females release *pheromones* which act as sex attractants for male.

Example. If in a female cockroach *corpus allatum* is surgically removed after moulting the female will not produce pheromone after it becomes sexually mature. But if corpus allatum of another adult female is transplanted to it then it again starts producing pheromones.

Now a question arises that a *Why has this system of sex-attractant evolved*? There are several answers.

1. Emission of a species sex-attractant may be a species isolating mechanism. It avoids gamete wastage.
2. Some pheromones may serve to synchronize the reproductive activities of the two sexes of a given species.
3. Series as a sexual excitatory function by bringing with members logically.
4. Attractant using that a male will find a female and mate with her.

(b) Eclosion

The process whereby the adult form of an insect emerges from the pupa after metamorphosis is called *eclosion* and is another activational effect controlled hormonally. Many moth species close at species *specific line* of adry.

The eclosion hormone which is produced by neurosecretory cells in the brain plays a critical role in this process. (Truman, 19/1; Truman and Riddifoord 1920).

Example. If the eclosion hormone is cirficted into pupae that are near the end of the metamorphosis eclosion behaviour, such as abdomen movements and being spreading after emergence *can be activated at any time of the deny.*

Moths that have their *brains removed* usually emerge; therefore, the presence of eclosion hormone is not an absolute requirement for eclosion to take place. The process, however, is not as coordinated in brainless subjects and some activities (e.g., wing spreading are usually absent. Thus, although the hormone may not be necessary

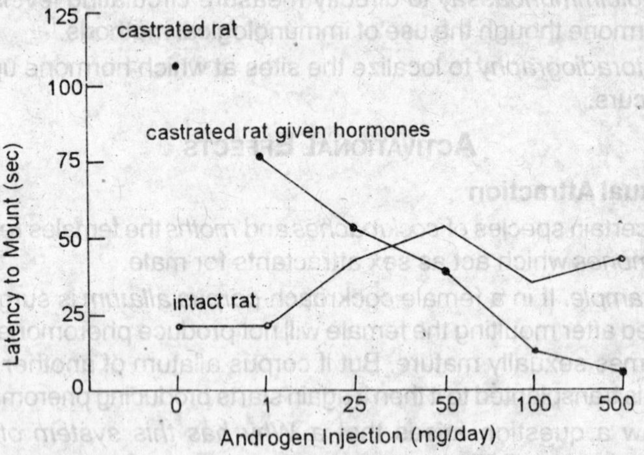

Fig. 7.2. Sexual behaviour in castrated rats.

for eclosion, it does appear to be necessary for proper coordination of the sequence.

(c) Development Phases

In adult male desert *locusts* sexual behaviour is exhibited when corpora allata are removed and when corpora allata is transplanted from other locust is restores its sexual behaviour (Lohrer 1961, Pener 1965), however, similar experiments have revealed that corpora allata are not needed for sexual behaviour in certain grasshoppers.

(d) Moulting

In some types of norms and molluscs and uncrustaceous investors have concentrated the presence of one or more neuro-

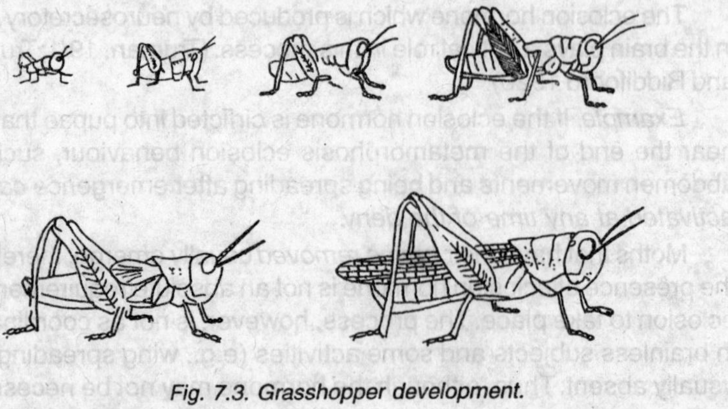

Fig. 7.3. Grasshopper development.

secretory or endocrine glands whose secretions effect several differentiation and malivation of gaindes and ethaulati reproduction.

Many crustacean moult periodically as they grind. *Removal of both eyestalk in these animals shortens the interval between moults.* If these are given the extracts of particular neurosecretory gland moulting is prevented clearly, the gland produces a moult-inhibiting factor.

(e) Colour Change

In short weasels which undergoes seasonal changes in pelage, orcont colour during spring and fall melts. In spring metamorphose stimulating hormone (MSH) secretion increases and new brown hairs replace the white coat colour during the full months MSH secretion is inhibited by the action of another hormone *melatonin* secretion by the sexual gland, the hair then is not argumented and returns to white.

(f) Aggression and Sexual Behaviour

Example. When male ring doves are *castrated* they show decreased levels of aggression courtship and copulation behaviour. When treated with crystalline *testosterone* the normal levels of behaviours are restored (*Banfield* 1971) *Mutchusin* 1969, 1971, 1978).

Testosterone effects both on sexual and aggressive behaviour.

(g) Secondary Sexual Characteristics

Example (i) The characteristics cock's becomes greatly decreased in castrated roosters.

(ii) Male cats, which spray urine probably as a marking behaviour, often cease to spray after their testes are removed.

ORGANIZATIONAL EFFECTS

Neonatal male rats pups were injected with estrogen. Histological examinations of the rats revealed some degeneration of the seminerous tubles, where sperms are produced. The investigations mated that although these males showed mounting behaviour when placed with a receptive female, the behaviour was irregular—the mounts were often incorrectly oriented and no ejaculation occured.

The *organizational effects* of hormones on female behaviour have been conducted on guinea pigs and Rhesus monkey. Female progeny of females that were treated with androgens when they were frequent have external genitalia that were masculinized and they exhibited male like sexual behaviour.

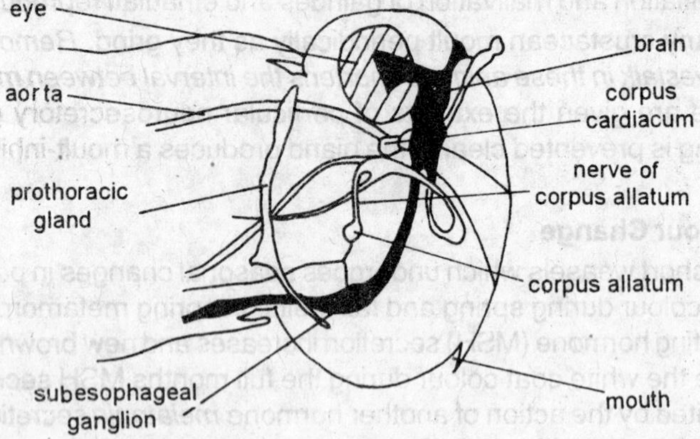

Fig. 7.4. Brain neurosecretory cells and glands of the head region in the grasshopper.

Endocrine Environment Behaviour Interactions

Some activational effects involve complex interactions among behaviours, hormones and specific environmental estimate. Will discuss in detail—one example reproductive sequencies in ring doves which illustrates this interrelationships.

A. Reproductive Sequence in Ring Doves

At each stage on this sequence, the internal state of each kind interacts with external variables to produce the observed behaviour pattern. The variables consists of —

1. The hormonal state of both the male and female dove, excluding feedback loops.
2. The behaviour of each member of the pair that stimulates changes in the hormonal levels and behaviour of its mate.
3. Environmental eves, such as nests and eggs, that influence hormonal and behavioural changes in both.

To delimine whether the presence of a mate or the nesting material affects incubation behaviours of female ring doves. *Daniel Lehrenan* and his colleagues used three experimental groups.

1. Control for males housed alone.
2. Females housed with a mate only.
3. Females housed with a male and nesting material.

They assessed the results of these pairings in terms of the percentages of test females in the each group that exhibited incubation behaviour when presented with a nest containing eggs.

Control female never incubated eggs. By days 6, 7, and 8 after pairing increasing percentages of females in the eccondgroup those housed with only a mate incubated eggs, by to day 8.100 per cent of females caged with both a male and nesting material incubated test eggs.

They concluded that presence of both a female and nesting material is necessary for complete incubation behaviour in a male.

Now a question may arise that why has such a fineiy tuned system of complex interaction among behaviour, hormones and external colours evolved in ring doves.

The answer to this question is that since the reproductive cycle involves the dual partical ration of both the birds and nececlatis that their activities be coordinated throughout, the reproductive success of the pain is guaranteed only if both perform certain acts in syutony e.g., Female that laid eggs before a nest was completed would contributed little to future generations. If the male developed a crop or began to produce crop milk before any rquats had happened etc.

B. Migratory Behaviour in Birds

Migration in behaviour in birds is also due to the result of the interaction between the environment and the hormones. *Reowan observed* that this peak of gonadal recurdiocence coincides with the peak of migration hormones activated the migratories urge.

Rowan's suggested that many factors influence migration. Increasing day length, increasing ambicint temperature and other environmental changes which affect the animal through pituitary activity and perhaps the nervous system directly.

MECHANISMS AND ONTOGENY OF HORMONAL EFFECTS

A. Stimuli and Mechanisms

To understand the action of a hormone, both the stimulation and its release and its effects upon behaviour must be known. The mechanisms by which hormones affect behaviour, is also complex. *Beach* (1948) separates four theoretical possibilities. Hormones may influence behaviour through their effects upon.

1. The whole organism (e.g., general activity level).
2. Morphologic structures employed in specific response patterns.
3. Peripheral receptor mechanisms.
4. Integrative functions of the C.N.S.

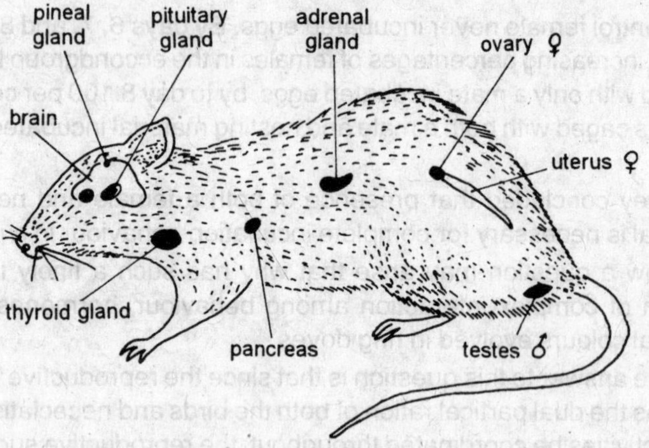

Fig. 7.5. Endocrine glands of the rat. Different glands receive either nervous system communication, circulatory system communication, or both from other parts of the body.

(a) Control of this development of nervous organisation.

(b) Control of periodic growth and regression of nervous elements.

(c) Control of sensitivity of stimulation.

B. Ontogenetic Development

If baby which is injected with testosterone, complete mating responses including covering, treating have been observed.

There are many examples which ultimately conclude that some interaction of hormonal action and behavioural development establishes a behavioural organization which then persists in the absence of hormonal influences.

Some Recent Development

Since 1950 research has centered increasingly upon the problem of untangling the delicate and complex interrelationship between hormones, the nervous system and behaviour.

1. *D.S. Lehrman* (e.g., 1964) has recently made three discoveries about the role of prolactin in parental behaviour in doves.

   First, Prolactin alone does not elicit parental behaviour, but requires estrogenic *priming.*

   Second, Low or hormone can act directly on a peripheral structure involved in behaviour. Without involving vagus effects on the C.N.S.

Third, The young birds provide the tactile stimuli that elicit regurgitation from the crop.

2. In somewhat similar program of research *Hinde* (e.g., 1958) has been exploring the hormonal bases and external stimulus conditions, controling nest building.
3. Another interesting approach to hormones and behaviour has been to emplified by the work of *R.P. Michael* regarding sexuality on the verge of *nymphonania.*
4. Active research continues to the field of migration too. The *Wolfsen theory* stated that the accumulation of fat sets off migratory behaviour has been challanged by the schools.

Farmer and *King* have shown by ingenious experiments that it captive birds were not allowed to accumulet for *Zugunrube* appears independently to response to long photoperiod.

Conclusions

Nearly, all behaviour is influenced by hormones in some way or the other. Hormone flow is elicited by a very broad range of physical and social external stimuli. Hormone may act directly upon central neural mechanism underlying behaviour but also certainly act upon pheripheral sensory reception.

The study of the ways to which hormones affect behavioural development through effects on peripheral organs as well as on differentiating neural tissue will require a vigorous new approach.

8

THE EVOLUTION OF SOCIETIES

Animals forms groups of enhance their foraging success and gain better protection against predators; but group living also exposes them to diseases, competition, and social interference. The costs and benefits of social living vary with the age, sex, experience, and physical condition of individuals. Social groups are maintained by chemical, visual, auditory, and tactile signals. Signals benefit both receivers and senders, but animals do not always communicate their intentions clearly and accurately. Complex social systems have evolved from offspring remaining with parents to help rear future broods and from cooperation among adults of the same generation.

This chapter deals with costs and benefits of group living, roles of the sexes, types of communication signals and their evolution, choice of associates, and the evolution of animal societies.

Even solitary organisms must interact to exchange gametes. Among many sessile marine organisms, the gametes do the associating by themselves in the water, but most mobile animals get together to copulate, even if they part immediately afterward. From these brief and simple interactions, social behaviour had developed into more and more elaborate systems, culminating in the complex societies of social insects and vertebrates. How these systems have evolved, why it was advantageous for animals to associate with one another, and how social groups are maintained are the subjects of this chapter.

The Costs and Benefits of Social Living

Humans tend to think that the highly complex societies of some vertebrates, ourselves included, represent a pinnacle of evolution.

This self-congratulatory view implies that "highly evolved" creatures, like ourselves, represent a form of life superior to the less highly social species. To adopt this attitude is to ignore the broad spectrum of potential disadvantages, as well as possible advantages that arises from social living. One of the more thorough explorations of the costs and benefits of living in a group is the study of bank swallow behaviour by *John Hoogland* and *Paul Sherman.* Bank swallows form nesting colonies of dozens to hundreds of pairs in sand quarries and river banks. *Hoogland* and *Sherman* showed that by nesting together the swallows are able to reduce the effectiveness of some of their diurnal predators. When a hunting jay or weasel approaches he colony, the birds mob the predator, much as gulls do and sometimes succeed in driving their enemy away. In addition, pairs seek to acquire a centrally located nest burrow and so to "use" the other members of the colony as a shield against nest-robbing predators that tend to attack the peripheral members of a group first. Moreover, because the birds link their reproduction with the pairs nesting around them, they avoid producing eggs or immature young when other birds are not reproducing. If the breeding period were extended, a predator

Table 8.1. Some Major Advantages and Disadvantages of Sociality*

Advantages
Reduction in predator pressure by improved detection or repulsion of enemies
Improved foraging efficiency for large game or clumped ephemeral food resources.
Improved defense of limited resources (space, food) against other groups of conspecific intruders.
Improve care of offspring through communal feeding and protection.
Disadvantages
Increased competition within the group for food, mates, nest sites, nest materials, or other limited resources.
Increased risk of infection by contagious diseases and parasites
Increased risk of exploitation of parental care by conspecifics
Increased risk that conspecifics will kill one's progeny.

* For solitary species certain of the disadvantages outweigh any benefits of social living; for social species the costs are more than matched by certain of the advantages of sociality.

could harvest the offspring of the swallows one by one, as they were generated. By synchronizing egg laying, a pair of swallows adds its progeny to a large pool of vulnerable offspring that exists for only a short period. Because predators can consume only a few nestling per day, synchronized breeders may have a better chance of getting their clutch through the dangerous period than a synchronous reproducers.

Some workers have suggested that an additional benefit of colonial nesting in this species might be improved foraging success. Birds that had failed to find much food might be able to follow other individuals to ephemeral concentrations of aerial insects these luckier or superior hunters had found. Although *Hoogland* and *Sherman* found little evidence to support this possibility in the colonies they observed, it might apply in other years or in other areas.

Offspring the benefits of colonial nesting in bank swallows is a number of documented costs. Nestings in large colonies suggests that the presence of many swallows in an area depletes food resources and actually makes it harder, not easier, for birds to find insects. Moreover, the swallows fight for nest sites and nesting materials; a solitary pair would not run the risk that a neighbour would make off with the hard-earned feathers it had collected for a nest lining. Nor would the male of a solitary pair run the same risk of cuckoldry, which would cause him to spend his breeding season caring for the progeny of an adulterous mate and another male. "Illicit" matings of this sort do occur in colonies. Finally, Hoogland and Sherman counted the bird fleas in bank swallow burrows, no doubt an edifying task, and found that the probability that a nest was infested was higher in large than in small colonies.

Because bank swallows never nest alone, but always in groups, one presumes that the benefits, primarily improved antipredator defense, outweigh the numerous disadvantages. However, the benefits of sociality vary from individual to individual, depending on the success of groups mobbing, the ability of the birds to claim central nesting sites, and their capacity to synchronize egg laying with other members of the colony. In other social species, the costs and benefits of group living vary so much from place to place or season to season that social units will change dramatically in size or even disband entirely when this is advantageous to individuals.

Animals in groups gain a number of advantages related to defence. While a group of animals is bigger than a single one, and

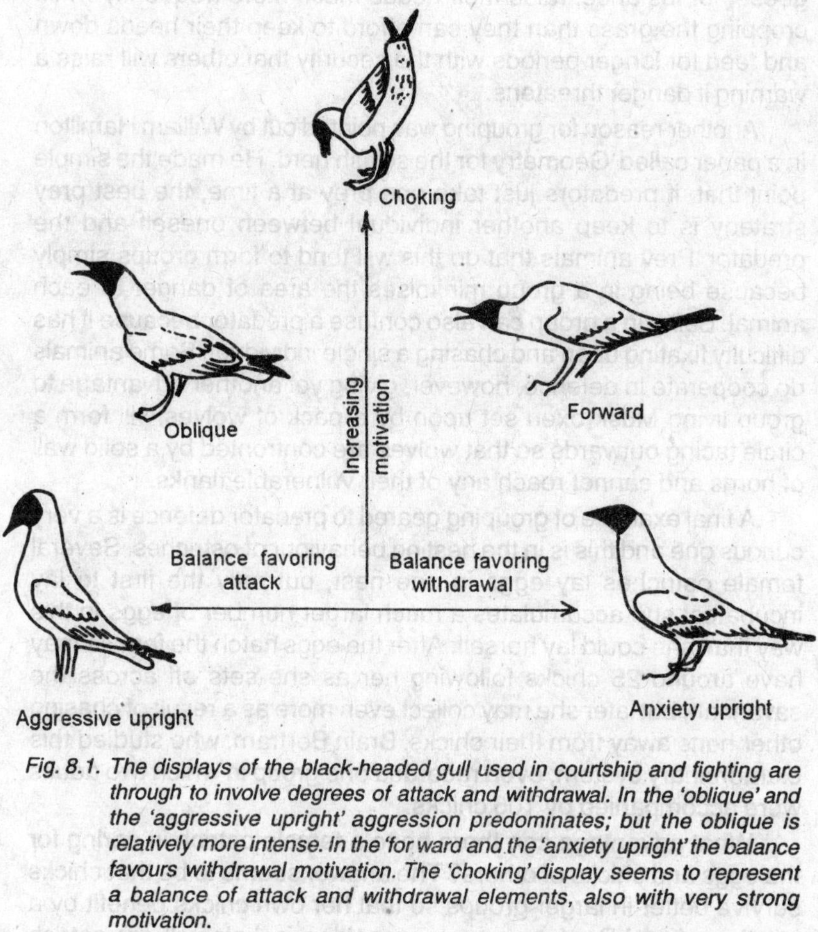

Fig. 8.1. The displays of the black-headed gull used in courtship and fighting are through to involve degrees of attack and withdrawal. In the 'oblique' and the 'aggressive upright' aggression predominates; but the oblique is relatively more intense. In the 'for ward and the 'anxiety upright' the balance favours withdrawal motivation. The 'choking' display seems to represent a balance of attack and withdrawal elements, also with very strong motivation.

so can be spotted from further off, if the predator only eats one prey at a time it will take longer to find meal if the prey are clumped than if they are evenly spread out. Furthermore, there are more pairs of eyes in a group, so the predator is less likely to creep up undetected. To some extent this is offset by the fact that animals in groups are able to be less vigilant than solitary individuals. Lone geese, for instance, raise their heads much more frequently while cropping the grass than they do when feeding in groups. Because there are more pairs of eyes in a group, so the predator is less likely to creep up undetected. To some extent this is offset by the fact that animals in groups are able to be less vigilant than solitary individuals. Lone geese, for instance, raise their heads much more frequently while cropping the grass than they can afford to keep their heads down and feed for longer periods with the security that others will raise a warning if danger threatens.

Another reason for grouping was pointed out by William Hamilton in a paper called 'Geometry for the selfish herd.' He made the simple point that, if predators just take one prey at a time, the best prey strategy is to keep another individual between oneself and the predator. Prey animals that do this will tend to form groups simply because being in a group minimises the area of danger to each animal. Being in a group can also confuse a predator because it has difficulty fixating upon and chasing a single individual. Some animals do cooperate in defence, however, giving yet another advantage to group living Musk oxen set upon by a pack of wolves will form a circle facing outwards so that wolves are confronted by a solid wall of horns and cannot reach any of their vulnerable flanks.

A final example of grouping geared to predator defence is a very curious one and this is in the nesting behaviour of ostriches. Several female ostriches lay eggs in one nest, but only the first to lay incubates: she accumulates a much larger number of eggs in this way than she could lay herself. After the eggs hatch the female may have around 25 chicks following her as she sets off across the savannah, but later she may collect even more as a result of chasing other hens away from their chicks. Brain Bertram, who studied this extraordinary system, even recorded one group in which two adults were accompanied by 105 chicks.

What advantage can there be to a female ostrich in caring for the eggs and chicks of others? The answer seems to be that chicks survive better in larger groups so that her own chicks benefit by a 'dilution effect.' Predators such as golden jackals will eat ostrich eggs and chicks, but the more of these that a female has with her the less likely is the one that is taken to be one of her own offspring. This cannot be the whole story, however, otherwise a female would gain whether she cared for her young herself or let another female do so. At the stage of incubation it seems that there are advantages to the hen that sits, for she can recognise her own eggs and she gives preference to them so that they remain in the centre of the nest and are more likely to hatch. She is not therefore begin a generous nanny helping out others, but is running the creche for her own advantage. It remains to be seen whether females with chicks

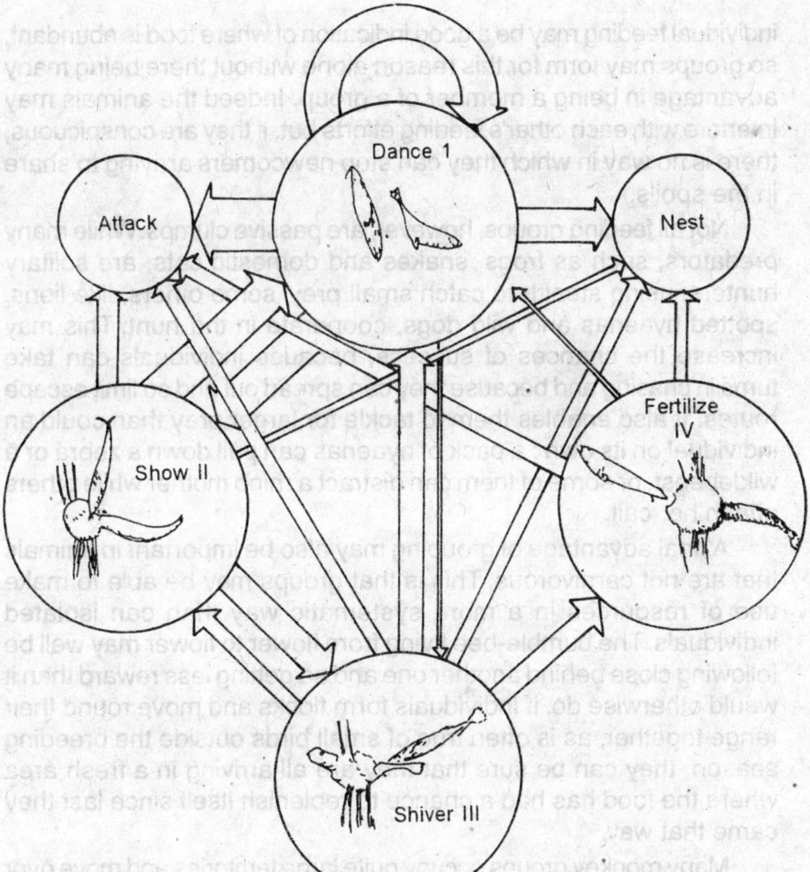

Fig. 8.2. The sequence of activities during courtship of ten-spined sticklebacks includes much aggression. The male is the darker of the two. The sequences of male courtship activities from dancing to fertilization are shown. The number of times that these behaviour patterns preceded or followed cther actions in represented by the width of the connecting arrows. Attacks may follow any part of the sequence.

also gain from giving care themselves rather than leaving them to be looked after by others.

Finding Food

A further range of benefits to group living is related to the exploitation of food supplies. If food is clumped, as is very often the case, animals that feed on it tend to become so too. Seeing another individual feeding may be a good indication of where food is abundant, so groups may form for this reason alone without there being many advantage in being a member of a group. Indeed the animals may interfere with each other's feeding efforts but, if they are conspicuous, there is no way in which they can stop newcomers arriving to share in the spoils.

Not all feeding groups, however, are passive clumps. While many predators, such as frogs, snakes and domestic cats, are solitary hunters, using stealth to catch small prey, some others, like lions, spotted hyaenas and wild dogs, cooperate in the hunt. This may increase the chances of success, because individuals can take turns in chasing and because they can spread out and so limit escape routes. It also enables them to tackle for larger prey than could an individual on its own : a pack of hyaenas can pull down a zebra or a wildebeest, or some of them can distract a rhino mother while others attach her calf.

A final advantage of grouping may also be important in animals that are not carnivorous. This is that groups may be able to make use of resources in a more systematic way than can isolated individuals. The bumble-bee flying from flower to flower may well be following close behind another one and so getting less reward than it would otherwise do. If individuals form flocks and move round their range together, as is often true of small birds outside the breeding season, they can be sure that they are all arriving in a fresh area where the food has had a chance to replenish itself since last they came that way.

Many monkey groups occupy quite large territories and move over them in a single partly searching for the fruit or buds on which they live. They too may gain from the systematic foraging that group living allows, but their groups are more integrated and the relationships of individuals within them more varied than in a flock of birds or a herd of deer. The advantages of food finding or predator avoidance may explain why they, like many other animals, come to live in groups, but to little to account for the richness of the interactions between the group members. It is to various facets of the behaviour of individuals within groups towards one another that we shall now turn.

Kinship

Not surprisingly, animals that live in social groups tend to be related to one another. Young ones are born into the group to which their mother and father belong and may often stay in it to become parents themselves. However, if this was always so, animals in the group would become more and more inbred. Too much inbreeding leads to young which are less viable, so it is discouraged by natural selection. What usually happens, therefore, is that, as maturity approaches, young animals of one sex or the other move away from the group in which they were born and so do not mate with close relatives. In birds it is usually the females that disperse to different areas, whereas in mammals the males are more likely to leave, although there are quite a few exceptions to this rule, including species in which both sexes move away.

Which sex disperse may have a profound effect on various aspects of social behaviour. Male white-crowned sparrows sing songs

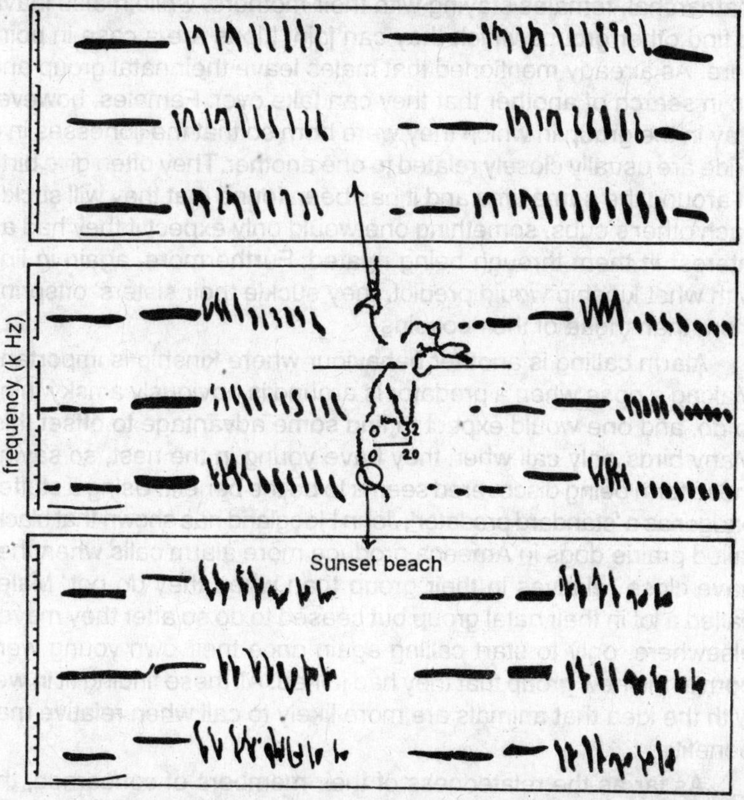

Fig. 8.3. The songs of each of six males white-crowned sparrows from three different areas around San Francisco. There are marked dialects, most obviously in the structure of the second part of the song, which is closely similar within an area but different between them.

which are almost identical with those of their neighbours and quite different from those of birds a few kilometres away. Song learning usually only occurs in the first three months of life and so it complete long before they set up their territories and start to sing. But they breed only a short distance from where they hatched and thus the song dialects are maintained. Although female birds do not usually sing, their choice of mate is probably influenced by the songs of prospective partners. In white-crowned sparrows, for example, there is evidence that females may prefer males from the same dialect area as themselves, though there is some controversy about this.

By contrast with birds, groups of many mammalian species are matriarchal, females staying with their mothers, while males leave to find other groups which they can join. Lions are a case in point here. As already mentioned that males leave their natal group and go in search of another that they can take over. Females, however stay in the group in which they were born so that the lionesses in a pride are usually closely related to one another. They often give birth at around the same time and it has been found that they will suckle each other's cubs, something one would only expect if they had an interest in them through being related. Furthermore, again in line with what kinship would predict, they suckle their sisters' offspring more than those of their cousins.

Alarm calling is another behaviour where kinship is important. Making a nose when a predator is around is obviously a risky thing to do, and one would expect to find some advantage to offset this. Many birds only call when they have young in the nest, so saving these form being discovered seems to be the benefit. Using a stuffed badger as a 'standard predator', John Hoogland has shown that black-tailed prairie dogs in America produce more alarm calls when they have close relatives in their group than when they do not. Males called a lot in their natal group but ceased to do so after they moved elsewhere, only to start calling again once their own young were born in the new group that they had joined. All these finding fit in well with the idea that animals are more likely to call when relative may benefit.

As far as the relatedness of their members of concerned, the most complex groupings are amongst the social insects. In the honey-bee, only the queen lays eggs and all other females in her hive are sterile workers. The eggs the queen lays are of two sorts: unfertilised ones, which will develop into fertile males (drones), and fertilised one, which usually develop into infertile females (workers) but, if nourished only on royal jelly, will form a new queen. This happens when the old queen dies or when the colony has grown to a point where it must split. A peculiarity of this breeding system is that drones have half the number of genes that queens or workers to and, when they fertilise an egg, all their genes pass to their offspring instead of just half of them. As a result of this system, which is known as haplodiploidy, workers share 3/4 of their genes with their sisters rather than 1/2, as most animals do. It is thought to be for this reason that it benefits them to raise sisters which will become queens rather than having daughters of their own. Through a quirk of their reproductive system, their sisters are more closely related to them than their daughters would be and are thus a better investment in the future of their genes.

The social system of bees is thus extraordinarily intricate, and this is probably because these animals are especially closely related to others in the hive to which they belong. However, having this unusual mode of inheritance is not a prerequisite for a social system of this sort. Termites are not haplodiploid yet they have a colony structure which is just as complex. More remarkable still is the naked mole rat, for this is a mammal which has colonies much more like those of a social insect. These animals live communally in groups of around 40 individuals which share a burrow. For must of the time they huddle together in one underground chamber, and it is here that the only female to breed has her young. She is large, and males approaching her size may mate with her, but most other large animals in the group neither breed nor forage : their main role seems to be to keep the colony warm. Smaller individuals are like the workers of social insects. They build nest and more around through the burrow system foraging for roots and tubers which they bring back for all the animals to feed upon. If there is a disturbance, all the members of the colony join in the task of carrying the young off to safety.

As well as the food that the workers bring back being shared, the young obtain nourishment form adults by eating their faces. Individuals also spread their urine round the colony and groom with it. These habits, unsavoury as they may seem, probably serve to pass information around the group for, if the breeding female dies or is removed, another one comes into reproductive condition very soon. This is just like the social insects. In these, queen substance is spread throughout the colony and stops the workers from rearing other queens. But, if the queen dies, this substance is no longer there, and the workers quickly start to rear a replacement.

Male rats are not very mobile and their burrow systems are isolated from each other. It is likely, therefore, that most of the animals within a colony are relatives and that their close social relationship stems from the common interest that this gives them in the offspring of the breeding female. Unfortunately we do not know just how closely related they are to each other, and this is very often the case in studies of social behaviour. It is easy to speculate that kinship is involved when we observe animals being altruistic to one another, but only in a few cases, such as that of the social insects, do we know in any detail exactly what the relationships are between the individuals involved.

Insect Societies

The Social Life of Insects

Ethologists honour four groups of insects—ants, bees, wasps and termites—with the description of 'social' insects. The crucial characteristic of these four groups is that, within their nests, they show a reproductive division of labour. Within the nest of a typical ant species there is a single queen who lays nearly all the eggs that are laid in the nest. The rest of the ants are sterile 'workers'. The workers carry out all the duties necessary to keep the colony going except for the laying of eggs.

There are more than 12,000 social insect species, and they show a fascinating range of ways of life. Within the ants, for example, there are 'army' ants with huge colonies of up to 22 million individuals, which bestride the jungle floor eating everything edible in their path. There are 'fungus gardening' species which grow fungus on specially prepared rotting leaves and live on the produce of the fungus. Other ant species live by milking honeydew from herds of little insects called aphids. In yet others, workers from living 'honeypots', as they hang upside down from the roof of their nest, their abdomens hugely distended with honey. Australian aborigines dig up the nests, take the ant's head between their fingers, and bite off the honeyed abdomen. The marvels of the social insects are almost endless, but we shall concentrate on some general properties of their social life that are closely related to their altruistic habits.

We must consider ants, bees and wasps separately from termites. Ants, bees and wasps belong to the insect order

Hymenoptera; termites make up the order Isoptera. A typical hymenopteran colony is founded by a single queen after her 'nuptial' flight. If the species is an ant, the foundress bites off her wings—she will never fly again. She excavates a small nest, lays her first brood, feeds and rears the larvae. She will never rear larvae again, because this first generation of workers themselves rear the next lot of eggs. It is an important fact that all the workers, in the first and later generations, are females. They are sterile and therefore in a sense sexless, but they contain the genetic make-up of females. Once the colony has reached a certain size the queen lays eggs which are reared as reproductives. The time when reproductives are first produced varies between species. In *Myrmica rubra,* a common garden ant in Europe, it is not until about 9 years after founding, when the colony has grown to about 1000 workers.

Altruism in Insect Societies

The distinctive property of social insects that cries out for explanation is the sterility of the workers. Why do these workers slave away to their death only to enhance the reproduction of another individual? An important part of the answer may be provided by *Hamilton's theory.* For Hymenoptera have an idiosyncratic genetic system, rather different from the usual Mendelian one. They are

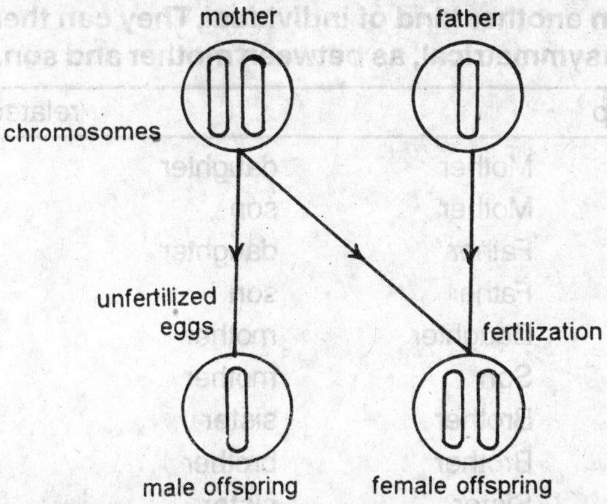

Fig. 8.4. Haplodiploid inheritance. Female have a diploid set of genes but males have only one ('haploid') set. Males develop from unfertilized eggs. This kind of inheritance is exceptional, but is found in hymenopteran insects and a few other groups of animals. The relatedness among individuals differ from those under diploid inheritance: a gene in a male, for instance, has a probability of one of being in his mother.

'haplodiploid'. The females, like most animals have two sets of chromosomes (they are 'diploid'); but the males have only one (they are 'haploid'). Male offspring contain no genes from their father; they develop from unfertilized eggs. Females contain genes from both their mother and their father. Now haplodiploidy, as *Hamilton* pointed out, will alter the normal pattern of relatedness. The sister of one family are exceptionally closely related because they all share exactly the same set of genes from their father; their father has only one set of genes to give. Many of the values of relatedness are changed under haplodiploidy. The relatedness between siblings under diploidy was calculated under the premise that if a father contributes a gene to one sibling there is only a chance of one half that he will give it to another. Under haplodiploidy that chance is one, not a half, and the total relatedness between sisters is therefore (½ × 1) + (½ × ½) = 3/4. Because the relatedness of mother to daughters, and one, other things being equal, 'breed' more efficiently by making sisters than by reproducing daughters. They may be why female hymenopterans have so often evolved sterility. Males, by contrast, have not evolved sterility. But then they are not exceptionally closely related to their siblings.

Table 8.2. Relatedness under haplodiploidy. The relatedness is the probability, given a gene is in one kind of individual, that it is in another kind of individual. They can therefore be asymmetrical, as between mother and son.

Relationship			relatedness (r)
*	Mother	daughter	½
*	Mother	son	½
*	Father	daughter	1
*	Father	son	0
*	Daughter	mother	½
*	Son	mother	1
*	Brother	sister	½
*	Brother	brother	0
*	Sister	sister	3/4
*	Sister	brother	¼

A large controversial literature has grown up around Hamilton's explanation of sterility in hymenopterans. It would be inappropriate to consider it here. No ethologist would doubt that kin selection has had some part in the evolution of hymenopteran social behaviour, but this statement leaves plenty of room for disagreement. The same exact theory cannot apply to termites. Termites are diploid, therefore the relatedness of siblings is normally the same as that of parents to offspring. But sterile castes have evolved in termites. As Hamilton's theory predicts, in termites both sexes of offspring become sterile workers; in termites, unlike Hymenoptera, there is no force of kin selection predisposing one sex to evolve sterility. However, although the theory can explain this difference from Hymenoptera, there is no widely accepted explanation of the evolution of sterile castes in termites. Suggestions have been made. The relatedness among siblings can be higher than that of parents to offspring if siblings are produced by highly incestuous matings but the offspring produced by outbreeding. A termite that 'went alone' might outbreed, whereas it could help to rear closely-related, inbred siblings if it stayed on as a helper. However, the idea is theoretical only, and is not well tested. I only mention it to illustrate that the theory of kin selection has been applied to termites as well as Hymenoptera. The theory is more easily applied to Hymenoptera.

Recognizing Relatives

We have not proved beyond doubt that kin selection has caused the evolution of altruism in avian helpers and social insects, but the theory probably contains a large measure of truth in these cases.

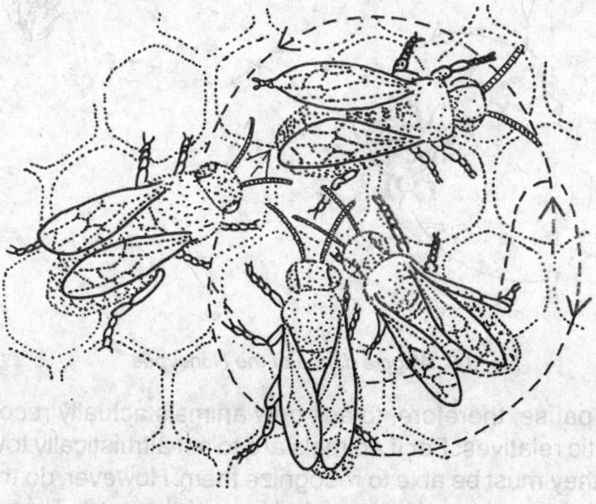

Fig. 8.5. Round dance of the Honeybee.

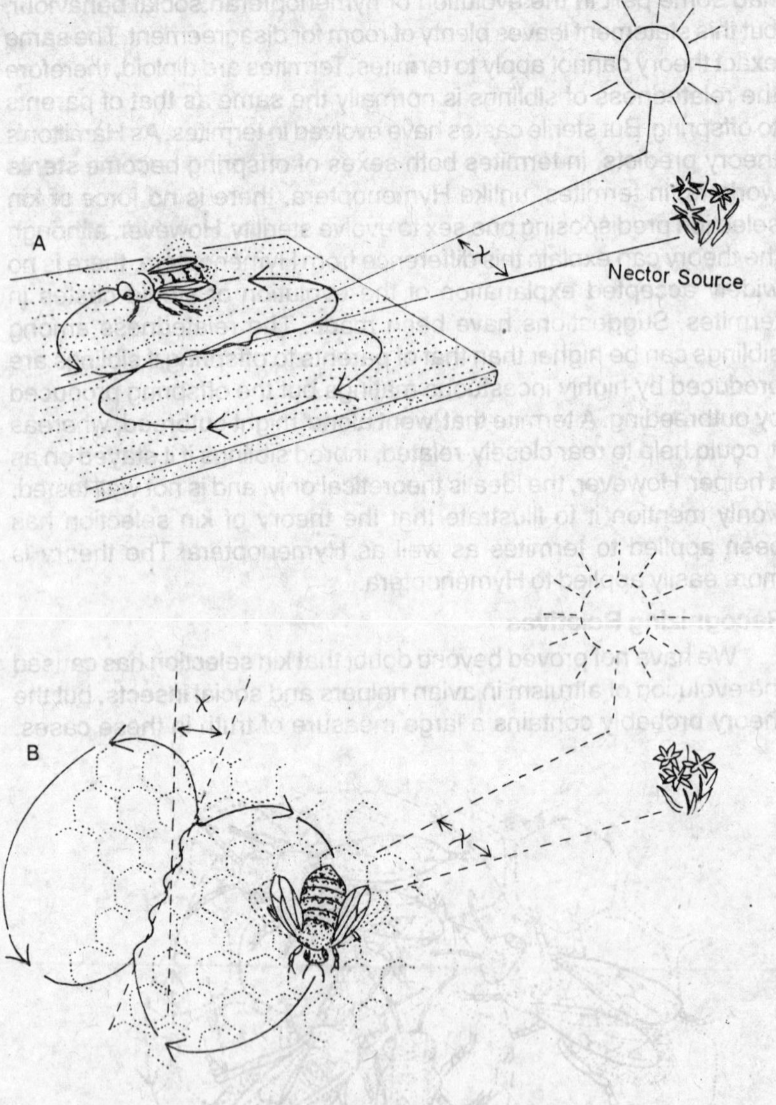

Fig. 8.6. Waggle dance of the Honeybee.

We might pause, therefore, to ask how animals actually recognize their genetic relatives. For if animals are to act altruistically towards relatives, they must be able to recognize them. However, do they do so? No work has been done on helpers at the nest. They could 'recognize' relatives merely as other young birds in the same nest as they were reared in as they would usually be siblings. In practice, recognition may be more sophisticated, but in the absence of facts speculation is unnecessary, because there is no difficulty of principle here. There has, however, been some work on insects merits mention. Colonies of social insects have been known for many years to possess individual colony odours. Individuals can tell members of their own colony (who are genetic relatives) from members of other colonies (unrelated) by their characteristic smell. The source of colony-specific odours is not completely known; but it probably develops from the diet of the colony, as individuals frequently regurgitate food to one another, and minor differences in the diets of different colonies would give them different characteristic odours. Different colonies perhaps secrete their own pheromones as well, but the importance of diet has been indicated by experiment. Kalmus and Ribbands moved two hives of honeybees from a typical honeybee environment to an isolated moor which had only one species of flower. The level of fighting between the two hives decreased, perhaps because they increasingly came to recognize each other as members of the same hive. Likewise, when parts of a hive were isolated and fed on different diets, the level of fighting within a hive increased.

Each member of the colony has to learn the colony odour. Young ants do not challenge other ants in the colony, regardless of their odour; but older ants challenge any ant that does not have the colonial smell. The learning may even take place during a sensitive phase of imprinting.

What I have said so far would probably have passed until quite recently as a short summary of how social insects recognize nestmates. It was supposed that relatives are recognized by cues of environment origin, by learning. A rush of recent evidence now suggests that story is incomplete. In several species, individuals have been found to be capable of distinguishing relatives even though they had no chance of learning to recognize them: they seem to possess an innate ability to distinguish relatives. The first important experiment was conducted on sweat bees, *Lasioglossum zephyrum*, by *Greenberg* in 1979. By various genetic tricks, he bred colonies in the laboratory that had twelve different kinds of relatedness to one another, varying from completely unrelated colonies (r~0) to different colonies produced from inbred lines (r*1). In his experiments he introduced a bee from one colony to the entrance of another colony.

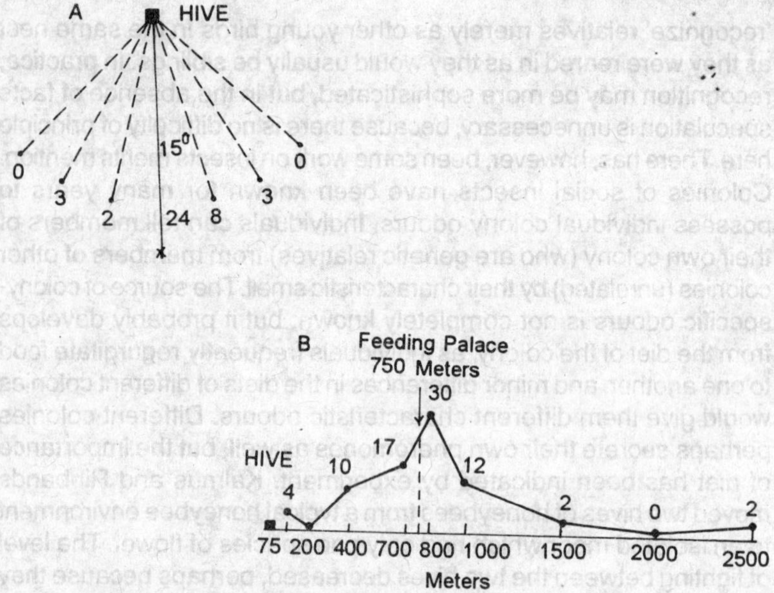

Fig. 8.7. Ability of worker bees to communicate with other workers about (A) the direction and (B) the distance to a food source.

At the colony entrance a 'guard' bee generally admits nestmates and challenges foreigners, and *Greenberg* observed the rate at which guard bees admitted introduced bees bearing differing degrees of relatedness to them. His result was strikingly positive; guard bees were more likely to admit bees the more closely related the introduced bee was. The important part of Greenberg's experimental design was that in nearly all cases the guard bee had never had any opportunity at all to learn the introduced bee's smell: in nearly all cases guard and introduced bee had spent all their lives in different colonies. For a minority of cases the bee would have been separated at the larval stage from the same colony, but no learning is thought to take place this early, and, in any case, the positive result still stands even if this minority of experiments is ignored. Sweat bees, it appears, can recognize degrees of relatedness among conspecifics of which they have no experience at all. Since Greenberg's work on sweat bees, moreover, a comparable ability has been demonstrated with varying degrees of certainly in other bees, and other animals, such as ants, mice, quails and tadpoles. In now seems reasonable to guess that genetic abilities to recognize relatives are widespread. Our original understanding must therefore be modified. Animals can

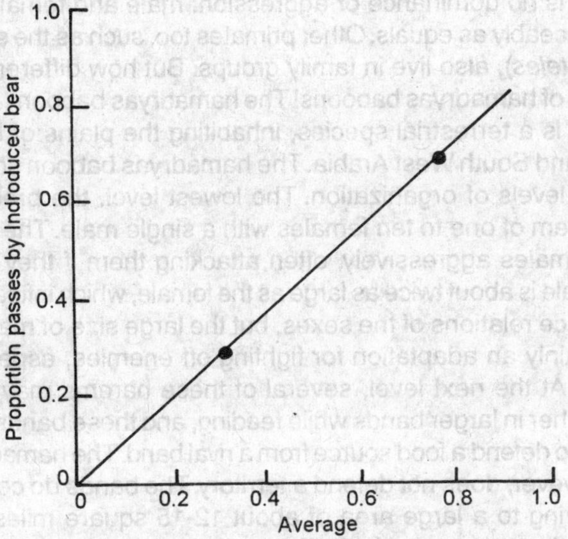

Fig. 8.8. Whether a guard bee of Lasioglossum zephyrum allows a conspecific bee to enter its nest depends on the genetic relatedness of the two bees: closer relatives are more likely to be admitted.

recognize their relatives by a mixture of innate and learned information, in ways that probably differ among species and are not, at this early stage of research, will understood.

PRIMATE SOCIETIES

The Social Life of Primates

Different species of primates live in different kinds of societies; indeed, the same species may form different kinds of social groups according to the conditions. The Hanuman langur (*Presbytis entellus*) forms both harems, with one male and several females, and multimale groups, with several males and a larger number of adult females. The reason is uncertain, but may be related to population density, for multi-male groups are commoner when the total population density of an area is lower. Before considering the altruistic behaviour of primates, let us consider briefly some of the different kinds of social groups.

The white-handed gibbon (*Hylobates lar*) lives in the trees of South East Asia in monogamous family groups. The male and female are similar in size and appearance, and live as a pair with 0-4 young. The pair defend a territory of about half a square mile. Within the family there is no dominance or aggression: male and female live together peaceably as equals. Other primates too, such as the spider monkeys (*Ateles*), also live in family groups. But how different are the societies of hamadryas baboons! The hamadryas baboon (*Papio hamadryas*) is a terrestrial species, inhabiting the plains of North East Africa and South West Arabia. The hamadryas baboon society has several levels of organization. The lowest level, the breeding unit, is a harem of one to ten females with a single male. The male treats his females aggressively, often attacking them if they stray away. The male is about twice as large as the female, which influences the dominance relations of the sexes, but the large size of males is probably mainly an adaptation for fighting off enemies, especially rival males. At the next level, several of these harems may walk around together in larger bands while feeding, and these bands may act as a unit to defend a food source from a rival band. The hamadryas baboon, however, does not defend a territory. The bands do confine their wandering to a large area of about 12-15 square miles, but different bands may overlap in their use of this area. It is, therefore, called a home range, to distinguish it from a defended area, which would be called a territory. At a higher level, several bands may join into larger groups for sleeping. If suitable shelters for sleeping are difficult to find, as many as 700 hamadryas may sleep together.

Whereas in the hamadryas baboon the mating unit is a single harem aggressively controlled by a single male, in the howler monkey (*Alouatta*), several adult males may live peaceably together in a single group. When Ray Carpenter watched howlers on Barro Colorado Island, in Central America, the group sizes were variable but contained an average of 3 males, 8 females, and 7 young. The male howler monkey is about 30% larger than the female, and has an enlarged voice box covered by a beard. The males roar daily, making the loudest animal noise in the American forests, a noise which carries for over a mile. This howling serves to space out the different groups. Within each group there is little aggression and no obvious dominance hierarchy.

These three species live in entirely different kinds of societies. One is monogamous, another polygamous within single male groups, another polygamous with multi-male group. One shows dominance and agression within groups, the other two do not. Two defend territories, the other does not. Other primate species live in yet other kinds of societies. Ethologists would like to be able to explain why different primate species live in different kinds of societies. Why are some species aggressive, others not? Some territorial, others not? These questions cannot yet be answered satisfactorily. Our understanding of the diversity of primate societies is still limited. One trend which can be explained is as follows. There is a tendency in species in which there are many females per male in the group for the males to be bigger than the females: in monogamous species the male is about the same size as the female; in polygamous species the male is larger. This has presumably arisen because sexual selection has favoured larger males in polygamous species, because larger males are more successful in fights over females.

Altruism in Primates

Many kinds of altruistic behaviour can be seen in primate groups. There is the feedings carrying and defence of young, not only by their mothers; cooperative searching for, hunting and exploitation of food; food-sharing; cooperative group defence against enemies; and most common of all, that favourite pastime of primates, grooming. Most, perhaps all, of these habits should probably be explained by kin selection. We, however, have considered the application of that theory in two examples already, and will therefore, use a primate example to illustrate another reason why altruism may evolve: the theory of reciprocal altruism.

The example concerns 'consorting', which is a habit of males, in many species that live in multi-male groups, whereby a male stays close to a female during the receptive phase of her oestrous cycle, and defends her from the advances of other males. It is an adaptation produced by the male competition component of sexual selection. In the olive baboon (*Papio anubis*), Craig Packer observed that two males may occasionally co-operate to fight off a single male: clearly, two will be stronger than one, which makes the advantage of co-operation clear. The advantage, however, is only for the one male that copulates with the female; he has gained a benefit of as much as one extra offspring: the other male has paid a cost of the risk of injury in a fight, but obtains no benefit. He has behaved altrustically. If what we have seen so far were the end of the matter natural selection should eleminate the altruistic habit. But it is not the end of the matter. The next stage arrives when another female in the troop comes into oestrus. The roles of the same two males may now be reversed. Packer saw ten cases in which a male who had previously been 'solicited' into co-operating to defend a female (but did not copulate with her) himself solicited a male into co-operative defence. In nine of the ten cases the solicited male was the individual whom he had previously helped. It looks, therefore, as if males form co-operating pairs to defend females and take turns in the copulating that is the end of the defence. If so, it would be an example of 'reciprocal altruism', Altruism can evolve under individual selection, without any need for the animals to be related, if the altruist is more than paid back later. The danger of any such reciprocal arrangement is that they will be cheated on. There is a clear short-term advantage to receiving altruism but then not paying it back; that way the cheat gains the benefit but does not pay the cost. This begin so, reciprocal altruism is expected mainly to evolve in species that form stable groups, with individual recognition. Then a cheat, after gaining a short-term benefit, can be recognized and excluded from future transactions; the cheating will then not pay. Without the opportunity and mechanism of discriminating against cheasts, reciprocal altruism is likely to break down. Because reciprocal altruism requires rather special conditions, it may be rare than kin-selected altruism. It is, however, a theoretical possibility; and in olive baboons it has been realized in fact.

Manipulated Altruism : The Control of Behaviour by Parasites

Natural selection, we have seen normally males animals behave in their own selfish interest. Even when if favour altruistic behaviour, it is only in the interest of some broader form of selfishness. However, conflicts of animals open many opportunities for exploiting other individuals, and animals may not therefore always behave in their own interests. There are probably numerous subtle forms of exploitation within a social group—a possibility we have considered before in relation to animal signals—but the idea of 'manipulation' in animal behaviour is a relatively recent one, and has not been the subject of much research. For clear examples we are forced to go to those unambiguous situations of conflict, parasite—host relations. Here are many examples in which animals do not behave in their own selfish interests, or the interests of their genetic relatives. Hosts under the influence of parasites do show altruistic behaviour, according to our adopted definition, but only because they are in some sense forced to, against their own interest. I should emphasize that I have picked parasite—host examples only because they unambiguously illustrate a process that may be expected to be of wider importance. Many cases of altruistic behaviour in nature may be due to still subtler forms of manipulation than those we are about to consider.

A young cuckoo (*Cuculus canorus*) being fed by its foster parent, such as a reed bunting (*Emberiza schoeniclus*), is a striking example of behavioural manipulation by a parasite. It is not in the reed bunting's interest to feed the cuckoo: natural selection favours reed buntings that rear their own offspring, not cuckoos. To our eyes, it is very strange that a reed bunting cannot tell when it is feeding a cuckoo rather than its own offspring. For the insatiable demands of the cuckoo are still met by is tireless foster parents even after the cuckoo has grown larger than its foster parent. It should be easy, we think, for a reed bunting to distinguish a great, ugly cuckoo from a reed bunting. Yet the parent does not. It continues to pour worms down the cuckoo's all consuming throat. Something about the continual gaping and begging of the cuckoo compels the reed bunting to continue to provide. The reed bunting is being forced by a parasite to do something against its won best interests. The behaviour of feeding the young, however, is not abnormal, it is just misdirected.

Let us now move on to some stranger examples, in which the parasite actually changes the behaviour of its host. The reasons are to be found in the life-cycles of the parasites. Many parasites grow up in a series of different host species. They may start off in one species, be transferred to an intermediate host, and then to a final host; but there can be a problem in effecting the transfer from an intermediate host to the next one. The normal behaviour of the intermediate host might not take it near the final host. A kind of fluke (flukes are small, flattened, worm-like animals) called *Dicrocoelium dendriticum,* for example, lives an ants as its intermediate host, and sheep as its final host. How is it to get from in ant to a sheep? The trick is to have the ant eaten by a sheep. Ants normally, understandably enough, avoid being eaten by sheep; they stay down in the soil away from grazing sheep. However, an ant harbouring a *Dicrocoelium* changes its behaviour. The infected ant typically contains about fifty *Dicrocoelium* individuals. One of these burrows into the ant's brain and somehow causes the ant to climb a blade of grass and fastens its jaws to the top of the blade; it clings fast until eaten, perhaps by a sheep. The parasite has then reached its goal.

There are many fascinating examples of this kind. Here are two more. Nematomorphs (small work-shaped animals), in live in insects as intermediate hosts, but in water as adults. The parasite has to

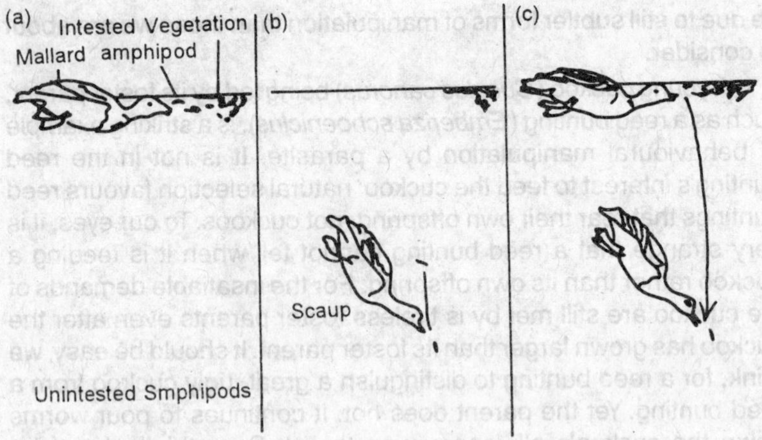

Fig. 8.9. Three different parasitic acanthocephalan worms have different effects on the behaviour of the amphipod Gammarus. Unparasitized Gammarus live in the bottom mud. But if parasitized by Polymorphus paradoxus (a), they swim towards the light and come to the surface where they may be eaten by dabbling ducks such as mallards (which are in turn parasitized by the Polymorphus). If parasitized by Polymorphus marilis (b), the Gammarus do not swim to the surface, but do come out of the mud; they are fed on by diving ducks such as scaups. (c) Gammarus parasitized by Corynosoma constrictum swim to the surface but dive when disturbed; they are fed on by both dabbling and dividing ducks.

bring its insect to water, and the nematomorphs do somehow bring this about. In one dramatic record, a bee was flying over a pond when suddenly, as it was about 6 feed above the water, it dived in. As soon as the bee splashed into the water, the worm exploded from the body they had so abused, and swam away, leaving it dead. A final example concerns the parasites of the amphipod *Gammarus lacustris. Gammarus* is a small shrimp-like animal which lives in fresh-water. Normally, *Gammarus* avoid the light, sheltering under pebbles on the bottom. However, when infected by the parasitic worm *Polymorphys paradoxus,* their reaction to light is reversed: they now seek the light and swim just below the surface. The parasite's motives is that the next host after *Gammarus* is a duck that feeds by dabbling at the surface; the *Polymorphus* changes the *Gammarus's* behaviour to make ti be eaten by its final host. The duck gets an easy meal, but it infects itself with a parasite when taking it.

Some biologists and psychologists would not include the manipulated altruistic behaviour in these parasitological examples as examples of real 'altruism'. They would prefer to count only kin selected and reciprocal altruism as real altruism, and exclude from the term behaviour carried out in the genetic interests of others. The process of natural selection does differ between the cases of kin selection, reciprocal altruism, and manipulation, and this might seem to provide a reason to separate them. But, following most authorities. I have defined altruism not be the process of natural selection that favours it, but in term of its reproductive consequences. An altruistic act is one that increases the number of offspring left by the recipient, and decreases the number left by the altruist. Manipulated behaviour fits the definition of altruism. The consequentialist, non-intention definition of altruism is far different from the normal everyday usage in which we thin of an altruist as intending to be kind. And perhaps some are tempted to include kin selection and reciprocal altruism as true altruism because the subjective intention (in the normal meaning of the word) can, by loose and wholly erroneous reasoning, be read into the genetic 'interest' of an individual. In kin selection and reciprocal altruism the individual does behave in its *genetic* self-interest (although not its own self-interest). However, it is not part of the definition of altruism that an animal should be behaving in its won genetic self-interest. The word was defined in terms of the effect on the number of offspring produced, regardless of whose genes those offspring carry.

Cooperation Through Group Advantage

W.C. Allee, in *Cooperation among Animals* (1951), synthesized much of the thinking about animal societies through the first half of this century. The ideas expressed are best understood in relation to contemporary development in the study of community ecology. *Clements* (1936) classified plant communities and started a school of "plant sociology", which emphasized the interrelationships of species and recognized the community as a "superorganism" with attributes transcending those of single-species populations. According to *Clements* living organisms reacted to the environment and so modified it that other species could not east there. The various assemblages of plants were so interdependent that they were distributed as single units. The idea of communities and ecosystems as superorganisms with emergent properties reached a peak with the writings of *Margalef* (1963), who discussed the qualities of mature and immature ecosystems. He treated the ecosystem as the functional biological unit that organizes itself so as to conserve and manage information. Although not explicitly stated, such reasoning led to the notion that the whole ecosystem was the unit of selection. Among animal behaviourists, the popularity of group selection reached its peak with Wynne-Edward's theory that social behaviour evolved mainly for the function of population regulating (*Wynne-Edwards* 1962).

Allee (1951) recognized the importance of natural selection at the level of the individual and did much work on egoistic, or selfish, behaviours such as competition and dominance. However, he felt that altruistic or cooperative forces were stronger, and he devoted much study to what he called *natural cooperation,* a force supposedly separate from natural selection. Allee believed that the phenomena he studied could be represented as several grades of social behaviour, listed here in generally increasing complexity:

1. *Invertebrate coloniality.* Allee thought the complexity and division of labour shown by such invertebrates as coelenterates in involuntary and, therefore, less advanced than the more voluntary cooperation seen in the vertebrates.
2. *Aggregation.* Organisms may be forced together by the actions of wind, tides, and other phenomena over which they have no control.
3. *Orientation to stimuli.* Animals may be brought together by their response to environmental gradients—for example, insects aggregate around lights. Such organisms must have at least a tolerance for each other.
4. *Locomotion to favourable locations.* When resources are in patches, animals may gather in taking advantage of them—for examples, birds in mulberry tree when the fruit is ripe.
5. *Clumping in the absence of substrate.* Mutual attraction and clumping occur in an attempt to find missing substrate. Brittle stars (Ophiuroidea) are typically dispersed in eel grass, but in aquarium devoid of objects they cling to each other. If small glass rods are planted as artificial substrate, they disperse and attach to the rods.
6. *Sleeping group.* In many species, individuals that are more or less isolated come together to sleep. Such behaviour may provide superior predator protection but may attract predators as well.
7. *Complex social life.* Allee considered highly developed social life an extension of sexual and family relations over a large

portion of the life span. He posited an unconscious tendency to cooperate that predisposes such species to develop complex societies.

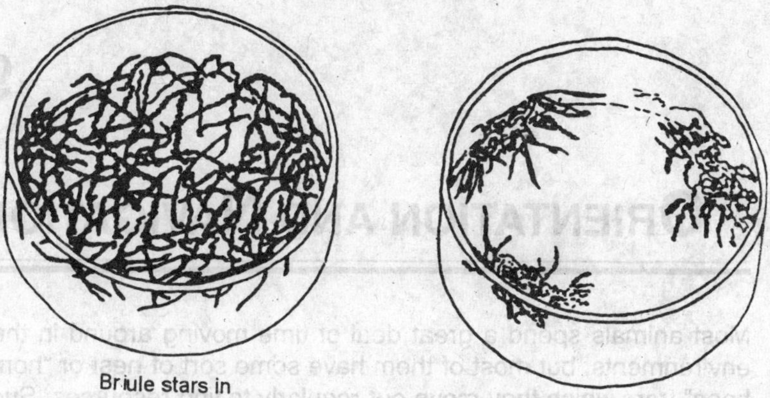

Fig. 8.10. Aggregation of brittle stars in Allee's laboratory.

Although he appreciated the negative effects of crowding, Allee also pointed out the unfavourable consequences of undercrowding. Typical of his experiments was the finding that goldfish (*Cyprinus* species) reared in a toxic colloidal suspension of silver survived longer in groups that they did alone; the slime produced by fish present in the "conditioned" water precipitated the silver to the bottom of the tank, thus protecting them. Other experiments showed that planarian flatworms (*Planaria* spp.) gain protection from ultraviolet rays by grouping and that the effect is not due simply to shading of one worm by another.

9

ORIENTATION AND NAVIGATION

Most animals spend a great deal of time moving around in their environments, but most of them have some sort of nest or "home base" from which they move out regularly to find resources. Such behaviour requires that the animal know where it is in relation to its home and how to get back there. Some animals move long distances from their homes and may have different summer and winter homes that are separated from one another by thousands of kilometers. Migration between different ranges requires very sophisticated navigational abilities, but many animals are able to get about with simple mechanisms.

The simplest orientation movements are known as *taxes* (singular: taxis) in which the animal assumes a particular spatial relationship to an orienting stimulus. The nature of the stimulus is denoted by adding the appropriate prefix to the word *taxis*: a *Phototaxis* is movement guided by a light; a geotaxis is movement guided by gravity; a *chemotaxis* is movement guided by detection of some chemical substance. A further distinction can be made between a positive taxis (movement toward the stimulus) and a negative taxis (movement away from the stimulut). The flight of a moth toward a light is a positive phototaxis, whereas the retreat of a cockroach from the same light is a negative phototaxis.

Many invertebrates possess a *light-compass reaction,* generally using the sun, in which the angle between the direction of movement and the direction of the moving to and from their nests, but they cannot compensate for movement of the sun. If an ant is held for 2½ hours in a dark box while it is on its way home, the sun will have traveled approximately 37°. When the ant is released, it shifts its direction approximately 37°, maintaining its original angle to the sun. Orientation to the moon has been demonstrated with intertidal amphipods of the genus *Talitrus* on beaches in Italy. These crustaceans escape seaward when disturbed; populations on coasts facing different directions have different but appropriate, escape directions even when tested in the laboratory. On moonless nights or under overcast skies, this orientation is lost. Unlike ants, however, the amphipods are able to compensate for the movement of the moon and maintain correct compass orientation of their escape.

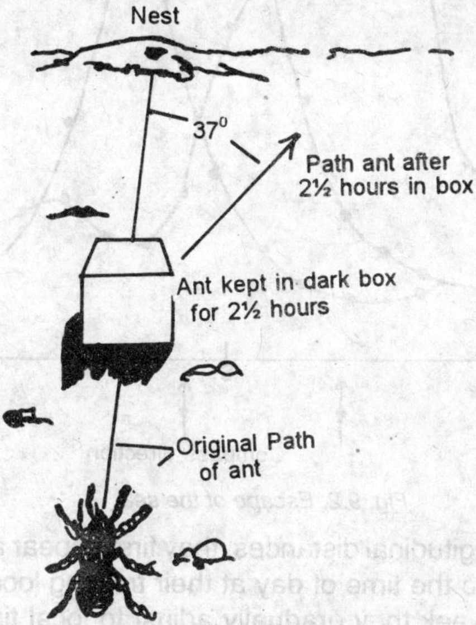

Fig. 9.1. Sun-compass orientation.

More intensive observation of orientation behaviour have been made with bees than with any other invertebrate. Bees can be trained to feed at a particular time at a particular spot in a certain direction from the hive. They have internal clocks that have periodicties of approximately 24 hours (circadian rhythms), which they use to tell time. Like amphipods, bees compensate for the movement of the sun. If they are held in lightproof boxes at a feeding tray they still fly directly to their hive when released. If trained bees are transplanted to an unfamiliar environment, they still appear at the appropriate direction at the correct time of day to find food. If they are shifted

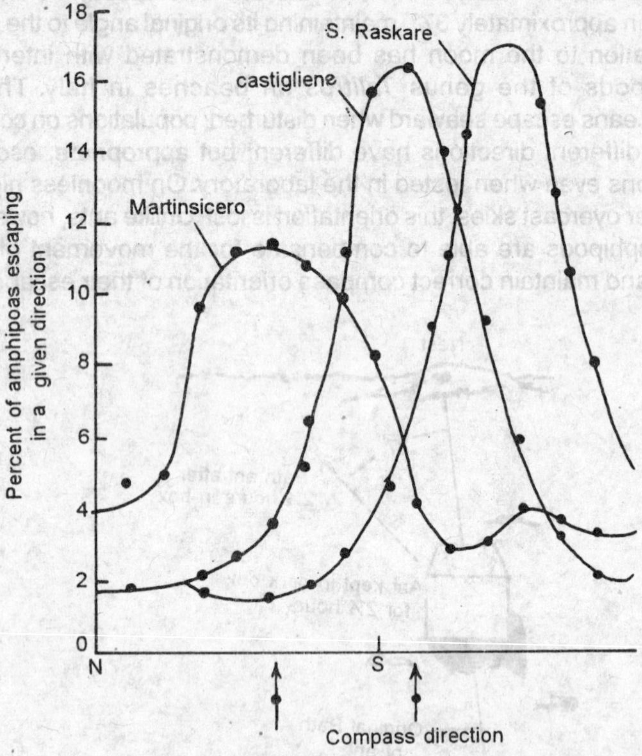

Fig. 9.2. Escape of the sea.

through great longitudinal distances, they first appear at the feeding sites according to the time of day at their training location. Over a period about a week they gradually adjust to local time, just as a person does after a transoceanic flight.

Navigation

The most spectacular long-distance migrations are undertaken by vertebrates. Pacific salmon hatch from eggs laid in gravel beds of rivers and, after varying amount of time in fresh water—intervals that depend on species and location—they migrate to the ocean, where they feed vigorously until they have reached maturity. At this time they migrate back to the coast, find the mouth of their home river, and swim back upstream, often to the very tributary in which they were born. They then spawn and die. The remarkable migration cycle has two components: navigation at sea to the general vicinity of the home river, and location of a correct river and tributaries.

Evidence gathered by a number of investigators reveals that salmon and other fishes can orient using a sun compass at sea but that they use a sense of smell to locate their home rivers. Salmon that are transported back downstream successful return to their home tributaries, but they cannot do so if their nostrils are plugged.

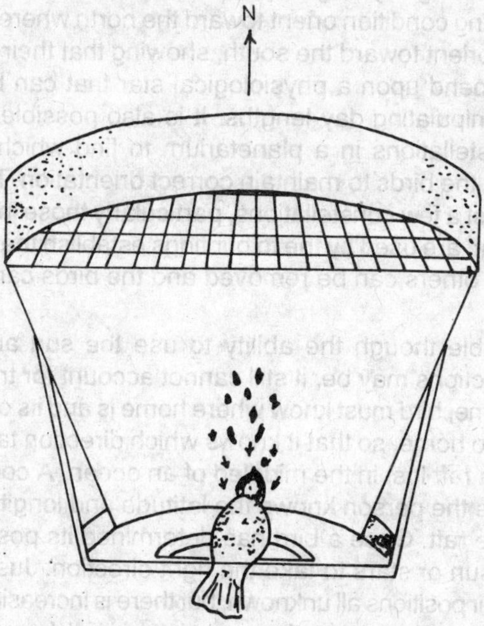

Fig. 9.3. Circular cage for orientation experiment.

Even in small lakes, fishes may use a sun compass to orient. White bass (*Roccus chrysops*) in Lake Mendota, Wisconsin, congregate in two small places for spawning. In one study, fish were displaced from the spawning grounds, and floats were attached to them so that they could be followed by observers in a boat. The fish showed good orientation toward the spawning grounds on sunny days, but they swam randomly on cloudy days.

The first conclusive evidence of navigational abilities in birds was not obtained until 1951, but for centuries it was known that some birds unhertake intercontinental migrations and yet return to specificities to breed. The first studies employed homing pigeons displaced from their lofts. The pigeons quickly orient homeward under clear skies but depart randomly under cloudy skies. Since the many species of birds have been shown to use the sun when navigating. Many small birds migrate at night, however, and they use the stars to orient. Indigo buntings will orient to "stars" in the artificialsky of a planetarium. The physiological condition of buntings may be changed by expositing some birds to increasing day lengths (as in spring and others to decreasing day lengths (autumn). In a planetarium, buntings that are in spring condition orient toward the north whereas those in fall condition orient toward the south, showing that their responses to the sky depend upon a physiological star that can be induced simply by manipulating day-lengths. It is also possible to blot out selected constellations in a planetarium to find which ones and necessary for the birds to maintain correct orientation. The relative positions of just a few constellations, particularly those in the vicinity of the Non-star, are used by the to buntings establish their migratory directions; all others can be removed and the birds can still orient correctly.

Remarkable though the ability to use the sun and stars to establish directions may be, it still cannot account for true homing. In order to home, bird must know where home is and its own position with respect to home, so that it knows which direction take. It is like a person on a raft lost in the middled of an ocean. A compass is of no use unless the person knows the latitude and longitude of both home and the raft. Once a bird has determined its position then it can use the sun or stars to take the right direction. Just how birds determine their positions all unknown, but there is increasing evidence that part of the information they use is derived from the earth's magnetic field. Homing pigeons can be disoriented on overcast days by attaching tiny to their necks; the magnets cancel the effects of Earth's magnetic field. Also, pigeons are unusually disoriented in the vicinity of sites on Earth where there are serious magnetic anomalies due to large non-deposits. There is yet no magnetic receptor dearly identified, but scientists are convinced that here must be one.

Invertebrates

Migrating locusts, flying in swarms, give the appearance of oriented directional movement. Studies of locust swarms have revealed that wind oxerts a displacing effect on individuals and thus on the swarm over time. Within the swarm there are groups of locusts that are flying in directions different from that of the swarm as a whole. When these sub-groups reach the periphery of the swarm, they turn back inwards toward the main flow of the swarm (*Rainey* 1959, 1962; *Waloff* 1958).

The means by which honey bee (*Apis mellifera*) communicate the location of food sources to one another in the hive and then make foraging flights to locate the food have studied by a number of investigators (*von Frisch* 1967; *Lindauer* 1961; *Gculd* 1975, 1976; *Wenner* 1971, 1975).

Bees and some other insects, including ants, can perceive polarized light (*von Frisch* 1967). The plane of polarization can provide an axis for orientation for these animals. It is also possible that in some animals the perception of polarized light may enable them to locate the position of the sun when it is obscured by partially overcast skies. The use of polarized light for orientation has also been suggested for a variety of other organisms, including fish (*Waterman* and *Hashimoto* 1974), octopodes (*Waterman* 1966), and salamanders (*Adler* and *Taylor* 1973).

Another examples is taken from the bee killing digger wasp *Philonthus triangulum.*

Let us consider first the homing problem of the bee-killing digger wasp *Philanthus triangulum.* Female wasps of this species dig burrows in the sand, to provide a nursery for their offspring. Each female digs several (about 6 or 7) cells off the side of her burrow, and lays an egg in each cell. When the egg hatches into a larva it will need food. That is where the bee-killing part of the wasp's name

Fig. 9.4. The bee-killing digger wasp uses physical landmarks to recognize its home nest.

comes in. The mother wasp goes out of her burrow, catches and kills a bee (by stinging), and brings it back to the burrow. She then opens the entrance to the burrow and takes the bee down to one of the cells. She continues to catch and bring back bees, one at a time, until each larva has about two bees to eat. The mother wasp, therefore, does not merely dig a burrow, and later leave it never to return: she departs from and comes back to it many times. On every return she has to find her burrow, distinguishing it from its surroundings. It is not an easy task to find a particular burrow, because these wasps can nest in quite dense groups; there might be more than twenty burrows within a circle of five yards radius. The problem of how the digger wasp locates her home has a special place in the history of ethology; it was one of the first questions about behaviour mechanisms ever to be asked, and experimentally answered. It was studied by *Niko Tinbergen* in 1929, on the heaths and sand dunes of Hulshort in Holland.

Tinbergen first confirmed, by individually marking all the nests and wasps in a particular area, that each wasp does indeed always return to her own burrow. He then performed some simple experiments to test whether the digger wasp recognized her own entrance by the distinctive array of odd objects (haphazardly fallen sticks, pine cones, stones etc.) around it, or by some stimulus emanating from the entrance itself. He placed around the entrance of each of several chosen burrows a neat circle of pone cones. He left them there for a few days, checking that the wasps still kept returning to their burrows. He then waited for the wasps to leave on bee-hunting expeditions, and while they were away he moved the cones a yard or so away from the entrances. When the wasps returned they landed where the entrance 'should' have been, in the centre of the circle of pine cones. Evidently they recognized the entrance by its surrounding landmarks, not by any stimulus from the entrance itself.

Many of the cues discussed in this chapter with regard to orientation in birds and mammals have also been demonstrated in some invertebrates. Among these are sun-compass orientation in spiders, beetles, and ants; geomagnetic force cues in bees; and landmarks in a variety of species (*Able* 1980). In addition, data have been reported which suggest beachhopper (*Talitrus*) use the moon (lunar orientation) to guide their nocturnal movements.

Fish

We have defined homing behaviour as the use of landmarks and navigation to return to a homesite. An animal may become displaced from its home by weather, or, as in the case of the Pacific salmon, it may return to a home location in the course of the life cycle. Alternatively, researchers may artificially move an organism from its homesite to another location in order to investigate the cues and means of navigation it employs to return home.

All species of salmon have similar life cycles. They are laid as eggs in river tributaries all over the northern hemisphere. They live their first few months in the river, then migrate downstream to the sea; they spend two or three years in feeding and growing out at sea, during which they cover thousands of miles; and when mature, they migrate back to exactly the same river, and same tributary, as they were born in. How do they find their way home? The answer is thought to be, mainly by smell. The attractive odour may come from either (or both) of two sources: the young salmon which are still in the stream, not yet having migrated to the sea, and any other characteristic odours in the stream. Salmons have been shown to be capable of the necessary olfactory discrimination, but the most direct evidence that they use their sense of smell comes from experiments, of the kind first performed by *W.J. Wisby* and *A.D. Hasler* actually plugged the noses of their salmon, which then homed less accurately than untreated controls recent experiments, the same result has been obtained by cutting the salmon's olfactory nerves. Experiments in which young salmons have been transferred from their stream of birth, to be released in another stream, have been used as evidence that salmon imprint on the smell of their river during a sensitive period just before they migrate downstream. In one such experiment, *L.R. Donaldson* and *G.E. Allen* took 72,000 young salmon at the 'fingerling' stage when they are about one year old from the Soos Creek Hatchery in Washington and divided them into two groups. In 1952 they released one group (identified by the removal of the right pelvic fin) at Issaquah Hatchery, and the other (which had their left pelvic fin removed) at the University Hatchery. The salmon returned in the winter of 1953/4; they homed on their release sites. None of them returned to Soos Creek, the site of their birth and first year of life. Of 71 marked salmon caught at Issaqua Creek, 70 lacked the right pelvic fins; all 124 marked salmon caught at the University Hatchery lacked their left pelvic fins. Results like that suggest that the young salmon the odour of their native stream (or of the young salmon in it), and later find their way home by seeking that scent. But there is another explanation. Salmon migrate in schools. The results of the transplantation experiment may only be due to the transplanted salmon following the lead of the native salmon. Recent commentaries, therefore, such as the review by *O.B. Stabell* maintain that the case for imprinting is at best not proved. But it is not in doubt that salmon rely critically on their sense of smell to guide them home. Whether they learn their home stream's distinctive smell during a discrete sensitive period is undecided.

Table 9.1. Numbers of coho salmon released and later recaptured in two rivers in Washington, with and without their olfactory sense impaired. Salmon with their noses plugged homed less accurately.

Stream of origin		No. released	No. recaptured	
			Issaquah	East Fork
Issaquah	Controls	121	46	0
	Nose plugged	145	39	12
East Fork	Controls	38	8	19
	Nose Plugged	38	16	

Amphibians

Orientation has been studied experimentally in southern cricket frogs (*Acris gryllus*) and northern cricket frogs (*Acris crepitans*) (*Ferguson, Landreth,* and *Turnispeed* 1965; *Ferguson, Landreth,* and *McKeown*, 1967). Experiments placed captured frogs in a large, circular plastic pen in the water, where the frogs could see no landmarks on the horizon. When southern cricket frogs were released in the pen under sunlit skies, they swam in the direction that would have taken them to land had they been released off the shore at their home pond. The results were the same when the experimenters carried the frogs to the pen with or without giving the frogs a view of the sky overhead, when they released the frogs under starlit skies, and when they released the frogs under moonlit skies. When researchers placed the frogs in the pen on a moonless night, with only stars visible in the sky, the frogs oriented themselves in two opposite directions; some headed toward the home shore and others headed directly away from home. When experimenters released the frogs during the twilight period, after sundown but before the stars or moon were visible, or under cloudy, overcast conditions, the frogs oriented in a random fashion.

Cricket frogs held in darkness for periods of thirty hours to seven days exhibited a decline in accuracy of orientation, which culminated in random orientation after seven days. "Dephased" frogs reestablished proper orientation when experimenters exposed them to normal dark/light cycles or to daily fluctuations of temperature and humidity. These frogs appeared to use a learned shore position, celestial cues, and a timing mechanism to orient themselves properly to the home shore. Similar results have been obtained for northern cricket frogs.

Reptiles

Sea turtles perform some of the most remarkable of all migratory feats. They may travel over a thousand miles to a small island to breed. The entire population of the Atlantic ridley turtle, *Lepidochelys kempi* apparently breeds on a small stretch of the Gulf Coast of Mexico. So for there is no evidence on the sensory basis of these journeys.

Carr (1967) has summarized a number of investigations on orientation by green turtles. Female turtles (*Chelonia mydas*) deposit their eggs on sandy beaches on certain islands; they lay about 100 eggs on each trip to the beach, and make from three to seven trips spaced twelve days apart. In non-reproductive periods the turtles live in areas with abundant turtle grass, generally along the coasts of continental land masses. In a specific study (*Carr* 1967), turtles were banded at Ascension Island. Young turtles that hatched on the island floated on currents that took them to the coast of South America. Adults had to swim against the same current to reach Ascension Island to reproduce. How did they find their way? The current may have certain features that allow the turtles to distinguish it.

Birds

In the last three decades, a variety of studies have been conducted or orientation in and navigation by birds. Many of the hypotheses that have been advanced for birds are also potentially applicable to other animals, such as insects and mammals (*Mathews* 1968; *Griffin* 1974; *Emlen* 1975; *Able* 1980).

Navigation requires both a compass to provide directional information and a map to provide the animal with information on its position relative to home or some other goal. As the following summaries of evidence for different cues illustrate, considerable evidence exists for the compass, but we still know very little about the "map" (*Keeton* 1969, 1970).

Topographic Features

Use of familiar topographic features is certainly a factor in navigation, but it is probably of secondary importance compared

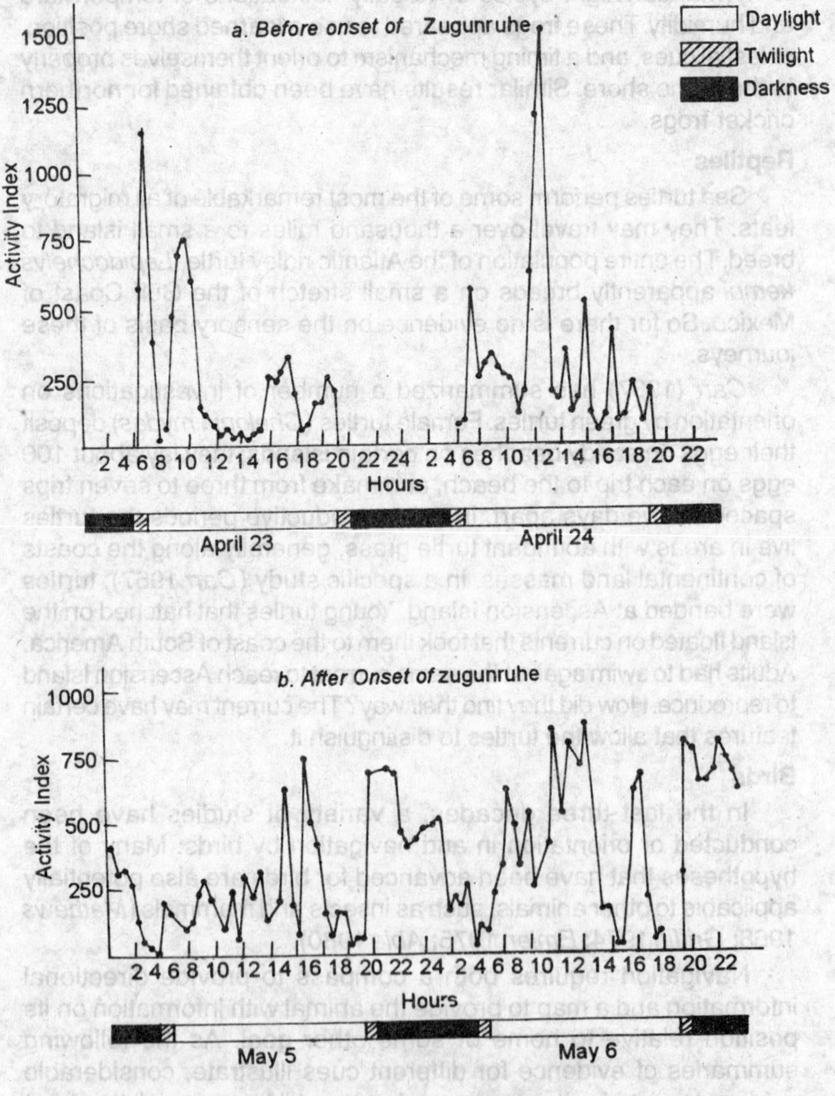

Fig. 9.5. Zugunruhe in male white-crowned sparrow.

with other cues. Nonetheless, many diurnal and nocturnal migratory birds are influenced by and may utilize topographic features. For example, birds congregate near and fly along coastlines or river valleys, and they may also use landmarks for piloting, particularly in areas at both ends of the migration, where they are more familiar in areas at both ends of the migration, where they are more familiar with specific topographic features. However, the use of topography as a guidance system has inherent drawbacks. How do first-time migrants who have never learned the landmarks find their way? What if a storm blows birds off the normal route into areas they have never traversed before? Also, visual landmarks alone cannot provide the necessary compass direction for proper orientation over long distances. Both landmarks and certain topographic features could be used in combination with some type of compass mechanism to provide sustained orientation during longer movements. Also, in familiar terrain, landmarks may suffice without any compass (*Able* 1980).

Sun

The most prominent regular cue for diurnal migratory birds, and possibly also for those that migrate at night, is the sun. The latter take a bearing at sunset and use that bearing of fly at night (*Able* 1982). The early studies of bird navigation devoted considerable attention to the sun (*Keeton* 1974). *Kramer* (1950, 1951) first showed that European starlings (*Sturnus vulgaris*), placed in an outdoor cage with the sun visible, exhibited migratory restlessness in the appropriate direction in the spring and fall. When *Kramer* used mirrors to alter the apparent position of the sun, the starlings' migratory restlessness shifted direction in a predict.

One way to demonstrate the use of a sun compass is to examine whether birds maintain a constant orientation at different times of the day even though the sun's position varies as it traverses the sky. Data from some diurnal migrants provides support for this alternative. Clock-shift experiments have provided a second, more rigorous test of sun-compass orientation. First, clock shifts of six hours, brought about by artificially shifting the birds' internal clock mechanism, resulted in predictable alterations of 90 degrees in the initial bearing of birds heading for the home site. This finding is consistent with the notion that birds use the sun as a simple compass. As a further test birds were clock-shifted six hours slow and released at a site 100 miles south of the home loft. If the birds

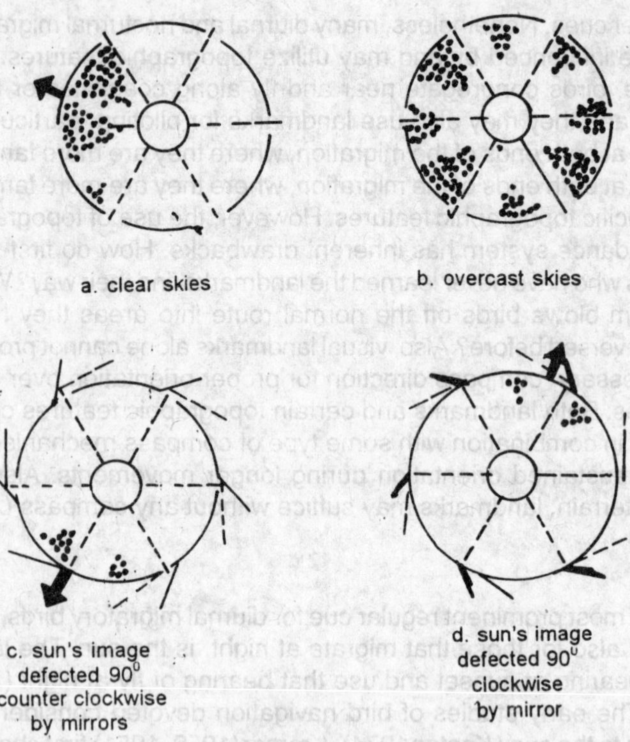

Fig. 9.6. Starling orientation experiment.

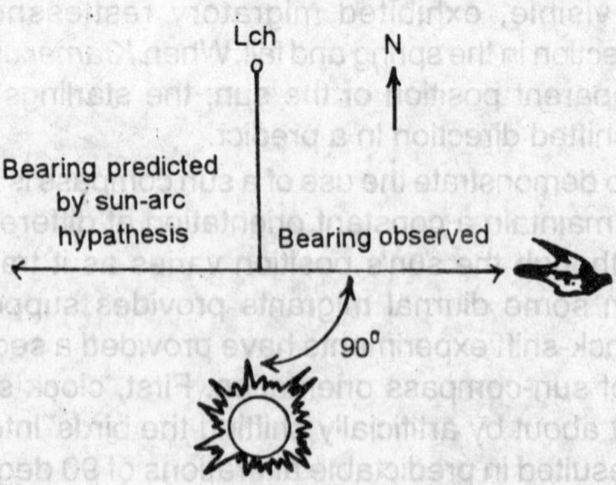

Fig. 9.7. Clock shifted pigeon experiment.

were using true sun navigation, they should have decided that they were 4,000 miles east of home and headed directly west. However, the birds headed directly east, as would be predicted if they were using the sun *only as a compass* and not as both a map and compass (*Keeton* 1974).

The sun, then, is an important directional cue for birds that are diurnal migrants and for homing pigeons. Evidence in support of the hypothesis that the sun can be used as both compass and map is lacking.

Stellar Cues

When nocturnally migrating birds are placed in outdoor cages with a view of the clear night sky, they exhibit migratory restlessness in a direction appropriate to the seasonal migration (*Kramer* 1949, 1951); *Sauer* (1975). *Sauer* exposed birds to planetarium skies, which permitted him to manipulate star patterns experimentally. Not only were the results of outdoor studies confirmed, but when he shifted star patterns 180 degrees in the planetarium, the direction of the *Zygunruhe* exhibited by the birds also shifted. Under cloudy skies or with no stars visible and with only diffuse, dim illumination in the planetarium, the birds exhibited random orientation. Sauer's studies on warbles (*Sylviidae*) have been confirmed in both field and laboratory by *Emlen* (1967a, 1967b) on indigo buntings (*Passerina cyanea*). Emlen's technique is innovative; the bird stands on an ink pad at the bottom of a funnel-shaped cage. The sides of the funnel are covered with paper, and the top is a wire mesh screen that permits the bird to see the overhead sky. When the bird jumps up against the sides of the funnel in its restlessness, it leaves marks on the paper. Investigators can turn the record of these marks into a vector diagram for purposes of analysis.

Meteorlogical Cues

Migratory activity in a particular region tends to be concentrated don a few days or nights. Birds are capable of responding to favourable weather conditions (*Drury* and *Keith* 1962; *Kreithen* and *Keeton* 1974). Data support the conclusion that birds often fly downward on their migratory flights (*Bellrose* 1967; *Richardson* 1971; *Bruderer* and *Steidinger* 1972).

Further experiments (*Able* 1982; *Able* et. al. 1982) monitored the natural nocturnal migratory flights of individuals of several *Passerina* species under various weather and wind conditions. When migrants could observe the sun near the time of sunset or the stars, they flew in the appropriate direction regardless of wind direction. If observations were made under totally overcast skies with no view of sun or stars possible, birds often flew downwind and thus sometimes in seasonally inappropriate directions. When migrant white crowned sparrows were fitted with frosted lenses and released from balloons aloft, they headed downwind, even though that direction was sometimes seasonally inappropriate. These results indicate the primacy of visual cues in birds for determining appropriate migratory direction and that, when deprived of visual cues, some species can determine wind direction.

Olfactory Cues

Several investigators have suggested that olfactory cues, primarily odors carried on the wind, may serve as an aid in orientation, particularly in homing pigeons (*Papi et. al.* 1972, 1973, 1974). The most direct test of such a hypothesis involve the removal of the olfactory capacity and the testing for expected orientation and homing behaviour. When the olfactory sense is blocked or when nerves are transected, behavioural effects that are not related to the homing ability being tested may result; and, therefore, unequivocal evidence of the possible role of olfaction in pigeon homing is difficult to provide. Indeed, studies by *Keeton* and *Brown* (1976), *Keeton, Kreithen,* and *Kiepenheuer,* and *Schmidt-Koenig* (1978) have all supplied data that demonstrate little or no effect on homing behaviour when the nasal passages are blocked or a local anesthetic is used on the nasal epethelium.

Geomagnetic Cues

Evidence from several sources has accumulated that supports a role for geomagnetic cues in orientation and navigation in some birds. After many failure by experimenters in attempting to determine whether birds could even sense the .05 Gauss geomagnetic field of the earth (e.g., *Emlen* 1970; *Kreithen* and *Keeton*, 1974), *Brookman* (1978), after a successful laboratory experiment, reported conditioning pigeons to magnetic fields. Measurement of *Zugunruhe* in European robins (*Erithacus rubecuta*) has provided a second line of evidence. Birds were placed in a cage around which a set of Helmholtz coils, which provide an artificial magnetic field, was constructed. *Wiltschko* and *Wiltschko* (1972; and *Wiltschko* 1972) have used this apparatus to show that the birds were not responding to the polarity of the horizontal component of the magnetic field; they did not shift the direction of their migratory restlessness when horizontal polarity was shifted. However, they did reverse direction when the vertical component was reversed; and they reversed the direction of orientation in a predictable manner. In an artificial magnetic field with a zero vertical component and strong horizontal component, the activity was random.

Other evidence for the role of magnetic fields in orientation comes from the work of *Southern* (1969, 1972) on ring-billed gulls (*Larus delawarensis*). Gulls in cages exhibited a strong tendency to walk in a southerly direction, except when storms produced temporary aberrations in the earth's magnetic field. Also when gulls with small magnets affixed to their backs were taken some distance from home and then released, they displayed a random pattern of dispersion. Control gulls without the magnets exhibited normal and consistent directional headings to the south. Additional confirmation has been reported by *Keeton* (1971) and *Larkin* and *Keeton* (1976), who used bar magnets attached to the wings of homing pigeons.

Walcott, Gould, and *Kirschvink* (1979) have reported the existence of an organ, located along the midline of the pigeon's brain, that contains 'magnetite granules. This organ may prove to be the source of sensitivity to geomagnetism in the pigeon.

It is clear from much of the foregoing discussion that may bird species possess multiple systems for orientation; the use of available cues apparently follows some type of hierarchical scheme. Which mechanism the organism employs depends on its preferences for using certain types of cues and the prevailing weather conditions at the location.

Mammals

In their studies on orientation and navigation in mammals, researchers have devoted particular attention to small rodents and bats. In many of the studies on rodents, experimenters have displaced the animal for a distance and observed the rate of return in terms of the number of animals that successfully returned home, the speed with which they returned, and aspects of the rodent's orientation as they left the displacement site. In some studies experimenters have attached reflector tape or radio transmitters to the animals to plot the rodents' course during homing.

Rodents have an area, called the home range, in which they carry out their daily functions. When experimenters have determined the home range, they can displace the rodent and monitor its homing behaviour until the rodent's return and recapture within its home range. Data from several species of mice indicate that, when displaced, these animals can find their home range again, some from a considerable distance (up to 30 kilometres) and with speeds ranging up to 300 metres/hour. In most studies the rodents are generally displaced for distances of only up to 500 metres and may take several days to return to the home range and reenter a trap (*Joslin* 1977). The return rate, or the percentage of mice that successfully return home, varies with the species and the nature of the habitat. Overall figures have indicated return rates ranging from 3 to 5 percent to over 80 percent.

Little work has been done on how rodents orient within their home ranges; however, *Drickamer* and *Stuart* (1984) investigated the movements of deermice (*Peromyscus*). In the winter the patterns of tracts left by the deermice of fresh show can be mapped. *Drickamer* and *Stuart* used traverses—each segment of the path from tree to tree or from tree to log—to examine the potential use of cues. They found that 86 per cent of all traverses were oriented from one tree to another tree and covered distances ranging from one to thirty metres. The concentration of tracks suggests that the vertical tree trunks are used for orienting. A significant positive relationship (r=0.51) between the length of the segment traveled by the mouse and the diameter of the tree to which it was headed further suggests that larger trees serve as more conspicuous objects on the horizon for orientation. Laboratory investigations (*Joslin* 1977) have supported the hypothesis that deermice may use conspicuous vertical cues on the horizon for orienting to goals.

The navigation and orientation systems of bats have received considerable attention. Bats taken from caves, barns, or buildings and displaced some distance often return. Sometimes the distances covered are great. A great deal is known about the homing of bats, but the difficulty of immediately recapturing them when they return and the possibility of missing many that do return make interpretations of the actual homing ability equivocal (*Cockrum* 1956).

Vision and familiar sounds at the home roost may provide terminal information detailing the precise location of the goal. But as in bird homing and migration, mechanisms involved in navigation may be different from those used in piloting. Determination of the initial directional course remains the primary unexplained phenomenon in bat homing. Preliminary experimental tests of the initial orientation of the fruit bat *Phyllostomus hastatus* seem demonstrate accurate initial orientation at the release site.

The possibility that search patterns provide an explanation of the orientation of migratory flight of bats can be dismissed. Direct migrations covering hundreds of miles have been recorded (*Griffin* 1958), and the seasonally appropriate directional flight of red bats, *Lasiurus borealis,* is a common place observation in the eastern United States. Yet it is remarkable that no author has suggested any explanation for how these travelers find their way! Bats often collide with towers in the same circumstances and often on the same nights as migrating birds of (*Tordoff* and *Mengel* 1956). Perhaps celestial orienting cues are a source of directional information for them.

Other Navigation Mechanisms

Orientation without Optical Cues : an Artifact? Several authors have reported oriented migratory fluttering in birds in the apparent absence of directional optical cues. *Merkel* and *Fromme* (1958) and *Fromme* (1961) worked with the European robin, *Erithacus rubecula*, and the whitethroat *Sylvia communis.* Birds were evenly and faintly illuminated, but they had no view of the sky. Various experimental locations were used, the most restricted of these being the placement of the test apparatus in a cellar without windows. Here directional orientation persisted, even in a steel cylinder with airtight doors, as long as one of the doors remained open. The magnetic field of the

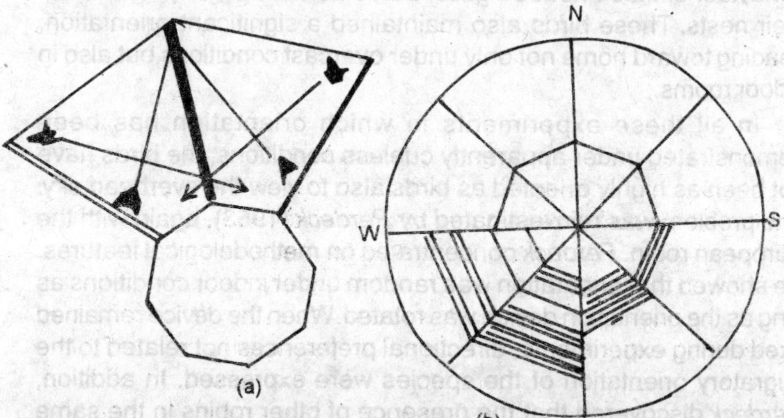

Fig. 9.8. (a) View of the test apparatus used by Merkel and Fromme. The movement from one radial perch to the next is automatically recorded. (b) A sample 30-minute record of the noctumal movements of orbin.

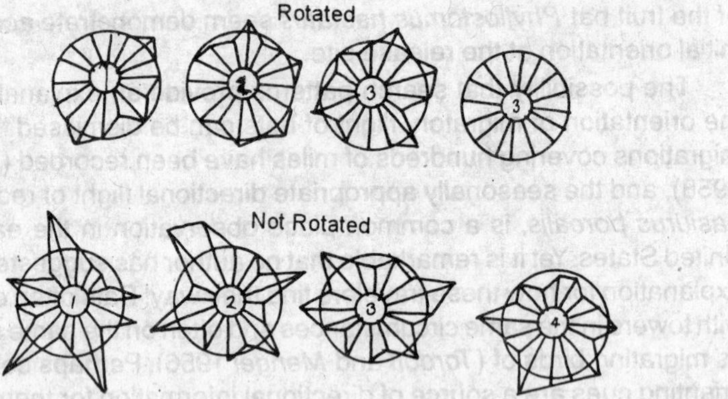

Fig. 9.9. Orientation of the robin, Ttithacus rubecala, in an experimental chamber lacking directional visual cues. The first row represents three individual experiments and the summary of thirteen such experiments. In the second row are four individual experiments which were performed without rotating the apparatus.

earth at the cage was undisturbed. This finding contrasts with Sauer's demonstration that garden warblers became disoriented when placed in a diffusely lit planetarium or under natural overcast. The conditions and recording methods used by *Merkel* and *Fromme* are different from those reported by *Sauer*, so that no direct comparison of results is possible.

Precht (1961) and *Gerdes* (1961, 1962) recorded the escape behaviour of black-headed gulls, *Larus ridibundus*, displaced from their nests. These birds also maintained a significant orientation, heading toward home not only under overcast conditions but also in indoor rooms.

In all these experiments in which orientation has been demonstrated under apparently cueless conditions, the birds have not been as highly oriented as birds also to view the everhead sky. The problem was reinvestigated by *Perdeck* (1963), again with the European robin. *Perdeck* concentrated on methodological features. He showed that orientation was random under indoor conditions as long as the orientation device was rotated. When the device remained fixed during experiments, directional preferences not related to the migratory orientation of the species were expressed. In addition, *Perdeck* discovered that the presence of other robins in the same building provided directional auditory cues which influenced the test chamber responses.

10

Behaviour Genetics

No aspect of an organism's structure of function is free of genetic control, and this includes behaviour. However, the analysis of the role of the genes in the fashioning of behaviour patterns is one of the most difficult and complex problems in the realm of biology and related sciences. The assessment of the relative importance of genetic and environmental factors in determining behaviour is not only difficult in itself, but when the question is asked of human behaviour, an emotional response sometimes bordering on the irrational is elicited.

In this chapter, some aspects of the inheritance of behaviour in animals are discussed. To begin with we consider some experimental results obtained with laboratory animals, which make it clear that certain behaviour patterns are inherited. We then proceed to consider aspects of behaviour in humans.

Genes to Behaviour

How are the processes that take place at the level of the gene and DNA translated into physiology and behaviour? These processes can affect behaviour in developing organisms as part of epigenesis, the interaction of the genetic program and the experiences and environment of the organism. Biochemical events involving genes can also effect behaviour in adult organisms through the turning on and off of specific genes or gene complexes, leading to the synthesis of particular proteins. A schematic representation of the major components in an organism's system and their feedback relationships is shown in Fig. The model attempts to show the pathways that connect genes and behaviour in both developing and adult organisms.

Some investigators have used *mosaics*, organisms whose tissues are of two or more genetically different kinds, to test aspects of the gene-to-behaviour sequence; mosaics permit us to observe both anatomical and behavioural anomalies combined in the same animal. *Hotta* and *Benzer* (1972, 1973) exposed fruit flies to chemicals that increased the frequency of mutations and produced various "abnormal" types of flies. They then used genetic techniques in mating the flies to produce mosaics, which had some normal and some mutant tissues—for example, one normal wildtype red eye and one mutant white eye (*Benzer* 1973)—as well as some affected behaviours—for example, courtship.

Benzer then sought to determine what tissue parts must be mutant for the abnormal behaviour to be expressed. He tested many fruit flies for the presence or absence of various mutant parts,

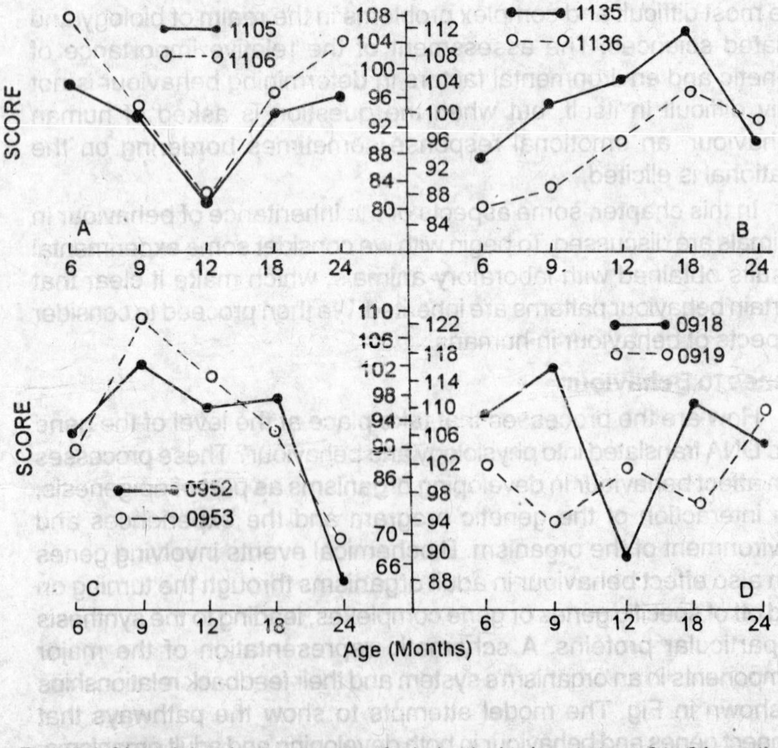

Fig. 10.1. Mental development scores for indentical twins at stage 6 to 24 months.

analyzed the resulting statistics, and made a map of the early embryonic fly on which he located the foci—the groups of cells that differentiate into specific structures and organs. We can use these maps to trace the effects of particular mutant genes on behaviour through the structures and physiological processes they affect.

GENETICS AND EVOLUTION

Gene Frequency

We have looked at ways gene affect behaviour, but there is another aspect to the gene-behaviour relationship. Behaviour may significantly affect the frequency and expression of certain genes in a population (*Ford* 1964; *King* 1967; *Oliverio* 1983). Changes in the gene pool, one of the outcomes of natural selection, can, in turn, affect behaviour.

For example, variations in courtship behaviour—the processes of mate selection and synchronization of activities that lead to mating—may alter gene frequencies. *Merrell* (1949, 1953) used wild-type and mutant strains of fruit flies to test this possibility. Four mutant strains—designated yellow, cat, raspberry, and forked—had sex-linked recessive alleles, expressed only in the male phenotypes and detected by examination of external characteristics. He started populations of flies in bottles with a mutant gene frequency of 0.5. Thus for each generation of flies, a departure from random mating would be indicated by an increase or decrease of the mutant gene in males. After several generations, the number of raspberry, cut, and yellow mutants decreased; the number of forked mutants showed no significant shift. Other tests demonstrated that although fertility or viability did not differ in any of the mutant types, mating success differed significantly. The test thus implicated some aspect of courtship and mating behaviour as the critical factor leading to differences in productivity and, hence, to shifts in gene frequencies.

Adaptations

Other investigations have demonstrated that gene-controlled behaviours may vary between groups of animals of the same species. Some of these differences are adaptive to habitat characteristics and others are adaptive to stages of a population cycle.

Habitat Adaptations

Prairie deermice (*Peromyscus maniculatus bairdi*) of the midwestern United States inhabit only fields are never caught in forested areas. In contrast, woodland deermice (*P. m. gracilis*) inhibit

Fig. 10.2. Complex responses by birds to simple stimuli. A—A territorial male European robin will attack a tuft of red feathers while ignoring a more complete model of a robin that lacks a red breast. B—Willow warblers attacking the stuffed head of a cuckoo.

only forested areas in south-central Canada and the extreme northern and northeastern regions of the United States. Mice of these subspecies exhibit several behaviours and preferences that are predictable from the habitats they occupy. *Woodland* deermice prefer a temperature of 29.1°C; grassland deermice select a temperature of 25.8°C in agreement with the cooler habitat of the letter subspecies (*Olgivie* and *Stinson* 1966). In addition, the two subspecies have different reactions to sand in a one-day test, grassland mice removed a median of 5.9 lb of sand (from a hopper) in contrast of 0.1 lb for woodland mice (*King* and *Weisman* 1964). Again this result seems logical because grassland deermice are terrestrial, living in the ground; woodland deermice and semi-arboreal, sometimes nesting in trees.

Population Cycles

The populations of meadow voles (*Microtus pennsylvanicus*) rise and fall in cycles that last three or four years (*Chitty* and *Chitty* 1962; *Krebs* 1966). During the low phase of the cycle in a particular area, voles are difficult to find. In contrast, at the time of peak populations, there may be over 200 voles per hectare. Trapping data indicate that many animals emigrate during the increase phase of the population cycle; at peak density, however, few animals emigrate, and losses in numbers must be attributed to deaths. In voles, blood proteins, whose production is regulated by specific gene loci, can be measured by electrophoresis. Electrophoresis is a technique in which substances are separated from one another on the bass of their electric charges and molecular weights.

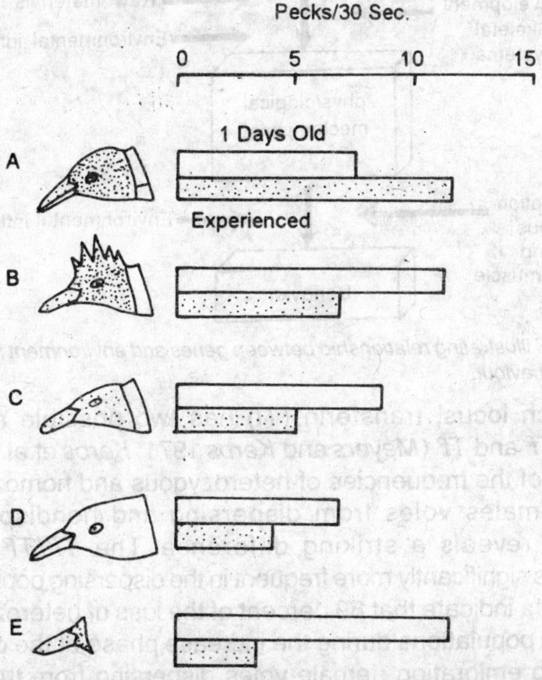

***Fig. 10.3.** Development of a preference in young laughing gull chicks for the model most closely resembling an adult of their species.*

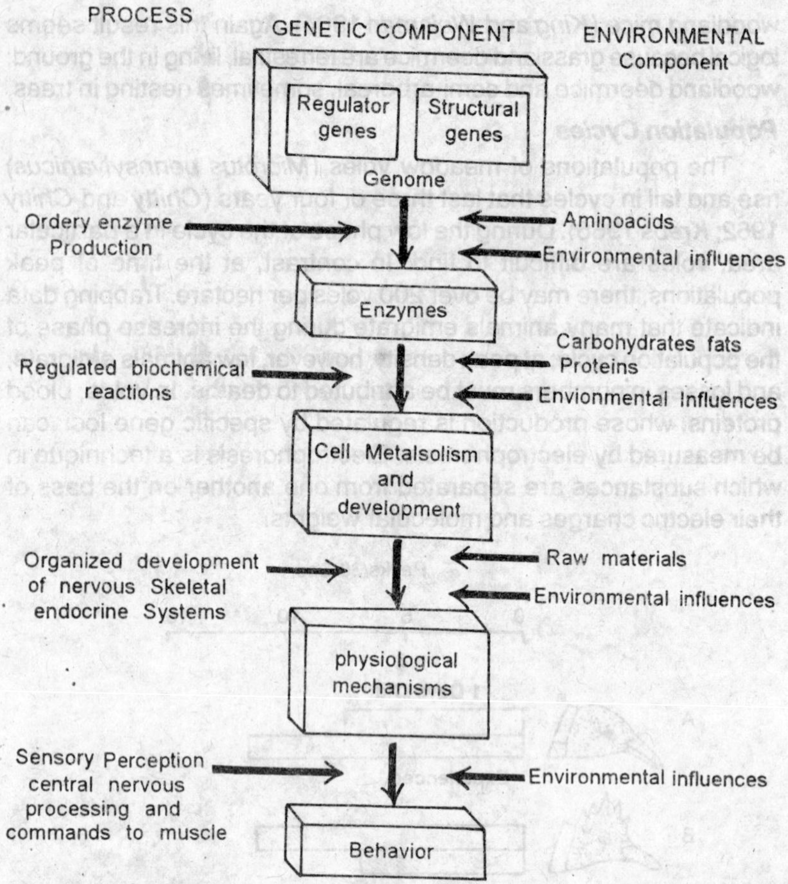

Fig. 10.4. Model illustrating relationship between genes and environment in control of behaviour.

One such locus, transferin (*Tf*) has two possible alleles, designated Tf^e and Tf^e (*Meyers* and *Kerbs* 1971; *Kerbs* et.al. 1973). Comparison of the frequencies of heterozygous and homozygous alleles in females voles from dispersing and nondispersing populations reveals a striking difference. The Tf^e/Tf^e allele combination is significantly more frequent in the dispersing population. Additional data indicate that 89 percent of the loss of heterozygous females from populations during the increase phase of the cycle is attributable to emigration (female voles dispersing from the local population). We should keep in mind that, in this instance, no direct gene-to-behaviour sequence has been clearly delineated; however,

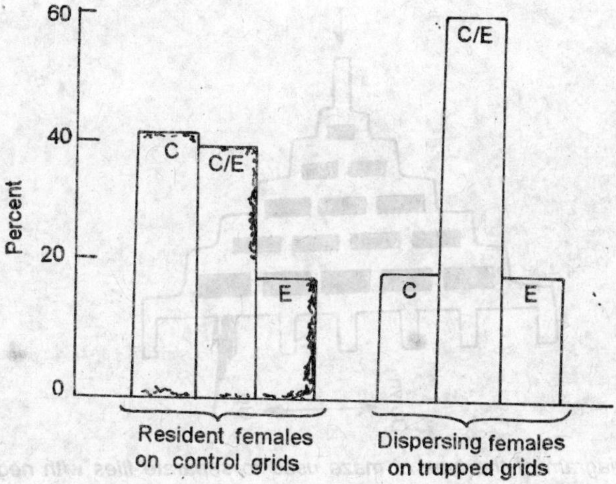

Fig. 10.5. Transferrin genotypes in female meadow voles. Genotypes of dispersing females are compared with those of resident females in the autumn during the increasing phase. C, C/E, and E represent the three transferrin genotypes.

changes in gene frequencies have been correlated with phases of the population cycle. The foregoing example also illustrates how rapidly the genotypic makeup of populations can change.

Breeding for Behaviour in Insects

Experimental work with *Drosophila* and *J.H. Hirsch* and *Th. Dobzhansky* and their associates has shown that fruit flies have different kinds of inherited behaviour patterns. Various techniques make it possible to select for specific types of behaviour patterns.

Selection Experiments

Normally, in laboratory-reared species of *Drosophila* most individuals are negatively geotactic; the flies walk upward against gravity when placed in a vertical tube. A maze has been devised to sort out from a population of flies those having different degrees of response to gravity. Individuals showing the most positive geotaxis were mated together, as were those showing the most negative geotaxis. The initial population of *D. pseudoobscura* used by *Dobzhansky* in his experiments contained individuals with positive, neutral, and negative responses in approximately equal proportions. Fig shows that after selection and inbreeding for 15 generations subpopulations were obtained all of whose members were either definitely negatively or definitely positively geotactic.

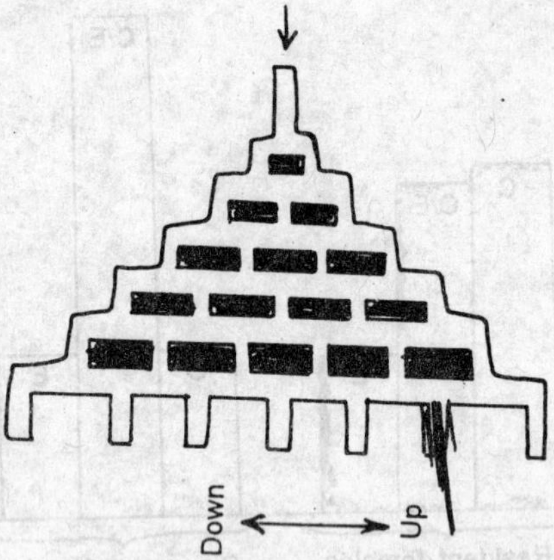

Fig. 10.6. Diagram of the type of maze used to separate flies with negatively genotactic responses from those with positive responses. Flies are introduced where indicated by arrow.

A similar result was obtained by using a maze that segregated flies according to the degree to which they are attracted to light. Fig. shows the results obtained after selection and inbreeding. The data

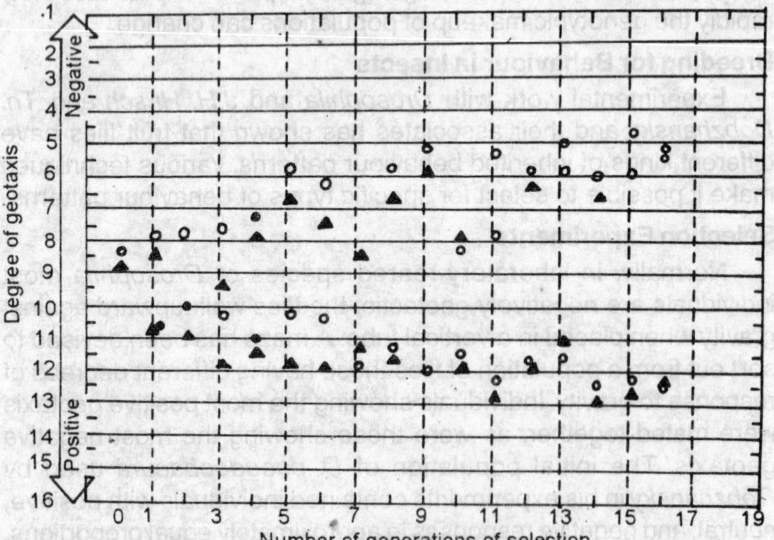

Fig. 10.7. Selection for negative (open symbols) versus positive (closed symbols) geotaxis in D. pseudoobscura. Circles represent females, triangles males.

supporting the genetic control of response to light are even more striking that those obtained for response to gravity.

Hirsch and *Erlenmeyer Kimling* have obtained very similar results with populations of *D. melanogaster* and have also been able to show that the response to gravity is determined by a number of genes.

The mating pattern illustrated in figure was used. It enables one to test the effects of chromosomes X, II, and III on geotaxis. The three chromosomes carrying the dominant markers B, Cy and Sb also carry inversions that suppress crossing over. The tester females were crossed to males from the following three different lines: (1) inbred for positive geotaxis, (2) inbred for negative geotaxis, and (3) an unselected control stock.

Only females flies carrying heterozygous B, Cy, Sb markers were used in the back cross. The eight types of expected female progeny are listed in figure. Flies heterozygous for each of the three chromosomes were compared to those homozygous for the chromosomes and the differences in response to light measured. The marked chromosomes were assumed to be unselected for geotaxis. The results are given in table, and they clearly show that

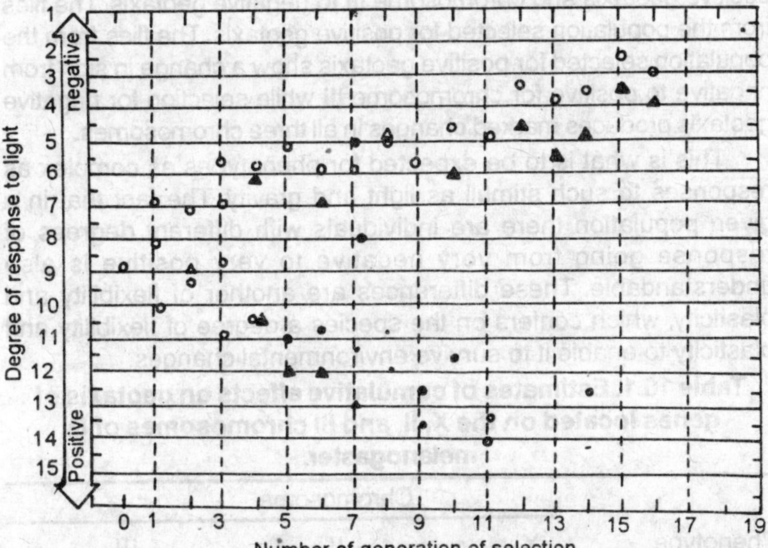

Fig. 10.8. Selection for negative (open symbols) and positive (closed symbols) photoxis in D. pseudoobscura. Circles represent females, triangles males.

all three chromosomes must have one or more genes affecting the geotactic response.

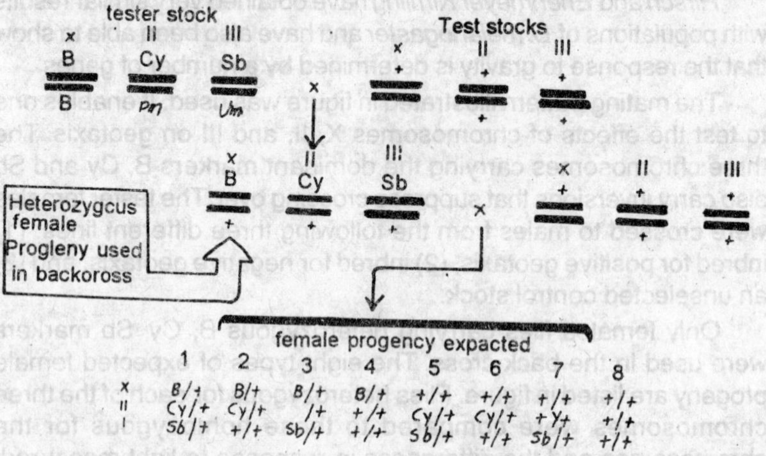

Fig. 10.9. Genetic analysis of geotactic response in Drosophila melanogaster.

In the unselected population, chromosomes X and II contribute to positive geotaxis and chromosome III to negative geotaxis. The flies from the population selected for positive geotaxis. The flies from the population selected for positive geotaxis show a change in sign from negative to positive for chromosome III while selection for negative geotaxis produces marked changes in all three chromosomes.

This is what is to be expected for phenotypes as complex as responses to such stimuli as light and gravity. The fact that in a given population there are individuals with different degrees of response going from very negative to very positive is also understandable. These differences are another of flexibility and plasticity, which confers on the species a degree of flexibility and plasticity to enable it to survive environmental changes.

Table 10.1. Estimates of cumulative effects on geotaxis of genes located on the X, II, and III chromosomes of D. melanogaster.

	Chromosome		
Phenotype	X	II	III
Positive	1.39	1.81	.12
Unselected	1.03	1.74	– 29
Negative	47	33	– 1.08

Mutations Affecting Behaviour in Drosophila

Although experimental results such as those described in the preceding section are helpful in demonstrating that certain types of behaviour have significant genetic components, they do not pinpoint the role of specific genes. It may be assumed that there re genes that have a primary effect in determining a behaviour pattern by their control of the development and functioning of some specific component of the nervous system.

A number of single gene mutants of *D. melanogaster* demonstrate anomalous behaviour. Some of them show marked deviations from the normal male courtship pattern; others will not fly, although their wings are well-developed; and some become paralyzed when shocked either mechanically or by changes in temperature.

A number of different kinds of behavioural mutants of *melanogaster* have been identified, some after treating wild type male flies with mutagens. Some of the mutants are listed in table. A wide range of different kinds of behavioural anomalies are represented. Some of these mutants have been analyzed anatomically and, as might be expected, divergences from the wild type condition have been found.

Table 10.2. Some behavioural mutants of *Drosophila melanogaster*.

Chromosome	Gene locus	Phenotypic description
X	*Hk*	Hyperkinetic leg shaking
	Sh	Vigorous shaking of body
	Eag	Shaking of body
	para	Paralysis at 29°C, but not at 22°C
	per	Disturbance of circadian rhythm
	drd	Drop dead beginning a day after eclosion
	tko	Become immobile when shocked
II	*sps*	Unable to walk or fly
	ap	Become immobile day after ecosion from pupa
	es	Hypersensitive to diethyl ether and chloroform
III	*fty*	Males court each other even in presence of females
	sc	Males have difficulty separating from females and copulation.

Mosaic flies with some mutant and some normal tissues have been produced. The effect of one tissue type on the other can be observed using these aberrant individuals. Mosaics have been constructed with the technique already described using X chromosomes carrying both morphological markers and behavioural mutations. The mutant tissues are readily identifiable because of the morphological marker on the X chromosome. One important finding made with behavioural mosaics is that the mutant tissue behaviour is generally unaffected by the presence of wild-type tissue immediately adjacent to it. This *autonomous* behaviour is not unexpected because it is also generally true for morphological mutant phenotypes in *Drosophila*. Autonomous action by genes refers to the fact that the effects of their activity are confined to the cell within which they reside. Thus a mutant gene in a single cell surrounded by cells with the wild-type allele will not affect the phenotype of the surrounding cells, nor will the mutant cell be affected by its wild-type neighbours. Sometimes, however, the gene products are diffusible through the cell membranes. In this case the action of the genes will not be autonomous. However, this is the exception rather than the rule.

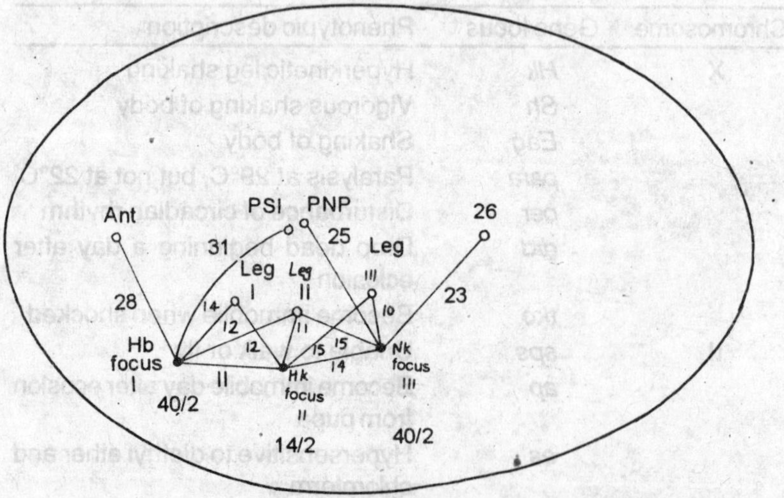

Fig. 10.10. The fate map sites of the behavioural foci for the hyperkinetic mutant HK. Each leg has a separate focus, each of which falls in the region of the blastoderm of the early embryo that becomes the central nervous system.

Y. Hotta and *S. Benzer* have extended the analysis of the behavioural mutants by mapping the embryonic blastoderm using a method original devised by *A.H. Sturtevant.* Basically the mapping depends on the supposition that the frequency with which any two parts of adult mosaics are of different sex-linked genotypes is directly related to the distance apart in the blastoderm of the sites that give rise to these parts. This generalization appears to hold after direct observation of many different mutant combinations, and "fate maps" have been constructed from them, as shown in figure.

Flies mosaic for morphological and behavioural mutant phenotypes have been scored for the coincidence and non-coincidence of the mutant phenotypes. For example, the mutant hyperkinetic (*Hk*) shakes its legs abnormally when anesthetized with either. A fate map located the site of the difficulty causing the hyperkinetic condition in the neurons of the thoracic ganglion of the flies' ventral nerve cord. Other behavioural mutant foci of activity have been established by this means, and some have been verified histologically by showing that the indicated cells are indeed morphologically mutant. Thus specific mutations causing abnormal development of certain parts of the insect's nervous system in turn cause specific changes in behaviour.

Mating behaviour in *Drosophila* also has genetic components. Discrimination of mates leading to non-random mating can greatly affect the array of genotypes in the population. *Pheromones,* chemicals that elicit courtship activity and mate selection, have been shown to allow differentiation of mates within a population, and to reduce the incidence of mating among close relatives of *D. melanogaster.* Furthermore, many instances have been demonstrated whereby individuals with a genotype in low frequency in a population will mate more frequently than effected under random mating. Some of these instances have been related to pheromone differences.

Behaviour modifications also may lead to reproductive isolation. For example, closely related species of some Hawaiian *Drosophila* use different markings on their wings to signal one's species identity between individuals. (Visual communication in species recognition is employed in many vertebrate species as well). Often sexual dimorphism reflects *sexual selection,* whereby courtship displays of the male are enhanced by the development of extreme phenotypes (including antlers and colourful plumage in the vertebrates). In the Hawaiian *Drosophila,* sexual dimorphism is frequently found, as one might expect, if males must visually attract the attention of receptive females. Courtship behaviour enhanced by morphological differentiation, may play an important role in sexual isolation and hence in evolution.

Each species of cricket has its characteristic pattern of chirps. *D. Bentley* and *R. Hoy* have shown that these distinctive song pattern are not learned, but are inherited. Hybrids made between two different species hirp with a pattern intermediate to the parents.

Several different kinds of song are common to males of each cricket species. These may be identified as (1) the "calling song" used by the male to call sexually receptive females, (2) the "courtship song" that facilitates copulation, (3) the triumphal song following copulation, and (4) a kind of song sung by fighting males; male crickets are very chauvinistic and will usually attack another male who invades their territory.

The sound pulses of the songs are made by the chirpers scraping their fore wings together. Figure shows the song patterns of two different species of cricket. *Teleogryllus oceanicus* and *T. commodus.* Also included in this figure are the patterns sung by F_1 hybrid males and back-cross hybrids made by crossing the F_1 hybrids back to the parental forms. The F_1 hybrids have songs resembling both parents, but there is a difference in song depending on how the cross is made. The reciprocal crosses indicate a maternal effect. In crickets, as in *Drosophila,* the male is heterogametic and the female homogametic. Hence the male receives his single X chromosome from his mother, and if some of the gene(s) on the X chromosome are involved in the determination of the song pattern, one might expect his song to resemble that of his maternal grandfather. However, it also is apparent that genes on other chromosomes are involved.

Genetic and Environmental Interactions in Dogs

Dogs have personality differences even within a breed. Some are timid and other are confident; some are gentle and others are aggressive. Those that are "socialized" (allowed to interact with people) early in life function as friendly and understanding companions of humans, whereas others of the same breed (or even of the same litter) that are not given similar experiences while young may become fearful or even hostile toward people.

J.P. Scott and *J.L. Fuller* have made extensive observations on *genetic and environmental factors* involved in the *building of behaviour patterns* in dogs. In some of their work, daily observations were continued until the dogs were 16 weeks of age. Every effort was made to observe the earliest manifestation of hereditary difference before the effects of experience were noted. Of course, during the first few days after birth, there was very little behaviour to observe in the pups.

As soon as recognizable behaviour became apparent, interaction between hereditary and environmental influences was already present. The puppies changed markedly in reactions from day to day in response to learning and increasing physical and mental maturity. The environment was optimal for learning, and innate faculties became active at *progressive developmental stages. Genetic* control was *acting on a very different animal* with each new developmental stage, and environmental influence was apparent.

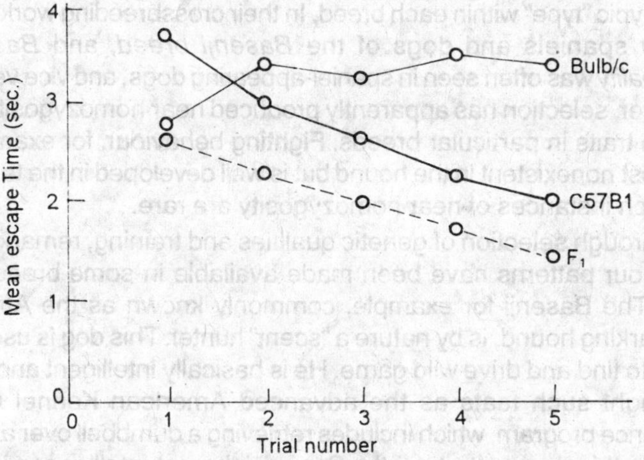

Fig. 10.11. Learning curves for a water escape response for two strains of mice and their F_1 hybrid cross.

Results obtained by *Scott* and *Fuller* were quite unexpected and significant. During the very early stages of development when behaviour is minimal, genetic differences had few opportunities to be expressed. When behaviour patterns did appear, however, genetically determined differences did not appear all at once early in development, to be modified by later experience. Instead, they developed under the influence of environmental factors. *Scott* and *Fuller* concluded that the raising and lowering of response thresholds is one of the most important way behaviour in dogs is affected by heredity. These studies showed that while *heredity* is an *important* factor in dog behaviour, details—and sometimes the actual appearance of specific patterns—depend to varying degrees on the *individual's experience.* Furthermore, they demonstrated that at least some genetic behaviour difference can be measured and compared as readily as hereditary physical differences.

Dog breeds generally have managed to retain a great deal of genetic flexibility, despite man's intensive selection over time. This was borne out by the further studies of Scott and Fuller in which 50 traits were examined in five pure breeds of dogs. Almost all the traits were significantly different among the various breeds. But a very few of them were found to breed true, as would be expected if they were controlled by single homozygous pairs of alleles. In addition, lack of correlation was observed between behaviour and phenotypic "type" within each breed. In their crossbreeding work with cocker spaniels and dogs of the *Basenji breed,* and *Basenji personality* was often seen in spaniel-appearing dogs, and vice versa. However, selection has apparently produced near homozygosity for certain traits in particular breeds. Fighting behaviour, for example, is almost nonexistent in the hound but is well developed in the terrier. But such instances of near homozygosity are rare.

Through selection of genetic qualities and training, remarkable behaviour patterns have been made available in some breeds of dogs. The Basenji for example, commonly known as the African non-barking hound, is by nature a "scent" hunter. This dog is used in Africa to find and drive wild game. He is basically intelligent and can be taught such feats as the advanced American Kennel Club obedience program, which includes retrieving a dumbbell over a high jump. In this demonstration, the Basenji inherent intelligence is put to a relatively artificial use.

Retrieving dogs have been selected and trained for more than 100 years in England by enthusiasts of the waterfowling sport. Several breeds exhibited some of the characteristics necessary for an excellent retriever, such as strength, moderate size, endurance, enthusiasm, aquatic ability, keen scenting ability, courage, favourable temperament, and trainability. By intercrossing the most favourable breeds—Newfoundland, setter, and spaniel—and selecting progeny, a litter of four puppies with a favourable combination of traits was obtained in 1868. From this beginning the golden retriever stock was developed. These dogs are light yellow in coat colour. They also have aquatic ability, pleasing temperament, keen nose, tracking ability, and tenacity to retrieve under severe conditions. It is natural for a pup from this stock to want to retrieve. Training merely perfects and polishes the performance.

Gene Mechanisms as Adaptations

The capacity of the cricket genotype to oversee the development of a nervous system whose compounds are designed for special functions is remarkable. It is even more surprising when you consider that each male cricket of a given species has a different genotype, a unique combination of genes drawn from the sample contained in his parents' bodies. Moreover, each individual is subject to a unique environment. No two male crickets eat exactly the same foods or experience the identical weather, social contacts, or anything else. The development of an individual cricket is, therefore, the product of an interaction between its unique genotype and a unique set of environmental influences. Yet almost all the males of the same cricket species develop essentially the same command interneuron in the right place with the right connections and the capacity to produce the appropriate score of impulses. Therefore, they all sing the same song.

This phenomenon is not restricted to a single species of cricket. It is axiomatic that members of the same species share certain basic features in common, especially those adaptation that are responsible for their ability to exploit key resources in their environmental and that are the basis for their reproductive isolation from other populations. Yet the ability of genetic mechanisms to compensate for the potentially disruptive effects of mutations and environmental deficits has been largely ignored by behavioural scientists until recently. Historically, the study of behavioural development has been the province of experimental psychology. As already noted, this discipline has been preoccupied with showing that environmental factors can influence the acquisition of behavioural traits. As *B.F. Skinner* has said: "I'm not the kind of behaviourist who denies that behaviour can be inherited. I accept an organisms as a highly complicated biological entity. What can you then do with it? That is the thing". The psychologists' interest in manipulating the behaviour of animals has resulted in many important discoveries, but it has also led to the widespread acceptance of the view of the developing animal as a billiard ball pushed hither and you by external forces. However, even studies designed to illustrate the role of the environment in structuring development have sometimes more or less inadvertently shown that the "billiard ball" knows where it should go and can resist at least some potentially disruptive influences on the direction of its development.

Environment and Development in Animals

The central hypothesis of the famous studies of *Margaret* and *Harry Harlow* has been that the social experiences of young primates may be critical for the development of normal social behaviour. According to this view, abnormal social behaviour in primates may be the product of social deprivation. The Harlow's standard experimental technique involves separation of the ir.fant rhesus from its mother shortly after birth. The baby is placed in a cage with an artificial "surrogate" mother, which may be a wire cylinder or terry-cloth figure with a bottle from which the baby is able to nurse.

The young rhesus gains weight normally and develops physically in the same way that non-isolated rhesus infants do. However, it soon begins to demonstrate signs of behavioural abnormality, which may eventually develop into continuous crouching in a corner, rocking back and forth, and various forms of self-mutilation. If confronted with a strange object or another monkey, the isolated baby is especially likely to react fearfully. It may withdraw as completely as possible from the object or playmate by huddling in a corner or by clinging tightly to a terry-cloth surrogate mother. Monkeys reared in total isolation for the first 6 months of their lives sometimes exhibit permanent disturbed behaviour in late life, when rocking, head banging, and self-biting may be almost the only activities of the adult animal.

The *Harlows* stress the significance of receiving maternal care for normal behavioural development. But they also point out that young rhesus monkeys reared alone with their mothers do not develop truly normal sexual, play, and aggressive behaviours. When the encounter animals their won age later in life, they are likely to behave very inappropriately, reacting with excessive fear or violence. Evidently contact and interactions with peers as well as a mother early in life are essential ingredients for normal behavioural development in the rhesus monkey.

How opportunities for social interactions might lead to changes in later behaviour has been examined, through use of the laboratory rat, by a team of University of California psychologists. They, too found that the way in which an infant rat was reared had substantial

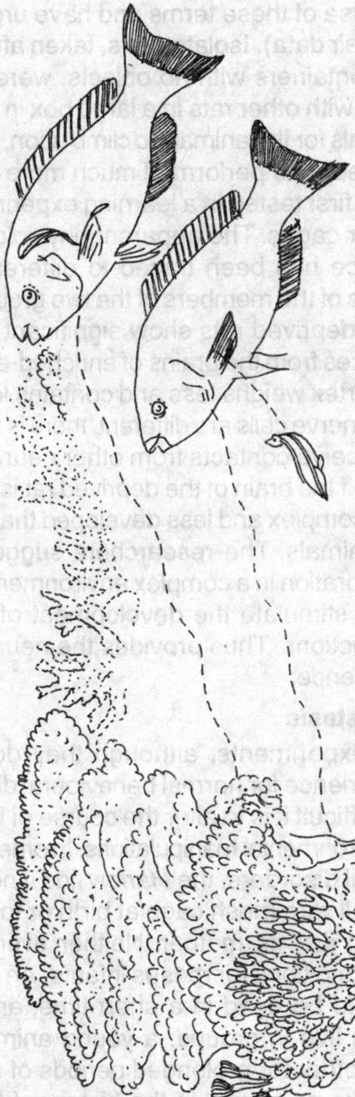

Fig. 10.12. Two Parrot fish at a cleaning station on a reef. A—The true cleaner fish approaches the parrot fish with an undulating dance. This behaviour induces the larger fish to adopt the tilted cleaning position shown in the figure. B—The false cleaner fish not only resembles the true cleaner in appearance, but also mimics its behaviour. Unlike the real thing, however, the mimic attacks the immobilized parrot fish and departs with a piece of its flesh.

effects on the development of its behaviour. They devised two basic rearing conditions: "improverished" and "enrich" environments. (By the use of these adjectives the researchers were clearly addressing their research to human problems and implying, as do the *Harlows*, that their findings can be applied to human beings. More recently they have qualified the use of these terms and have urged caution in the interpretation of their data). Isolated rats, taken after weaning and put in small bare containers with no objects, were compared with weaning rats raised with other rats in a large box in which were placed a variety of materials for the animals to climb upon, manipulate, and explore The "deprived" rats performed much more poorly than the "enriched" rats when first tested in a learning experiment after a number of weeks in their cages. The apparent physiological basis for this poor performance has been traced to differences in the development of the brains of the members of the two groups. Studies of the brain structure of deprived rats show significant anatomical and biochemical differences from the brains of enriched-environment rats. For example, the cortex weighs less and contains less protein, the connections between nerve cells are different, there is less surface area on many cells to receive contacts from other neurons, and so on. In general, it looks as if the brain of the deprived rat is structurally and physiologically less complex and less developed than the brains of the other group of animals. The researchers suggest that the early experiences of exploration in a complex environment and social interactions among rats stimulate the development of brain cells and neuronal interconnections. Thus provides the neural basis for later behavioural competence.

Developmental Homeostasis

Paradoxically, the experiments, although they do show the importance of early experience for normal behavioural development also demonstrate how difficult it is to alter the course of behavioural development through environmental manipulations. In order to produce a rhesus with behavioural disorders, the *Harlow* find it necessary to isolate an infant in a small wire mesh cage at birth or to keep it in a similar environment with just its mother. Neither event has ever occurred under natural conditions for rhesus infants. An abandoned (i.e., isolated) baby would be dead in a short time; and because rhesus monkeys always live in groups, a young animal is never entirely alone with his mother for prolonged periods of time, to say nothing of 6 months. Thus no animal in the history of the species has had to cope with the bizarre environmental situations devised by the *Harlows*, and other researchers in this field. It is not surprising that the animals' developmental system break down under these conditions. What is surprising and revealing is that otherwise totally isolated infants will develop relatively normal behaviour if they introduced into a cage with three other infants for just 15 minutes

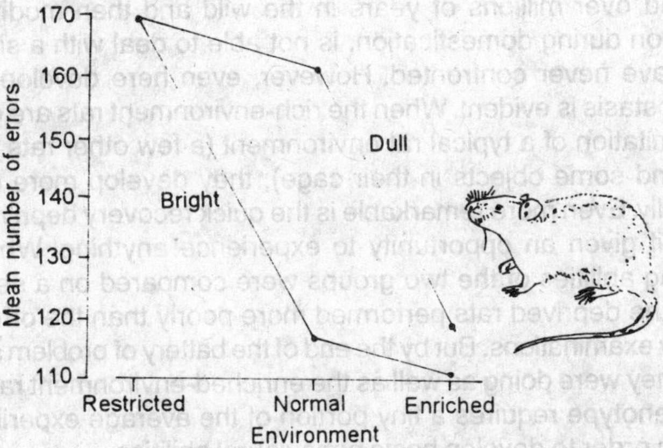

Fig. 10.13. Mean number of errors made in a maze by 'bright' and 'dull' rats reared in enriched, normal and impoverished environments.

each day. At the outset, the young rhesus monkeys simply cling to one another, but later more time is spent in play. In their natural habitat rhesus babies, beginning at about 1 month, start to play more and more with others about their age. By 6 months they are spending practically every waking moment in the company of their peers. Yet in the laboratory, despite the barren existence of isolated infants, a few moments of playtime a day is sufficient to counteract the effects of extreme deprivation. Rhesus monkeys, therefore, exhibit Developmental Homeostasis, the ability of developmental systems to compensate for environmental (and genetic) deficits and so produce key adaptive traits in animals exposed to less than optimal conditions.

Another unexpected demonstration of the resilience of the developmental system of the rhesus is provided by tests of isolated infants reared on a swinging, instead of immobile, surrogate mother. These infants develop more typical behaviour in many regards than those reared on non-moving surrogates. Again, stimulation that only approximates what a baby rhesus would receive from its active mother nevertheless provides the basis for the relatively normal development of some behaviours.

Exactly, the same points emerge from the Norway rat studies. In order to disrupt normal development, experiments have had to resort to extremely abnormal environments. No weanling Norway rat has ever been placed in a tiny featureless cage and isolated there for weeks, prior to the last 100 years. The rat's genetic program, evolved over millions of years in the wild and then modified by selection during domestication, is not able to deal with a situation rats have never confronted. However, even here developmental homeostasis is evident. When the rich-environment rats are given a pale imitation of a typical rat environment (a few other rats to play with and some objects in their cage), they develop more or less normally. Even more remarkable is the quick recovery deprived rats make if given an opportunity to experience anything. When the learning abilities of the two groups were compared on a series of tests, the deprived rats performed more poorly than the others on the few examinations. But by the end of the battery of problem solving tests they were doing as well as the enriched-environment rats. The rat's genotype requires a tiny portion of the average experimental input in order to develop basic behavioural abilities.

If we were to extrapolate from these studies to human beings, an evolutionary approach would suggest that young human children are probably not high susceptible to various environmental deficits. The human genetic-developmental system should be as protected against environmental deprivation as a white rat's or a rhesus monkey's developmental program. Yet a large majority of developmental psychologists and the general public, at least in the United States, take it for granted that experience young children have in the first few years of life have an enormous impact on the development of their intelligence, personality, and other aspects of their behavioural repertoire. This viewpoint is reminiscent of the philosophy of the Englishman *John Locke* as expressed in 1686 in his "Essay on Human Understanding." (This paper was written in Amsterdam, *Locke* having been expelled from England 2 years before for unpatriotic behaviour. It was rumored he chose Holland because of a Dutch mistress, but he himself claimed he did so because "there was but little beer in France." Either reason seems sufficient). *Locke* proposed that the human infant's mind is like a blank sheet of paper on which experience is written and that humans are, in effect, entirely products of what happens to them. The advice often given to young parents presumes that *Locke's* view is correct and, moreover, that the human infant is highly impressionable and sensitive to its initial experiences in life.

There is growing evidence, however, that human genetic programs can compensate for severe environmental deficits even if these occur early in life, as studies of the mental performance of a group of Dutch teenagers have shown. These adolescents were born

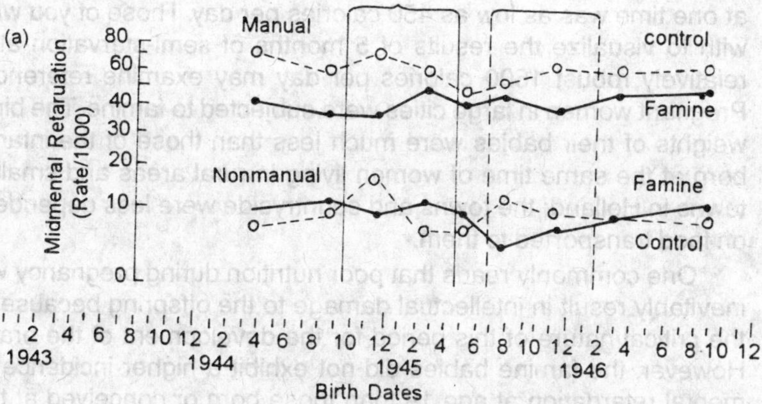

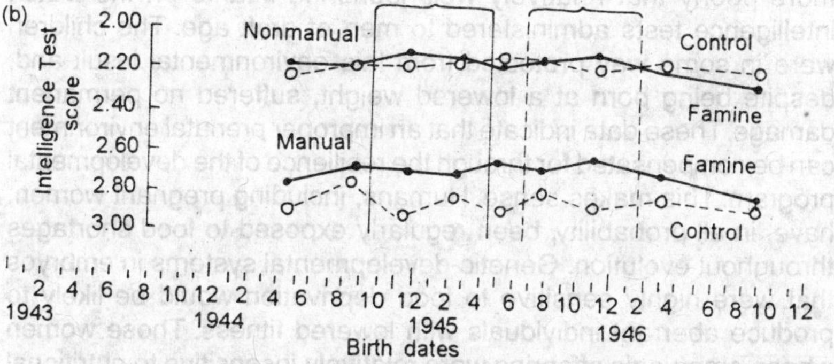

Fig. 10.14. Evidence for developmental homestasis in humans. A—Rates of mild mental retardation in 19-years-old Dutch men divided into two groups on the basis of their fathers' occupations (manual or non-manual). All the subjects were either born during the period of urban famine or conceived during this time and born later. The famine group was born in urban areas where starvation occurred; the control group was born in rural areas, where food shortages were much less severe. B—The intelligence test scores of 19-year-old from control and famine groups. Infants born of conceived under famine conditions were no more prone to mental retardation or low intelligence test performance than those conceived or born at the same time in rural, non-famine conditions.

or conceived at a time when their mothers were being starved during the Nazi transport embargo, which prevented food from reaching the large Dutch cities during the winter of 1944-45. Deaths from starvation were common, and many persons lost one-fourth of their body weight. For most of the famine period the average caloric intake was about 750 calories (roughly one-half the absolute minimum standard) and at one time was as low as 450 calories per day. Those of you who with to visualize the results of 5 months of semi-starvation at a relatively robust 1600 calories per day may examine reference. Pregnant women in large cities were subjected to famine. The birth weights of their babies were much less than those of the infants born at the same time of women living in rural areas and smaller towns in Holland; the towns and countryside were less dependent on food transported to them.

One commonly reads that poor nutrition during pregnancy will inevitably result in intellectual damage to the offspring because of the critical nature of this period for the development of the brain. However, the famine babies did not exhibit a higher incidence of mental retardation at age 19 than those born or conceived at the same time in on-famine areas. Nor did these deprived babies score more poorly that relatively well-nourished infants on the Dutch intelligence tests administered to men of draft age. The children were in some way protected from this environmental insult and, despite being born at a lowered weight, suffered no permanent damage. These data indicate that an improper prenatal environment can be compensated for through the resilience of the developmental program. This makes sense. Humans, including pregnant women, have, in all probability, been regularly exposed to food shortages throughout evolution. Genetic-developmental systems in embryos that were highly sensitive to food deprivation would be likely to produce aberrant individuals with lowered fitness. Those women whose embryonic offspring were relatively insensitive to nutritional deficits would tend to produce normal functional individuals with normal to superior fitness.

I am not saying that human are impervious to all environmental influences or that it is a good idea for pregnant women to starve. Still, just because there is an environmental factor that potentially could affect development is no guarantee that it will, in fact, have an irreparable effect. Our genotype as a general rule can withstand a good deal. *Jerome Kagan,* a child psychologist, recently went to *Guatemala* expecting (according to this own account) to find that the child-rearing techniques practiced in that country would permanently stunt the intellectual growth of children. The expectation was based on the knowledge that *Guatemalan* villages keep young infants from birth to about 1 or 2 years of age in conditions that strike American observes as amounting to extreme deprivation. The young baby is kept in a dark hut, experiences only limited social interactions (it is fed but talked to), and is denied toys. Despite all this *Kagan* says that had he seen American children treated this way he would have called the police, the children's developmental program compensates for the first year or so of confinement. By 11 they score as well on Kagan's test as middle-class Americans of the same age. They are socially well-adjusted capable human beings.

If the developing animal really were a passive billiard ball designed to respond to every environmental impact, the end product would often be a disaster. Individuals would run the risk of being permanently warped by a transitory environmental deficit, such as a temporary shortage of food or lack of social stimulation. They might fail to develop the traits that historically have led to survival and reproduction. The genes of these developmentally "sensitive" individuals would surely be less likely to survive than those that had the capacity to guide development over or around obstacles in a developmental path that had in the past produced animals with high fitness. If behavioural scientists were to look for the adaptive phenomenon of developmental homeostasis in behavioural development, they would probably find many new examples of it.

Currently, little is known about how genetic-developmental systems are able to keep on even keel in the face of great genetic and environmental variation. But it seems likely that as more is learned about the regulatory properties of the genotype more will be learned about developmental homeostasis. Perhaps there is considerable redundancy in genetic control mechanisms. Just as the designers of manned space vehicles provide these machines with a number of backup systems so that failure is not fatal, so too the genotype may have the capacity to provide developing cells with alternate metabolic pathways for the production or disposal of key substances. Should a mutant gene's product interfere with the manufacture of degradation of a material, other genes may become activated and provide substitute enzymes for these tasks. The maintenance of metabolic homeotasis within cells should promote the regulation of cell function and development. In this way genes can maintain control of the course of development of nervous systems and so steer behavioural development along certain lines. The detailed analysis of the role of genetic redundancy and other properties of the genotype that contribute to developmental homeostasis is a formidable but enormously excited challenge for behavioural scientists.

11

THE PHYLOGENY OF BEHAVIOUR

Two important characteristics of behaviour should be apparent forms the discussions that have appeared thus far in this book—its diversity and its adaptiveness. The tremendous diversity of the behaviour of different species, not only among distantly related species but often among closely related species as well, has been apparent in virtually every chapter. However, this diversity is by no means random variation. Animals of each species have become adapted to live a particular "life style" that is appropriate for a given environment; that is, they have become adapted. The most parsimonious explanation for the diversity and adaptedness of behaviour, and indeed of all biological phenomena, comes from the theory of evolution. Seldom has the development of a theory so revolutionized and synthesized a scientific endeavour as the theory of evolution has the life sciences.

The driving force of evolution has been natural selection. It will be recalled that natural selection was defined as a process that resulted in the differential reproduction by individuals of different genotypes. There is good evidence in support of the view that behaviour is, at least in part, dependent upon genotype. It follows that as different genotypes reproduce differentially, the behavioural patterns that are dependent upon those differentially reproducing genes will become differentially represented; that is, behaviour evolves.

Prerequisite to an understanding of the evolution of behaviour is an understanding of evolutionary processes in general. Some of these processes are being discussed here.

Microevolution

Evolutionary biologists have generally assumed that natural selection produces large-scale, qualitative change through the accumulation of small, quantitative increments. These small changes, referred to as *microevolution,* may come about through changes in thresholds in the sensory, motivational, or motor systems that control behaviour patterns (*Manning* 1971). Changes in thresholds lead to modifications in frequency or amplitude of displays. For example, the long calls of common and herring gulls (genus *Larus*), which function to advertise territory, are similar, but the two species emphasize different parts. The basic pattern involves jerking the head down, then throwing it back while calling. The herring gull has increased the amplitude of the downward jerk of the head; the common gull has increased the amplitude of the throwing back of the head (*Tinbergen* 1959). Such differences probably function in species recognition. Microevolution also seems important in enabling organisms to adjust to changing environments, but can it lead to the evolution of new species, as Darwin assumed? *Gould* and *Eldredge* (1977) think not, and have argued that new species have arisen in peripherally isolated populations through rapid change, rather than by slow transformation in the central ancestral area.

One of the cornerstones of research by ethologists, which began in the early 1900s, has been the idea that behaviour patterns can be treated like anatomical features. They "dissected" behaviour patterns to learn their "anatomy". They also demonstrated homologies in behaviour, which are patterns shared by species through descent from a common ancestor. For instance, *Van Tets* (1965) studied birds of the order Pelecaniformes, scoring species for the presence or absence of behavioural traits and comparing the pattern with the phylogenetic tree as inferred from morphology. Some behaviour patterns, such as the pre-landing call, are common to all members of the order; others, such as head wagging, as seen in gannets and boobies, are restricted to one family. The general conclusion from this and other comparative studies (for example, *Lorenz* 1972) is that the distribution of behavioural similarities and differences in a group of species tends to be correlated with phylogenetic relationships as disclosed by morphological or other means (for example, the fossil record or protein analysis).

There is a widespread tendency, particularly among nonbiologists, to talk about "progression" in evolution and to refer to some traits or species as "advanced" and others as "primitive". These terms are holdovers from orthogenetic interpretations that view evolution as proceeding in a predetermined direction independent of external factors. We should talk, however, or progress in evolution only in reference to either conformity of phyletic trends or to an approach to some *arbitrarity* designated final stage (*Williams* 1966): thus forms that share traits with that final stage could be said to be advanced. The evidence that groups "progress" in a more general sense or that adaptations become more perfect can usually be countered by examples of ancient, primitive forms that are highly successful (for example, sharks) and by the fact that adaptations in one direction usually mean trade-offs in some other area; improvements in terrestrial locomotion, as an example, are likely to be matched by decrements in swimming ability. Nevertheless, as well become evident form the examples that follow, many researches speak of "improvements" in traits through evolutionary time.

The Nature of Genetic and Micro-Evolutionary Changes

It is worth while trying to discover if there is a common thread running through the diversity of genetic and micro-evolutionary changes just described. We have emphasized the quantitative nature of the changes and many of them can, in behavioural terms, be directly related to changes in thresholds. If, in the same situation, two species or strains differ in the relative frequency with which they perform patterns. A and B from a common repertoire, it is reasonable to interpret this as due to threshold differences in the mechanisms controlling A and B in each case. There are numerous sites at which the control mechanisms could be affected. There might be changes of threshold in the sensory system, in systems affecting motivation, in the motor system, or in any combination of these. Threshold changes could be produced by genes acting on the nervous system itself as in the hyperkinetic mutants of *Drosophila* or more indirectly, by genes affecting metabolic rate or hormone secretion which will in turn affect the nervous system. In no case do we have proof, but sometimes the circumstantial evidence for direct action on the nervous system in strong. Thus the differences in the performance frequency of various sexual patterns between inbreed lines of guinea pigs remain even after injections of sex hormones or thyroid hormones, which increase the basic rate of metabolism. It seems most probable that the lines differ in the threshold properties of the sexual mechanisms in the brain.

It is not difficult to explain changes in the qualitative form or speed of patterns in terms of threshold changes also. Each fixed action pattern must have a co-ordinating 'centre' of some kind whose structure and properties are inherited and which, when activated, produces a stereotyped pattern of output to lower centres controlling muscles and groups of muscles. The centre calls into play each muscle group in the correct order, at the correct time and for the correct duration. The result must depend on the subtle series of threshold relationships both within the co-ordinating centre itself and between the various motor centres which control the muscles. The output of the co-ordinating centre may be completely 'pre-set', or it may be modified by feed-back from the muscles as the fixed action pattern is actually being performed. In insects we know centres, probably small groups of neurons, which control singing in crickets and flight in locusts. These produce quite normal output when isolated, although they can be modified by feed-back. *Hoyle* discusses the different types of control in more detail in relation of his own work on insects, which have proved excellent material for such studies.

Genetic changes affecting thresholds either within the co-ordinating centre or one of the subordinate muscle groups will change the form of a fixed action pattern. For instance, muscles might be brought into action earlier or later in a sequence; they might be held on for a longer or a shorter time, or they might change the intensity and speed of their response as more or less or their constituent fibres are activated. A beautiful example of the behavioural results of such changes from *Hunsaker's* work on lizards. Males of the genus *Sceloporus* show rhythmic head bobbing movements during courtship and also whenever they meet other lizards as a kind of species-identity signal. Each species has a characteristic pattern of bobbing which results from a series of contractions in the muscles which extend the front legs and thereby raise and lower the head and shoulders. Micro-evolution has produced several distinct versions by a series of changes of the type outlined above. This example is particularly clear because only two main groups of muscles are involved, but it is not difficult to envisage the differences in emphasis illustrated in Fig. evolving in just the same way.

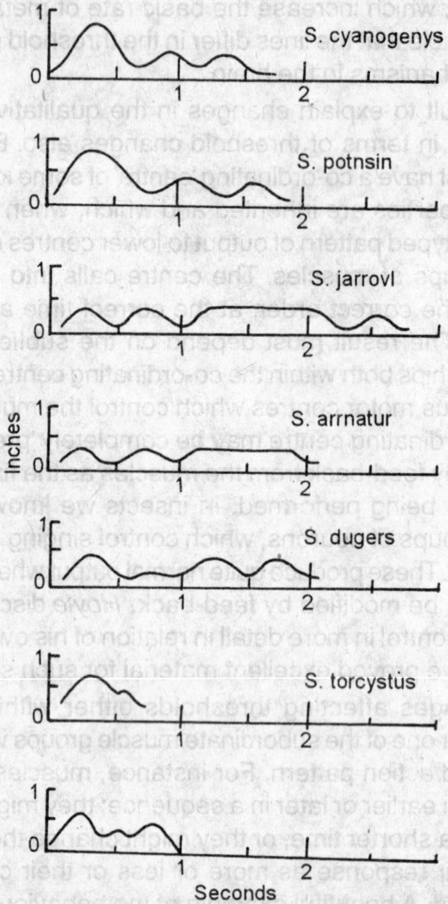

Fig. 11.1. Diagrams illustrating the head-bobbing movements in the courtship of several species of lizards of the genus. Sceloporus. The elevation of the head is portrayed on the ordinate as a function of time, which is shown on the abscissa.

We may conclude that gens affecting threshold within the nervous system have been most important in the evolution of behaviour. Most of the amazing diversity of fixed action patterns we observe will have evolved by the accumulation of small quantitative changes, in just the same way as the body form of animals have evolved.

Evidence for the Evolution of Behaviour

In this section we examine the types of evidence used to demonstrate how behaviour changes in a more or less permanent

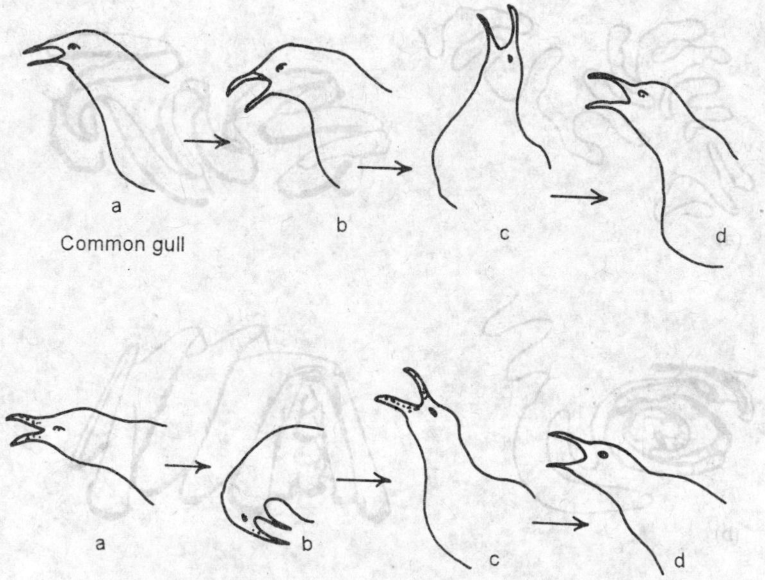

Fig. 11.2. Variators of emphasis in the 'oblique long call' sequence of two gull species. The sequence reads from left to right. At the start the head is in the oblique posture (a), it is jerked down as the bird begins calling (b), and is then thrown back (c) as the calls continue. The head individually lowered (d) as the calls die away. The common gull shows little emphasis of (b) and much of (c), the converse is true of the herring-gull.

way through generations. The evidence is of three general types: (1) phylogeny, the tracings of the patterns of evolutionary change through time; (2) adaptation, the way natural selection actually works; and (3) speciation, the formation of new species (*Alexander* 1978).

Phylogeny

The most direct evidence about phylogeny is provided by the fossils which, for all their inadequacies, represent the only genuinely historical record of the course of evolution. However, since actions normally leave no permanent trace to be fossilized, this is not a very fruitful source of information about the evolution of behaviour. There are some exceptional cases where the consequences of behaviour have survived as fossils. Fig shows a series of fossil tracks left by feeding organisms from the Cambrian to the lower Carboniferous. From their form and position in the fossil series, the following sequence of events in the evolution of foraging behaviour in this group can be proposed; the animals originally fed in loosely

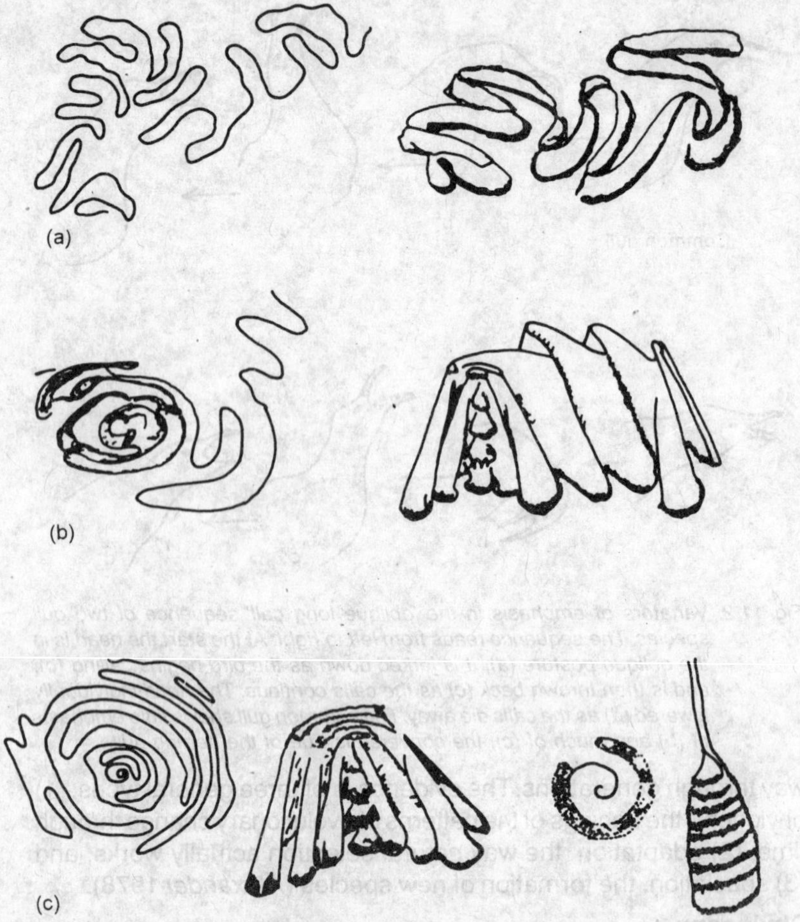

Fig. 11.3. Tracing the course of behavioural phylogeny from fossils; fossilized tracks left by an unknown sediment feeder (Dictyodora) in deposits from the Cambrian (a) to the lower Carboniferous.

meandering paths just below the surface but gradually showed a greater tendency to move sideways and to burrow deeper until the tracks finally took the form of extremely regular spirals. This series of chains probably increased the efficiency of foraging by reducing the changes of covering the same ground (*Seilacher* 1967).

The frequency and type of bone fractures observed in excavated human skeletons can be used to provide information about their way of life; the incidence of past cranial bone fractures in American

Indians of different periods correlates with their mode of life, deduced from other evidence; post-cranial fractures become less common with the transition from a nomadic, hunting lifestyle to life in more or less modern towns. This correlation can be used to assess the lifestyle of populations for which only skeletons are available (*Steinbock,* 1976).

Table 11.1. Making deductions about behaviour from fossilized remains; the Incidence of postcranial in Amerindian populations differing in age and way of life.

Date	Way of life	% of skeletons with postcranial fractures
4000—1000 BC	Nomadic hunters	10.7
1000 BC—1000 AD	Permanent villages	5.4
1000—1600 AD	Agriculture/towns	3.9

Even when the consequences of behaviour are not preserved in this way, we can get some information about the behaviour of fossilized animals from structures which regularly correlate with a particular behaviour in similar forms alive today. For example, fossil ants resembling modern worker ants have been found in the cocene; it is reasonable to conclude that these animals showed at least some of the worker-like behaviour of their modern counterparts and therefore that they lived in complex social groups with division of labour (*Lin* and *Michener*, 1972). The information obtained in this way can be very precise; for example, study of the vocal tract and the way muscles are attached to bones in modern humans has demonstrated that these place constraints on the phonetic repertoire. Given this information, measurements of the fossilized hominid skulls can be used to make deductions about exactly what sounds these people could and could not produce. For example, neither *Australopithecus* nor primitive Neanderthals were able to pronounce the vowels i, u and a or the constants g and k; more advanced Neanderthals did not have these structural restrictions on speech (*Lieberman,* 1977).

Domestication

Artificial selection by humans has changed the behaviour of numerous animals with which humans have associated over thousands of years. Domestication involves more that just taming or socializing animals; humans also control the breeding, care, and feeding of domesticated animals. Some ethologists have argued that behavioural plasticity is greater in domestic animals than in their wild counterparts and that neoteny, the persistence of juvenile characteristics in adults, in typical in domestic animals. Others have emphasized the degeneracy of domesticants; they cite the laboratory rat. But few of these generalizations hold true when the entire spectrum of domesticated animals is considered (*Ratner* and *Boice* 1975; *Price* 1984). Certain species are preadapted to domestication. For instance, most carnivores maintain a nest, and urinate and defecate some distance away from it; they are thus suited to living in human habitations. Other factors facilitating the process of domestication include large social groups, the presence of males in the group year round, precocial young, and an omnivorous diet (*Hale* 1969).

The evolution of dog behaviour provides the most striking example of selective breeding by *Homo sapiens* (*Scott* and *Fuller* 1965). The wolf (*Canis lupus*), ancestor of the dog, and the principal carnivore of the northern hemisphere prior to the arrival of humans, has a complex, cooperative social system. As the wolf was domesticated and transformed into the dog, it spread all over the world in conjunction with humans and underwent adaptive radiation. Dogs (*Canis familiaris*) were used for protection, for hunting, and for herding other domesticated animals. The relative isolation human tribes 8,000 to 10,000 years ago, with only occasional genetic interchange, probably encouraged the evolution of distinct dog breeds. As human contract increased with improved transportation, the different breeds of dogs mixed, and many breeds, such as those in South Africa and North and South America lost their identity. Only in the last few hundred years have humans practiced rigid artificial selection, which has resulted in increased separation of dog breeds. The great diversity of form and behaviour of current breeds illustrates the potential behavioural effects of selection, whether natural or artificial, or certain phenotype.

Scott and *Fuller* (1965) concluded that the behavioural patterns of the dog and wolf are essentially the same. Selection has modified primarily the agonistic and investigatory systems. For example, the bulldog breed, in the interest of an English sport, was selected for its tendency to attack the nose of a bull and hang on, in contrast with the more typical slashing attach from the rear used by wolves. Terrier breeds have been selected for their tendency to attack prey relentlessly, regardless of any injury suffered; the usual wolf pattern is to snap and then withdraw to avoid injury. Other breeds have been selected for opposing reasons; scent hounds and bird dogs are examples of peaceable animals that can be kept in groups in a kennel.

Wolves and dogs search for prey rather than lie in wait and therefore have a well-developed investigatory repertoire. Some dogs, such as scent hounds, are bred for the ability to follow a trail. Others, such as bird dogs, use their visual and olfactory sense much more equally; after rapid quartering of ground, they locate prey by scent only when they are a few paces away, and then the freeze. Dogs must be trained for all of these tasks, but in each case artificial selection has produced a phenotype that is predisposed to specific behavioural traits.

Comparative Series. Perhaps the best available evidence for the evolution of behaviour has come from comparative studies of closely related forms. The presence of a series of intermediate forms between two extremes suggests what the evolutionary sequence was, but it may not represent the exact route because the living intermediates may have changed from the ancestral type. It is usually assumed that behaviours evolve from simple to complex, and such series are often referred to as progressions; but inferring goal-directed change in evolution in unwise, as we noted at the beginning of this chapter.

One comparative series is that of balloon files, members of the family Empididae (order Diptera). According to *Kessel* (1955) the story began in 1875 when *Baron Osten-Sacken* was visiting in the Swiss Alps and noticed bright silvery flashes among the shadows of the fir forest. Believing that they were silvery insects he used his net to capture some, but ended up with dull-coloured flies. Later he noticed that along with male flies, he had netted packets of filmy material. *Osten-Sacken* assumed that the males carried the material on their backs to attract females during the mating swarm. Another observer later discovered that the male flies carry the packets underneath their bodies and suggested that the somewhat flattened balloons serve as aeronautical surfboards on which the flies cavorted among the sunbeams. Another suggestion was that the devices serve as warning signals to birds and predaceous insects; however, no evidence exists that flies are distasteful to predators. The male flies seem to use the balloon for attracting females, stimulating mating, and reducing the likelihood that the female will try to eat the male once they have coupled. *Kessel* (1955) described the evolutionary stages that seem to have led to this situation.

Stage 1. In the majority of empidids, both sexes capture insects independently, and no presentation of prey is associated with mating. These flies sometimes prey on conspecifics; when the male attempts to copulate, it may be eaten by the female.

Stage 2. The male capture a prey item and presents it to the female. He copulates with her as she eats the prey instead of him. Many other insects follow this procedure, and natural selection theory predicts that intraspecific deception should occur if the deceiver gains a reproductive advantage over a conspecific. In an unrelated insect, the scorpionfly *Hylobittacus apicalis* (order Mecoptera), the male with a prey item advertises to females by means of a pheromone. But sometimes a male assumes the behavioural posture of a female; he approaches a male that has a prey item, lets the male try to copulate, then steals the prey item and either eats it or uses it to attract a female (*Thornhill* 1979).

Stage 3. Rather than taking the prey and searching for a female, some male empidids join other males, each with a prey item, in an aerial dance. The prey, Kessel posits, is now a stimulus for mating rather than a distracting to avoid mate cannibalism. The female enters the swarm and selects one of the males.

Stage 4. In many species of he genus *Hilara*, the male wraps the prey loosely with some silken threads, and action that seems to quite the prey.

Stage 5. In several species of he genus *Empis*, from the western United States, the male applies elaborate silken wrappings to the prey, which then resembles a balloon. When male and female meet in midair, the male transfers the balloon to the female and climbs on her back. The pair alights on a plant, and the female rolls the balloon about, probes it, and eventually consumes the prey item while the male copulates with her.

Stage 6. The male catches a small prey and may consume its fluid so it is no longer edible; he then constructs a complex balloon. The female accepts the balloon and plays with it during copulation, but gets no meal from it.

Stage 7. The prey item is very small, of no food value to either sex, and pieces of it are plastered at the front end of the balloon. The balloon is now the sole stimulus for copulation.

Stage 8. The *Hilara sartor* male give the female a balloon that has no prey at all. This behaviour is thought to be the final stage in the series.

Kessel argues that the nuptial feeding of the female initially functioned to reduce the chances of the male's being consumed by the female during mating. *Thornhill* (1976) suggested an alternative function based on parental investment. Male insects may invest in their offspring by providing nutrition to their mates in the form of glandular secretion, prey captured by the male, regurgitative food offering or the male himself. Females may select mates on the basis of the quality of food the males offer. In all cases, the substitution of a balloon for prey represents deception of the female by the male.

It is useful, if not critical, to have an independent assessment of phylogenetic relationship with which the behavioural series can be compared. For instance, many wasps (order Hymenoptera) capture prey to provide food on which they lay their eggs and on which the larvae subsequently feed. Some species carry the prey anteriorly with the mandibles; other species carry the prey posteriorly, sometimes on a terminal abdominal segment called an "ant clamp" (*Evans* 1962). The series, with many intermediate forms, correlates well with other evolutionary evidence, and we can be confident of the direction of change. The selection pressure directing the change from anterior to posterior carriage is thought to be cleptoparasitism. When carrying prey with the mandibles, the wasp must put the prey down to open or dig a burrow. Other flies or wasps can then lay eggs on the prey as well and thus diminish the food resource.

Among vertebrates, the evolution of pigmentation in the mouth-breeding cichlid fish (*Haplochromis* spp.) was studied by *Wickler* (1962). Presumably under the selection pressure of egg predation, some species of cichlid females pick up eggs after spawning (i.e., after the eggs have been laid and externally fertilized) and "incubate" them in the mouth. The faster the females pick up the eggs after laying, the less the chance of predation. In some species the females take the eggs before they are fertilized by the male. The male presents his anal fin, which bears yellow spots resembling eggs; the female snaps at these spots and takes in sperm, which fertilize the eggs in their mouth. Comparative studies have shown that these yellow spots evolved from less conspicuous pearly spots on the vertical fin that have no signaling function. *Eibl-Eibesfeldt* (1975) has pointed out that the following preadaptations existed :

1. Female readiness to pick up eggs,
2. Pearly spots on the vertical fins to the male,
3. A lateral display by the male during courtship and spawning.

The case in which intermediates are not available is more common and we must draw evidence from related species that have evolved more or less individually from some common ancestor. A weak inference can be made on the basis of the assumption that those behaviour patterns common among the various living species represent ancestral traits.

Tinbergen (1959) compared the behaviours of different species of gulls (family Laridae); in particular he examined those patterns that are relatively constant throughout a species and of the "fixed pattern" type, that is, they are developmentally stable and resistant to environmental change. The basic similarities in displays throughout the family strengthen the conclusion, previously based on morphology, that the family is monophyletic, that is, it originated from a single ancestral type. If a species is judged to be closely related to other species because it shares many of their characteristics, any differences can be considered to have been recently acquired.

A variety of animals construct objects for prey capture, shelter, or mate attraction-for example, spider webs, caddisfly cases and nets, bee hives, bird nests, and bowers. It is relatively easy to examine these semipermanent structures and to compare closely related species. *Collias* and *Collias* (1963) studied the nests of African weaverbirds of the family Ploceidae, tracing the evolution of the nest from the ancestral cup nest to intricate domed structures with bottom entrances. Primitive members of the family began the nest as a simple cup or platform; in some species a dome is added later.

In the true weaves (subfamily Ploceini) a ring is made from which the roof is built, followed by the floor. In other species a communal thatched roof is constructed, with brood chambers of many individual pairs underneath. The researches developed a key to the nests of the true weaves and actually suggested some taxonomic revision in the family based on rest construction.

The techniques used to weave the nest from grasses also vary among the weaves, and *Collias* and *Collias* noted trends toward more regularity in the pattern, tighter weaving, and finer stitches within the subfamily. In general, species making the more "advanced" nests used the more refined pattern of weaving. In many species

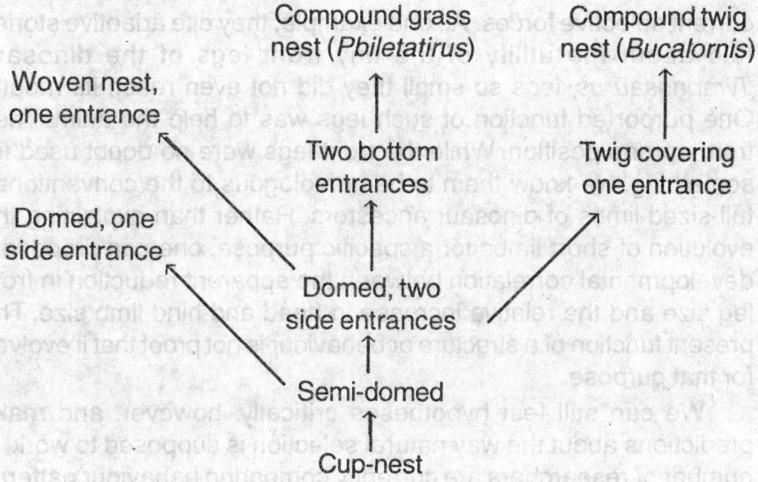

Fig. 11.4. Main trends in evolution of nest building in the weaverbird (family Ploceidae).

the male build the basic nest and, along with neighbouring males, advertises it to the female, who makes the final choice. Males may make many nests before one is chosen for completion by the female, implicating sexual selection in the evolution of nest construction.

Study of Adaptation

Many studies have attempted to illustrate the adaptiveness of behaviour by emphasizing the ways in which behaviours ensure an organism's survival and reproduction. The courtship of ring doves is adaptive because it coordinates the physiological processes necessary for reproduction in the male and female. But does such evidence really tell us anything about evolution by natural selection? Natural selection leaves only the best adapted or optimal phenotypes for our inspection.

A common practice is to study an organisms's traits or behaviour pattern and to propose an adaptive story for each trait considered separately. Behaviour patterns or structures found to be suboptimal are considered to be the results of trade offs among competing demands within the same organism and thus also the result of adaptation. The possibility that present traits are not adaptive to present conditions is usually not considered. *Gould* and *Lewontin* 91979), among others, have criticized this approach; they argue that present traits are constrained by phyletic heritage, developmental pathways, and general architecture and are not just the result of current selective forces. As one example, they cite adaptive stories told about the utility of the tiny front legs of the dinosaur *Tyrannosaurus,* legs so small they did not even reach its mouth. One purported function of such legs was to help the animal rise from aprone position. While the front legs were no doubt used for something, we know them to be homologous to the conventional, full-sized limbs of dinosaur ancestors. Rather than explaining the evolution of short limbs for a specific purpose, one can propose a developmental correlation between the apparent reduction in front leg size and the relative increase in head and hind limb size. The present function of a structure or behaviour is not proof that it evolved for that purpose.

We can still test hypotheses critically, however, and make predictions about the way natural selection is supposed to work. A number of researchers are currently comparing behaviour patterns with the theoretical optima that might be achieved in the best of all possible worlds. The models of optimal foraging already discussed were constructed to predict strategies based on maximizing some parameter, such as rate of energy intake. The degree to which such models fit data collected in the field demonstrates the importance of natural selection and selective processes in causing changes in gene frequency. The most useful models are those that are falsifiable, enabling us to reject hypothesis.

Observing Natural Selection in the Field. The best evidence for the evolution of behaviour is documentation in a natural population. Perhaps the strongest case we have so far is the work of *Kettlewell* (1965) on morphological change in moths, where melanism, the development of dark pigmentation, was a result of the darkening of the substrate by soot. As the Industrial Revolution led to increased burning of coal, the bark of trees used as resting places by the moths became darker. Those moths whose darker colour matched the background were subject to less predation than were the lighter-coloured moths. A behavioural change in back-ground preference must also have occurred so that the moths remained cryptic.

Intense natural selection has been demonstrated in one of the ground finches (*Geospiza fortis*) of the Galapagos (*Boag* and *Grant* 1981). On one of the islands, *Daphne Major,* regular rainfall throughout the early 1970s produced abundant seeds, and seed-eating finches thrived. In 1977 a drought occurred and the finches failed to breed, declining in numbers by 85 percent. The corresponding decline in seed supply was non-random, with small seeds becoming scarce and large, hard seeds remaining relatively common. The result was strong selection for birds with large beaks that could handle the seeds, and within a short period of tome the average beak size increased dramatically. *Boag* (1983) went on to show that beak size is heritable and that offspring showed a strong phenotypic response to natural selection of their parents.

Speciation and Reproduction Isolation

An important question in the study of the evolution of behaviour is that of how species are formed. As already defined a species as a group of organisms that are capable of interbreeding freely under natural conditions but which do not interbreed with organisms other groups. In the study of the evolution of behaviour, it is differences of at least the species level that are generally of interest. There are many quantitative and even qualitative differences among members of different populations of the same species. In considering the problem of the formation of species (*speciation*), we consider how such differences come to produce reproductive isolation and the formation of species.

Allopatric Speciation

Two groups of animals are said to be *sympatric* if they occur together in the same geographical region and *allopatric* if they are geographically separated. The dominant view among biologists is that speciation requires geographic isolation; that is, that speciation must take place in allopatry. However, some biologists maintain that sympatric speciation occurs. Thus, while *Mayr* (1963) maintains that allopatry is necessary for speciation. *Scudder* (1974) maintains that alternative forms, including sympatric speciation, are possible. We shall limited our discussion to allopatric speciation.

The process of allopatric speciation begins with a single population that clearly represents but one species. Somehow this population must become divided by a geographical and ecological barrier, so that animals from what are now sub-populations are no longer able to interbreed freely. Populations may be separated by the development of a mountain range during a time of glaciation, by being swept up on different islands, or by being isolated in different lakes. During the period that gene exchange between the two sub-populations is prevented, there will be changes in the gene pool within each sub-population. Many of these changes will be related to

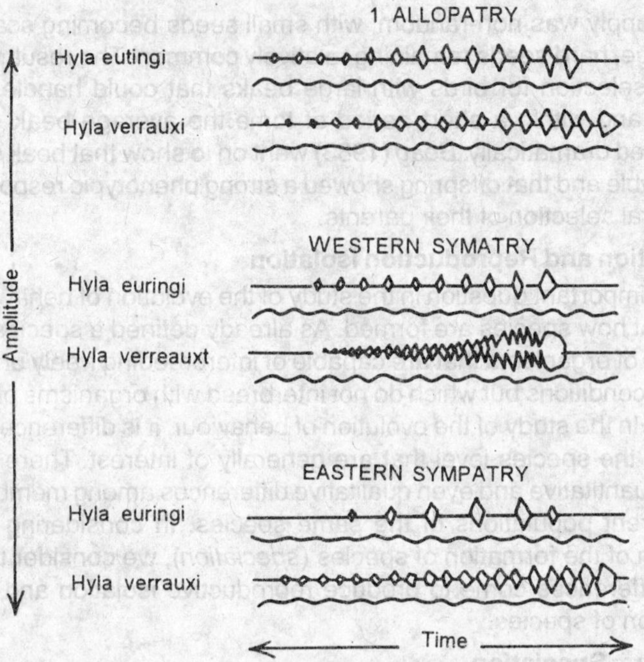

Fig. 11.5. Oscillograms of songs of two Australian frog species. The amplitude of each note of the song is shown on the vertical axis, and time is shown on the horizontal axis. Under each song recording is a 50-cycles-per-second reference line. The songs of the two species are similar when comparing individuals caught in areas where the two species do not overlap (alloparty). When the two species do overlap (sympatry), song differences are pronounced. This divergence of behaviour between closely related sympatric species is referred to as character displacement.

natural selection as the sub-populations adapted to slightly different habitats. The extent and rate of these changes will be a function of phylogenetic inertia and ecological pressure. Still other genetic changes may results from random processes, such as genetic drift. The latter factors may be particularly important in small sub-populations.

At some point the geographical barrier is crossed or removed and the two sub-populations interact. What happens then depends on the degrees and kinds of changes within the two sub-populations that have occurred during their period of allopatry. If changes in the gene pool have been minimal, the two sub-populations will freely interbreed and eventually merge and become a single population once more. Alternatively, the two sub-populations may have been separated for so long that no breeding can occur; they have already formed species. However, divergence may be less extreme than this. The two sub-populations may have diverged to an intermediate degree, so that hybrids could occur but would be at a competitive disadvantage. If hybrid animals are not as healthy or as fertile as those conceived from mating within sub-populations, the fitness of animals breeding with partners from outside their sub-population would be lower than that of animals choosing mates from within the sub-population. In that case, there would be selection against matings between animals from different sub-populations.

It is at this point, after two divergent sub-populations have come together, that reproductive isolating mechanisms confer selective advantages. Reproductive isolating mechanisms are those biological properties of individuals that prevent the interbreeding of sympatric groups. They develop under the present circumstances because of the competitive disadvantage of hybrids between the two subpopulations. Individuals that breed with members of the foreign sub-population would be at a reproductive disadvantage relative to those that breed with members of their own sub-populations. Thus the reproductive isolating mechanisms are reinforced and mating between the two sub-populations becomes infrequent or impossible. They now have become two different species.

Typically, sympatry will occur among previously isolated sub-populations only in part of their geographical range. It is in the areas of sympatry that selection will act against interpopulation breeding and foster the development of isolating mechanisms. The animals from the two sub-populations that occur in the region of sympatry should diverge more from each other than those living in areas of allopatry. This is the phenomenon of *character displacement*. Character displacement is the phenomenon whereby animals from two newly evolved species interact so that further divergence between the species results in the area of sympatry.

Darwin's Finches

An excellent example of the manner in which a presumed single population can diversify in adapting to new habitats called (*adaptive radiation*) is provided by a group of finches living in the Galapagos Islands off the coast of Ecuador (*Lack*, 1947, 1953). Because Darwin studied these birds during his voyage on the *Beagle*, they are called

"Darwin's finches". There are currently 14 species of finches on the Galapagos Islands. They have radiated into a number of very diverse forms. This diversification can be seen most easily in an examination of their feeding habits and bills. The bills of some species are adapted for crushing seeds of various sizes; some are adapted for grasping insects and buds; and others are adapted for probing after insects. The appearance of the bill may also function in displays and hence foster reproductive isolating.

Lack proposed that a flock of ancestral finches may have arrived as the first passerine birds on the Galapagos. They found great potential for exploitation of resources because of an absence of both competition from other species and of predators. The result was an adaptive radiation in which the original animal became isolated on different islands, adapted to local conditions, and formed the total of 14 species from the single founding population. Many of these species have evolved into distinctly "unfinchlike ecological niches" as a result of the unique conditions on the Galapagos. "The geographical isolation from both mainland birds and from each other, together with the opportunity for rapid diversification in adapting to new habitats, probably was critical to the adaptive radiation of Darwin's finches.

Experimental Studies of Reproductive Isolation

A number of researchers have attempted to recreate artificially the kinds of interactions that must occur when two previously sympatric subpopulations are reunited (*Koopman,* 1950; *Knight, Robertson* and *Waddington,* 1956; *Kessler,* 1966; *Crossley*, 1974). All studies used fruit flies of the genus *Drosophila,* the "geneticist's while rat". Flies of two closely related species or of two similar genotypes differing in some prominent marker gene are intermixed and permitted to mate. Such flies can either mate with partners of their own genotypes or form hybrid matings. The offspring of hybrid matings are removed and not permitted to reproduce. If this procedure is followed over several generations, reproductive isolation between the population of matings in the first few generations and in control lines may be hybrid matings, the per cent of matings in the selected lines that produce hybrids decreases over 20 to 40 generations.

Crossley (1974) analyzed the details of the reproductive interactions between her ebony and vestigial flies after selection against hybrids. She found that the courtship interactions of her two populations had been changed over the course of her experiment.

There occurred (1) an increase in the tendency of females to resist courtship from males of the strange (i.e., heterogametic) strain, and (2) a greater tendency of males to respond to the repelling movements from females of the foreign strain by terminating courtship than to respond similarly to the repelling movements of females of their own strain.

These studies differ from the selection studies in that the experimenters did not select for particular behavioural characteristics. Rather, as natural selection appears to do on occasion, the experimenters selected against hybrids. As a result of selecting against hybrids, increased reproductive isolation and evolutionary changes in behaviour developed. Events that occur when two previously allopatric subpopulations first interact may not be too much different from those that occur under experimental conditions with these *Drosophila*.

Tradition

Although the evolution of persistent changes in form and behaviour usually has a genetic basis, natural selection may act to produce both plasticity of expression and the tendency to transmit behaviour to offspring through non-genetic means. For example, experiences of female rats can bias the behaviour of offspring and even grand offspring (*Denenberg* and *Rosenberg* 1967). The physiological mechanisms that cause these changes in behaviour are poorly understood, but the effects of environment on the behaviour of subsequent generations probably are more widespread and important in producing persistent changes in behaviour than are generally recognized.

We have a better understanding of the role of *tradition*, the passing of specific forms of behaviour from generation to generation by iearning. Compared with genetically controlled behaviour, tradition transmitted behaviour can spread through a population, rapidly, sometimes in less than one generation. One common form is dialect in animal communication; birdsong has received the most attention. Males learn song during a critical period, when they imitate phrases heard in their neighbourhood. Male immigrants into a local area may adopt the local dialect. The adaptive significance of this phenomenon could be to enhance communication between male and female in pair bonding. It may also maintain the separateness of locally adapted populations.

Migratory animals tend to use the same routes and return to the same places as their ancestors. Presumably the young learn the routes from their elders. Water-fowl, for example, use traditional routes and stop at the same places each year during migration, and grazing mammals use the same trails for many generations. Breeding grounds may also be used year after year; in lek species the mating areas may persist for centuries. Tradition is particularly important in species living in closed social groups with prolonged socialization of young.

Most examples of the social transmission of acquired behaviour are from insects, birds, and mammals (reviewed by *Galef* 1976 and by *Bonner* 1980). Among coral reef fish, French grunts (*Haemulon flavolineatum*) exhibit social traditions of daytime schooling sites and twilight migration routes (*Helfman* and *Schultz* 1984). Individuals transplanted to new schooling sites and allowed to follow residents used the new migration routes and returned to the new sites even in the absence of resident fish. Naive, control grunts released at the same new sites failed to use the new routes or to return to the new sites.

Nonhuman primates can develop new behaviour patterns through invention and incorporate them rapidly and rather permanently into their social systems. The techniques for washing sand from potatoes and wheat grains introduced by a Japanese macaque (*Macaca fuscata*) and copied by others in the group.

Another type of learned behaviour that occasionally spreads through populations is tool use. Tools enable otherwise generalized species to specialize in particular tasks without evolving morphological adaptations; for example, animals may use sticks or similar objects, rather than specialized beaks, to extract insects from crevices in tree bark.

Although the mechanisms of behavioural change involved in tradition and human culture usually is assumed to be quite different from the mechanism of genetic change brought about by natural selection and drift, learned responses can be highly stereotyped and may persist for many generations. The big difference between learned and genetically controlled traits is the speed with which learned traditions can spread through the population, most notably in human societies. An old controversy, which socio-biology has recently kindled a new, concerns the extent to which differences in human social behaviour are influenced by genetics and by culture.

Durham (1979), among others, has attempted a synthesis of human cultural and genetic evolution and has argued for the coevolution of genes and culture. During the organic evolution of *Homo sapiens*, according to Durham, there was selection at the level of the individual for the capacity to modify and extend phenotypes on the basis of learning and experience. As the capacity for culture evolved, the developing culture in each population was adaptive for the members of that population and increased their inclusive fitness. The process was further accelerated by group-level cultural selection, which operates analogously to natural selection and independently of it. Competition among cultures results in selection of the "more fit" cultures, those that best exploit the environment to their own, advantage (*Dawkins* 1976). *Durham* emphasizes, however, that cultural "fitness" is dependent on the reproductive success of the individuals that make up a culture.

Deriving General Principles of Behavioural Evolution

A quite plusible reconstruction of the evolution of a particular behaviour maybe derived from the application of studies such as those described above. Broad surveys of complex behaviour in a wide range of contemporary species can suggest the intermediate for as through which a modern behaviour may have evolved. If the animals compared are closely related and if the behaviour concerned is fine grained, the plusible became the probable; it also becomes possible to discover general principles about how behaviour changes during evolution.

Sound Production in Crickets

Sound production and reception in crickets have been intensively studied and the neurophysiological mechanisms and their genetic bases are well known. Crickets produce sounds by rubbing their elyta together, and behavioural observation and experiment show that temporal patterning is a significant aspect to the song. Analysis of wing movements and spike patterns in motor neurons shows that wing beat frequency during flight is the same as that used during stridulation and that the same muscles and motorneurons and used in the same ways during these two activities. It seems likely that stridulation movements evolved from flight movements, with minor alterations in auxiliary muscles changing the up-and-down movements the song patterns of six sympatric species of cricket; a likely sequence of changes in the production of the song is that the continuous trill of the primitive forms (traces 1 and 2) was gradually broken up into chirps by amplitude modulations. Initially these varied irregularly in duration and in the number that occurred in a single pulse of sound (3) but in more advanced forms, these parameters were fixed (5 and 6) and systematic variation of amplitude within chirps occurred (4). These changes are presumably brought about by alteration of the properties of the identified inter-neurons, motor neurons, and neuromuscular junctions which bring about this behaviour.

There is much less information about the evolution of sound reception in females; sense organs which detect airborne vibration appear to have evolved from proprioceptors monitoring body movement. Further evolution took the form of increasingly complex higher order sensory processing in parallel with the production of more sophisticated songs by males. This comparative study has identified a possible phylogenetic sequence of the song types, demonstrating the kind of changes which have taken place in the overt behaviour and suggesting possible physiological bases for these (*Elsner* and *Popov*, 1978).

Courtship in Peacocks

A constellation of similar evolutionary changes are believed to have occurred in many cases where related species show homologous display movements. These are well-illustrated in a classic study of the evolution of courtship displays in the peacock. The proposed course of evolutionary events is as follows; in the ancestors of the peacock (presumed to resemble the modern domestic fowl *Gallus domesticus*), during courtship the male gave a call like that used to attract companions to a new food source, while scratching the ground, stepping back and picking up pebbles, which served to attract females. The movements became exaggerated as in the impayan pheasant (*Lophophorus impejanus*) in which the male bows rhythmically before the hen with wings and tail expanded and in the peacock pheasant (polyplectron bicalcaratum) which scratches the ground the bows with spreading wings and tail, offering any available food to the female. Finally, feeding disappears altogether in the peacock itself, which erects its elaborately, coloured tail feathers and shakes these in front of the female, taking several backwards step and bowing (*Schenkel*, 1956).

In such proposed phylogenies a series of changes in the signal tends to occur, accompanied by appropriate changes in the sensory systems of the receiver. The movements may change in form, often becoming more stereotyped; thus the irregular pecking of the cock and stridulation of primitive crickets give way to the regular movements of the Impayan pheasant and the repetitive calls of more advanced crickets. Many such changes in form and temporal patterning can be interpreted as the result of an alteration in some threshold for performance of different components of behaviour. The functional context in which the action is shown may change; thus, movements of flight become movements of calling and behaviour which accompanies feeding is used in a sexual context. This is associated with a change in the causal relationships of the behaviour patterns. As discussed above, displays often have their evolutionary origins in the behaviour shown by animals in motivational conflicts; in peacocks, displacement feeding movements have been incorporated into the courtship ritual, and performance of the display in adults is no longer dependent on the original motivational conflict. The fact that juvenile male peacock have been observed pecking at food and scratching while showing the courtship movements supports this idea. Finally, morphological alterations evolve which enhance the conspicuousness of the movements. Classically, this process was seen as the consequence of a mutually beneficial improvement in the efficiency of information transfer. In the modern view. It results from successive attempts to manipulate and avoid manipulation as both participants seek to maximize their inclusive fitness.

Thus, notwithstanding the limitation of the techniques for studying the phylogeny of behaviour, these have suggested possible intermediate stages in the evolution of highly complex behaviour, have allowed deductions to be drawn about the probable phylogenetic history of simpler actions among closely related species and have identified some of the kinds of evolutionary change to which behaviour is subject.

12

PHEROMONES

Different animals understand their own signals to communicate ideas. However, for this they also depend on chemical signals. The hunting dogs reach their prey by smelling the earth. The female cats show sexual play and give noise to such males which are not seen but are heard. These are few examples where we find chemicals as a means of communication. Such chemicals which carries or spreads the information are called *Semiochemicals* (Greek *Semeion* - mark or signal). These can be of two types:

1. Intra specific—Communication within the species (pheromones).
2. Inter specific—Communication between different species.

When the chemicals are used as inter specific and it helps to the producers than it is designated as *Allomones.*

When it helps to the receivers they are termed as *Kairmones.*

Pheromones are those chemicals which bring coordination in the same species. Pheromones are generally regarded as carriers of excitation and the term means chemical communicator of individual through internal release of chemical messengers. Now the evidences are pouring the literature there are individuals which liberates materials externally which sets are, initiate behavioural or developmental responses in the recepients. The term pheromone has been apply to such external secretion that serves to integrate conspecific individuals of a group (*Karlsow* and *Luscher* 1959). Even symbiotic relationship (*Devenyort* 1955) involving animals of different species may be influenced by similar materials, *Parks* and *Bruce* 1961 coined a new term as *exocrinology* for these studies.

The term pheromone originally applied to the sex attractants of insect previously but now it includes various kinds of agents released into the environment and functioning in all major groups to integrate members of the population. These may be ingested, absorbed by body surface or perceived by olfaction. They evoke specific behavioural, developmental or reproductive responses which are very significant for ecology and survival of the species.

In early literature on the subject chemicals used in communication were usually referred to as "*Ectohormones*". Since 1959 the less awkward and *Etymologically* more accurate term pheromones has been widely adopted.

Numerous instances are known among Protistants in which the determination of sex and attraction of gametes to permit fertilization are accomplished by specific chemical substances. Raper's observation of mold *Achiya* indicate that atleast four ectohormones are involved in causing (i) Differentiation of male and female element. (ii) Their delimination, and (iii) Their eventual fusion.

The chemical which stimulate gametogenesis are often referred as *Termones,* those which are responsible for the coming together are referred as *Gamones. Tylers* has used both the term in his egg-sperm experiment. Fertilizin which is diffuse from egg causes activation and aglutination of the sperms. It is obvious that termones and gamones are diffusion ectohormones.

Difference between Pheromones and Hormones

Sometimes it has been seen that some endocrine product may be active both within an individuals (hormone) and between individuals of a colony (pheromones) according to *Lischer* 1963 hormones from corpora allata of termites function in dual manner.

However, there are some difference between hormones and pheromones.

Hormone	Pheromone
1. Internal secretory product	1. Externally secretory product.
2. Act within the individual	2. Act between the individuals of colony or population.
3. Not species specific (broadely)	3. These are species specific.

It is also known that the production of pheromone by exocrine glands may be under endocrine control as has been demonstrated by *Barth,* 1962 in the cockroach and the *Martan* 1962 in guina pig. Activity of human sebaceous glands is regulated by endogenic substances (*Strauss* and *Pochi* 1963).

Mode of Action of Pheromones

Pheromones act generally by the following modes :

1. Olfaction : Sex attractants-Insect and Mammalian reproductive pheromones.
2. Ingestion : Queen substances secreted by mandibular gland of queen (honey bee) is α ketodecanoic acid. It is ingested by the workers and act to inhibit the development of their ovaries.
3. Contact or absorption : In desert locust male adult produces secretion from their glands, these are absorbed by young nymphs with the result that the young ones have accelerated growth through metamorphosis.

   In lucosid spiders the male's cutaneous gland secrets chemical substances which is absorbed by female's skin and induces and sexual stimulation.

Classification of Pheromones

Classification is based on action path way.

1. Releaser effect: Pheromones exert there effect or influence by directly affecting the recipients central nervous system to produce rapid and reversible changes in behaviour.

   Example : (i) Sex attractants. (ii) Alarm pheromone. (iii) Agression pheromone.
2. Primer effect : By affecting central nervous system an endocrine mostly (pituitary gland) glands and initiate a chain of physiological adjustments which eventually modifies the behavioural patterns, they are slow in reaction.

   Example—Sex pheromones invertebrates queen substances.

   Locust substances.

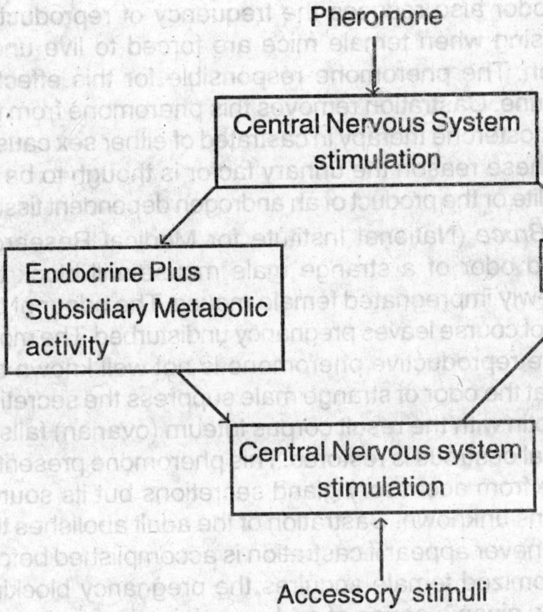

Pheromones influences behaviour

Pheromones influence behaviour directly or indirectly. If a pheromone stimulates the recipient central nervous system into producing an immediate change in behaviour it is said to have a releaser effect. If it alters a set of long-term physiological conditions so that the recipient's behaviour can subsequently be influenced by specific accessory stimuli. The pheromone is said to be have a primer effect.

Primer Pheromones

Pheromones of the categories which have primer effect causes important physiological changes without an immediate accompanying behavioural response *S. Van der Lee* and *L.M. Boot* 1955 (Mammalian Endocrinologist) give Lee Boot effect— "Female mouse placed in groups of four, show an increase in the percentage of pseudopregnencies". Normal, pattern of reproduction can be restored by removing the olfactory bulb of the mice or by housing the mice separately. When more and more female mice are forced to live together their oestrous cycles become highly irregular and in most of the mice the cycle stops for long period. *W.K. Whitten* (Australian National University) give the Whitten effect—The odour of male mouse can initiate and synchronize the oestrous cycles of female mice. The male odor also reduces the frequency of reproductive abnormalities arising when female mice are forced to live under crowded condition. The pheromone responsible for this effect is present in male urine. Castration removes this pheromone from the urine whereas testosterone therapy in castrated of either sex causes it to appear. For these reason the urinary factor is though to be an androgen metabolite or the product of an androgen dependent tissue.

Miss Helen Bruce (National Institute for Medical Research, London) observed odor of a strange male mouse will block the pregnancy of a newly impregnated female mouse. The odour of the original stud male of course leaves pregnancy undisturbed. The mode of action of mouse reproductive pheromone is not well known but the evidence is that the odor of strange male suppress the secretion of hormone prolactin with the result corpus luteum (ovarian) fails to develop and normal oestrous is restored. This pheromone present in bladder urine free from accessory gland secretions but its source and identity remains unknown. Castration of the adult abolishes the pheromone and it never appear if castration is accomplished before puberty. Ovariectomized female acquires the pregnancy blocking capacity if they are given a series of androgen injections.

Mice remain vulnerable to the pregnancy blocking pheromone for about 4 days after coitus, while the zygotes are developing in the oviducts. The blastocysts are moved into uteri but implantation is impossible because the ovaries do not produce the hormone required to build up the kind of an endometrium required for nidation. With pregnancy at an end, the female promptly returns to estrus, ovulates and again becomes sexually receptive. Adjustments of great complexity involving both nervous and endocrine systems intervene between olfactory stimulation and endometrial failure. Hypothalmic stimulation leads to a diminished release of a prolactin by the anterior pituitary and a consequent failure of the ovarian corpora to secrete progesterone. The return to estrus would involve the production of two other pituitory hormone F.S.H. and L.H. The sequence is as :

Stimulated olfactory epithelium	→	Ordinary neurons
of CNS (Neuro humors)	→	Neuro secretory
Cells of Hypothalamus (3 Neuro-hormones as releasing factor)		3 anterior
Pituitary gonadotrophins	→	Ovaries
(Progesterone and Estrogens)	→	Responses to accessory sex organs and behavioural changes.

Releaser Pheromone

Among the pheromones which causes simple releaser effect sex attractants constitute a large and important category.

A. Sex Attractant Pheromone

At present time, four insect and two mammalian sex attractant pheromones are recognised. Their exact role is never been rigorously established experimentally in living animals. In fact mammals seem to employ musk like compounds alone or in combination with other substances to serve several functions, to mark home ranges to assist in territorial defense and to identify the sexes.

The nature and role of 4 insect sex attractants is well known.

Adolf F.J. Butenandt and *Karlson* et. al. (Max Plant Institute of Biochemistry in Munich) extract 12 mg of esters of Bombykol from 250,000 female silk worm moths.

Martin Jacobson, Morton Beroza and *William Zones* (U.S. Department of Agriculture) get 20 mg of Gyplure from 500,000 female Gypsy moths.

Bombykol and Gyplure obtained by extracts of insects and separation is done by chromatographic method. *T. Yamamoto* (U.S. Deptt. of Agriculture) use another technique on American Cockroach. Virgin females were housed in metalcans and air was continuously drawn through the cans and passed through chilled containers to condense any vaporized materials. By this method 12.2 mg obtain from 10,000 females.

Insect attractant pheromones has a great power of diffusion. If volatility and diffusion are taken into consideration, it can be estimated that the threshold concentration is no more than a few hundred molecules per cubic cm. If .01 micro gram of Gyplure (approximate of one female) is distributed with maximum efficiency it excite more than billions male moths.

How the male moth gets the source of tinting? *Iseschwinck* (Munich University) done the experiment, when male moths are activated by the pheromones they simply fly up wind and thus inevitably more towards female. If by accident they pass out of active zone they either abandon the search of fly about at random until they approach the female, there is a slight increase in concentration of pheromone and this serve as a guide for remaining distance.

Female cockroaches produce sex attractant pheromone from one or more anal, sternal and tergal glands. Experiment on cuban cockroach (*Byrootria fumigata*) the corpora allata is necessary for the production of pheromone. In *Periplaneta americana* the oocyte development is controlled by corpora allata so the concomitant endocrine control of pheromone production provides a mating signal during correct period (ootheca is being formed and extruded). The same result is obtained with pycnoscelus-surinemensis.

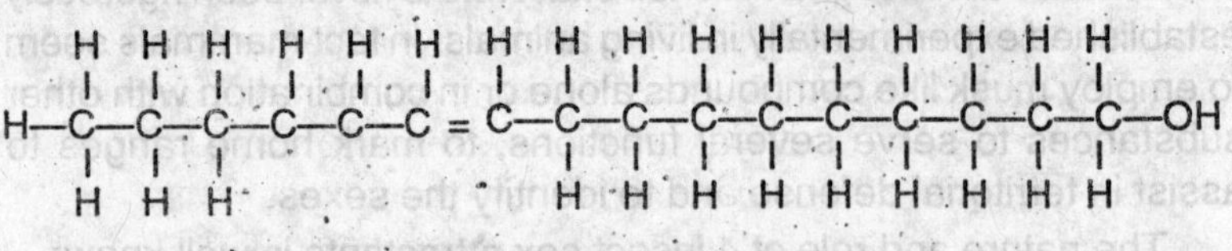

BOMBAY KOL (Silk Worm Moth) $C_{16}H_{30}O$

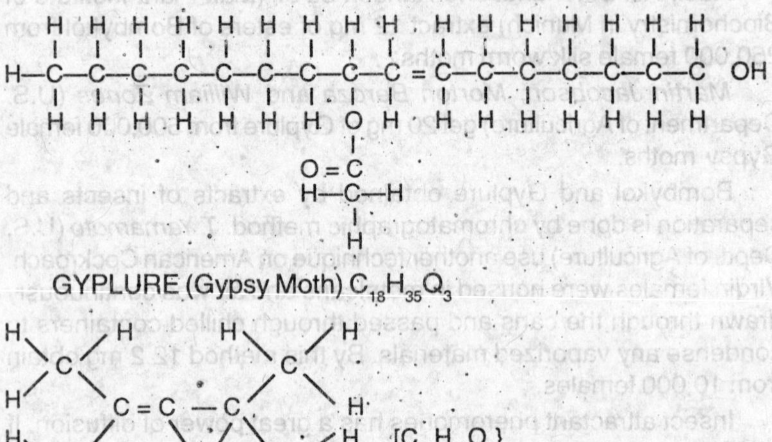

GYPLURE (Gypsy Moth) $C_{18}H_{35}O_3$

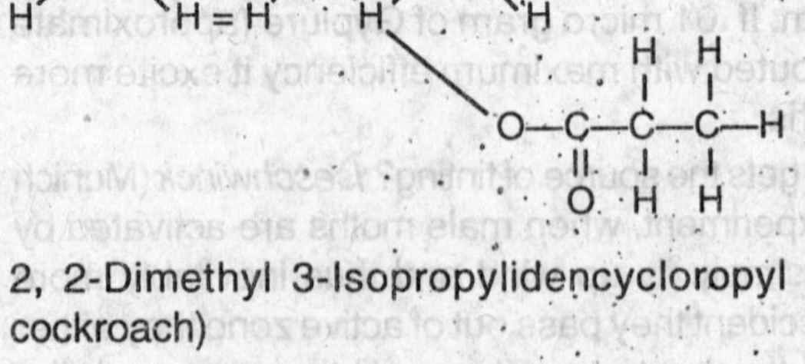

$[C_{11}H_{18}O_2]$

2, 2-Dimethyl 3-Isopropylidencycloropyl propinate (American cockroach)

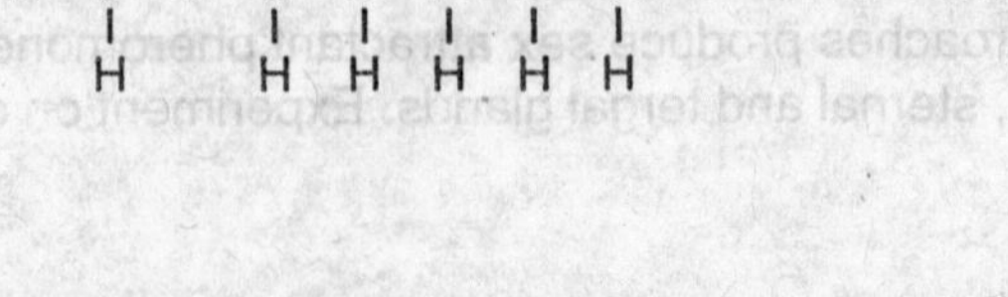

In *Bombyx mori, Antheraea pernyl* (the oak silkworm) *Galleria mellonella* (wax moth) etc. the adult endocrine system controls neither sex pheromone production nor oocyte development. So it is likely that only those insects which have regularly repeated reproductive cycle, with intermittent periods when mating is possible will possess an endocrine mechanism co-ordinating the two events (olfactory communication mechanism).

In many orthoptera (male or female both) produces stridulatory song which is auditory communication mechanism. The signal is sent out at the appropriate time. Chemical signal can often acquire greater versatility than communication by sound.

In the honey bee's (*Apis mellifera*) sex attractant is 9-oxodecenoic acid, in addition to its sex attractant activity, stimulating the Drone's olfactory receptors. The same substance play an important development and behavioural roles within hive, primarily affecting the female worker bees.

$$CH_3—\overset{\overset{\displaystyle O}{\|}}{C}—(CH_2)_5—CH—COOH$$

9-Oxodecenoic acid

$$CH_3—\overset{\overset{\displaystyle OH}{|}}{CH}—(CH_2)_5—CH=CH—COOH$$

9-hydroxy decenoic acid

B. Maturation Pheromone

In the hive only one female is able to reproduce and other females (workers) the oogenesis in inhibited. If the queen is removed from hive.

1. Workers modify one or more of the worker cells into large queen cell.
2. Ovaries of many of workers begin to develop. There is thus behavioural and developmental responses by the worker, to the disappearance of their queen. Queen substance largely composed of 9-oxodecenoic acid it combined with inhibitory scent and inhibits oogenesis. The combined substance is identified as odour of 9-hydroxydecenoic acid.

Maturation pheromones activity can be seen in various animals as termite (*Kalotermes flavicollis*), Desert locust (*Schistocerca gregaria*) etc.

Source of Pheromones

Animals which produce pheromones (bees, wasps, ants termites etc). have the reservoirs.

All kinds of ants have well developed exocrine glandular system. Many of most prominent of these glands, whose functions have long been a mystery to entomologists have now been identified as the source of pheromones.

Trail Substance

Trail substance is a limited guide for the worker ants. It is secreted by the *Dufour's* gland via the sting of insect. It helps as a guide to worker in search of food or nest. The artificial trail can be made by using the extract of Dufour's gland of freshly killed worker. Trail substance is a volatile substance and the range of activity is about 2 minutes.

Alarm Substance

It is also a component of ant pheromone when ant is disturbed than a chemical is liberated from the special glandular reservior. The chemical soon excite other workers with whom it comes in contact. To the human nose the alarm substance is mild or pleasant but to the ant they represent an urgent toxin that can propel a colony into violent and instant action.

Some Examples of Pheromones

Most of compounds are fatty derivative. They conveyed from individual to individual by various ways. The olfactory active substance need air as medium. Alarm substance released from the skin of an injured fish needs water as medium. Orally active pheromones needs hormones like factors among colonial insects, termite substance conveyed by faces.

Pheromones Among Human

Primer pheromones might be difficult to detect since they can affect the endocrine system without producing overt specific behavioural responses. Striking sexual differences have been observed in the ability of humans to small certain substances.

Following 4-Alarm substance are now analysed.

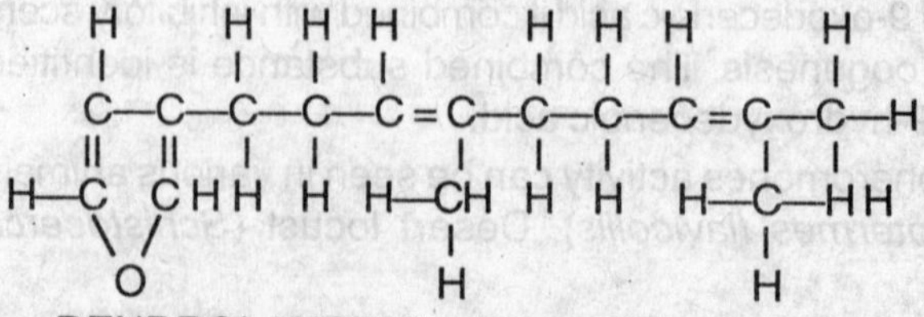

DENDROLANSIN [Lasinus Fuliginosus] $C_{15}H_{23}O$

Table 12.1. Some Examples of Pheromones Among Insects

Species	Manner of effect	Source	Function	Chemistry
Bombyx mori (Silk worm moth)	Olfactory	Lateral gland at the tip of abdomen of female	Attract male of same spp. releases male sex behaviour	$C_9H_{10}O$
Porthetria dispar (gipsy moth)	"	"		$C_{16}H_{34}O_2$ 10-acetoxy 1-hydroxy cis 7 hexadecene
Periplanta americana (cockroach)		Pouch like organ in female	Attract male of same spp. releases male sex behaviour	?
Belostoma indica (tropical bug)	"	Special ducts in male	Sex attractant ?	2-Hexenol acetate
Formic rufa (ant)	"		Marking scent to establish trails	Formic acid.
Pogonomyrmex badius (ant)	"	Mandibular gland	"Schreckst off" communicates alarm	?
Apis mellifera (honey bee)	Oral	Head gland of queen bee	Queen bee substance inhibits ovarian development in workers and prevents development of additional queen	$C_{10}H_{16}O_3$ 9 Keto, 2-decenoic acid
Kalotermes flavitollis (termites)	"	Head of functional female	Inhibits development of supplementary reproductives possibly through endocrine systems.	?

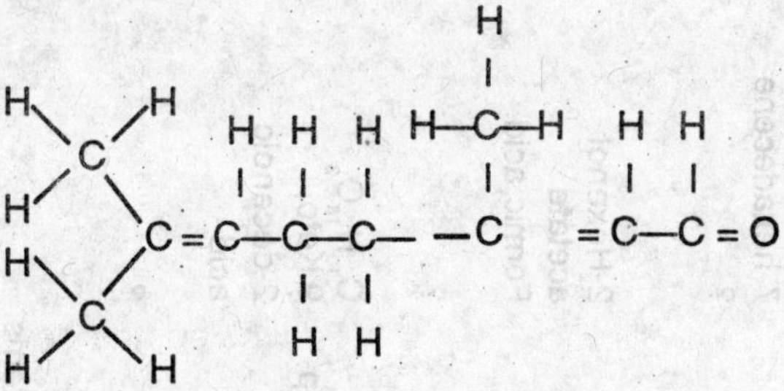

CITRAL [Atla Sexdens] $K_{10}H_{16}O$

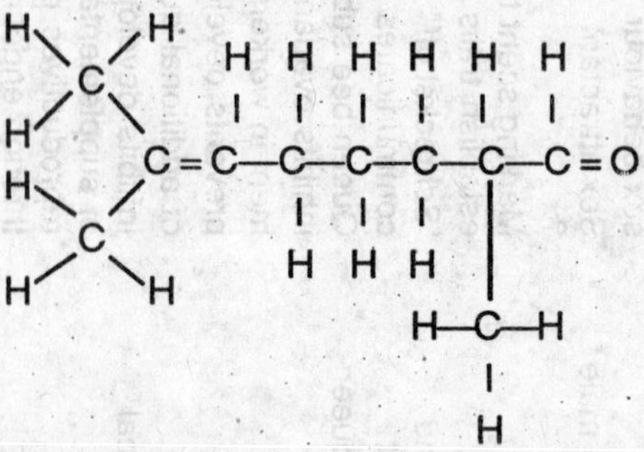

CITRONELLAL [Acanthomyops Claviger] $C_{10}H_{18}O$

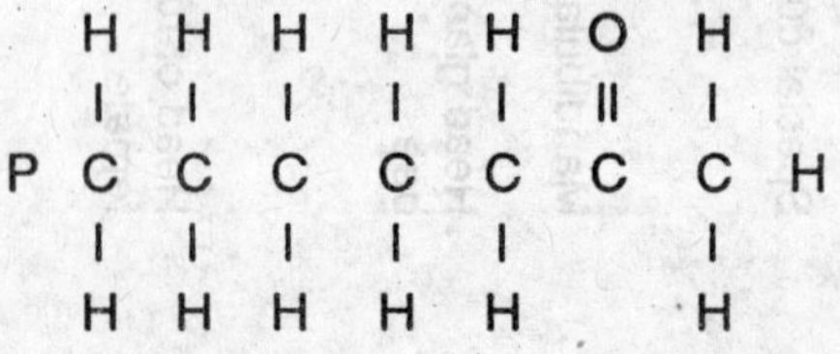

2. HEPTANONE (Iridomyrmex Pruinosus) $C_7H_{14}O$

J. Le Magnen (French) reported that odor of exaltolide (synthetic lactone of 14 hydroxy tetradecenoic acid) perceived clearly by sexually mature mostly at the time of ovulation. Males and young girls were found to be relatively insensitive but a male subject became more sensitive following an injection of estrogen.

Exaltolide is used commercially as a perfume fixative. *Lemagen* also reported that the ability of his subjects to detect the odor of certain steroids paralleled that of their ability to smell exaltolide. These observations hardly represent a case for the existence of human pheromones.

13

SEXUAL SELECTION

Biological fitness is a function not only to an animal's ability to survive but also of its ability to contribute to the gene pool of the next generation and subsequent generations—to reproduce. An organism that copes successfully with its problems of obtaining food and water, finding shelter, and avoiding predators, yet fails to reproduce, still may have a fitness of zero. Reproductive behaviour is of obvious importance to fitness and lies very close to the very definition of species. Because of its obvious importance, and because animal behaviourists generally treat reproductive behaviour as either individual nor social behaviour reproductive behaviour is begin taken under a separate head. Reproductive behaviour possess a set of related questions about mating and the rearing of offspring. Before conception, there may be a struggle among males, and a characteristic (often strange) sequence of behaviour patterns in which the sexes court each other. The term courtship, as used by the ethologists, refers to all the behavioural interactions of the male and female which come before, and lead up to, the fertilization of eggs by sperm. Its form and ostentation vary among species. In some species it does not even exist. In others, such as the stickleback (which we shall consider below) it lasts a few minute. In others it may last for months. Male and female waved albatrosses (*Diomede irrorata*), which live on Isla Espanola of the Galapagos Islands, may court each other, with an extensive repertory of stereotyped movements of the neck and bill, for several hours a day, day in day out for much of the year. The question courtship possess is why it exists, and why, in some species, it has taken on so extravagant a form.

One answer is that courtship ensures that animals male with other individual, of the correct species, sex and condition, Its extreme development, however, of numerous energetic and colourful displays, and extreme structural modification in males, cannot be so easily explained. It is difficult to believe that all the paraphernalia of sex is needed merely to ensure that males are not confused with females or males of another species: that could be achieved much more quickly and less colourfully. Moreover, the most bizarre sexual characteristics, such as the amazing plumage of male birds of paradise, probably decrease the chances of survival of their bearers. They use up energy, hamper flight, attract predators. They are therefore something of a puzzle. They must possess some hidden function to compensate their obvious disadvantages, because if they did not they would have been eliminated by negative selection. Darwin invented a special theory to solve the problem, his theory of sexual selection.

Fig. 13.1. The extinct giant deer Megaloceros giganteus.

Choosing a Member of the Right Species

Matings between members of different species are very rare in nature. For example, in a sample of 725 copulating pairs of two species of cicada, which are distinguishable but similar in appearance to the human eye, only seven contained members of two species; the other 718 were matings of he same species. Different species often do not get a chance to interbreed because they live in different geographic areas, or in different localities within an area (one species at the tops of trees, the other on the ground, for example), or they might be active at different times of day. Of the few species they do have the opportunity to interbreed, none do so frequently. They usually do not because the one species does not respond to the sexual lures of the other.

It is adaptive for animals not to mate with members of other species. Hybrid offspring are usually inferior to those produced by matings of members of the same species, often because they are sterile. The best known example of a sterile hybrid is the mule. The mule is the offspring of a he-ass and a mare; if a she-ass mates with a stallion the result is a 'hinney', which is also sterile. Natural selection will act to prevent animals from mating with members of other species if the offspring so produced would be sterile. Natural selection will favour animals that produce normal healthy offspring by choosing to male with members of their own species, rather than producing sterile hybrid offspring. There have been many investigations of the factors animals use to ensure that they mate with members of the same species. Let us take crickets as an example. They are approximately 3000 species of cricket in the world. They do not all live in the same place, but in any one place there may be more than one, and perhaps half a dozen, species. They songs are produced by males when they rub their wings together, and are heard by the female through the ears on her front legs; they are familiar to us from night-time walks, or the vicarious experience of the cinema. The chirrups of he crickets are calling male. Females are attracted only by the songs, as can be demonstrated by putting out a loudspeaker broadcasting the tape-recording some of a male cricket: females of that species will approach it. Females, moreover, only approach loudspeakers playing song of their own species. They are therefore using the song to choose a mate of the correct species. (Parasitic flies are also attracted to the loudspeaker. Parasitic flies are one of the hazard of the male cricket's sex life; they are attracted to calling males, on whom they lay their eggs. The growing parasitic larvae soon inactivate the male. It is a common risk in many species that by broadcasting to females, a male attracts predators and other enemies).

The all-importance of song in the species recognition of crickets can be shown by another experiment. Female crickets can be trained to walk along a 'Y-maze' on which they have to turn to the left or right. If, for instance, the loudspeaker on the left were playing the song of the female's own species, while the one on the right played the song of some other species, the female turns to the left. The side of the songs can be reversed to control for any bias the female may have for turning in one direction or the other.

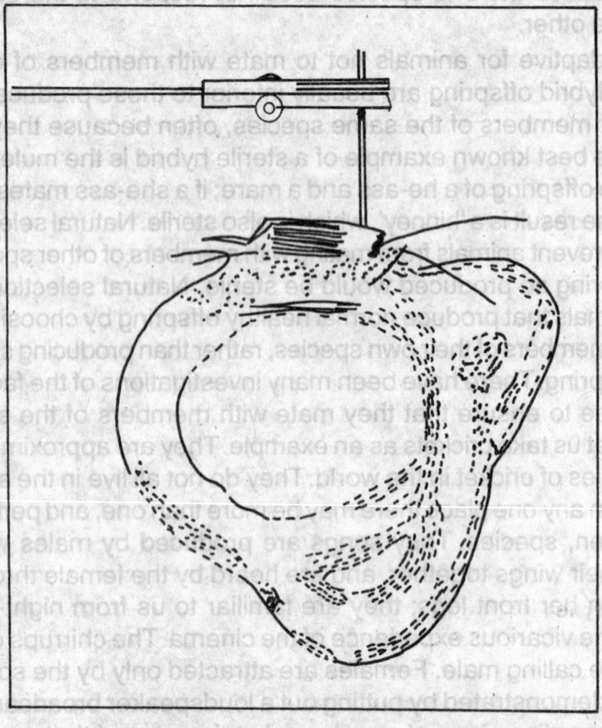

Fig. 13.2. Sexual song in crickets.

The Y-maze has also be used to see how hybrid female crickets behave. If a male and female cricket of two closely related species are forced to mate, some hybrid offspring will be produced when *R. Hov* put these hybrid females on a Y-maze he found that they preferred the songs of hybrid males to those of either parental species. The result is interesting because it bears on the following problem. When a new cricket species evolves, the male song and the female receptor must change in a co-ordinated fashion: a change in one without a change in the other would be disastrous. The fact that the song and the receptor change together in the hybrids suggests that there is some control mechanism which prevents the two form becoming uncoupled form each other; such a mechanism would have the effect of making it more likely that, whenever a male or female cricket changed its song or song-preference during evolution, some other individual of the other sex will have made the complementary change.

Sexual Selection

The success of an individual is measured not only by the number of offspring it leaves but also by the quality of probable reproductive success of those offspring. Thus, it becomes important who its mate will be. *Darwin* (1871) introduced the concept of *sexual selection,* a special process that produces anatomical and behavioural traits that affect an individual's ability to acquire mates. Sexual selection can be divided into two types: one in which members of one sex choose certain mates of the other sex (*intersexual selection*), and a second in which individuals of one sex compete among themselves for access to the other sex (*intrasexual selection*). One result of either type is that the sexes come to be different—that is, *dimorphic.*

Intersexual Selection

Intersexual selection individuals of one sex (usually the males) "advertise" that they are worthy of an investment; then members of the other sex (usually the females) choose among them. Most naturalist after *Darwin* discounted the importance of mate choice in evolution, but it has recently become a popular topic of study. *Fisher* (1958) used birds as an example. Suppose a plumage characteristic in males is lined to some survival or reproductive advantage for the bearer. If females prefer those males as mates, they will have higher reproductive success and will likely produce sons with the trait and daughters that prefer the trait when choosing a mate of their own. Further development of the trait will proceed in males, as will the preference for that trait in the females, resulting in a runaway process. A possible result of such a process is seen in the peacock. Sexually selected traits may become so exaggerated that survival of the males is reduced. Counter selection in favour or less ornamented males will occur—due, for example, to greater predation on he more ornamented males—and sexual selection is brought to a steady state.

As an example of intersexual selection in action, consider the bowerbirds of southeast Asia. Bowerbirds are unique in that they build a structure whose only function seems to be to attract mates. The male stain bowerbird (*Ptilonorhynchus violaceus*) of Australia and New Guinea stations itself alone in the forest, clears a space, weaves the bower from twigs, and decorates it with brightly coloured objects *Gilliard* (1963, 1969) reasoned that bowers function in place of showy plumage, and noted that those species with the fanciest bowers had the drabest plumage. He argues that such behaviour arose to synchronize mating between males and females.

Intrasexual Selection

Intrasexual selection involves competition within one sex (usually males), with the winner gaining access to the opposite sex. Competition may take place prior to mating, as with ungulates such as deer (family Cervidae) and antelope (family Bovidae). Typically, males live most of the year in all-male herds; as the breeding season approaches, males engage in highly ritualized battles, using their antlers or horns. The winners of these battles gain dominance and do most of the mating. Antlers are better developed in those cervid species where males compete strongly for large groups of females (*Clutton-Brock, Guiness,* and *Albon* 1982).

It is often difficult to determine which type of sexual selection is operating to produce an observed effect, since members of both sexes may be present during courtship. The "roughout display" of the male brown-headed cowbird (*Molothrus ater*) consists of a "glunk-chee" sound, made as the bird spreads its tail, lowers its wings, topples forward, and nearly falls over. This display seems to intimidate other males and to attract females. Similarly, the chirp of male field crickets (family Gryllidae) is used to exclude other males from an area and to attract females (*Alexander* 1961).

Females may incite competition among males and thus may gain some control over the choice of mate. For example, female elephant seals (*Mirounga angustirostris*) vocalize loudly whenever, a male attempts to copulate. This behaviour attracts other males and tests the dominance of the male attempting to mate (*Cox* and *Le Boeuf* 1977). In response to the female's sounds, the dominant harem master will drive off low-ranking, potentially inferior mating parents.

Males may fight directly at the time of copulation, as exemplified by the male of the yellow dung fly (*Scatophaga stercoraria*). A male will sometimes attack a mated pair, displacing the male and mating with the female.

Nor does competition among male to sire offspring cease with the act of copulation. Females of many species store sperms and may remate before sperms from the previous mating are used up, creating the possibility of sperm competition (*Parker* 1970b). Sperm competition can be thought of as a selection pressure leading to two opposing types of adaptation in males: those that reduce the chances that a second male's sperm will be used (first male advantage), versus those that reduce the chances that the previous male's sperm will be used (remator male advantage) *Gromko Gilbert,* and *Richmond,* 1984).

Fig. 13.3. A male Bellbird advertising his genetic quality (?) by vocalizing at full volume in the ear of a female that has come to his calling perch in his territory.

First male adaptations include mate-guarding behaviour and deposition of copulatory plugs, both of which reduce the chance of sperm displacement by a second male. In fruit flies (*Drosophila melanogaster*) first males may also transfer to females "anti-aphrodisiac" substances that inhibit courtship by other males (*Jallon,*

Antony and *Benamar* 1981). Female spiders store sperm for long periods and often mate with several males. In laboratory studies of the bowl and doily spider (*Frontinella pyramitela*), *Austad* (1982) found that the first male's sperm had priority in fertilizing eggs over sperm of subsequent males, as is the usual case in spiders.

Among insects, however, the advantage usually seems to accrue to the sperm of the remator male (*Parker* 1970b). In the yellow dung flies he studied, the last male to mate sired 80 per cent of the offspring subsequently produced. The ultimate remator made adaptation may be the penis of the demselfly (*Calopteryx maculata*) which has a dual function: it removes sperm deposited in the female by a previous male via a special "sperm scoop" and then replaces it with its own (*Waage* 1979).

Moths and butterflies (Lepidoptera) produce two kinds of sperms, one kind is the usual type (*eupyrene*) that fertilizes the eggs. The second type (*apyrene*) contains no nuclear material but may comprise more than half the sperm complement. Why should males waste energy on these "dud" sperm? *Silberglied, Shepherd,* and *Dickinson* (1984) suggest that they are the result of sperm competition, possibly

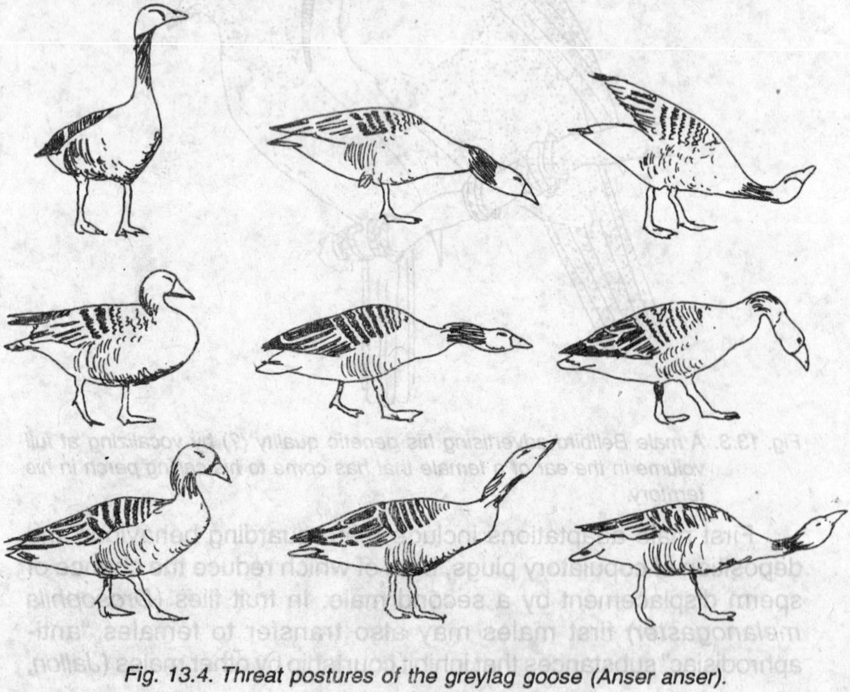

Fig. 13.4. Threat postures of the greylag goose (Anser anser).

displacing eupyrene sperm first males or delaying remating by the female.

Following conception, male-male competition may take a different form. In mice the *Bruce effect* operates early in pregnancy: a strange male (or his odor) causes the female to abort and become receptive (*Bruce* 1966). Among langur monkeys (*Presbytis entellus*) strange males may take over a group, driving out the resident male. The new male may then kill the young sired by the previous male (*Hrdy* 1977). Females who have lost their young soon become sexually receptive, and the new resident male can inseminate them. Infanticide by adult males thus may be reviewed as a "remator male" adaptation.

Timing or Seasonality

Most of animals breeds in a particular time of their life or season, that is why they are usually known as seasonal breeders. In lower animals the timing or seasonality is less understood. Timing in the higher animals is a more complicated affair. Something is known of fish, birds, and mammals of the Northern temperate zone. Reproduction of most of these begins in spring. The first phase is migration towards the breeding grounds. This done by all individuals at approximately the same time, though there may be weeks between the arrival of the first and the last comers. This rough timing is again due not to social behaviour but to reactions to an outside factors. The main factor here is the gradual lengthening of the day in late winter. Various mammals, birds, and fish have been subjected to artificial day-lengthening. The result was that the pituitary gland in the brain began to secrete a hormone which is its turn affected the growth of the sex glands. These then began to secret sex hormones, and the action of these sex hormones on the central nervous system brought about the first reproductive behaviour pattern, migration. Often a rise in the temperature of the environment has an additional effect.

In some cases, the timing process is not very accurate. The different individuals do not all react to the lengthening of the day with the same promptness. There may be a considerable different between the male and the female of a pair. It has been found, in pigeons and in other animals, that if the male is further advanced than the female, his persistent courtship may speed up the female's development. This has been found in the following way. When a male and a female are kept separately in adjoining cages so that

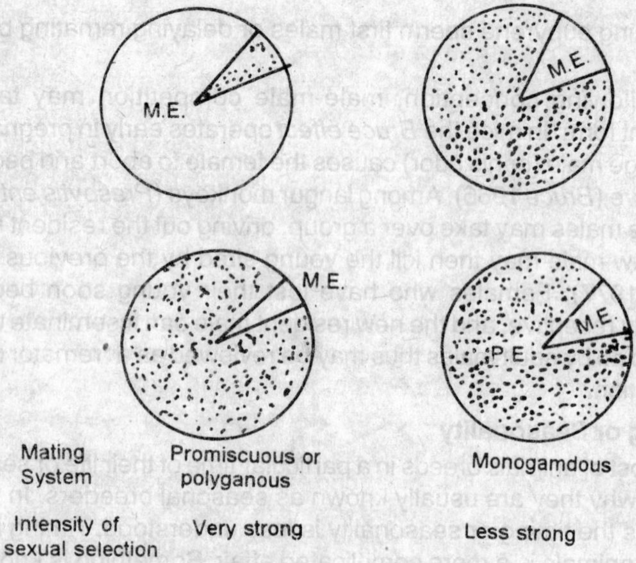

Fig. 13.5. The total resources of time and energy used by an animal in reproduction is referred to as reproductive effort. This is represented by a circle. Reproductive effort can be partitioned into parental effort (provisioning and rearing offspring) and mating effort (acquiring mates). These are represented by the stippled and coloured areas of the circles respectively. In general, males put relatively more into mating effort than do females, but this varies between species. The intensity of sexual selection therefore also varies. The differences in relative parental effort of the sexes is often related to the mating system. In monogamous species male and female effort is more similar than in polygamous and promiscuous species.

they can see and even touch each other, but are prevented from copulating, the persistent courtship of the male will finally induces the female to lay eggs. These of course are infertile. It may occur in captivity, when no males are available, that two female pigeons form a pair. Of these two, one then shows all the behaviour normally shown by the male. And although their reproductive rhythms may have been out of step at the beginning, the final result is that they both lays eggs at the same time. Somehow, their mutual behaviour must have produced synchronization, not merely of behaviour but also of he development of eggs in the ovary.

The Sex Ratio

If one male can fertilize the eggs of dozens of females why not produce a sex ratio of, say, one male for every 20 females? With this ratio the reproductive success of the population would be higher than with 1:1 ratio since there would be more eggs around to fertilize. Yet in nature the ratio is usually very close to 1:1 even when males do nothing but fertilize the female. The adaptive value of traits should not be viewed as being 'for the good of the population', but 'for the good of the individual', or mor precisely 'for the good of the gene.' As *R.A. Fisher* first realized the 1:1 sex ratio can readily be explained in terms of selection acting on the individual; his argument is simple but subtle.

Suppose a population contained 20 females for every male. Every male has 20 times the expected reproductive success of a female (because there are no average 20 mates per male) and therefore a parent whose children are exclusively sons can expect to have almost 20 times the number of grandchildren produced by a parent with mainly female offspring. A female biassed sex ratio is therefore not evolutionarily stable because a gene which causes parents to bias the sex ratio of their offspring towards males would rapidly spread, and the sex ratio will gradually shift towards a greater proportion of males than the initial 1 in 20. But now imagine the converse. If males are 20 times as common as females a parent producing only daughters will be at an advantage. Since one sperm fertilizes each egg. Only one in every 20 males can contribute genes to any individual offspring and females therefore have twenty times the average reproductive success of a male. So a male biassed sex ratio is not stable either. The conclusion is that the rarer sex always has an advantage, and parent which concentrate on producing offspring of the rare sex will therefore be favoured by selection. Only when the sex ratio is exactly 1:1 will the expected success of a male and a female be equal and the populations table. Even a tiny bias favour the rarer sex: in a population of 51 females and 49 males where each female has one child, an average male was 51/49 children. The *average* value is the same whether one male does most of the fathering of whether fatherhood is spread equally among the males.

The argument that the sex ratio should be 1:1 can be refined by re-phrasing it in terms of investment. Suppose sons are twice as costly as daughters to produce because for example, they are twice as big and need twice as much food during development. When the sex ratio is 1:1 as son has the same average number children a daughter. But since sons are twice as costly to make they are a bad investment for a parent: each of its grandchildren produced by a son is twice as costly as one reduced by a daughter. It would therefore pay the parents to concentrate on making daughters. As the sex ratio swings towards a female bias, the expected reproductive success of a son goes up until at a ratio of two females to every male an average son produces twice the number of children produced by an average daughter. At this point sons and daughters give exactly the same return per unit investment: a son costs twice as much to make but yield twice the return.

Table 13.1. In polygamous or promiscuous species males have a much higher potential reproductive rate than females. The data for man came from the Guinness Book of Records: the male was Moulay Ismail the Bloodthirsty, Emperor of Morocco, the woman had her children in 27 pregnancies. The data for elephant seals are from Le Bouef and Reiter, for red deer from Clutton-Brock et. al, (1982). In the monogamous kittiwake, where male and female invest similarly in each offspring, the difference in maximum reproductive output is negligible (Clutton Brock 1983).

Species	Maximum number of offspring produced during lifetime	
	Male	Female
Elephant seal	100	8
Red deer	24	14
Man	888	69
Kittiwake gull	26	28

This means that when sons and daughters cost different amounts to make, the stable strategy in evolution is for the parent to *invest* equal in the two *sexes* and not to produce equal numbers. An example to illustrate this point is *Bob Metcalf's* (1980) study of the sex ratio of two species of wasp; *Polistes meticus* and *P. Variatus.* In the former females are smaller than males, while in the latter they are similar in size. As predicted, the *population sex* ratio is biassed in *P. metricus* and not in *P. variatus*. In both species the investment ratio is 1:1.

The prediction that parents should invest equally in sons and daughters does not always hold, and demonstrations of these deviations from 1:1 investment are among the most convincing pieces of evidence that the sex ratio has evolved in the way suggested by *Fisher.* We will pick out some examples in the following paragraphs.

(a) Local Mate Competition

Fisher's theory predicts a different outcome when brothers complete with each other for mates (so-called '*local mate competition*'). Suppose, for example, that two sons have only one chance to mate and that they compete for the same female. Only one of them can be successful in mating, so from their mother's point of view one of them is *wasted.* This is an extreme example, but it illustrates the general point that when sons compete for mates their value to their mother is reduced. The mother should therefore bias her ratio of investment towards daughters. The exact degree of bias predicted by Fisher's theory depends on the degree of local mate competition. Extreme competition is to be expected in species with limited powers of dispersal (because brothers will stay together in the same place) and therefore, it tends to be associated with inbreeding. In the extreme case of inbreeding, a mother *knows* that all her daughters will be fertilized by her egg sons. The best sex ratio in this instance is to produce just enough sons fertilize the daughters, since any other males will be wasted. The crucial difference between this and the earlier argument for a 1:1 sex ratio is that here the ratio of males to females in the rest of the population does not matter. A female biassed ratio within a brood will not give other parents a chance to benefit by concentrating on sons. An example which supports this prediction is the viviparous mite *Acarophenox* which has a brood of one son and up to 20 daughters. The male mates with his sister inside the mother and dies before he is born.

Jack Werren (1980) has tested the prediction that the degree of bias depends on the extent of local mate competition. He studied the parasitoid wasp *Nasonia vitripennis* which lays its eggs inside the pupae of flies such as *Sarcophaga bullata.* If one female parasitizes a pupa, her daughters are all fertilized by her sons and as predicted the sex ratio of her clutch of eggs is biassed towards females. Only 8.7 per cent of the brood is male. If a second female lays her eggs in the same pupa, what should her sex ratio be? If she lays few eggs she should produce mainly sons, since the first female has laid predominantly female eggs. But as the proportion of the total number of eggs in the pupa that come from the second female increases, the chance that sons of the second female will compete for mates also increases. Therefore, her brood should have a female biassed sex ratio. Werren found exactly this pattern: when the second female's clutch was 1/10 the size of the first female's it contained only males, but when it was twice as large as the female's it contained only 10 per cent males, and the quantitative details of the change in sex ratio with relative brood size were much as predicted.

(b) Local Resource Competition or Enhancement

Anne Clark found that the South African prosimian *Galago crassicaudatus* has a male biassed investment ratio among its offspring. She pointed out that this could be explained by the species' life history. As with most mammals, female *Galago* disperse less far than males, and often end up competing both with their mother and with each other for rich sources of food such as gum and fruit trees in the mother's home range. This local resource competition among females reduces their value as offspring in the extreme case only one daughter might be able to survive on the food available near home, and so investment in other daughters would be wasted.

Exactly the opposite effect could arise if the sex that stays at home, rather than hindering one another or their parents, actually helps. In some bird species, males but not females stay at home and help. The consequence of this is to make males slightly more valuable than females as an investment (since they help the parent in its future reproduction) and hence a male-biassed investment ratio might be expected (*Emlen* et. al. 1986).

(c) Maternal Condition

It has been stated already that male red deer compete for females by prolonged roaring and antler wrestling contests. In these contests it is an advantage for a male to be big, and size depends among other things on how well fed the male was as a youngster, which in turn depends on his mother's ability to compete for good sources of food and hence produce a plentiful supply of milk. In other words, there is a direct link between a mother's competitive ability while lactating and her son's expected reproductive success. Now if a mother *knew* that her son's would be highly successful harem holders it would pay her to invest heavily in sons rather than daughters: the pay-off in terms of grandchildren would be much greater. Similarly, a mother who *knew* that her sons would not grow up to be big and strong would do better to have daughters, since a daughter's future reproductive success does not depend so much on her mother's milk. Exactly this pattern has been found in red deer: dominant females

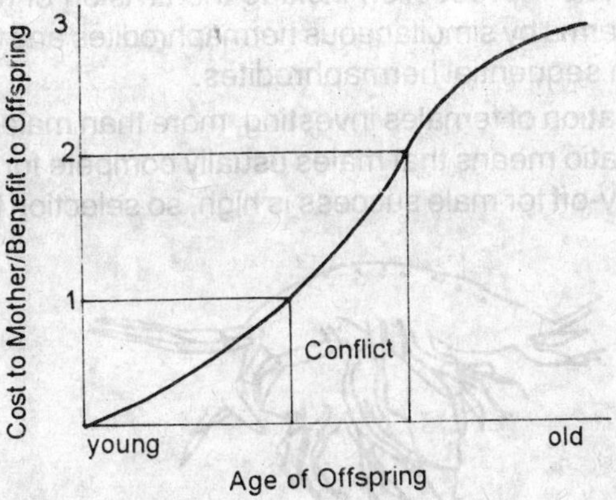

Fig. 13.6. Ratio of cost to mother and benefit to offspring as function of age of offspring.

who are able to gain access to food feeding sites while lactating and hence produce strapping sons. Tend to have sons, while subordinate females have daughters. It is not known how the sex ratio of adjusted by the mother in the red deer or the *galagos* studied by *Clark*, but the fact that they are adjusted is striking because agricultural geneticists have failed to select for adjustment of the say ratio or to separate male and female sperm of domestic mammals (imagine the value of a female biassed sex ratio to the milk farmer!) and it has often been concluded that mammalian sex ratios are very inflexible. Adjustment of the sex ratio in hymenoptera such as the wasps studied by Warren is not at all a problem because the mother can determine whether an egg becomes male or female by whether or not she fertilize it.

(d) Population Sex Ratio

When the population ratio of investment deviates from 1:1 compensatory bias in favour of the rarer sex should occur. In Metcalf's study of *P. metracus* he found that some nests produced only male offspring. These offspring are the product of unfertilized eggs and are produced by workers in nests where the queen has died. In the remainder of the nests in the population *metcalf* found a female biassed sex ratio, so that the ratio of investment in the population as whole is 1:1.

Finally, it is worth pointing out that the theory of sex ratios discussed here is one example of a more general theory of sex *allocation*. Other examples of the problem of allocation or resources to male and female reproduction include the division of resources into eggs and sperms by simultaneous hermaphrodites and the timing of sex change in sequential hermaphrodites.

The combination of females investing, more than males and 1:1 population sex ratio means that males usually compete for females. The potential pay-off for male success is high, so selection for males

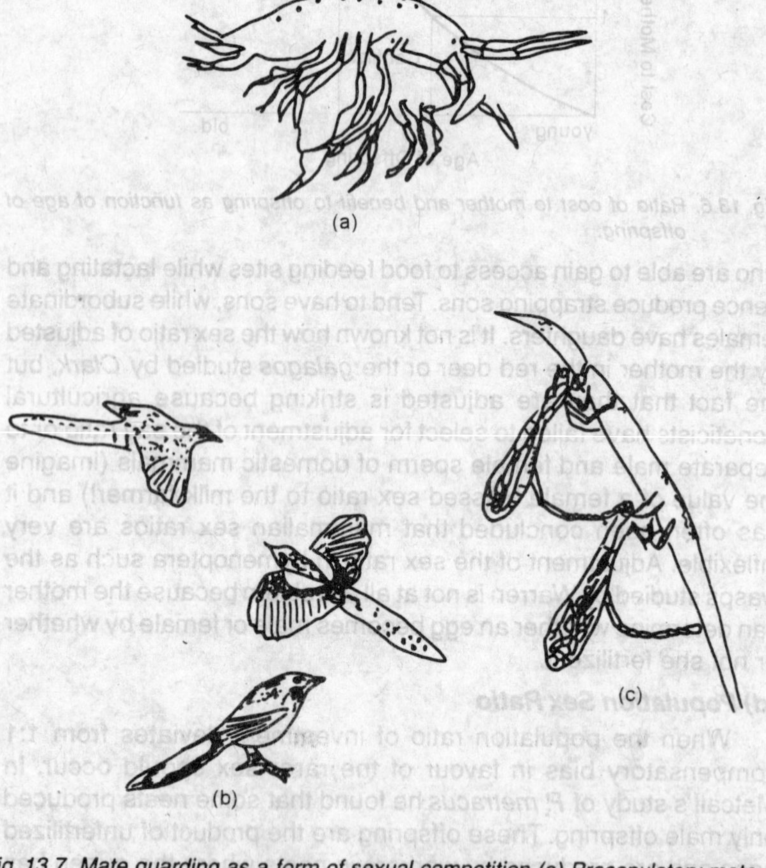

Fig. 13.7. Mate guarding as a form of sexual competition (a) Precopulatory mate guarding in the freshwater amphipod Gommarus. The mature female in this species is ready to be fertilized immediately after she moults. Males guard females in the few days preceding this mould. (b) Male European magpies (Pica pica) assiduously guard their mates against intruding males just before and during the period of egg lying. (c) After copulation the males damselfly guards the female while she lays her eggs, by clasping her thorax with the tip of his abdomen in the 'tandem' position.

ability to acquire matings is very strong. Selection for traits which are solely concerned with increasing mating success is usually referred to as *sexual selection*. It can work in two ways: by favouring the ability of one sex (usually males) to compete directly with one another for fertilizations, for example by fighting (*intrasexual selection*) or by favouring traits in one sex which attract the other (*Inter-sexual selection*). Often the two kinds of selection act at the same time.

The intensity of sexual selection depends on the degree of the difference in parental effort between the sexes and the ratio of males to females available for mating at any one time (referred to as the operational sex ratio). When parental effort is more or less equal, as for example in monogamous birds where both male and female feed the young, sexual selection is less intense than in species will very different levels of parental effort. This follows from the point made earlier that the sex making little investment competes for the sex which makes a large investment. If equal numbers of the two sexes come into breeding condition at the same time, the degree of sexual selection is reduced because there is less chance for a few males to control access to very large numbers of females. In contrast, when females come into breeding condition asynchronously there is a chance for a small number of males to control many females one after the other. With such high potential pay-off, sexual competition is very intense.

Ardent Males

The most dramatic and obvious way in which males complete for mates is by fighting and ritualized contests, and often males have evolved weapons for fighting. Males may dispute over direct access to females or over places where females are likely to go, as for example, when male speckled wood butterflies defend patches of sunlight. Fighting is often a risky business, as illustrated by the injuries sustained by red deer stages. The most intensive fights in many species occur when females are ready to be fertilized and once a male finds a female he often guards her.

Males of the compete in ways which are less conspicuous than fights, but are no less effective and often more bizarre. The invertebrates are a particularly rich seam of examples. Females dragonflies, as with many other insects, mate with a number of males and store the sperm in a special sac (the spermatheca) in the body for use at a later date. The males compete for fertilizations by trying to ensure that previous sperm is not use by the female. The penis of male *Ortheturm cancellatum* is equipped with barbed whip at the end which is used to scrape out of the female any sperm left by previous males before he inject new sperm into the sperm previous males before he inject new sperm into the sperm previous males before he injects new sperm into the sperm sac. *Croethemis erythraea*, another dragonfly, uses an inflatable penis with a horn-lime appendage to peak the sperm of previous males into corners of the spermatheca.

In some invertebrates (especially insects) the male cements up the female's genital opening after copulation to prevent other males from fertilizing her. The male of *Moniliformes dubius*, parasitic acanthocephalan worm in the intestine of rats, produces a chastity belt of this kind but in addition to sealing up the female after copulation, the male sometimes 'copulates' with rival males and applies cement to their genital region to prevent them from mating again. No less remarkable are the habits of the hemipteran insect *Xylocoris maculipennis*. In normal copulation of the species the male simply pierces the body wall of the female and inject sperm, which then swim around inside the female until they encounter and fertilize her eggs. As with the acanthocephalan worms, males sometimes engage in homosexual *copulation*. A male *Xylocoris* may inject his sperm into a rival male. The sperm then swim inside the body to the victim's testes, where they wait to be passed on to a female next time the victim mates.

Competition between males to prevent each other's sperm from fertilizing eggs is sometimes referred to as 'sperm competition.' The sperm of a second male displaces that of the first male to mate with a female dungfly. Sperm competition also occurs in vertebrates. For example, during courtship male salamanders and newts deposit little sperm-capped rods of jelly (spermatophores) on the bottom of the pond and then try to manouever the female onto the spermatophore to achieve fertilization. In the salamander *Ambystoma maculatum* males compete by depositing their spermatophores on top of those of other males. The top spermatophore is the one that fertilizes the female's eggs.

The fourth example of the arcane methods of male-male competition found among invertebrates is the use of anti-aphrodisiac smells. *Larry Gilbert* noticed that female *Heliconius erato* butterflies always smell peculiar after they have mated. He was able to show experimentally that the scent does not come from the female herself, but is deposited by the male at the end of mating. *Gilbert* also found that the scent discourages other males from mating with the female, perhaps because it resembles a scent used by males to repel one another in other contexts.

Reluctant Females

Since females in the great majority of species are the chief provides of resources for the zygote; they might be expected choose their mates carefully in order to get something in return. To put it another way, each egg represents a relatively large proportion of a female's lifetime production of gametes when compared with a sperm, so the female has more to loss if something goes wrong. Mating with the wrong species could cost of a female frog her whole year's supply of eggs, but would cost the male very little apart from lost time—he could still go on to mate successfully with a member of the correct species the next day. Not surprisingly therefore, females are on the whole choosier than males during courtship. Choosiness extends not only to discriminating between species, but also to discriminating between males within a species. Females offer and perhaps sometimes to obtain good genes for their offspring.

(a) Good Resources and Parental Ability

In many animals species males defend breeding territories containing resources which play a crucial role in the survival of a female's eggs or young. For example male North American bullfrogs (*Rana catesbiana*) defend territories in ponds and small lakes where females come to lay their eggs. Some territories are much better for survival of eggs than other and there are the ones which females prefer. One factor which has an important influence on survival of eggs is prediction by leeches (*Macrobdella decora*). Two environmental features of a territory influence leech predation: if the water is warm the eggs develop faster and are therefore exposed to predation for fewer days, and if the vegetation in the water is not too dense the eggs can form into a ball which the leeches find hard to attack. In territories with a dense mat of vegetation the eggs lie in a thin film on top of the plants and more easily attacked. The bullfrogs also show that female choice and male-male competition may go hand in hand. The preferred territories are hotly contested by males and the largest, strongest frogs end up in the best sties.

Food is a resource which often limits a female's capacity to make eggs and during courtship females may choose whether or

Fig. 13.8. Sexual selection in male bull frogs. Males complete by wrestling and calling (left and middle) for good territories, in which the females prefer to lay their eggs (right). The good territories have high survival of eggs because they are warm and because the vegetation is not tool dense.

not to mate with a male on the basis of his ability to provide food. In some birds and insects for example, males may provide food for the female during courtship ('courtship feeding') which makes a significant contribution to her eggs. Female hanging flies (*Hylobitacus apicalis*) will mate with a male only if he provides a large insect for her to eat during copulation. The larger the insect, the longer the male is allowed to copulate and the more eggs he fertilizes. The female gains from a large insect by having more food to put into her eggs. In birds, the male usually helps to feed the young and courtship feeding may play the additional role of indicating to common term (*Stern fuscata*) there is a correlation between the ability of the male to bring food during courtship feeding and his ability to feed the

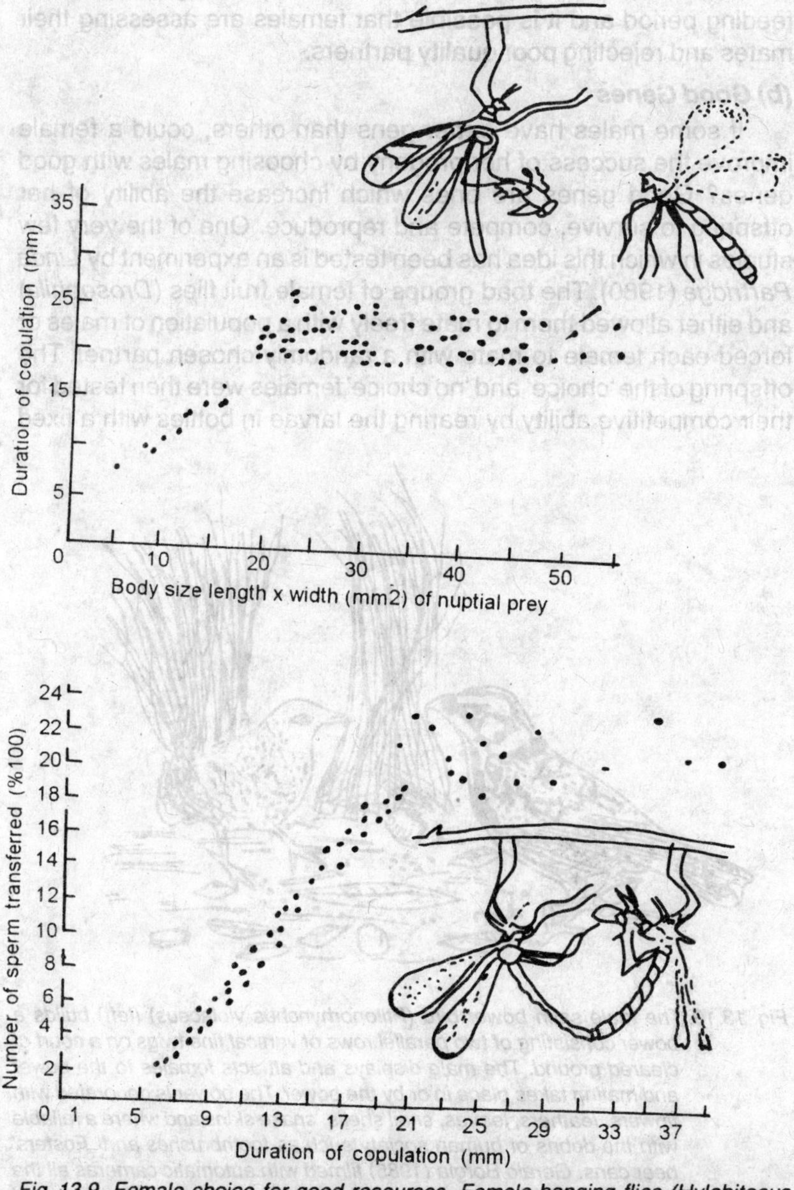

Fig. 13.9. Female choice for good resources. Female hanging flies (Hylobitacus apicalis) mate with males for longer if the male brings a larger grey item to eat during copulation. The male benefits from long copulation because he fertilizes more eggs.

chicks later in the season. Pairs often break up during the courtship feeding period and it is possible that females are assessing their mates and rejecting poor quality partners.

(b) Good Genes

If some males have *better* gens than others, could a female improve the success of her progeny by choosing males with good genes? Good genes are ones which increase the ability of her offspring to survive, compete and reproduce. One of the very few studies in which this idea has been tested is an experiment by *Linda Partridge* (1980). The toad groups of female fruit flies (*Drosophila*) and either allowed them to mate freely with a population of males or forced each female to mate with a randomly chosen partner. The offspring of the 'choice' and 'no choice' females were then tested for their competitive ability by rearing the larvae in bottles with a fixed

Fig. 13.10. The male satin bower bird (Ptilonorhynchus violaceus) (left) builds a bower consisting of two parallel rows of vertical fine twigs on a court of cleared ground. The male displays and attracts females to the bower and mating takes place in or by the bower. The bower is decorated with flowers, feathers, leaves, snail shells, snake skins and where available with the debris of human society such as toothbrushes and Fosters' beer cans. Gerald Borgia (1985) filmed with automatic cameras all the matings at 22 bowers and found that males with more decorations get more females (snail shells, blue feathers and yellow leaves were especially attractive). In part a male's decoration reflects his competitive ability, since males steal decorations from each other's bowers.

number of standard competitors (these were distinguishable by a genetic marker). *Partridge* found that the offspring of the *choice* group did slightly but consistently better than those of *no choice* females in the larval competition experiments. This experiment suggest that females are able to increase the survival of their offspring by choosing good genes in their mates, but it must be born in mind that the results could also be in part explained by intrasexual competition: in the *choice* experiment, the males that mated may have been superior competitors against other males.

The theory of sexual selection is most famous as an attempt to explain the evolution of excessively elaborate adornment and displays of male peacocks, pheasants, birds of paradise and so on. Some elaborate displays may have evolved for use in contests between males, but the most widely accepted (and still controversial) account of how such traits evolve depends on selection by females for good genes. In this case, however, the genes are not ones which increase the ability of a female's offspring to survive and compete (in fact the bizarre adornments usually seem to be a hindrance to survival) but rather for genes which increase the sexual attractiveness of her sons.

R.A. Fisher was the first to clearly formulate the idea that elaborate male plumage may be sexually selected simply because it makes males attractive to females. This may sound circular, and indeed it is, but that is the elegance of Fisher's argument. At the beginning, he supposed, females preferred a particular male trait (let us take long tails an art example) because it indicated something about male quality. Perhaps males with longer tails were better at flying and therefore collecting food or avoiding predators. An alternative starting point is to suppose that longer tails were simply easier for females to detect and that they therefore attracted move mates. For example, male natterjack toads *Bufo calamita* with louder croaks attract more females simply because females locate males by moving passively down sound gradients. If there is some genetic basis for differences between males in tail length the advantage will be passed on to the female's sons. At the same time, a gene which causes females to prefer longer than average tails will also be favoured since these females will have sons better able to fly or more readily detected by potential mates. Now once the female preference for longer tails starts to spread, longer tailed males will gain a double advantage: they will have sons that are both good fliers and attractive to females. As the positive feedback between female preference and longer tails develops gradually the benefit of attractive sons will become the more important reason for female choice, and the favoured trait might eventually decrease the survival ability to males. When the decrease in survival counter-balances sexual attractiveness, selection for increasing tail length will grind to a halt.

Fisher's verbal argument has been formalized and extended in population genetics models by *Russell Lande*. His analysis underlines the importance of the assumption, implicit in Fisher's argument, that the genes causing variation in preference and in tail length become associated with one another, or in other words show covariance.

Although 'tail length' is the classic textbook trait in discussion of sexual selection it is in fact only recently that evidence for sexual selection of tail length in real life has been collected. *Malte Anderson* showed that females of the long-tailed widow bird *Euplectes progne* in Kenya prefer males with long tails. This highly polygynous species is an ideal candidate for sexual selection; the male is a sparrow-sized bird with a tail up to 50 cm long. The female's tail is about 7 cm long, presumably close to the optimum for flight purposes. *Anderson* studied 36 males which he divided into four groups. In one group he docked the tails to about 14 cm, while in another group he attached the severed bits of group 2 males by an average of 25 cm. The remaining two groups were controls: one lot were left untouched and the others had their tails cut and glued without altering the length. By counting the number of nests in each territory, *Anderson* showed that before his experimental manipulations there was no difference in mating success of the different groups, while afterwards the long tailed males did significantly better than the controls or the shorter tailed birds.

Another nice experimental study of a sexually selected elaborate display is that of *Clive Catchpole* on the song of the European sedge warbler. The song consists of a long stream of almost endlessly varying trills, whistles and buzzes and is sung by the male after arriving back on the breeding territory from the winter quarters: as soon as a male pairs, it stops singing. Catchpole's measurements showed that the males with the most elaborate songs are the first to acquire mates, as expected from Fisher's hypothesis of sexual attractiveness. Further, when female warblers were brought into the

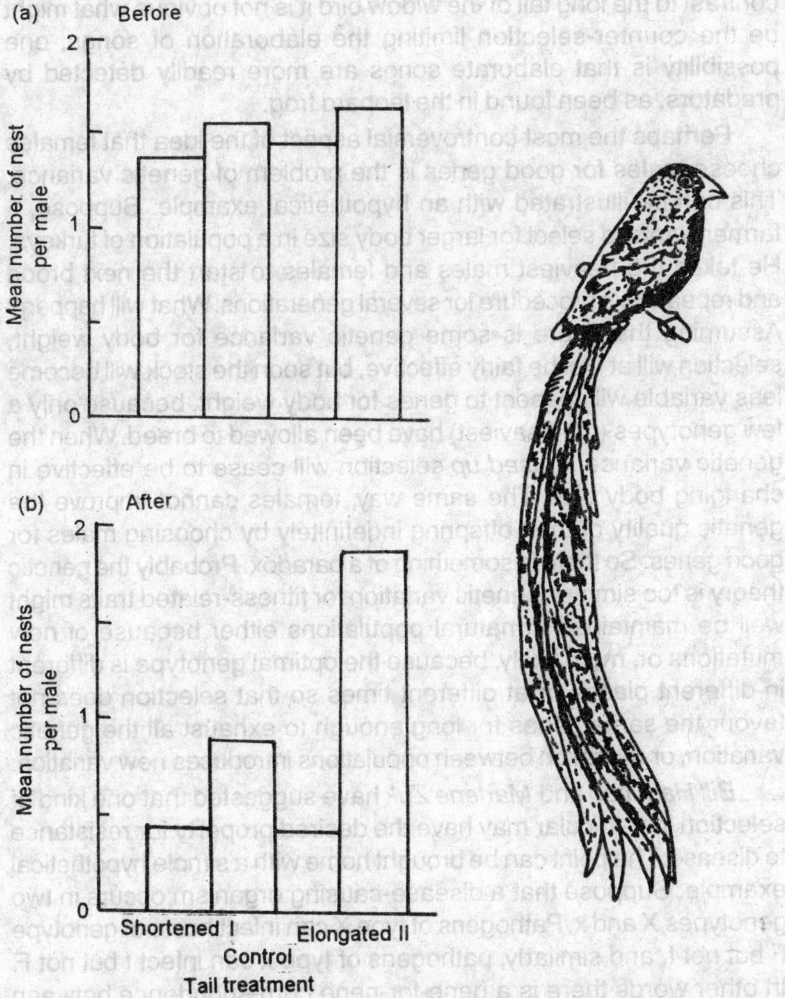

Fig. 13.11. Sexual selection for tail length in long-tailed widow birds. The top line shows that there was no difference between the four group before the tails were altered. The bottom lines shows that after the tails were cut and lengthened the mating success went down and up respectively. The two kinds of control birds were (I) unmanipulated, and (II) cut and glued back without altering length. Mating success is measured as the number of active nests in each male's territory.

laboratory and treated with oestradiol to make them sexually active, they were more responsive to large than to small repertoires. In contrast to the long tail of the widow bird it is not obvious what might be the counter-selection limiting the elaboration of songs; one possibility is that elaborate songs are more readily detected by predators, as been found in the leopard frog.

Perhaps the most controversial aspect of the idea that females choose males for good genes is the problem of genetic variance. This can be illustrated with an hypothetical example. Suppose, a farmer wants to select for larger body size in a population of turkeys. He takes the heaviest males and females to start the next brood and repeats this procedure for several generations. What will happen? Assuming that there is some genetic variance for body weight, selection will at first be fairly effective, but soon the stock will become less variable will respect to genes for body weight, because only a few genotypes (the heaviest) have been allowed to breed. When the genetic variance is *used up* selection will cease to be effective in changing body size. The same way, females cannot improve the genetic quality of their offspring indefinitely by choosing males for good genes. So there is something of a paradox. Probably the genetic theory is too simple. Genetic variation for fitness-related traits might well be maintained in natural populations either because of new mutations or, more likely, because the optimal genotype is different in different places or at different times so that selection does not favour the same genes for long enough to exhaust all the genetic variation, or migration between populations introduces new variation.

Bill Hamilton and *Marlene Zuk* have suggested that one kind of selection in particular may have the desired property for resistance to disease. The point can be brought home with a simple hypothetical example. Suppose that a disease-causing organism occurs in two genotypes X and x. Pathogens of type X can infect hosts of genotype F but not f, and similarly, pathogens of type x can infect f but not F. In other words there is a gene-for-gene correspondence between pathogenicity and susceptibility to the disease. Now let us start with X as the common form of the disease and F as the common host genotype. Because of its resistance to X, f is at a strong selective advantage and will become commoner, but as f increases in frequency so disease type x gains an advantage relative to X. This in turn swings the pendulum back in favour of host F and thence pathogen X and so on. It is easy to imagine (and imagination has been made respectable by computer simulations) that the host-pathogen interaction will produce endless cycles of genetic change with constant pressure on both sides to keep ahead or catch up with the other. What does this have to do with sexual basis of their resistance to disease, resistance which would be passed on to progeny and help in the never-ending race to keep ahead. As *Zuk* (1984) points out aviculturalists and poultry breeders can readily tell a sick bird by looking at its plumage (for example, a turkey suffering from erysipelas has a swollen purple snood), so why should females have not evolved a similar ability? Seen in this light, courtship displays and their associated body structures are a sort of medical examination by which one partner assesses the other.

As yet there is no evidence that females of any species are skilled diagnostic veterinarians, but *Hamilton* and *Zuk* have some evidence that is consistent with their idea. They argue the sexual selecting for external features indicating vigour and disease resistance will have been strongest in those species that are exposed to the most disease (simply because in these species it pays the female to scrutinize her partner closely). Therefore, if traits like the showy plumage and elaborate courtship songs of male passerine birds have evolved as indicators of disease resistance, there traits should be most highly developed in the most diseased species. In a survey of the blood parasites, plumage and song of 109 North American passerine bird species that is generally what *Hamilton* and *Zuk* found.

The Handicap Principle

Amotz Zahavi has suggested an alternative view of elaborate males displays. He points out that the long tail of the peacock is a handicap in day to day survival, a fact which few would dispute. He then goes on to suggest that longer tails are preferred by females precisely because they are a handicap and therefore act as an advertisement of male quality. The tail demonstrates a male's ability to survive in spite of the handicap, which means that he must be extra good in other respects and is able to pass on any genetic component of this quality to his sons and daughters. The difficulty with this argument is that the offspring inherit not only the *good genes* but also the handicap, and under most conditions they will be better off without either than with both. For this reason, Zahavi's idea is unlikely to work in its original form, but if we modify it to say that the handicap is not a fixed trait but is simply conditional upon male vigour its major problem disappears. In this modified version, the size of male's handicap is a direct index of his current vigour, as in *Hamilton* and *Zuk's parasite hypothesis* where a male's plumage or other displays indicate his degree of disease resistance. In fact, the idea that the intensity of a male's sexual displays are not genetically fixed (as in Fisher's hypothesis) but are simply a phenotypic expression of his current condition, can be extended beyond the handicap principle. Any display or trait (antler length, croaking rate, and so on) whether or not it is a handicap, could be a reliable index of a male's well-being and be used in assessment by females or rival males.

Male Investment

We have so far assumed that females are investors and males are competitors. While this picture describes most animal species, thee are exceptions. In many birds, some amphibians, and arthropods, both male and female invest about equally in the eggs or young by feeding, guarding or brooding. Sometimes the usual sex roles are completely reversed so that males do the investing and females the competing. The ideas about sexual conflict and sexual selection can still be applied in modified form to species with equal or primarily male investment. When both sexes are equally for the offspring, for example, courtship may involve assessment and choice by males as well as by females. Males of species with internal fertilization can never be absolutely sure that they have fathered the children of their mate, and one role of courtship may be as an insurance against cuckoldry. A prediction of this idea is that courtship allows males to assess whether or not females have previously mated with others. This was tested by Erickson and Zenone. They found that male barbary doves (*Streptopelia risoria*) attack a female instead of courting her if she performs the *bow posture* (an advanced stage of courtship) too quickly. Since the females which responded in this way had been pretreated by allowing them to court with another male, the reaction of the test males in rejecting eager females is adaptive if courtship plays a role in assessing certainly of paternity, before investing in offspring. It would not have been predicted by the older view that male courtship serves to sexually arouse the female!

In species with high male investment, females tend to be the competitive sex and males may be choosy. In moorhens (*Gallinula chloropus*) males do almost three-quarters of the incubation and females play an active role in competing for the chance to mate with good incubators. These ideal husbands are small and fat: well equipped to survive on their reserves during long incubating stints.

In other species, investing males may actually reject low quality females.

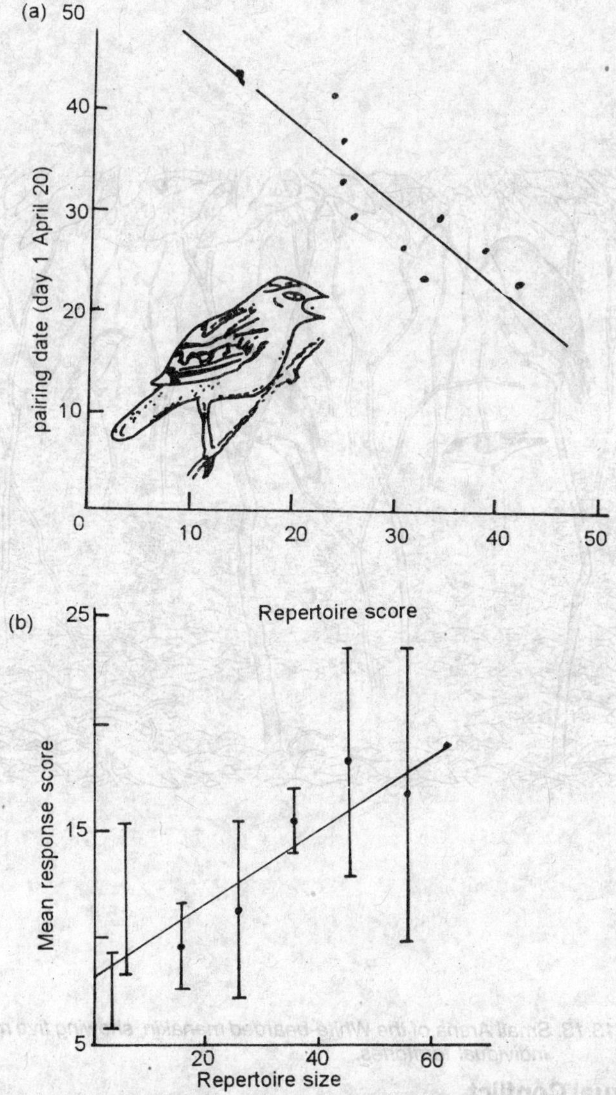

Fig. 13.12. Male sedge warblers with the largest song repertories are the first to acquire females in the spring. The size of song repertoire is estimate from sample tape recordings of each male. The results were collected in such a way as tc control for the possibilities that older males, or males in better territories, both mate first and have larger repertoires. (b) The mean ± s.e. response score of 5 females to repertoires of different sizes. The response score measures sexual behaviour.

Fig. 13.13. Small Arena of the White-bearded manakin, showing five males at their individual territories.

Sexual Conflict

Recall the view of the origin of anisogamy as the primeval example of sexual conflict. The conflict was one about mating decisions. Macrogametes might have done better had they been able to discriminate against microgametes, but in the evolutionary race microgametes won. Similar, but more directly observable, conflicts of interest between the sexes are still apparent today, not only with respect to mating decisions but also in the context of parental investment, multiple matings and infanticide.

(a) Mating decisions

As we have emphasized, females have more to lose and therefore tend to be choosier than males. Thus for a given encounter it will often be the case that males are favoured if they to mate and females if they do not. An extreme manifestation of this conflict is enforced copulation as exemplified by scorpionflies (*Panorpa* spp.).

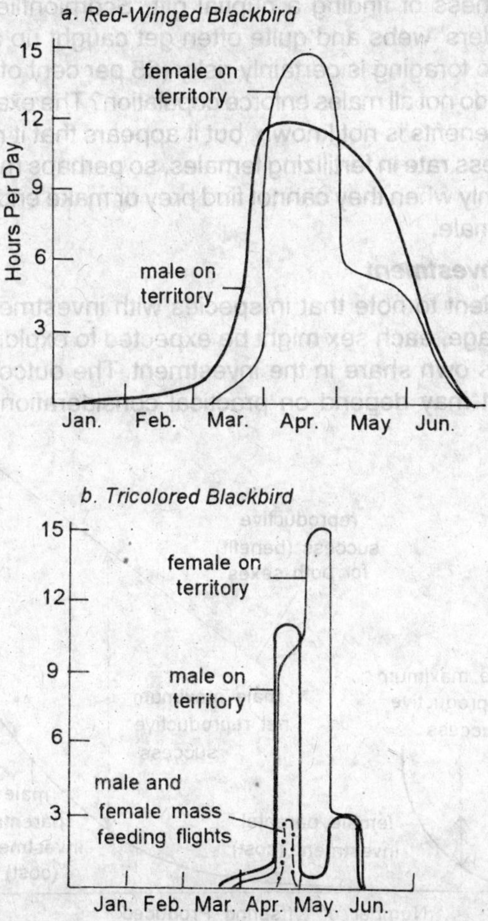

Fig. 13.14. Time expenditure for pair of red-winged blackbirds and tricolored blackbirds during breeding season.

Male scorpionflies usually acquire a mate by presenting her with a nuptial gift in the form of a special salivary secretion or a dead insect (this is very similar to the *Hylobittacus* described earlier). The female feeds on the gift during copulation and turns the food into eggs. However, sometimes a male enforces copulation: he grasps her with a special abdominal organ (the notal organ) without offspring a gift. Enforced copulation appears to be a case of sexual conflict. The female loses because she obtains no food for her eggs and has to search for food herself, while the male benefits because he avoids the risky business of finding a nuptial gift. Scorpionflies feed on insects in spiders' webs and quite often get caught up in the web themselves, so foraging is certainly risky (65 per cent of adults die this way). Why do not all males enforce copulation? The exact balance of costs and benefits is not known, but it appears that it results in a very low success rate in fertilizing females, so perhaps males adopt this strategy only when they cannot find prey or make enough saliva to attract a female.

(b) Parental investment

It is sufficient to note that in species with investment beyond the gamete stage, each sex might be expected to exploit the other by reducing its own share in the investment. The outcome of this sexual conflict may depend on practical considerations such as

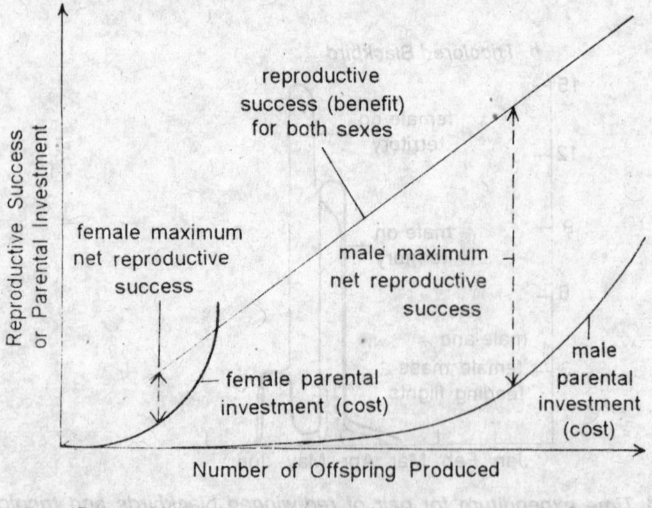

Fig. 13.15. Parental investment and reproductive success as a function of number of offspring produced.

which sex is the first to be in a position to desert the other. When fertilization is internal, for example, a male has the possibility of deserting the female immediately after fertilization and leaving her with the egg or young to care for.

(c) Infanticide

Male lions may slaughter the cubs in a pride shortly after they take over as group leaders. The behaviour (which is also seen in some primates) probably increases male reproductive success, and clearly decreases female success. This seems to be a case of sexual conflict in which the males have won, but it is perhaps surprising that females have not evolved counter-adaptations. They could, for example eat their own young once they have been killed in order to recoup as much as possible of their losses.

(d) Multiple matings

As Bateman's experiments with *Drosophila* showed females often gain little by mating with more than one male. However, because of sperm competition males gain by mating with already fertilized females. Multiple matings are likely to be costly to the female at the same time as being advantageous to the male. This dramatically illustrated by the dungflies. When two males struggle for possession of a female, the female is sometimes drowned in cowdung by the fighting males on top of her!

Conflicts of interest between the sexes will lead to an evolutionary race of the sort envisaged by *Parker et. al.* for sperms and eggs. There is no simple answer to the question "Which sex is more likely to win the chase?" As we discussed earlier, factors such as the strength of selection and the amount of genetic variation will determine how fast the two sexes can evolve adaptation and counter-adaptation, but it is not possible to make any more specific statements about the outcome of sexual conflict races.

The Significance Courtship

Some aspects of courtship behaviour can be interpreted in terms of sexual conflict and sexual selection. However, this is not true of all courtship signals; many are designed for species identification, and here the interests of the two sexes are similar because both benefit by mating with a member of the same species. Some of the clearest examples of this role of courtship come from studies of frog calls. When several species of frogs live in the same pond, each has a characteristic and distinct mating call given by the male, and females are attracted only to calls of their own species. In some frogs (e.g., the cricket frog *Acris crepitans*) it has been shown that the female's selectivity of response results from the fact that the auditory system is turned to the particular frequencies in the male call.

Courtship displays may also play a role in competition between males within a species for mating opportunities. Often the same displays simultaneously serve to repel other males and attract females. An example for which this has been demonstrated experimentally is the mating call of the pacific tree frog (*Hyla regilla*). Males are repelled and females attracted by loudspeakers broadcasting the mating call and females select out of a group of loudspeaker the one which calls for the longest bouts. Females may choose between displays purely on the basis of sexual attractiveness, as explained by Fisher's theory of sexual selection, but there is also the possibility that differences in courtship between males many indicate habitat quality, for example males with territories containing a lot of food might be able to afford to spend more time displaying.

A third role of courtship to which we have referred is assessment. In a species with male parental care females may assess the ability of the male to look after young and males may assess whether a female has previously been fertilized. Early work by ethologists on birds and fish showed that at the beginning of courtship males are often aggressive and females are coy or reluctant. Courtship was seen, therefore, as serving to synchronize sexual arousal of the partners. A possible explanation of why it should be necessary to overcome aggression and reluctance is that the early phases involve assessment by both partners before investing in offspring.

Mating Systems

Despite the argument, based upon relative parental investment by the sexes, that polygyny should be the normal mating system in animals, monogamy is common in some groups for example, in 90% of birds. Parental investment in many birds, however, is almost as great in the males as it is in the female. Young altricial birds in nests are very vulnerable to predation and survive best if they can develop rapidly. If the female alone can seldom rear two young, it is advantageous for males to help with parental care. *Armstrong* (1955) suggested that wrens (*Troglodytes troglodytes*) are monogamous, when both parents are needed to collect enough food for the young to develop. In long-lived birds, long-term mate fidelity seems to result in great efficiency of reproduction due to rapid mating and effective co-operation between mates (*Coulson* 1966, *Davis*, 1976). *Crook* (1964), who compared 90 species of weaver bird, *Ploccinae,* found that the primarily insectivorous species which inhabit forests are monogamous whereas savannah-dwelling seed-eaters are colonial and polygynous. The monogamous species defend and forage in territories while the seed-eater living in flocks and probably obtain food more readily hence the males do not have to feed the young.

Habitat quality may affect mating systems. *Verner* (1964) found that female long-billed marsh wrens (*Telmatodytes palustris*) would mate with a male which was already mated, rather than with an unmated male, if the quality of the territory of the first male was much better than that of the second. *Verner* and *Willson* (1966) therefore proposed that there is a polygyny threshold in habitat quality. This idea was developed by *Orians* (1969) who quotes supporting evidence from his own studies of red-winged *Agelaius phoeniceus* and yellow headed *Xanthocephalus xanthocephalus* black birds. *Orians* proposes that polygyny should be commoner in habitats where food is locally abundant, especially if the number of nest-sites is restricted and that it should be prevented in precocial birds. In such situations, genes which have the effect of favouring polygyny if the male himself, or the territory which he controls, is of high enough quality, should spread in the population. It seems that this has happened in habitates where there is a period of rapid resource increase. In addition to marsh and savannah dwellers, such as those

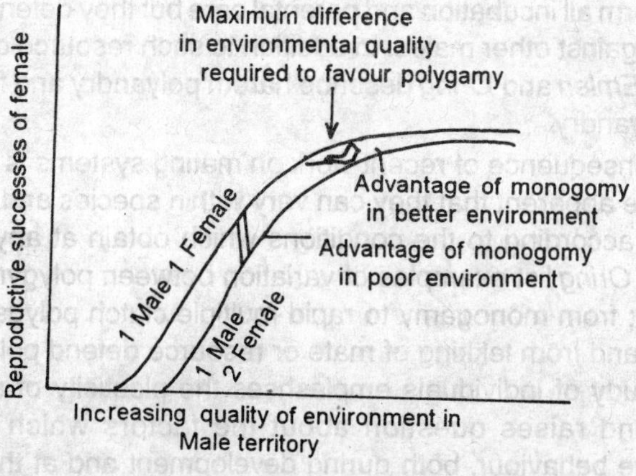

Fig. 13.16. Graphical representation of reproductive success and territory.

mentioned, polygyny occurs in many upland game birds (*Wiley* 1974), in the ruff *Philomachus pugnax* and a few sandpipers (*Lack* 1968).

Emlen and *Oring* (1977) summarise the conditions necessary for the occurrence of polygyny in terms of the relative economics of defending one or several males. This depends upon the spatial distribution of resources, the sex ratio and the time at which mates become available. If females show synchrony in their sexual receptivity, polygyny is more difficult. *Emlen* and *Oring* describe three major types of polygyny, involving resource defence by males, harem defence, or selection by females from male assemblies such as leks. It is nor surprising that polygyny is the normal mating system in mammals for the female's parental investment is always large. *Eisenberg* (1966) lists a few cases of monogamy, for example, marmosets (Callitrichidae), gibbons (Hylobatinae), beavers (Castoridae), one seal and a few terrestrial carnivores.

An intermediate between polygyny and polyandry, in which females lay a second batch of eggs which the male incubates or guards but where the male may also mate with other females was called *rapid multiple clutch polygamy* and *Emlen* and *Oring*. This is common in fish and occurs occasionally in birds. Polyandry has not been reported for mammals (*Eisenberg* 1966) but *Jenni* (1974) describes it in several species of birds. Most are in the order Chondriiformes and all have precocial young. The females of the American jacana (*Jacana spinosa*) defend large territories in march areas which include the smaller territories of several males. The males perform all incubation and parental care but they defend their territories against other males. In addition to such resource defend polyandry, *Emlen* and *Oring* describe harem polyandry and female access polyandry.

One consequence of recent work on mating systems is that is has become apparent that they can vary within species and within individuals according to the conditions which obtain at any time. *Emlen* and *Oring* list examples of variation between polygyny and monogamy; from monogamy to rapid multiple clutch polygamy to polyandry, and from lekking of mate or resource defend polygyny. Detailed study of individuals emplashses the plasticity of mating systems and raises question about the factors which affect reproductive behaviour, both during development and at the time that decisions about the behaviour are taken.

The Struggle of Males and Females

Because the males reproductive success is probably limited by the number of females they can attract and defend from other males rather than their sperm *supply* natural selection will favour are property in a male that enables him to mate with more females. This general principle finds its natural realization in the fascinating diversity of adaptations by which the males of different species seek to increase their share of the mating. Let us examine some examples.

The most obvious form of competition for females is straightforward fighting. We have considered the subject before, in the previous chapter, and we only need notice here the kinds of characteristics it leads to in males. It will favour strength, and fighting in its simplest form will favour increased size. In toads, for instance, the males sit on the backs of females for a few days before the female lays her eggs. Other males try to dislodge the sitting males from the females, by pulling them off. In experiments larger males are much better at dislodging smaller males from females than vice versa. There is an advantage in being large in males, and large males are more likely to mate with females than are small males. In other species males have evolved special weapons with which they fight over females. The narwhal's tusk is an example. It is found only in males, who use their tusks to fight each other. Most males suffer severe wounds from these fights and in one sample from a narwhal population over 60% of the males had broken tusks. Other strange weaponry can be seen in the males of some kinds of beetles, and the antlers of deer have evolved for the same reason. In species that fight, a male can increase his effective strength by forming a coalition with another male, and ganging up on competitors. Males lions do just that groups of two or three male lions try to take over harems of females by forcibly evicting the existing male owners. Bigger coalitions of males are more successful in taking over harems.

However, physical fighting is not the only means by which males compete with each other. When a coalition of male lions successfully takes over a harem of females, the first thing the males do is to kill at the young lion cubs in the pride. The cubs, of course, were fathered by the previous males, and are no loss to the new owners. There is even a gain to them. While a lioness is lactating for her cub she will not produce another cub; but when her cub is killed (or is weaned) she soon becomes ready to reproduce again. By killing the cubs, the males bring forward the time when they can start reproducing.

Infanticide by males which have just taken over a harem is probably common in nature in many species: it has been observed several times in the Hanuman langur (*Presbytis entellus*, species of primate) and has been anecdotally recorded in many other mammals.

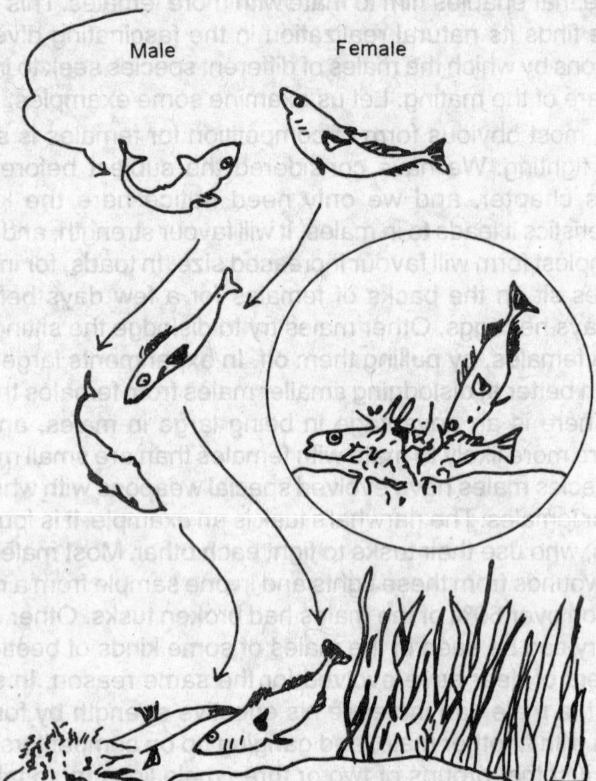

Fig. 13.17. Mating behaviour in Stickle back.

A more subtle, less gory, form of male competition is the microscopic battle fought among the sperms of different males. In many species, if a female mates with two males, the second male by some means or other manages to fertilize many more than half he eggs. In the fruitfly *Drosophila*, for example, the second male fertilizes from 83% to 99% of the eggs. Males have evolved other kinds of counter-adaptations to prevent sperm competition. Some males damselflies stay with their mates after copulation and fight off any other males who come near, only after the female has laid her eggs does the male leave her. Likewise, the acanthocephalan worm *Moniliformis moniliformis* male sticks a 'chastity belt' on a female after mating, probably to prevent other males from copulating with her. The males are also know to 'repe' other males, cementing up the victims' genital openings to render them incapable of copulation. An extraordinary kind of sperm competition has been found in the hemipteran insect *Xylocoris maculipennis* by J. Carayon. Copulation in this species is achieved by injection, the male simply punctures the side of the female with his genitalia and squirts his sperm in. However, a male sometimes compulates with another male. His sperm then migrate to the victims tests, and when the second male comes to compulate with a female, the first male's sperm will be injected in her.

Courtship

Courtship behaviour functions to bring together two animals of different sexes of the same species under conditions where mating is likely to occur and to be successful. The first problem is that of simply locating a potential mate. Obviously, the conspecific must be of the opposite sex if reproduction is to be successful. Courtship often entails a complex sequence of interacting signals, that function to ensure that an animal mates with an appropriate partner. Timing is an important part of successful reproduction; both male and female must be in appropriate physiological condition for mating. This is ensured by the synchronization of cycles that results from the interactions of environmental stimuli and of the two animals themselves. The classical ethologists pointed out that many courtship patterns contain elements of conflict, often because the initial response of an individual to a stranger in its vicinity may be aggressive.

Mating systems vary greatly from species to species. Some species, such as some swans and geese, are truly monogamous and pair for life. Many migrant birds from pair bonds that last for a single season. Some primates are serially polygamous, in that they pair with several individual partners, each for a definable period of time. In simultaneous polygamy, an individual maintains simultaneous pair bonds with several individuals of the opposite sex. Many mammalian species appear to be completely promiscuous, with many copulations and no pair bonds (*Brown*, 1975).

Arthropods

Initial detection often occurs via olfaction. For example, silk moths, *Bombyx mori,* are noted for the extreme sensitivity of the male to a sex attractant, bombykol, produced by the female. Males are attracted from great *Drosophila* (*Spieth,* 1974). Various combinations of behavioural patterns—such as wing vibration, leg vibration, wing seamaphoring, circling, and licking—make up the courtship patterns of different species.

Fishes

Different species of fishes display diverse courtship patterns. The behaviour of tropical aquarium fishes has been most carefully described. Blackchin mouthbreeders, *Tilapia melanotheron,* display four "pure" courtship patterns—patterns that occurs in virtually no other contexts. These are the nod (rapid titling down and forward), quiver (a pattern motionless near the nest) *Barlow* and *Green,* 1970). The most prominent display in the courtship of guppies is the adoption of a "sigmoid" or S-shaped posture by the male. Male platys approach females with a sidling movement and also display patterns of backing toward the female, quivering of the body, and swinging of the gonopodium (*Clark, Aronson,* and *Gordon,* 1954).

Amphibians and Reptiles

Male bullfrogs adopt territories from which they emit their familiar loud choruses. Females appear attracted by such choruses. Many crocodilian species emit loud roads.

The courtship behaviour of anoles (American chameleons), *Anolis carolinensis*, is familiar to many Americans with backyards. The male displays a rhythmical up-and-down bobbing of his body co-ordinated with the exposure of his dewlap, a bright red flap of skin beneath the chin (*Crews,* 1975). Courtship in most species of snakes is based on tacttie stimulation of the female and olfactory stimulation of the male. The dramatic so-called "courtship dances" in which two snakes become closely intertwined have generally proved to be male-ale interactions, presumably aggressive in nature (*Porter,* 1972).

Birds

Birds have provided some of the most dramatic examples of courtship behaviour. The complex displays of grebes, gulls, ducks, herons, and other birds have been favoured subjects of study by ethologists. The attention paid to bird songs and the legends surrounding them is indeed substantial.

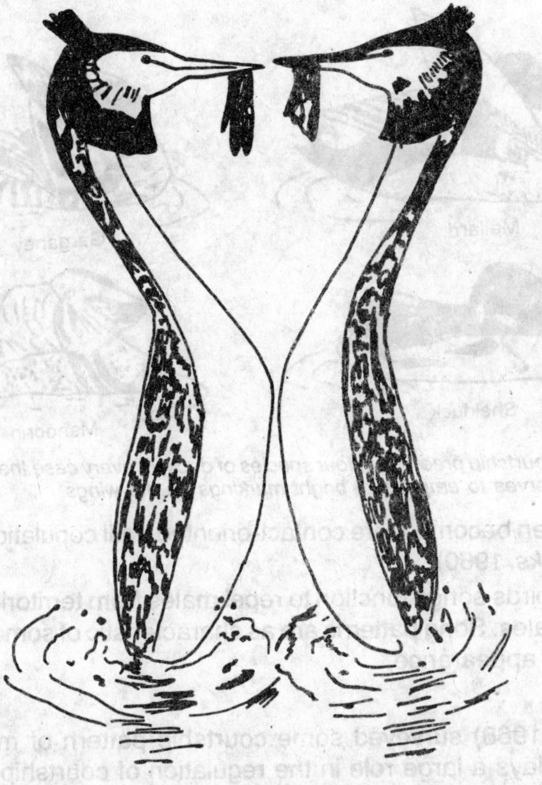

Fig. 13.18. Male and female great crested grebes, Podiceps cristatus, performing one of their elaborate mutual displays. This one involves presentation of nest material.

Lorenz studies various courtship patterns of male ducks. These can be seen in many duck pons. In the *grunt-whistle*, the male lowers its bill to the water, arching the body upward, flicking the bill and emitting a loud whistle followed by a grunting sound. The *head-up-tail-down* display in accompanied by a loud whistle. In the *down-up* display, the breast is dipped into the water and the bill is jerked upward and outward, flipping a column of water as it goes (*McKinney*, 1969).

In green herons males set-up territories. The full-forward display functions to drive away other males from a resident male's territory. The calls of the males attract females. Females are initially repelled with full-forward displays, but females persist and courtship is eventually initiated with the appearance of the snap display and the stretch display. After pairing, the two birds fly about the territory, occasionally showing the more intense flap-flight display. Display

Fig. 13.19. Courtship preening in four species of duck; in every case the movement serves to emphasize bright markings on the wings.

patterns then become more contact-oriented until copulation occurs (*Meyerriecks,* 1960).

Many birds songs function to repel males from territories and to attract females. Song patterns are as characteristic of some species as are their appearance.

Mammals

Ewer (1968) surveyed some courtship pattern of mammals. Olfaction plays a large role in the regulation of courtship in many species. Patterns of anogenital investigation and urine testing are common. After smelling a frame, the males of many species display a *Flehmen* response, in which the neck is extended and the upper lip is curled. This pattern appears to function more in facilitating perception of the odor than as a display.

Female mammals often solicit mountings, sometimes by approaching a male, nuzzling or licking him, and often by darting from him. Much of the running from the male done by females appears more as solicitation behaviour than escape.

Courtship in bottle-nosed dolphins entails vocalization, mouthing of the partner nuzzling of the partner's genitalia, rubbing of bodies, stroking with flukes or flippers, displaying of the white underside, leaping, chasing, and head butting (*Puente* and *Dewsbury,* 1976).

Mating

The consummation of courtship activity is the occurrence of actual mating, wherein fertilization of ova is accomplished. Patterns of mating are almost as diverse as are courtship patterns, including both internal and external fertilization via a great variety of process. A select sample of mating patterns will be described.

Arthropods

In many species, sperm are "packaged" in a *spermatophore,* a kind of bag or sack containing the sperm. Perhaps the heights of impersonality are reached in some species of mites, pseudoscorpions, millipedes, and springtails, in which the male deposits a spermatophore on the substrate and the female comes along later to pick up the package. The pair may never meet. Many aquatic species have external fertilization (e.g., the male horseshoe crab, *Limulus,* hooks himself on behind the female and remains there until eggs are released). Most terrestrial form have some kind of internal fertilization, wherein the spermatophore is placed directly into the female using some kind of appendage (*Alexander,* 1964).

Fishes

Fertilization in fishes can be either internal or external. In anabantid fish, such as gouramis and bettas, a bubble nest is built and spawning generally takes place beneath it. Eggs and sperm are released simultaneously and eggs are fertilized as they float up to the nest. In other species—such as guppies, platys, and *Gambusia*—a gonopodium (modified and fin) is used to transfer sperm to the female. Male nurse sharks grasp the hind edge of one of the females' pectoral fins, turning her on her back, and insert their specialized claspers (*Budker,* 1971).

Amphibians and Reptiles

Most species of salamanders accomplish internal fertilization through use of spermatophore, although external fertilization and parthenogenesis occurs in some species. Most frogs and toads have external fertilization, with eggs fertilized as they are laid. Exceptions include tailed frogs, which have an intromittent, organ, and African toads, in which cloacal glands of the male and female are brought together. Typically, a ripe female frog orients toward a calling male and is clasped by him in a state termed "amplexus." Eggs are released and fertilized in several bouts. If a male clasps another male or is clasping a female that has finished spawning, a release call by the recipient results in release by the clasping male (*Rabb,* 1973).

Reptiles display internal fertilization through use of intromittent organs. Among anoles, the male delivers a neck grip, twists his tail around that of the female, and inserts his hemipenis.

Birds

In contrast to their courtship activities, the copulations birds generally show little variability. In most species there is no penis, and thus sperm are transferred from the cloacal gland of the male to that of the female. In copulation, the two cloacal glands are brought together in the "cloacal kiss." In Japanese quail, for example, the male grasps the feathers of the female's neck in his break, mounts her by standing on her back, and orients until the cloacal glands can be brought together (*Sachs,* 1969).

Mammals

Much of the research on mammalian copulatory behaviour has been done using laboratory rats, *Rattus norvegicus,* as subjects. Their copulatory behaviour will be described in some details. If a receptive female is introduced into a cage containing a vigorous male, a predictable sequence of events ensues. After initial courtship behaviour (sniffing, chasing, soliciting by the female), the male pursues the running female. As the male mounts the female from behind, the female adops a stereotyped posture, termed "lordosis". The head and hindquarters are elevated to produce a concave arching of the back, while the tail is deflected to the side. Two kinds of copulations typically occur. In some, called "intromissions," the male's penis is inserted into the female's vagina for a period of about 1/4 second before the male displays a rapid dismount. Although there may be extravaginal pelvic thrusting by the male prior to insertion, there is no intravaginal thrusting.

After about ten such brief intromissions spaced about a minute apart from each other, the male displays a different pattern, called "ejaculation." The male shows a kind of convulsive thrusting during ejaculation and clasps the female for a few second before he dismounts (in a stereotyped vertical posture). A complete group of intromissions ending with an ejaculation is termed a "series." Ejaculation and the associated sperm transfer never occur without prior intromissions. Ejaculation is followed by a temporary cessation of sexual activity. After about 5 minutes, activity resumes and a second series occurs. Normally, rats will complete about seven ejaculatory series before they may be considered satiated, according to a criterion of 1/2 hour with no intromissions (*Beach* and *Jordan,* 1956).

14

Sexual Behaviour in Insects

Reproductive behaviour involves first the location of a mate, followed by courtship and mating, oviposition, and sometimes brood care. Territoriality may arise out of competition for mates or food. Reproductive behaviour is typically complex and highly variable. The evolutionary role of this complexity and variability is generally assumed to be reproductive isolation.

Mate Location

A wide variety of mechanisms function in bringing together the sexes. Initially, over relatively long distances, these mechanisms usually involve the visual, olfactory, and auditory modes of communication, singly or in combination. The visual stimuli associated with mate location and the responses they elicit vary in complexity. The simplest pattern, observed among water striders (Gerridae, Hemiptera), involves a male insect approaching any moving object of appropriate size that happens to enter its visual field. Other insects require more specific stimuli to induce approach. For example, male damselflies in the family Lestidae (Odonata) approach any insect that has transparent wings and flies in a manner similar to that of damselflies. Damselflies in the genus *Calopteryx* (Calopterygidae) requires even more specific stimuli, the males of different species being able to recognize members of their own species by the amount of light allowed to pass through the wings.

More complex visual stimuli involved in mate location include the use of luminescent organs in signaling for a mate. This is best exemplified by the fireflies in the beetle family Lampyridae. In some species both the males and females produce light and have a very complicated signaling system; in other species only the male produce the light signals. In either case the light-signaling systems are species-specific. In fact, several species have been discovered through observation of differences in signaling systems. The mating behaviour of *Photinus pyralis,* is described in the following quote.

At dusk the male and female emerge separately from the grass. The male flies about two feet above the ground the emits a single short flash at regular intervals. The female climbs some slight eminence, such as blade of grass, and waits. Ordinarily she does not fly at all, and she never flashes spontaneously. If a male flashes within three or four yards of her, she will usually wait a decorous interval, then flash a short response. At this the male turns in her direction and glows again. The female responds once more with a flash, and the exchange of signals is repeated—usually not more than five or 10 times—until the male reaches the female, waiting in the grass, and the two mate.

In some tropical species of fireflies, several males congregate in a single tree and flash synchronously *McKlroy* et. al. (1974) describe this phenomenon as follows.

In Burma and Siam and other eastern countries, all the fireflies on one tree may flash simultaneously, while on another tree some distance may this same synchronous flashing would be apparent, but out of step with those of the first tree. Observers have been particularly impressed by the display, which is one of the interesting sights of the Far-East. In some cases visual markers in the environment may be involved in mate location. For example, among the true flies males may swarm or sit near a marker and wait until a female approaches the marker.

The use of olfactory cues in mate location is widespread among the insects. Sex pheromones are produced by specialized glands in males, females, or both sexes of a given species. Sex pheromones are very potent substances; "Detected by the insect in fantastically minute amount, these attractants are undoubtedly among the most potent physiologically active substances known today". The sex pheromone *bombykol,* produced by the female silk moth, *Bombyx mori,* can elicit a response from a male in a concentration of 1000 molecules per *cubiccentimeter.* Electrophysiological studies of antennal receptors have revealed that a single molecule of bombykol can initiate a single nerve impulse. Most sex pheromones are apparently species-specific however, there are several known examples of nonspecificity.

The production by female insects of pheromones that attract males has been observed and described for several species in a number of orders. The combined lists of *Jacobson* (1965), and *Jacobson* (1974) are composed of 12 species of cockroaches (Orthoptera); 4 species of true bugs (Hemiptera; Heteroptera); 4

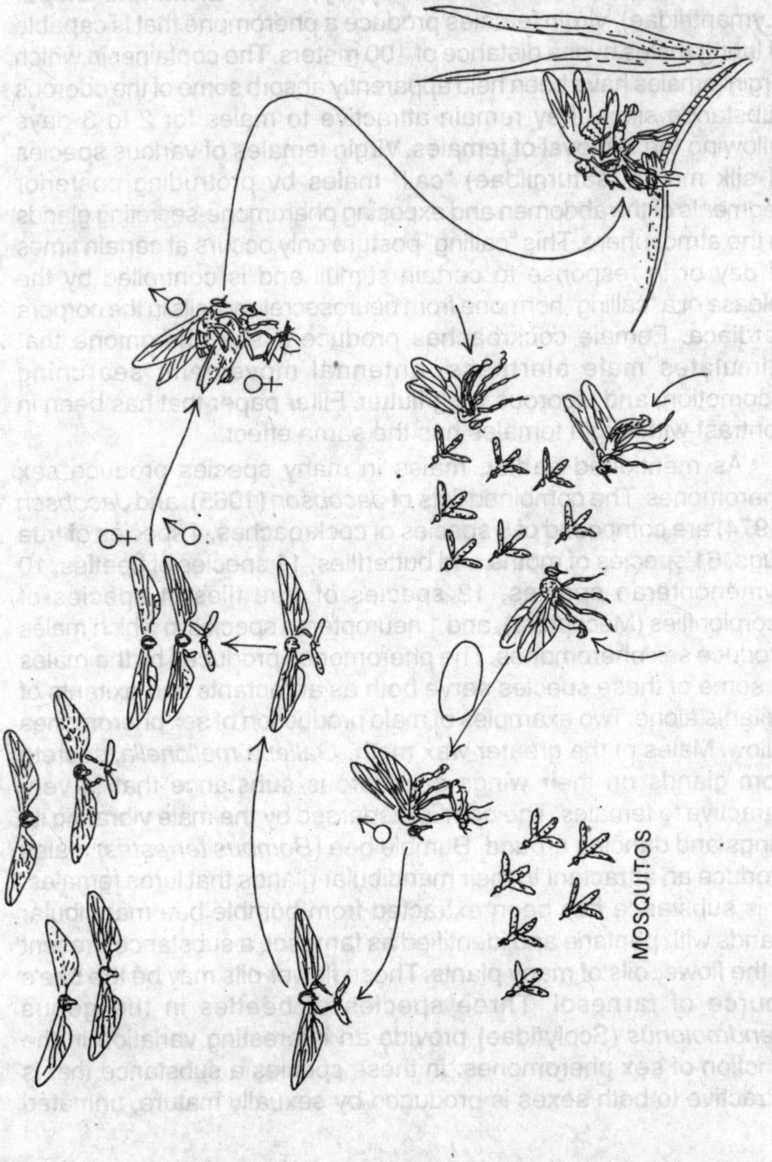

Fig. 14.1. Complete courtship sequence of an empid fly, Rhamphomyia nigripes.

species of homopterans (Hemiptera, Homoptera) 184 species of moths and butterflies (Lepidoptera); 34 species of beetles (Coleoptera); 31 species of bees, wasps, sawflies, and ants (Hymenoptera); 8 species of true flies (Diptera); and 3 termite species (Isoptera) in which the females are known to produce sex pheromones. Some examples follow. In he Gypsy moth, *Lymantria dispar* (Lymantriidae), virgin females produce a pheromone that is capable of luring males over a distance of 100 meters. The container in which virgin females have been held apparently absorb some of the odorous substance since they remain attractive to males for 2 to 3 days following the removal of females. Virgin females of various species of silk moths (saturniidae) "call" males by protruding posterior segments of the abdomen and exposing pheromone-secreting glands to the atmosphere. This "calling" posture only occurs at certain times of day or in response to certain stimuli and is controlled by the release of a "calling" hormone from neurosecretory cells in the corpora cardiaca. Female cockroaches produce a sex pheromone that stimulates male alertness, antennal movement, searching locomotion, and vigorous wing flutter. Filter paper that has been in contrast with virgin females has the same effect.

As mentioned earlier, males in many species produce sex pheromones. The combined lists of *Jacobson* (1965), and *Jacobson* (1974) are composed of 5 species of cockroaches, 4 species of true bugs, 61 species of moths and butterflies, 11 species of beetles, 10 hymenopteran species, 12 species of true flies, 4 species of scorpionflies (Mecoptera), and 1 neuropteran species in which males produce sex pheromones. The pheromones produced by the males of some of these species serve both as attractants and exitants of exitants alone. Two examples of male production of sex pheromones follow. Males of the greater wax moth, *Galleria mellonella,* secrete from glands on their wings an adorous substance that is very attractive to females. The odor is dispersed by the male vibrating its wings and dancing around. Bumble bee (*Bombus terrestris*) males produce an attractant in their mandibular glands that lures females. This substance has been extracted from bumble bee mandibular glands with pentane and identified as farnesol, a substance present in the flower oils of many plants. These flower oils may be the bee's source of farnesol. Three species of beetles in the genus *Dendroctonus* (Scolytidae) provide an interesting variation in the function of sex pheromones. In these species a substance that is attractive to both sexes is produced by sexually mature, unmated females feeding on fresh Douglas fir phloem. Since both sexes are attracted by this odorous substance, it serves to bring them together, which eventually leads to courtship and copulation, *Trypodendron lineatum,* an ambrosia beetle, produces a substance that has a similar effect. Such substances have been called *assembling scents.*

Acoustic signals serve to bring the sexes together for a large number of insects—may Orthoptera. Hemiptera-Homoptera, and Diptera; a few Lepidoptera, Hemiptera-Heteroptera, and Coleoptera. Calling songs may be quite elaborate and are usually species specific. In general, a response to one of these songs depends in part on the state of internal readiness for courtship and copulation. In the majority of cases, specialized structures are involved in the production of sounds, but several insects produce sounds as a direct result of wing movement in flight. Examples of sounds produced in this fashion that serve to bring the sexes together have been found in members of the Diptera, Hymenoptera, and Orthoptera. For example, the flight sound of a mature female mosquito is very attractive to a male and elicits compulatory behaviour. The males react similarly when the frequency of the flight sound is produced by a tuning fork. In some insect groups the acoustical counterpart of the "*assembling scent*" has been found. For example, cicadas in the genus *Magicicada* sing an aggregating song is chorus that is responsible for assembling both males and females. In many cases, aggregation of males and females may occur in association with biological activities other than sex, for example, at emergence, oviposition or feeding sites. Such aggregation can subsequently lead to sexual activities.

Once an insect has been induced to approach a possible mate, other stimuli are usually required to release further pursuit. These are generally olfactory, but they may also be a characteristic behaviour on the part of the pursued. An insect that recognizes a member of its own species may or may not be able to distinguish the sex of the individual it pursues. Males of *Pyrrhocoris* and *Gerris* (Hemiptera-heteroptera) and *Leptinotarsa* (Coleoptera, Chrysomelidae) recognize members of their own species by specific odors but will attempt to copulate with either sex. On the other hand, some fruit flies are able to recognize the sex of an individual by its odor.

Courting and Mating

Following mate location, variable and often elaborate behaviour is released within ultimately leads to copulation. Courtship displays function is species and sex identification, the meeting of solitary individuals, and in the stimulation and maneuvering involved in copulation.

During courtship, escape and attack responses are usually inhibited, at least to the extent that copulation may occur. However, they may not be inhibited completely. For example, as mentioned earlier the female praying minted very often devours the head and part of the thorax of a male that is attempting to copulate. This is not as tragic as it seems since removal of the head, in particular the subesophageal ganglion, increases the vigor of the male's compulatory movements. Signals from the intact subesophageal ganglion are though to inhibit various endogenous nervous patterns in the rest of the body. Thus decapitation removes inhibition of copulatory movements. In addition to releasing vigorous compulatory movements, the female mantid also obtains a highly nutritious meal.

Many of the cues that serve in mate location act further as releasers of courtship and copulation. For example, the flight sound of the female mosquito not only attracts males but also releases compulatory behaviour. In many insects the sex pheromone produced by one of the sexes also serves as an excitant (aphrodisiac), stimulating courtship or attempts at copulation. For instance, the sex pheromone produced by female American cockroaches (*Periplaneta americana*) not only attracts males but induces a wing-raising display and attempts to copulate. In the absence of a female, males may attempt to copulate with one another if the female odor is present. This female sex pheromone is also capable of eliciting courtship behaviour in males of other *Periplaneta* species and males of *Blatta orientalis* (oriental cockroach). In some insects, the sexually excited male or the female may produce a substance that acts as an excitant for the opposite sex. Such is the case in *Lethocerus indicus*, a giant water bug, and a family of beetles.

During sexual excitement the male is easily recognized by its odor, for this abdominal glands secrete a liquid with an odor reminiscent of cinnamon. This substance, produced in two white tubules 4 cm long and 2-3 mm thick, occurs to the extent of approximately 0.02 ml per male and is used in southeast Asia as a spice for greasy foods. The female does not secrete the substance, which is believed to act as an aphrodisiac to make her more receptive to the male. Males of Malachiidae, a family of tiny tropical beetles, entire females first with a tarty nectar and then epose them to an

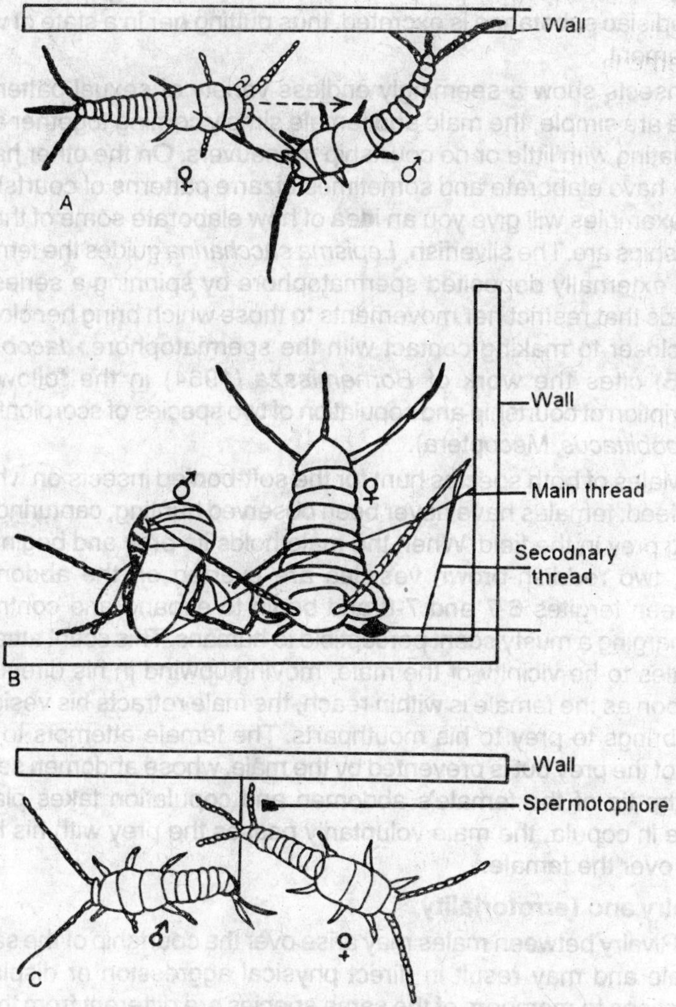

Fig. 14.2. Courtship in the silverfish, Lepisma saccharina. A—Approach. B— Male affixes threads to wall and floor and deposits spermatophore. C—Female is guided to the spermatophore by the threads.

aphrodisiac. The males possess tufts of fine hair growing out of their shells (in some species on the wing covers, in others on the head). These hairs are saturated with a glandular secretion that the females cannot resist. During the mating season, the male searches for a female; when he finds one he offers his tuft of hair, which the female then accepts and nibbles upon. In so doing, her antennae come in contact with microscopic pores in his shell, through which the aphrodisiac substance is excreted, thus putting her in a state of wild excitement.

Insects show a seemingly endless variety of sexual patterns. Some are simple, the male and female simply coming together and copulating with little or no courtship maneuvers. On the other hand many have elaborate and sometimes bizarre patterns of courtship. Two examples will give you an idea of how elaborate some of these courtships are. The silverfish, *Lepisma saccharina* guides the female to an externally deposited spermatophore by spinning a series of threads that restrict her movements to those which bring her closer and closer to making contact with the spermatophore. *Jacobson* (1965) cites the work of *Bornemissza* (1964) in the following description of courtship and copulation of two species of scorpionflies (*Harpobittacus*, Mecoptera).

Males of both species hunt for the soft-bodied insects on which they feed; females have never been observed hunting, capturing, or killing prey in the field. When the male holds its prey and begins to feed, two reddish-brown vesicles are everted on the abdomen between tergites 6-7 and 7-8 and begin to expand and contract, discharging a musty scent perceptible to humans. This scent attracts females to he vicinity of the male, moving upwind in his direction. As soon as the female is within reach, the male retracts his vesicles and brings to prey to his mouthparts. The female attempts to get hold of the prey but is prevented by the male, whose abdomen seeks out the tip of the female's abdomen and copulation takes place. Once in copula, the male voluntarily passes the prey with his hind legs over the female.

Rivalry and Terrotoriality

Rivalry between males may arise over the courtship of the same female and may result in direct physical aggression or displays. Reactions to members of the same species are different from those involved with escape or defence, which are stimulated by threats of danger. When a male grasshopper approaches another male that is serenading a female with a courtship song, they face one another and sing a characteristic "*rivalry*" song. Eventually one of the male leaves, and the other continues the courtship. Male silverfish, *Lepisma* (Lepismatidae, Thysanura), will fight over a female.

Territorial behaviour is not common among insects, but there are some very definite examples. Males of certain odonates in the genera *Calopteryx* (Damselflies) and *Pachydiplax* (dragonflies) defined territories against other males. A male entering another's territory is met with an attack. Females are recognized by sight since there is distinct sexual dimorphism in these particular species. Instead of aggression, these females are met with attempts to copulate. In some other genera of dragonflies (e.g., *Aeshna* and *Libellula*), when the sex of an approaching member of the same species is not recognized males will attempt to copulate with males. The tendency for males to avoid such encounters results in their spreading out into individual territories. The suggestion has been made that such sexual encounters between males preceded the evolution of territorial behaviour. Male crickets are territorial; when one male enters another's territory, he is greeted by the "*rival*" song of the other. This is followed either by the exit of the intruder or holder of the territory or by fighting. A rank order may become established among males whose territories are in close proximity to one another. A female entering a male's territory is also greeted with the "*rival*" song, and will either leave or remain quietly. Price (1975) provides a list of examples of territoriality is several insect orders.

Competition in mating often occurs even after insemination. Since sperm are commonly stored in a spermatheca, it would seem likely that more than one male could successfully inseminate a given female. A number of mechanisms that help prevent additional inseminations have evolved. For example, male dung flies "protect" females from the copulation attempts of other males. Other examples include the tandem flight of dragonflies (Odonata); prolonged copulation characteristic of the love bug, *Plecia nearctica* (Diptera, Bibionidae); and mating plugs formed in the genital tract of the female from secretions of the male accessory glands.

Males of some species—certain *Drosophila* (Diptera, Drosophilidae), dragonflies, and others—congregate and display at a particular place. These aggregations of males, called *leks,* are presumably more effective in attracting females than isolated males.

Courtship

Once conspecifics of differing sex have been drawn close to each other there would be a selective advantage to behaviour patterns that would suppress aggressive or escape responses and lead to sexual receptiveness. Courtship serves both of these purposes and includes an array of behaviours that ranges from a slightly modified attach response to ritualized attack, appeasement, sexual stimulation, and finally copulation. Courtship involves all forms of communication. Pheromones are used in numerous species for both recognition and sexual excitation. Males of some beetles will attempt to mate with a piece of paper treated with female sex pheromone, while ignoring an actual female held captive under glass nearby. During the courtship of the grayling butterfly, *Eumenis semele*, the male spreads its wings and bows toward the female so that her antennae come in contact with scent patches on his forewings. Many grasshoppers move about the female in an excited manner while singing a courtship song prior to attempting copulation. Visual displays are also common and sometimes spectacular, as are the aerial displays of some butterflies and the bioluminescent flashing of fireflies. Tactile stimulation, though often not obvious, is also important and widespread. In some species the male offers the female an object or some food, which apparently increases the female's

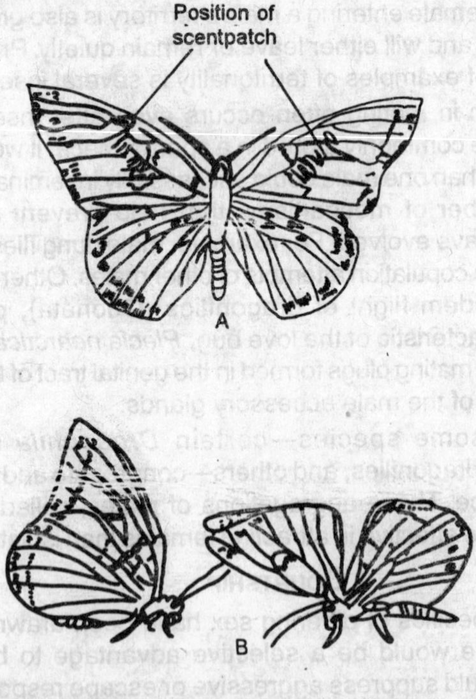

Fig. 14.3. Chemical communication in the grayling butterfly, Eumenis semele. A—The position of the scent patches on the dorsal surface of the male's forewings. B—The male bowing to the female during courtship to expose the scent patches, which the female contacts with her antennae.

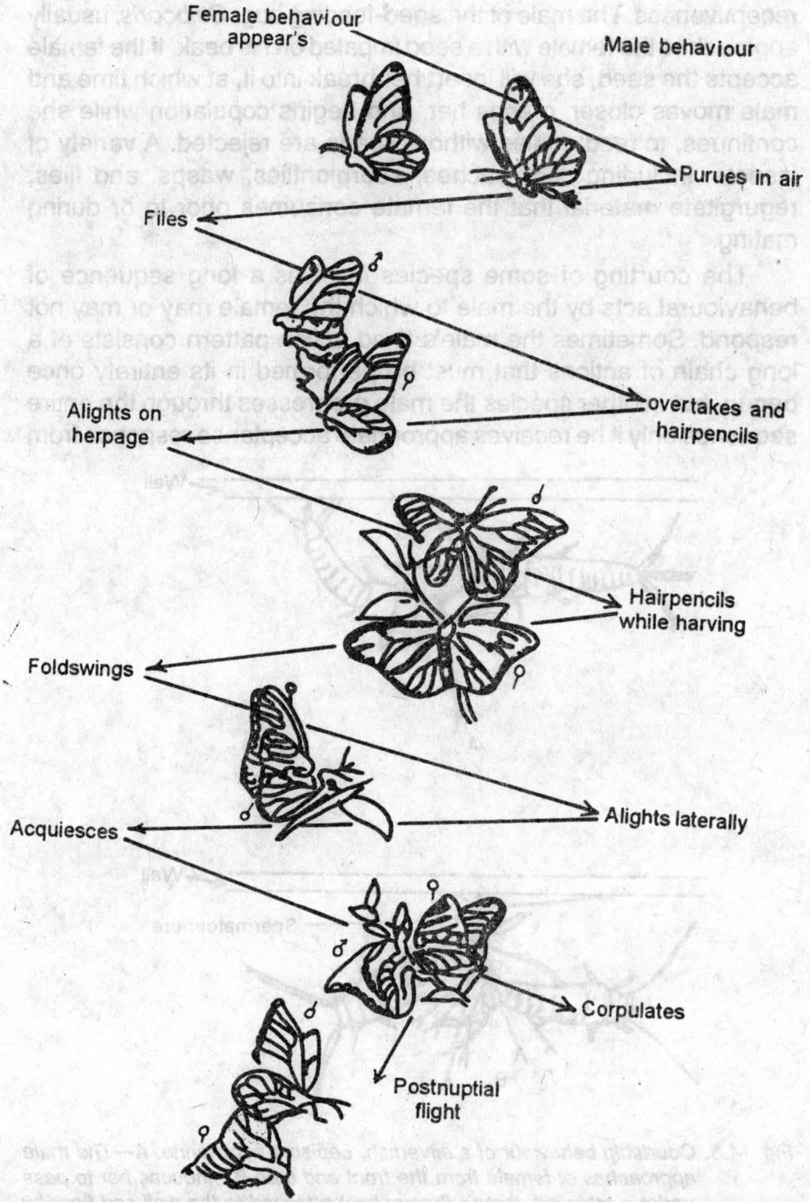

Fig. 14.4. Courtship behaviour of queen butterflies, showing the response chain resulting from an encounter between a male and a female.

receptiveness. The male of the seed-feeding bug, *Stilbcoris,* usually approaches the female with a seed impaled on his beak. If the female accepts the seed, she will insert her break into it, at which time and male moves closer, grasps her, and begins copulation while she continues, to feed; males without seeds are rejected. A variety of insects, including cockroaches, scorpionflies, wasps, and flies, regurgitate material that the female consumes prior to or during mating.

The courting of some species involves a long sequence of behavioural acts by the male to which the female may or may not respond. Sometimes the male's fixed action pattern consists of a long chain of actions that must be performed in its entirety once begun, but in other species the male progresses through the entire sequence only if he receives appropriate acceptance response from

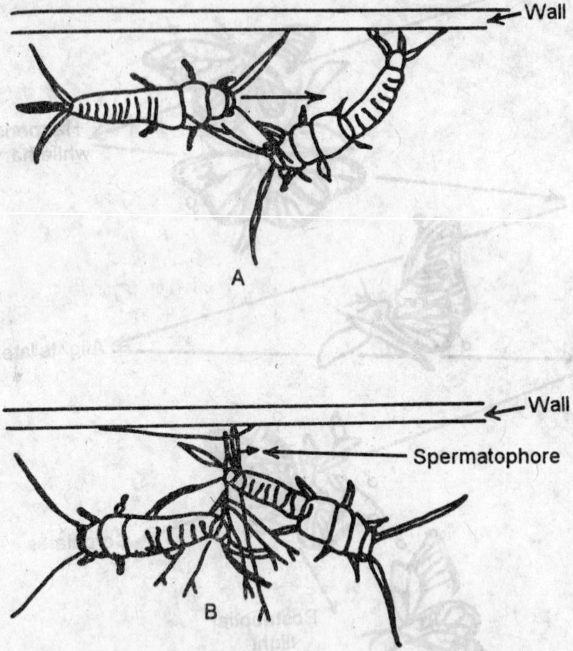

Fig. 14.5. Courtship behaviour of a silverfish, Lepisma saccharina. A—The male approaches of female from the front and then B—induces her to pass under a main silk thread (heavy line) attached to the wall and floor by establishing a number of secondary guiding threads (fine lines). The system of threads leads the female to the spermatophore deposited earlier by the male.

the female at key points along the way. Such sequences have been studied in a number of insects, but two examples from opposite ends of the evolutionary spectrum of the insects will serve as examples.

The silverfish, *Lepisma saccharina,* engages in complicated precopulatory behaviour in which the male produces a package of sperm (spermatophore) and then leads the female to it and with a series of silk threads that restrict her movement. This so-called leading and bridling behaviour culminates with the male pushing the female's ovipositor into contact with the spermatophore.

The courtship behaviour of flies belonging to the genus *Drosophila* involves the most elaborate set of signals of any of the mating sequences studied thus far. A simplified catalogue of signals commonly used by male *Drosophila* is presented in table.

Table 14.1. Major courtship elements of *Drosophila males*

Orientation	o turns toward another individual, slight raising body.
Tapping	o lifts, straightens foreleg, and strikes downward against other individual; almost invariably occurs at start of courtship.
Wing vibration	Oriented o extends wing nearest o's head, then vibrates it rapidly; interspecific variation in degree of extension, amplitude and speed of vibration, and angle of extended vane with respect to substrate, i.e., horizontal titled, vertical.
Wing flicking	o flicks one wing sharply out, then back to resting position.
Wing waving	o extends one wing, then slowly waves up and down.
Wing semaphoring	o alternately and repeatedly flicks wings sharply outward, then back a resting position; one wing is moved outward while the other is returning to the resting position.
Wing scissoring	o repeatedly and rapidly extends both wings horizontally outward and back to the resting position.
Leg vibration	o displaying at rear of o extends and vibrates forelegs against abdomen of usually against venter.

Leg rubbing	o displaying at rear of o extends and rubs forelegs to and fro against abdomen of o, usually venter, but some species against sides or dorsum.
Circling	o periodically circles about o, facing her as he moves, often from rear to front and back, sometimes completely about female.
Licking	o opens labellar lobes, extends proboscis, and licks o genitalial o may either lick intermittently and repreadly or continuously for a prolonged period; some Hawaiian species have modified labellar lobes which grasp o's genitalia.
Mounting	o curls tip of abdomen under and forward, rear upward, thrusts head under o's wings or between her spread wings, grasps her body with his fore and mid legs, and attempts intermission.
Countersignalling	Leg striking, kicking, wing-vibration signals resulting from one male attempting the court another male.

Table 14.2. The major elements of the courtship behaviour of *Drosophila* females.

Female acceptance behaviour	
Wing spreading	Female spreads both wings outward and upward and holds them extended until male mounts.
Genital spreading	Female slightly droops tip of abdomen slightly extrudes genitalia, and spreads oviposition apart.
Ovipositor extrusion	Female extrudes ovipositor posterior; restricted to Hawaiian species.
Female repelling behaviour	
Decamping	Female breaks contact with courting male by running, jumping, or flying.
Kicking	Female kicks vigorously backward with hind legs, striking face when he courts a harrier.

Fluttering wings	Female rapidly flutter wings in movements of small amplitude; often occurs when male initiates tapping action
Abdomen elevation	Female elevates abdomen, thus raising tip high above substrate and inhibits male courtship such as kicking and leg vibration; often observed when male attempts to court feeding female; may be accompanied by extrusion.
Abdomen depression	Female depresses tip abdomen and wing tips against substrate; inhibits male courtship actions such as kicking and leg vibration
Extrusion	Female extends and elongates tip of abdomen, thus exposing articulating membranes directs tip of her abdomen toward male face, usually causing male to turn quickly away and engage in cleaning behaviour.

Drosophila adults are attracted by odor to decomposing and fermenting plant material, particularly fruits, where they feed on the bacteria and yeast and in many case mate and oviposit. The communality of this food odor response helps to bring conspecifics of both sexes together, but also results in the formation of mixed species aggregations. Although there is some separation of species resulting from species-specific responses to time of day and environmental conditions, several species are likely to be present at the same place at the same time.

According to the *Spieth* (1974) the females must feed steadily to fulfill the nutritional demands of egg production, whereas the males feed for only short periods, after which they turn their attention to mating. Since the males cannot visually distinguish between similar appearing species or conspecific male from females, they approach any passing individual with roughly the right conformation. Consequently, a system of signals has evolved that Spieth concludes has the advantage of providing the males with a means of ascertaining the receptiveness of females with a minimum expenditure of time and energy, the unreceptive females with a way to repel males without interrupting their feeding, and receptive females with an opportunity to sample several males thereby bringing sexual

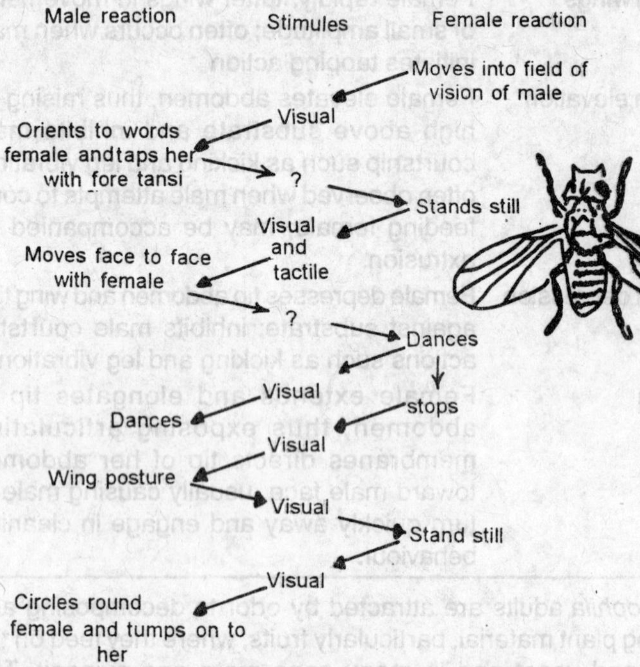

Fig. 14.6. The sequence of stimulation and response steps that characterize the courtship behaviour of Drosophila suboscura.

selection into operation. In *Drosophila suboscura,* for example, the courtship ritual consists of a series of sexually alternating stimulation and response steps. The male begins by facing the female and tapping her with his forelegs. If the male perceives the appropriate visual and tactile stimuli from the male perceives the appropriate visual and tactile stimuli from the female, he will extend his proboscis and more to a face-to-face position. The male then taps the female's head and both began a dance that consists of a series of side steps while still aligned face to face. As the dance progresses, the male raises his wings at right angles to the body with the leading edge tilted downward. The female responds by ceasing to dance. The male then circles his mate and finally jumps upon he and attempts copulation.

In most insects, successful courtship usually culminates in copulation. Various species-specific positions are employed, and the duration of copulating may range from seconds to hours. Postcopulatory behaviour is also highly variable. In many species

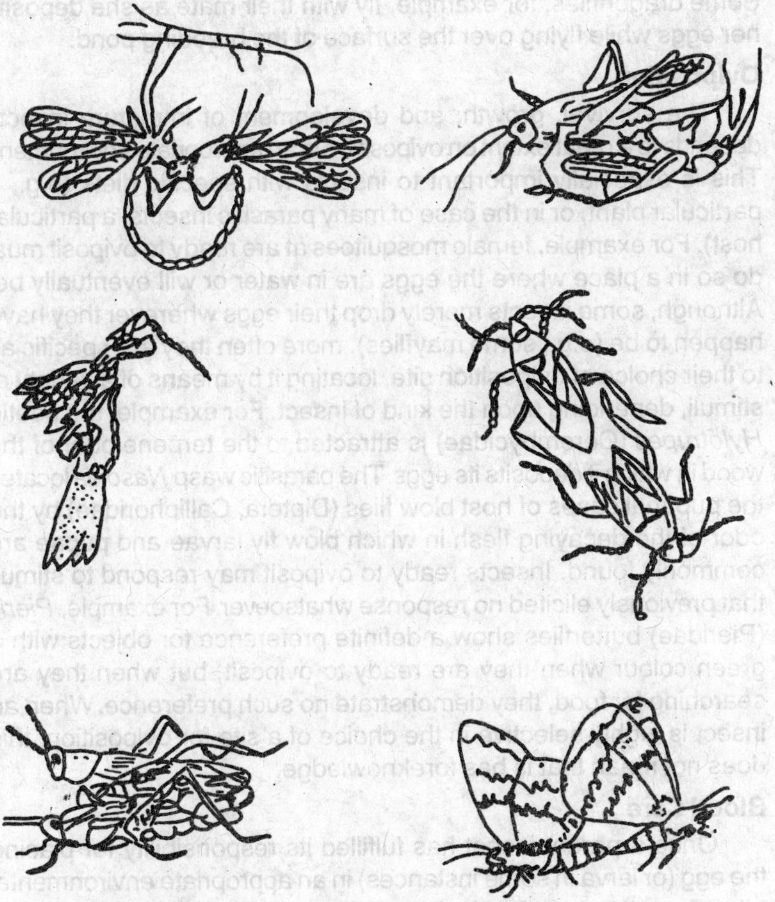

Fig. 14.7. Various positions assumed by insects during copulation.

the pair separates immediately, and the female begins to search for an oviposition she whereas the male goes off in search of another mate. In others the pair remains together for some time. In some species the male continues to provide a food offering, which prevents the female from eating the spermatophore he has deposited, whereas in others, like the mantids, the female may actually devour her mate. The females of a few species such as a screwworm, *Chchliomyia hominivorax,* mate only once, but most male several times although they are usually unresponsive to the advances of males for some time after each copulation. In a variety of species belonging to different orders, the male may remain the female during oviposition.

Some dragonflies, for example, fly with their mate as she deposits her eggs while flying over the surface of the breeding pond.

Oviposition

The survival, growth, and development of immature insects depends to a great extent on oviposition in an appropriate environment. This is especially important to insects with specific diets (e.g., a particular plant, or in the case of many parasitic insects, a particular host). For example, female mosquitoes at are ready to oviposit must do so in a place where the eggs are in water or will eventually be. Although, some insects merely drop their eggs wherever they have happen to be (e.g., some mayflies), more often they are specific as to their choice of oviposition site, locating it by means of a variety of stimuli, depending upon the kind of insect. For example, the beetle *Hylotrupes* (Cerembycidae) is attracted to the terpene odor of the wood in which it deposits its eggs. The parasitic wasp *Nasonia* locates the puparial cases of host blow flies (Diptera, Calliphoridae) by the odor of the decaying flesh in which blow fly larvae and pupae are commonly found. Insects ready to oviposit may respond to stimuli that previously elicited no response whatsoever. For example, *Pieris* (Pieridae) butterflies show a definite preference for objects with a green colour when they are ready to oviposit, but when they are searching for food, they demonstrate no such preference. When an insect is highly selective in the choice of a site for oviposition, this does not mean that is has foreknowledge.

Blood Care

Once a parent insect has fulfilled its responsibility for placing the egg (or larva in some instances) in an appropriate environmental situation, it may simply leave. However, some continue an association with the eggs and immature stages. This association is most highly developed in the social insects (ants, bees, wasps and termites) in which the blood are "reared" from egg to adult by workers. Social insects differ from social vertebrates in that a vertebrate colony is composed of a number of mating pairs and offspring, whereas an insect colony is usually the product of a single female or single male-female pair.

Many nonsocial insects also do more for their offspring then merely deposit the egg. Earwig females (Dermaptera) lay their eggs in burrows in the ground and guard them until they hatch. Female beetles in the genus *Omaspides* protect their brood from the ravages of invading ants. Some nonsocial insects go to the extent of preparing elaborate nests and stocking them with food for their young. For example, a female solitary wasp in the genus *Bembix* digs a nest in the soil with her legs and mandibles. She then capture an adult fly (or sometimes another insect) and brings it to the nest. *Evans* (1957) describes the prey capture.

The capture and stinging of the prey occur with great rapidly. The female wasp proceeds slowly through the air or hovers over a source of flies, pouncing upon the flies either in flight or at rest; then she descends to the earth or to some other solid object, where she quickly bends her abdomen downward and forward, inserting the sting on the ventral side of the thorax or in the neck region of the fly.

Following capture, the wasp return to the nest and deposits an egg on the fly. Thereafter, a number of flies are brought to the nest as the larva grows. Eventually, the larva spins a cocoon, pupates, and emerges from the ground as an adult. Depending on the species and the time of year, it may remain over the winter in the cocoon.

Insects in Groups

Insects, display many gradations between solitary behaviour and complex, organized social behaviour. However, comparatively few insect species are truly social (*eusocial*) a few thousand perhaps, and these are found in only two orders, Isoptera (termites) and Hymenoptera (ants, bees, wasps and relatives).

Matthews and *Matthews* (1978) provide a useful classification of intraspecific insect associations other than sexual interludes. The divide interactions into aggregation, simple groups, primitive societies, and advanced societies (*eusocial* insects).

Group behaviour is thought to provide a number of different benefits; including protection a result of such things as collective displays and more efficient detection of potential predators, increased efficiency in detection and utilization of food, and moderation of adverse physical environmental factors. Insects that product defensive secretions and/or display warning colouration on doubt derive increased protection by pooling their defensive capabilities. Wools apple aphids, *Eriosoma lanigerum* (Hooptera, Eriosomatidae), secrete waxy material, which gives their bodies a whitish.

Parental Behaviour

For both males and females, the act of mating may be only part of their investment of time and energy in the perpetuation of their genes. *Parental investment* is defined as the behaviour that increase the probability of some offspring surviving to reproduce at the cost of the parent's ability to produce more offspring. Minimal behavioural investment tends to be compensated for by maximum physiological capacity to produce offspring (for example, in house flies). Conversely, maximum investment (such as brood care) tends to be offset by reduced production (but higher survival) of offspring. As a general rule, the investment of the female is much greater than that of the male; she not only converts most of her nutrients to the production of eggs, but she seeks a suitable site for oviposition and in some cases guards her eggs and even feeds her offspring. The male, on the other hand, produces large numbers of much smaller sperm, with less energetic investment, and attempts to use these to inseminate as much females as possible. But it is not always quite as simple at this, and it will pay us to look briefly at male parental investment before returning to that of the female.

Male Investment and Assurance of Paternity

Male sperm, as we have seen, are stored in the female's spermatheca and released at the time of oviposition. When females mate more than once, it is generally the sperm from the last mating that are released to fertilize the eggs. This has been termed *sperm precedence.* In instances of sexual selection involving male territoriality or nuptial feeding, it is to the male's advantage, once he has mated, to ensure that the female does not mate again before she lays her eggs. Thus a male dragonfly guards the female while she lays her eggs, ensuring his paternity of the resulting offspring. Indeed, Jonathan Waage, of Brown University, showed that the penis of certain

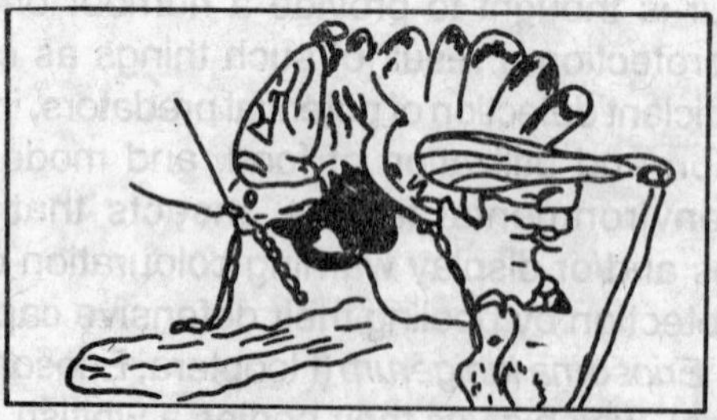

Fig. 14.8. A freshly mated female Mormon cricket feeding on proteinaceous material in the spermatophore deposited by the male.

damselflies serves a dual function: By means of a scooplike extension of the penis, any sperm present in the female genital tract is removed before the male introduces his own sperm.

Male field crickets, having attracted a female via a calling song; switch to a "courtship song" then, after mating, to a *staying together*" song, which ensures that she will not mate again before laying eggs. In the case of the hangingflies we discussed earlier, the male has invested efforts in a "nuptial gift," and it is to his advantage to ensure that the resulting offspring carry his genes; if the gift is large and the female fully inseminated, she enters a refractory period and will not mate again until her eggs have been laid.

The male hangingfly, in fact, contributes indirectly to egg production by supplying nutrients to the female. It has recently been shown by the use of radioactive tracers that male butterflies of certain species contribute to egg production via nutrients supplied in the spermatophore. The spermatophore of some of the Orthoptera are particular large. In the Mormon cricket 20% of the male's weight is lost in a single mating. Much of the spermatophore consists of a mass of proteinaceous material this is consumed by the female following mating. Females compete vigorously for calling males, and males reject smaller females, who have been shown to be less fecund. Since rejection usually occurs after the female has mounted the male, it appears that it is this time that the male assesses the weight of the female. This unusual example of "sex reversal"—competition among the females for males and male selection of females likely to produce the most offspring—has recently been elucidated by Darryl Gwynne, then at the University of New Mexico.

Sex reversal in which the male cares for the eggs is also known in a few instances. In giant water bugs there are repeated bouts of copulation interspered with egg laying these bouts are dominated by the male, who receives several eggs on his back following each mating. On one occasion a pair were seen to copulate over 100 times in 36 hours, resulting in the transfer of 144 eggs to the back of the male. While carrying the eggs, the male subjects them to necessary aeration, and he assists the young as they emerge from the eggs. Such a system ensures that the male carries eggs that bear his genes. But a male can be "cuckolded" if he mates with a female who carries sperm from a previous mating. Robert Smith, of the University of Arizona, vasectomized a male and mated him to a female that had previously mated to a male homozygous for a dominant genetic marker, a dorsal stripe. The vasectomized male received 75 eggs as a result of this mating; most were infertile, but several were fertile and give rise to striped offspring, demonstrating that this male had been "cuckolded" by another male. Thus, despite this elaborate mating system, a male does run a risk when he mates with nonvirgin female.

Brood Care by Females

Since, in general, it is the female that invests the most in term of physiological commitments to her offspring, it is not surprising that when brood care occurs, it is usually the female that is involved. Brood care has evolved in a variety of insects that have developed strategies opting for maximum protection of the offspring as opposed to production of large number of offspring that are "left to their own devices," Extreme examples are provided by certain cockroaches and by tsetse flies, which retain and nourish their larvae internally, in a uteruslike structure. More commonly eggs are "brooded" by the female, although brooding does not usually imply the transfer of heat as it does in birds. Rather it serves to protect them from predators, egg parasites, mold, or other factors that might destroy them. Female stink bugs of several species cover their eggs and small larvae much as a hen will cover the chicks. Such eggs suffer high mortality from ants and other predators if the female is removed crickets in which the female prepares a burrow that she guards vigorously from intruders. The eggs are laid in a cell at the bottom of the burrow and then the young hatch, they are fed by the mother, at first with small, infertile eggs that she lays, later wit food brought in

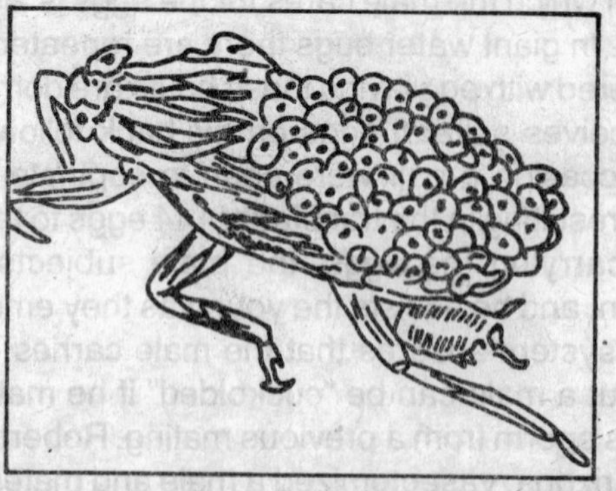

Fig. 14.9. A male giant water bug bearing eggs, in this case in part hatching to produce striped offspring known to have been fathered by another individual.

from the outside. Beetles of several families exhibit brood care that in its initial stages may involve both sexes. Dung beetles (Scarabaeidae) prepare balls of dung, roll them to a suitable site, and bury them as food for the larvae. Often male and female work together to make the dung ball and bury it. In some cases the female remains the burrow after laying her eggs, standing guard and keeping the pellets clean and well formed. She may remain with her offspring until they are fully developed, then emerge from the ground with them. In carrion beetles (Silphidae), male and female work together to bury a dead animal, then prepare a ball of decaying flesh in which the eggs are laid. As the larvae grow, they are fed by the female with regurgiated food, much as a bird fees its nestlings.

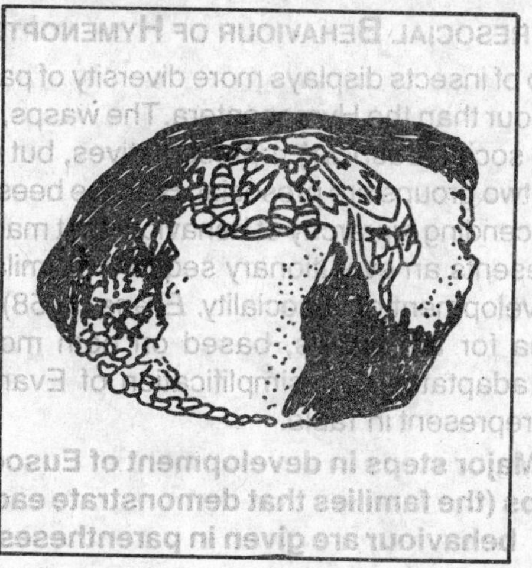

Fig. 14.10. Female carrion beetle (or burying beetle) feeding her larvae.

Brood care reaches its greatest development among the Hymenoptera, particularly in social groups such as the ants and some of the bees and wasps, which we shall consider in the next chapter. In solitary wasps—from which the social Hymenoptera are believed to have evolved—the female commonly prepares a nest, provisions it with paralyzed insects or spiders, and after oviposition seals it off in such a way that it is well protected against predators and physical factors in the environment. We have already discussed some of these wasps briefly. Of special interest are those species in which the nest cell is not sealed off immediately but is visited repeatedly by the female, who brings in prey over a period of days as the larva grows. *Progressive provisioning,* as this is called, involves such contact between parent and offspring as well as further protection of the larva as a result of the mother's continued presence in the nest. Of even greater interest are cases of *communal nesting,* that is, instances in which several females are active in the same nest, preparing and provisoning individual cells without aggression with other females in the nest. In many cases these females are sisters or mother and daughters, and the "extended family" they represent is perhaps an important progenitor for the colonies of social species. In both cases—progressive that the selective advantages relate to reduction in the opportunities for entry by various nest parasites and predators.

Presocial Behaviour of Hymenoptera

No group of insects displays more diversity of parental care and social behaviour than the Hymenoptera. The wasps, bees, and ants all have truly social (*eusocial*) representatives, but most members of the former two groups are presocial. Both the bees and the wasps display an ascending hierarchy of behaviour that many investigators believe represents an evolutionary sequence similar to that which led to the development of eusociality. *Evans* (1958) presented one such schema for the wasps, based on both morphological an behavioural adaptations. A simplification of Evans' behavioural sequence is represent in Table.

Table 14.3. Major steps in development of Eusocial behaviour of wasps (the families that demonstrate each level of behaviour are given in parentheses).

Step of Behavioural Sequence	Comments
1. Prey-egg (Pompilidae)	The female locates a prey, temporarily paralyzes it with her sting, lays eggs on the prey, and departs; the prey recovers and carries the wasp larvae which feed as external parasitoids.
2. Prey-natural crevice-egg (Pompilidae, Sphecidae)	The female drags the paralyzed prey to an available protective crevice where it is left with an egg attached; the female thus provides young with a level of protection.

3. Prey-nest-egg (Pompilidae, Sphecidae)	Female paralyzes a prey and then constructs a nest in which it is placed along with an egg; this is slightly advanced level of parental care.
4. Nest-prey-egg (Pompilidae, Sphecidae)	Same as step 3 except the nest constructed before the prey is captured; this introduces homing in that the female must return to a previously selected nest site.
5. Nest-prey-egg-prey (Sphecidae)	Similar to step 4, but the addition more prey egg is laid introduces mass provisioning as a more advanced form of parental care.
6. Nest-prey-egg-prey (Sphecidae, Eumenidae)	Instead of mass provisioning, the nest is provisioned progressively with fresh prey; this bring the female into contact with her developing offspring; in some species, the female remains in the nest when not provisioning, thereby reducing predation, and may also clean the nest of partially consumed food.
7. Prey macerated by female (Eumenidae, Vespidae	In the process of progressive provisioning, the fresh prey are macerated by the female and fed to the larvae; this brings the female into direct contact with her offspring and provides opportunity for trophallaxis and the transfer of pheromones.
8. Female life prolonged and offspring remain with nest (Vespidae)	The prolonged female life result in overlap with the first generation offspring, which remain and lay eggs in cells they add to the nests; this results in small colonies consisting of the mother and a group of undifferentiated daughters.

9. Trophallaxis and division of labor (Vespidae)	Mother and daughters cooperate in nest building and the care of young, but there is no permanent division of workers and egg-laying castes, trophallaxis paves the way for queen dominance.
10. Queen dominance (Vespidae)	The original offspring are all females that are incapable of producing their own female offspring, thus separating the reproductive and workers castes; intermediates may be common.
11. Differential larval feeding (Vespidae)	Differential feeding of the larvae and trophallaxis lead to the production of a well-defined workers caste strongly differentiated from the queen, and a reduction in the number of intermediates.

The aculeate (stinging) Hymenoptera are believed to have evolved from parasitic forms, and the behaviour of many solitary, predatory wasps is clearly related to that of their parasitic ancestors, except that the ovipositor is used to narcotize their prey. In the more primitive species, the female searches for an appropriate prey, paralyzes it with her sting, lays an egg on its surface and departs. These prey, left unprotected, are clearly subject to consumption by other organisms or to accidental destruction before the wasp larva can complete its development. The tendency of other species to conceal their prey, either before or after they have deposited their egg on it, is a form of parental care that has obvious survival benefits. Placing a single prey in a brow in the manner of some *Ammophila* species is the full extent of parental care displayed by a number of wasps, yet it requires a great deal of effort. These species utilize single prey such as caterpillars that are large enough to fulfill the nutritional needs of their young. The progressive provisioning displayed by the sphecids, for example, not only permits the utilization of smaller prey that can be carried back to the nest in flight, but also has the potential for improving larval nutrition. The repeated visits of the female to the nest also provides an opportunity for her to come in contact with her young and opens the way to the exchange of

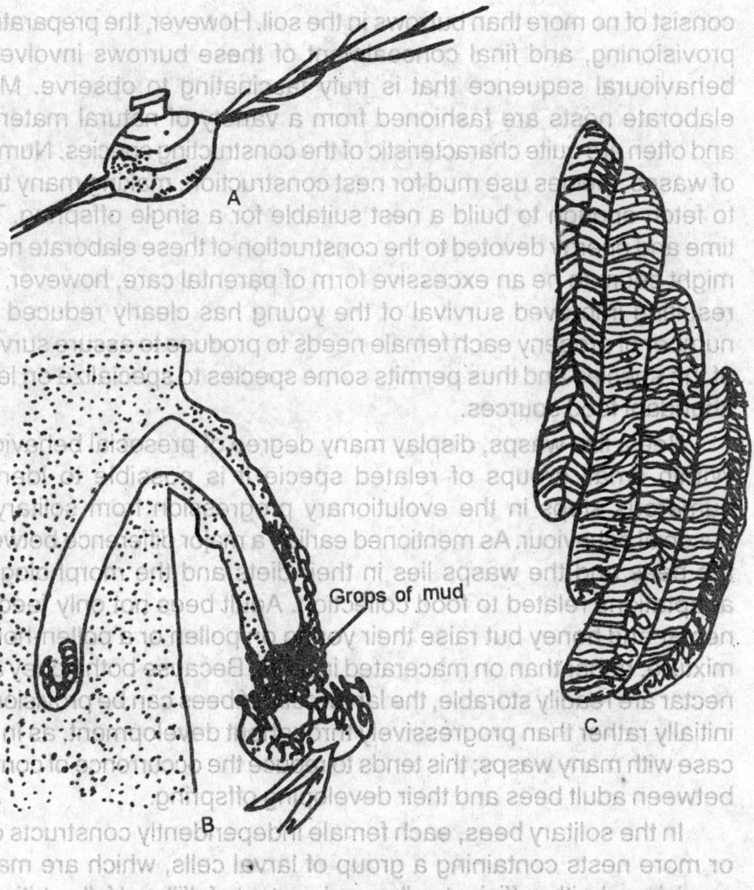

Fig. 14.11. Some examples of nests constructed by solitary hymenopterans (A) Juglike nest of the potter wasp Eumenes. B—Section through the nest of the solitary wasp Oplomerus showing the stronge tunnel provisioned with a caterpillar and the down-curved entrance tube constructed of mud. C—Group of mud nest units of the pipe organ and dauber Trypoxylon.

hormones, pheromones, and gut symbionts, along with partially digested food (*trophallaxis*) and the possibility for parental control over the development of offspring, as seen in the burying beetle *Necrophorus*.

The development of patterns of behaviour that increased provisioning seems to have been accompanied by the construction of fairly elaborate nests. Among the wasps, the simplest nests consist of no more than burrows in the soil. However, the preparation, provisioning, and final concealment of these burrows involves a behavioural sequence that is truly fascinating to observe. More elaborate nests are fashioned from a variety of natural materials and often are quite characteristic of the constructing species. Number of wasps species use mud for nest construction, making many trips to fetch enough to build a nest suitable for a single offspring. The time and energy devoted to the construction of these elaborate nests might seem to be an excessive form of parental care, however, the resulting improved survival of the young has clearly reduced the number of progeny each female needs to produce to assure survival of the species and thus permits some species to specialize on less-abundant food sources.

Bees, like wasps, display many degree of presocial behaviour. Within small groups of related specie it is possible to identify numerous steps in the evolutionary progression from solitary to eusocial behaviour. As mentioned earlier, a major difference between the bees and the wasps lies in their diets and the morphological adaptations related to food collection. Adult bees not only feed on nectar and honey but raise their young on pollen or a pollen-honey mixture, rather than on macerated insects. Because both honey and nectar are readily storable, the larval cells of bees can be provisioned initially rather than progressively throughout development, as in the case with many wasps; this tends to reduce the occurrence of contact between adult bees and their developing offspring.

In the solitary bees, each female independently constructs one or more nests containing a group of larval cells, which are mass-provisioned with sufficient pollen and nectar to fulfill and full nutritioned requirements of the larvae. Such is the way of the leaf-cutter bee, *Megachile rotundata*, which nests above ground in preexisting holes of appropriate diameter and depth. The female, after locating a suitable nest site, constructs a series of larval cells in the tubular hole by lining the tunnel with pieces cut from leaves and petals. Each cell is provisioned with 7 to 12 loads of pollen and then separated from the next by several discs of leaf material. In her life span of about 6 weeks, each *M. rotundata* female will construct 5 or 6 such nests, totaling approximately 30 larval cells.

Other solitary bees, like the soil-nesting alkali bee, *Nomia melanderi*, form nesting aggregations. The female constructs a more or less vertical burrow and, like the leaf-cutter bee, constructs a series of larval cells mass provisioned with pollen. The suitability of certain patches of soil often results in the construction of a number of rests in one area but there is no cooperation between individual females.

Perhaps because of the tendency to mass provision made possible by their food type, the bees do not show the variety of parental care behaviour displayed by the wasps. However, species of *Nomia* construct composite nests consisting of a communal entrance with cells off on side branches that are constructed and provisioned by individual females. *Michener* (1969) classified these species as communal and suggest that they benefit from an economy of nest construction labour. The regular back-and-forth movement of a group of females using a single entrance probably enchances the defense of the brood as well.

The parental care and brood-rearing behaviour of some presocial wasps and bees is clearly only slightly less elaborate than that of eusocial species. However, the formation of large social groups, involving a division of labour and queen dominance, requires a higher level of behavioural coordination.

15

HUMAN BEHAVIOUR

Of all the organs of the body, by far the most complex is the brain. The evolution of the intricacies of the human brain has enabled *Homo sapiens* to become the dominant species on the planet Earth. The human is the animal which is best equipped to overcome challenges imposed by the environment. Largely as a result of his superior mental abilities, the human is less restricted than other species by environmental problems. In this sense, his greater independence of environment, the human is the most highly developed of all creatures. Not only does he cope with existing environment problems, but he is also capable of changing the environment to suit his needs. In so doing, he is also influences the course of his own evolution as well as that of other species. The ultimate benefits to be reaped from tempering with natural selection poses many profound questions bearing on the very future of life on earth. How the human uses his superior brain to manipulate the environment is a direct reflection of human intelligence. Intelligence is reflected in the many facets to human behaviour which has the potential to influence the very existence of all life forms.

Any information which can be obtained to provide an insight into the nature of human intelligence and behaviour has a decided value. Since the human nervous system is so complex, it should come as no surprise that analysis of the hereditary and environmental components which influence human intelligence and behaviour is the most difficult to all gentle investigations. However, many facts related to the problem are known. Sufficient data may in time accumulate to permit a breakthrough into a fuller appreciation of the human mind and the reasons why the human acts as he does.

Let us begin by noting several obvious facts about one aspect of human behaviour, speaking a language. The ability to speak entails a decided advantage for the species. It enables members of one generation to pass the assembled knowledge, the culture of the society, to the next generation with no loss of time and no need on the part of the later generation to learn everything anew. Without this ability, each new generation of a species loses the advantage to profit from the knowledge accumulated by the efforts of previous ones. The ability to communicate through speech and the written works associated with oral language makes it possible for each generation to build on the culture of its ancestors, to utilize it, and improve it, thus becoming more and more independent of environmental problem.

These points about language indicate something about behaviour in general. Just like any other feature of an individual, behaviour is a phenotypic characteristic. We speak of *species-specific behaviour* when we refer to the assemblage of behaviour patterns which is considered typical of a given animal. We expect members of a given species to behave or conduct themselves in a certain way. A particular species of bird goes through its expected courting behaviour, builds and nest incubates its eggs, and feeds its young. The young, in turn, respond to the mother in a typical fashion. If the bird is a duck, the young will follow her shortly after hatching and will soon learn to swim.

The behaviour of human parents in rearing their young has specific features which are expressed during the long period of development required by the very helpless human infant. While incapable shortly after birth of performing feats such as walking, swimming, or feeding himself, the potential of the human infant exceeds by far that of any other animal. The human's genetic endowment has provided him with a central nervous system of the utmost complexity. The countless nerve connections and relays in the brain itself provide him with the machinery to accomplish something no other animal can do. This is the ability to deal with abstractions. To receive symbols, arrange, and interpret them in a meaningful coherent fashion. There is evidence from studies with chimpanzees that these animals possess a certain ability to deal intelligently with symbols, but it does not approach the facility of the human in this respect. Moreover, the human is unique in the ability to communicates symbols freely through a spoken language. Again, chimpanzees have been taught after labour to utter a few simple words. The inability to the ape to speak freely may not be a consequence of the fact that the ape is not equipped anatomically to handle spoken words. Speech demands a voice box, palate, nose, and mouth cavities of a certain construction. No other animal possess the type of anatomy required for the free formation of words.

Also essential for proper speech is a normal hearing apparatus. This is the receiving part of the system which delivers the sound stimuli to the brain. It is here that these stimuli are arranged and interpreted into a meaningful whole. Before any human can speak properly, he must be provided with these sound stimuli. These enable him in turn to form his own sound by association of symbols in the complexes of his brain. Ability to speak a language, therefore, is a complex characteristic. It depends in large part on anatomy, the hearing apparatus to receive the sound stimuli, the brain cells to interpret them, and the vocal apparatus to form words. The anatomical features necessary. Certainly, every fertile human mist carry in the sperm or egg information required for the embryo to develop these anatomical requirements. A defective gene may cause upsets in the differentiation of one of these structural parts. Without sound reception, a human will never be able to speak clearly and freely, even though the brain and vocal apparatus have developed perfectly.

But proper hereditary information is by no means the sole requirement. An environment trauma (viral attack during fetal development, birth accident) may deprive an infant of sound reception. Again, without the appreciation of sound, even though the proper genetic endowment is present, such a person will lack the facility with spoken words possessed by a more fortunate individual.

The ability to deal with symbols is a reflection of the intangible known as *intelligence*. The increase in intelligence which the human possesses over other animals is seen in the greater ability of memory retention, to think, and to arrange information stored in the brain in such a manner that foresight is possible. This foresight enables the human to act in ways which permit him to avoid potentially harmful environmental influences. With foresight, the human can plan and transmit his knowledge and culture. This thinking, learning and foresight depend, of course, on a brain with billions of neurons (nerve cells). Increase in intelligence thus depends to a large degree on increase in the number of brain cells and thus increase in the size of the brain. It is well knowr. that differences in intelligence among different animal species is related to differences in brain size. However, *within* a species, there is no evidence to indicate that a larger brain reflects greater intelligence. Nevertheless, the species-specific behaviour of the human in regard to the ability to think, plan, and speak depends on an important genetic component which provides the information to construct a brain of a certain size as well as the other anatomical features essential for the expression of the brain's full potential. However, the environment in which these genetic factors operate is a paramount importance may carry genetic factors for normal speech or above-average intelligence, the environment can deprive him of the full benefits which they provide. We see that behaviour, like any other phenotypic characteristic, is thus the result of the interaction of many genes and the environment. It should also be evident that species-specific behaviour has been subjected to the operation of the forces of evolution. Natural selection has provided each species with a particular behaviour which enables it to cope best with its environmental challenges, behaviour. Therefore, has an advantage just as any other characteristic of a species of individual. The human's behaviour, as noted here, has enabled him to become less restricted than any other species, to be able to make use of the greatest number of environmental niches, and to manipulate the environment for his own end.

While much of human behaviour depends on heredity, it must not be forgotten that a great deal of it depends on learning. Human behaviour is distinguished by its plasticity, its ability to be changed and molded. A normal child is born with genetic factors which make it possible for him to receive environmental stimuli. Among the most important of these are the one provided by other humans. An infant is associated with a family, and as he continues to develop, he is exposed to more and more members of his particular cultural group or society. Certain stimuli will determine what language he will speak. He carries the genetic information which provides he ability to speak, but whether he will speak and what language he will become expert in, depends on other humans about him. The amazing inherent talent which a child possesses to learn a language is seen by the youngster's ability to become expert in several tongues if he receives he appropriate sounds during his formative years. No one must tell him what is English, Chinese, or French. He sorts them out automatically without being told a thing about rules of grammar. But alas, this genetic ability is lost about the age of puberty, so that no adult will ever have the facility of the child who can so effortlessly master two or more tongues. Not will most adults ever speak a language perfectly if they have learned it after they have reached sexual maturity. There is nothing that dictates what language a child must speak. The great plasticity of this aspect of human behaviour is quite apparent, as is the interaction of its hereditary and environmental components. Few human behavioural traits are based exclusively on genetic factors. This plasticity provide by the environmental interaction with the genetic component is seen in a child's general conditioning as he becomes integrated into his particular society. He learns the customs of his society and tends to behave according to the expectations of his specific cultural group. Reared in a completely different society with different values and expectations, another child with the same genetic potential as the first one will come to behave very differently as he acquires different habits, skills, and beliefs prescribed by the second group.

GENES, CULTURE AND BEHAVIOUR

The fact that sociobiologists understand that animals do not have within them a gene or genes ordering the individual to be automatically aggressive or to copulate once a day no matter what does not satisfy many social scientists. These students of human behaviour often argue that there is not even an indirect connection between human genes and human behaviour.

For *Sahlins* and many others, two aspects of human behaviour are taken as convincing evidence that human behaviour has not evolved to promote the survival of the genes of individual human beings. First, the critics not that if you were to ask any human why he did something, the person would *never* respond that he performed the action to elevate his inclusive fitness. But if a body cuckoo cold talk, it too surely would not tell an interrogator that it rolled its host's eggs out of the nest "because may genes made me do it". No animal need possess a conscious awareness of the ultimate reasons for its activities. It is enough that its proximate mechanisms predispose it to behave in ways that help its genes survive. On the proximate level we enjoy eating sweet foods, fail in love, want to be liked by our friends and relatives, and learn a language because we possess physiological mechanisms that make it easy for us to do these things. We do not have to be aware of the caloric content of honey and its contribution to cell growth and maintenance in order to eat it. It is sufficient that our brains tell us that honey tastes good. Thus evolutionary theory does not require that an animal know the genetic consequences of its acts, only that animals with certain abilities or preferences will tend to replace those with other aptitudes.

The second objection is much more challenging. Social anthropologists have documented that human cultures are amazingly diverse. To them this suggests that we are the completely flexible product of whatever social environment we choose to invent. How can it be said that humans tend to behave in ways that elevate their fitness when humans around the world behave in such different ways? There are polyandrous, polygynous, and monogamous societies; cultures in which females make important political decisions and others in which males dominate the females. There are human groups for which warfare is a constant fact of life and other groups that never fight openly with one another. The list of cultural peculiarities is seemingly endless; depending on where you were born, you might be allowed to marry your first cousin or not; you might be allowed to look at your mother-in-law or not; you might be forced to have your penis cut from stem to stem at adolescence or not. Moreover, cultures are not only enormously variable, but they the capacity of change rapidly as seen in the transformation of human societies over the past 10,000 years from predominantly hunter-gatherer groups to the industrial high-density, high technology civilizations of today.

The indisputable diversity and flexibility of human behaviour potentially provide test after test of the evolutionary approach. We are an animal species. We do have within us genes. Therefore, evolutionary theory predicts that human behavioural traits, no matter how diverse of flexible, rest on evolved abilities that will tend to promote the genetic success of individuals. In fact, opponents of sociobiology could demolish not only sociobiology as applied to human beings but evolutionary theory in general if they could show that any basic human ability consistently reduces the inclusive fitness of individuals! *Darwin* recognized this general point, writing: "If it could be roved that any part of the structure of any one species had been formed for the exclusive good of another species, it would annihilate may theory, for such could not have been produced through natural selection". In modern terms, Darwin's challenge could be rephrased to read that current evolutionary theory would be annihilated if it could be shown that a trait originated exclusively for the genetic benefits of individuals that did not share genes with the practitioners of the trait. There are certainly some possible candidates for genetically altruistic human traits that by evolutionary terms should not have evolved if they really do result in lowered inclusive fitness; risking death in warfare, the use of birth control measures, and the practice of infanticide, for example, however, to show how these and other similar phenomena are all explicable by an evolutionary approach and that we can, at least for the movement, retain our confidence in evolutionary theory as applied to animals in general and humans in particular. But first let us consider why it is that cultural diversity may have evolved.

The Evolution of Cultures

Earlier chapter in this book presented evidence in support of two principles that are essential for an understanding of the possible evolutionary bases of human cultures. We begin with the realization that cultures are the product of the neural machinery within the human skull. Cultural practices and learned behaviours, transmitted from one generation to the next through tradition (teaching). One basic biological principle relevant to an understanding of cultures is that all learned abilities have a genetic foundations of some sort and, therefore, can and to evolve. In order for cultural practices to be acquired by each new generation, young human must have brains that are capable of storing the information contained in tradition. We can predict that to the extent culturally acquired information elevates individual fitness there will be selection in favour of humans who are able to absorb this information more completely, rapidly, or with less expenditure of effort than others. Conversely, if a cultural practice reduces inclusive fitness, individuals who were *unable* to learn it should enjoy a selective advantage.

Human brains appears to have properties that facilitate certain kinds of learning. The classic example is language learning; there is considerable evidence that the brain contains neural mechanisms that facilitate language acquisition and use. Simply because there is cultural variation in the languages spoken around the world is not proof of the independence of this behaviour from genetic influence. Genes are deeply involved in the production of the neural foundation of language, and it seems inescapable that alleles have been selected on the basis of how well "their" brains have been able to contribute to language use. If selection can produce humans with great skill at language, it should also be able to shape the design of the human brain so that humans can readily learn which individuals are relatives, how to use tools, how to behave in certain social settings, how to extract economic grains from the environment and so on. As in the case of language, the successful genes need not produce neural mechanisms that are restrictively designed to permit only the learning of a specific cultural trait (such as the Turkestan of Jivaro or English language). Instead, genes may endow their brains with more open learning abilities, permitting a more flexible (but still not completely plastic) accommodation to whatever cultural variables dominate an individual's social environment. By this view, cultural evolution involves selection for various learning abilities that permit individuals to adopt that cultural practices of their societies. Thus genetic and cultural evolution complement one another rather than compete with each other.

But why have individuals with genes that indirectly helped them acquire a culture enjoyed a selective advantage in the history of our species? To answer this question it may be helpful to consider a second basic behavioural principle discussed earlier in the book: that social organization is a reflection of the ecology of the species. If a species confronts varied ecological conditions, it may evolve the ability to alter its social structure to match the conditions of the movement. If this is true for humans, cultures may have evolved because the adoption of cultural practices enabled humans to invade a wide variety of environments each with its own special ecological pressures. Many authors have noted that cultural adaptation permits a more rapid adjustment to variable ecological circumstance than biological adaptation. If this hypothesis is correct, we should predict hat cultural variation is not random; instead, the cultural practices of one group should permit individuals to exploit their environment more successfully that other cultural options. We should also expect that cultural changes elevate the inclusive fitness of the individuals that invert or adopt the change. People that practice the wrong option (one that results in lowered relative reproductive success) will be replaced by those who practice the superior option, transmitting this option culturally to their offspring. It seems obvious that the spread of tool use and the perfection of tool manufacture that occurred during human evolution were related to improve resource acquisition and the ability to produce and support more offspring. Moreover, the two major cultural innovations of the past 10,000 years—the invention of agriculture and the development of the use of fossil fuel as an energy source— have (1) spread extremely rapidly around the world, and (2) been associated with a phenomenal increase in the human population. Both inventions, in effect have given participating individuals access to more calories and nutrients for conversion into offspring than non-participating individuals.

But what about those many peculiar culture practices that actually appear to reduce the fitness of individuals rather than raise it? In some American culture it is customary to cook cornmeal in a 5 per cent solution of lime, pour off the solution, cool the mash, wash it, and finally use the prepared material to make tortillas. This involves a considerable expenditure of time and energy as well as he loss of some of the thiamine, niacin, nitrogen, and fat content of the corn. Although it would be easy to interpret this method of corn preparation as maladaptives, *S.H. Katz* and his colleagues have shown that alkali cooking actually improves the nutritional quality of cornmeal. It does so by breaking down the indigestible portions of the corn kernel and thereby freeing for human use substantial quantities of certain essential amino acids, including lysine, which is deficient in those parts of the corn seed that can be digested without this system of preparation. A diet that is dependent on cornmeal as a staple would be dangerous without alkali cooking; we lack the metabolic pathways to manufacture lysine, an amino acid that appears in many vital proteins. Moreover, alkali-treated corn has a lower ratio of the amino acids leucine to isoleucine, which improves its nutritional value. It is hard to imagine how someone invented this cooking procedure, but it is striking that all South and Central American cultures that rely on maize as the basic foodstuff have adopted the practice.

Table 15.1. Effect of Alkali cooking on the availability of certain essential amino acids.

	Milligrams of amino acid per gram of nitrogen in	
Amino Acid	Uncooked com	Alkali-treated com
Histidine	0.012	0.028
Isoleucine	0.074	0.158
Leucine	0.309	0.358
Lysine	0.045	0.126
Threonine	0.145	0.381
Ratio: Leucine/Isoleucine	4.2	2.3

Another cultural trait that has baffled outside observers is the sacred-cow tradition, which forbids or greatly restricts the sale or slaughter of cows for meat. In India the Hindu reverence of cows is such that they are allowed to roam freely, interfering with traffic, and they are almost never killed, even in times of famine. Why are not emaciated beasts sold or killed and eaten by their owners? *Marvin Harris* has argued persuasively that far from being a liability cow love is almost certainly to the economic benefit of the Indian

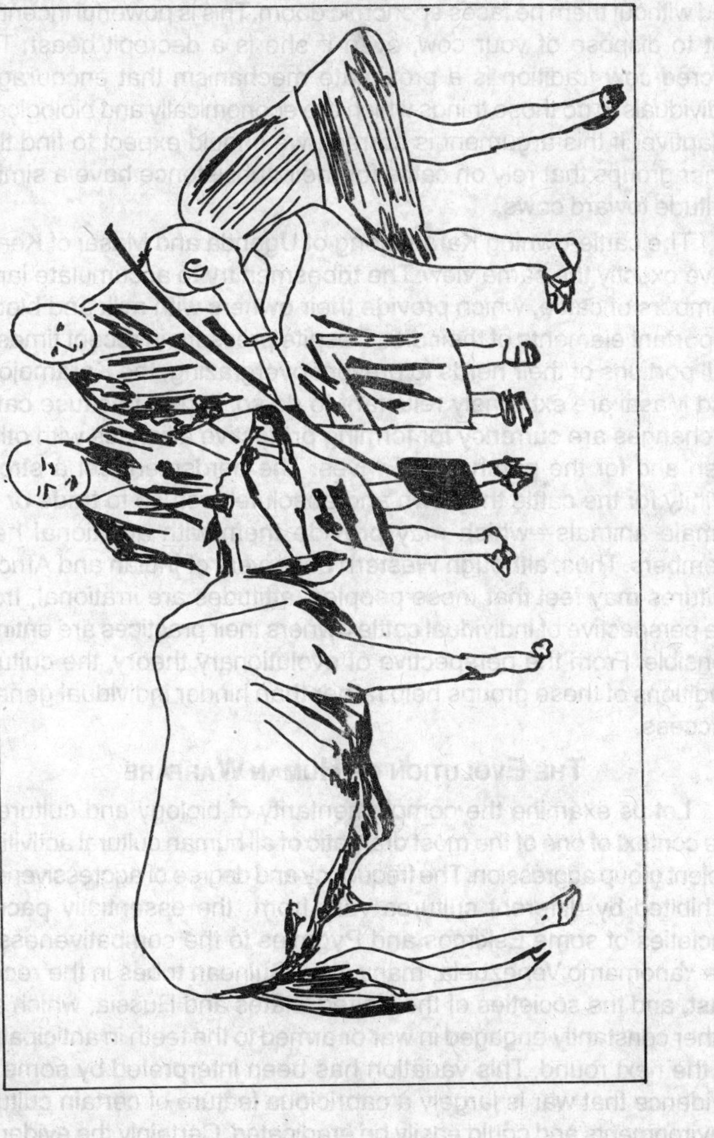

Fig. 15.1. Group of Masai Men.

peasants who maintain the tradition. Emaciated or not, cows provide their owners with a certain amount of milk and with quantities of dung, which is used for fertilizer, cooking, and construction; most importantly they occasionally produce offspring, the males of which can be converted into oxen. Oxen are the Indian farmer's tractor, and without them he faces economic doom. This is powerful incentive not to dispose of your cow, even if she is a decrepit beast. The sacred-cow tradition is a proximate mechanism that encourages individuals to do those things which are economically and biologically adaptive. If this argument is correct, we should expect to find that other groups that rely on cattle for their subsistence have a similar attitude toward cows.

The cattle-owning Karamojong of Uganda and Masai of Kenya have exactly the same view. The tribesmen try to accumulate large numbers of cattle, which provide their owners with milk and blood, important elements of their diet. Despite pressure in recent times to sell portions of their herds to prevent overgrazing, the Karamojong and Masai are extremely reluctant to do so, in part because cattle exchanges are currency for forming protective alliances with other men and for the purchase of wives. The herdsmen feel a strong affinity for the cattle they own and absolutely refuse to trade or kill female animals—which may provide them with additional herd members. Thus, although Western observers of Indian and African cultures may feel that these peoples' attitudes are irrational, from the perspective of individual cattle owners their practices are entirely sensible. From the perspective of evolutionary theory, the cultural traditions of these groups help rather than hinder individual genetic success.

The Evolution of Human Warfare

Let us examine the complementarity of biology and culture in the context of one of the most dramatic of all human cultural activities, violent group aggression. The frequency and degree of aggressiveness exhibited by different cultures vary from the essentially pacifist societies of some Eskimos and Pygmies to the combativeness of the Yanomamo Venezuela, many New Guinean tribes in the recent past, and the societies of the United States and Russia, which are either constantly engaged in war or armed to the teeth in anticipation of the next round. This variation has been interpreted by some as evidence that war is largely a capricious feature of certain cultural environments and could easily be eradicated. Certainly the evidence indicates that the genotypes of humans do not blindly program them to engage in warfare. But this does not mean that our genes have not endowed our nervous systems with special properties that provide humans with the *capacity* to engage in group aggression under *certain conditions.* From a evolutionary perspective it is possible that some ecological circumstances may so elevate the benefits of warfare to individuals that they exceed the costs to individual inclusive fitness. An argument that the abilities that underlie the capacity for warfare are generally advantageous does not imply either that warfare is inevitable or that it is morally desirable. This does not prevent me from accepting as a hypothesis that the traits that may be employed in group aggression can help maximize the inclusive fitness of the humans that possess these traits. Here we shall examine the plausibility of the hypothesis, taking this example to illustrate how (1) evidence about our history as a species, (2) comparative behavioural data from other living animal species, and (3) ecological comparisons of human cultures may be brought to bear upon the question.

The Evidence from Human History

Group warfare is an organized and usually premediated aggression with the goal of destroying or maiming other human beings whom the aggressors may not even know. Perhaps paradoxically, it is promoted by the cooperative abilities of members of each warring group. The great capacity or cooperative among the members of human clans, tribes, villages, and nations can be seen as the product of powerful selection forces operating on our ancestor. Although experts disagree on some details of the interpretation of the fossil record, there is nearly universal acceptance of the following points. Our hominid forebears were group-living primates, closely related to the great ape lineage, whose key ecological *invention* was to become a hunter of living animals in addition to gathering vegetable foods. The use of tools became a central adaptation that enabled protohominids to capture and process game efficiently. The importance of this culturally transmitted adaptation may be related to the dramatic increase in the brain size of members of the human line over time and is surely reflected in alteration in the anatomy of the hand. The relation between the thumb and other fingers of *Australopithecus africanus* (a fossil hominid thought to be in the direct human evolutionary line or close to it) differs from that of modern humans. The changes that had occurred by 500,000 years

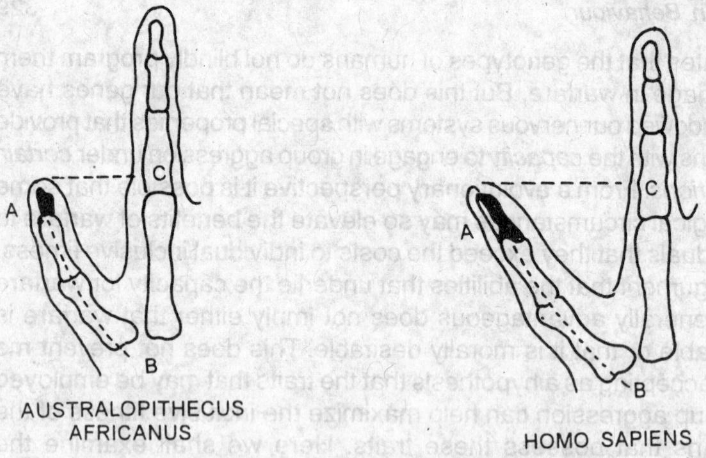

Fig. 15.2. Comparison of the Thumb and Forefinger.

ago permitted better *power* and *precision* grasping, capabilities of the greatest importance in tool making and tool using. The evolution of a longer straighter thumb with a broader tip meant that humans could apply more surface area to an object held between the thumb and other fingers, thus producing area a firmer grip. There is a corresponding increase in the abundance diversity, and quality of the tools found in hominid campsites over their evolutionary time.

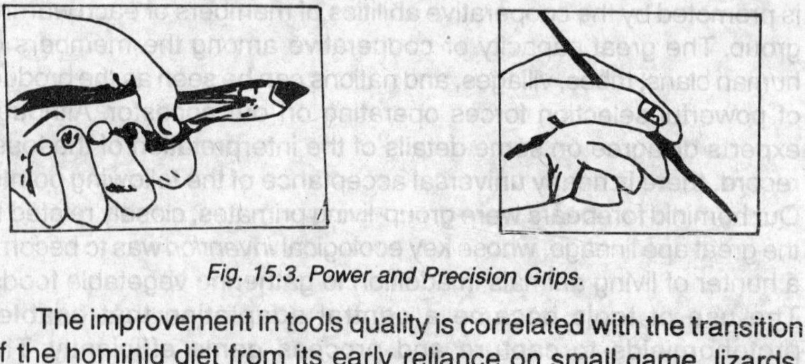

Fig. 15.3. Power and Precision Grips.

The improvement in tools quality is correlated with the transition of the hominid diet from its early reliance on small game, lizards, birds, fishes, and antelopes to a focus on really large animals. Campsites of our ancestors living as long as 300,000 years ago have been found that contain great quantities of the bones of the largest mammals then alive—mastodons, rhinoceros, and cave bears. Bit-game hunting many have been the central adaptation of most human populations form that time until the recent past. *Paul*

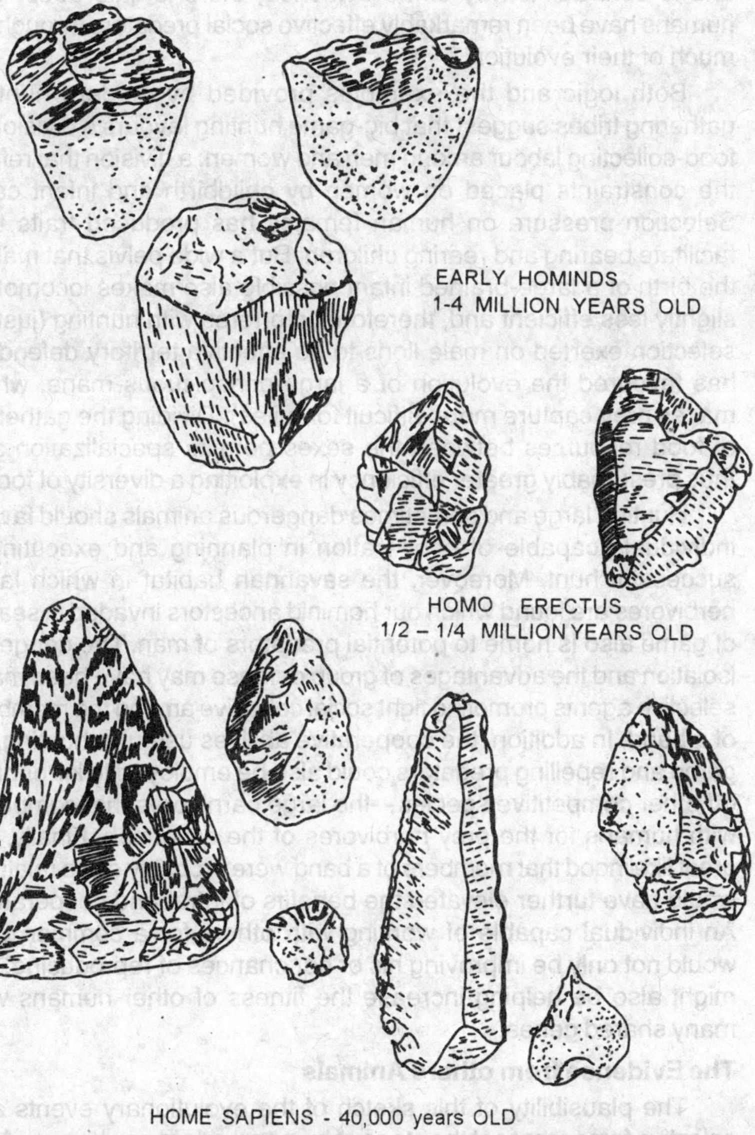

Fig. 15.4. The Evolution of Tools.

Martin has advocated the hypothesis that humans were such successful hunters that they may have been responsible for the mass extinctions of bit-game species that occurred first in Africa and then later in North America. Although his position is controversial and is contradicted by some evidence, there is little doubt that humans have been remarkably effective social predators throughout much of their evolution.

Both logic and the examples provided by modern hunting gathering tribes suggest that big-game hunting favours a division of food-collecting labour among men and women, a division that reflect the constraints placed on women by childbirth and infant care. Selection pressure on human females has produced traits that facilitate bearing and rearing children. But a wide pelvis that makes the birth of a large-brained infant possible also makes locomotion slightly less efficient and, therefore, interferes with hunting (just as selection exerted on male lions to be effective territory defenders has favoured the evolution of a large conspicuous mane, which makes prey capture more difficult for miles). Dividing the gathering of food resources between the sexes permits specialization and thus presumably greater efficiency in exploiting a diversity of foods.

Hunting large and sometimes dangerous animals should favour individuals capable of cooperation in planning and executing a successful hunt. Moreover, the savannah habitat in which large herbivores are found which our hominid ancestors invaded in search of game also is home to potential predators of man. The danger of isolation and the advantages of group defense may have been major selective agents promoting tight social cohesive among the members of a band. In addition, the cooperative abilities useful in hunting big game and repelling predators could also be employed to kill or drive off other competitive species—the large carnivores that competed with humans for the prey herbivores of the savannah. Finally, the great likelihood that members of a band were probably close relatives would have further elevated the benefits of intraband cooperation. An individual capable of working with others for a common goal would not only be improving his or her chances of reproducing, but might also be helping increase the fitness of other humans with many shared genes.

The Evidence from others Animals

The plausibility of this sketch of the evolutionary events and selective forces that set the stage of human warfare can be examined by making some careful comparisons of human behaviour with that of other animals. A common complaint of critics of an evolutionary analysis of human behaviour is that sociobiologists make caplicious use of comparative material, selective whatever species help support the hypotheses favoured by sociobiologist. However, as has been outlined in previous chapter, there are species to gain insight into the ecological and historical bases of a behavioural characteristic of interest. First, we can look at *closely related* species to determine which of their behaviour are similar to our own. This information can help us determine the history of a behaviour pattern. Second, we can examine *unrelated* species that experience similar ecological pressure to determine if they have evolved convergent behavioural solutions to these shared problems. For humans, therefore, the species whose behaviour may be most useful for an understanding of our own abilities include : (1) other primates, especially out closest relatives, the chimpanzee, gorilla, orangutan, and gibbon, and (2) the mammalian social carnivores, which are not closely related but which face the same obstacles associated with capturing big game.

Human Sex Role in Society

The relatives roles of the hereditary and environmental components in the expression of human behaviour has been a matter of heated debate in relation to the topic of human sex roles in society. Many persons expect a female to behave in a certain manner and a male to respond to a particular situation in a characteristic way which is distinct from that of a female. Generally, the female is expected to be more retiring, the male more aggressive. For a male or female to behave in a fashion considered inappropriate to his or her sex is often labeled *unnatural.* According to the ideas of some students of behaviour, characteristic male and female behaviours are fixed in the species. There is a genetic basis, according to this thesis, which is responsible for the expression of male social behaviour and of female social behaviour. The argument of the proponents of this idea is based mainly on studies of the fossil record, observations made on the behaviour of various species of primates (the order to which the human belongs), and observations of behaviour patterns in various human societies today. Proponents of the concept attribute the fixation of sex roles to the advantage this behaviour imparted to the human species during its evolution in the relatively recent past. Social behaviour in which sex roles are distinct would have given a decided benefit to the human when he become a hunter. Once the human species assumed hunting activities, so the argument goes, an assignment of roles to each sex would have given insurance to the success of these activities and hence would have increased the survival of the group. Any genes or genetic combinations which would magnify the differences in sex role behaviour would be selected under the force of natural selection. As the hunter, the human male would require strength, endurance, speed, alertness, and prowess among other attributes. Such hunting effects would benefit from improvement of these traits, and so any genetic determinants in this direction would be selected. However, such traits would benefit directly only the active hunter, the male. Interactions of genes for excellence in these traits with the hormonal environment would *fix them* in the male. On the other hand, any hormonal controls to make the female more different from the male would have value. The changes in the female during this part of human evolution would have occurred only in relations to the requirements of the male in his role as the hunter. In her role as childbearer and rearer, intelligence would not be at a premium. Cunning, strength, and swiftness would not be needed for food gathering activities. Genetic controls to make the female subordinate and docile would benefit the needs of the male after the demands of the hunt and would in turn make him a more efficient hunter. Hormonal regulations in the female to make her sexually receptive at any time would have value. They would have been selected in relation to the need of the hunter who would be assured that the female would be available to him whenever, he so desired after return from the hunt. Strong bonds would form between males. Such bonding would be of advantage since it would assure cooperation during hunting activities and would reduce hostility among the males. No such bounding among females would be required, since the female's activities did not depend on major group activities. Therefore, sex roles become more and more distinct as the success of the hunt became more and more a certainty.

According to this idea, we see today in our human societies behavioural traits which characterize the male and those which are typically female. These exist because they are set in the genetic program where they become established by natural selection during the time the human emerged as a hunter. So today, mainly the men perform decision-making activities. Women accept these decisions. The male is aggressive, whereas the female is the homebody concerned primarily with child rearing and pleasing the male. All of the behaviour patterns which we today consider male or female are largely a direct result of the genetic factors which control these activities and which became established in the past in relation to the hunting activities of *Homo sapiens.*

There are, of course, many arguments which can be offered to counter these ideas. Not the least is the demonstration that human behaviour is highly plastic, a feature which distinguishes it from that of other species. There is good evidence to show that an individual, regardless of sex, may be just as intelligent, docile, aggressive, or strong as any other, depending on the experiences of the individual during development. No evidence exists for some sort of maleness of femaleness in behaviour which is set and cannot be modified. There is no clear-cut evidence from the fossil record to reveal details of the social life of the human when the species emerged as a hunting society. It is possible that some of the participants in the hunt were actually females. The studies of behaviour in other primate groups should be viewed with suspicion when conclusions are drawn from them and applied to the human. First of all, one should remember that the human is a human and the monkey today is a monkey. To draw conclusions on the basis of an animal's behaviour today and to apply it to humans who once lived in the past are misleading tactics. All of our biological evidence indicates that no animal as it exists today is ancestral to any other which exists today. While they may have had a common ancestor in the distance past, this ancestor may have behaved very differently from either one of the modern groups today. In short, today's human did not rise from today's monkey or ape. Moreover, conclusions are often drawn from groups in which the male *happens* to be dominant. In some primate groups such as the gorilla, there are no sharp distinctions which set off the male as the strictly dominant sex. In addition, many conclusions are drawn on the behaviour of animals in captivity, a factor known to distort an animal's behaviour. Studies on so-called primitive human societies today provide no concrete evidence to the genetic assignment of sex role. The primitive populations of the human species cannot simply be assumed today to be exactly like those populations in existence millions of years ago. We don't know that much about the history of these groups today nor of the cultural changes in them which are very recent adaptations.

While a genetic components certainly exists for species-specific behaviour, we must recognize the advantage the human has in the flexibility which his genetic endowment permits in his behaviour. It would seem less wise, on the basis of highly debatable arguments.

To assert that certain types of behaviour are programmed on inherent than to appreciate the *range* of human responses made possible by the very plasticity which the genetic endowment allow. A fuller understanding of the subtle interplay between genetic factors and environmental influences in the establishment of human behaviour can lead to a deeper appreciation of differences among groups and individuals. It may also offer clues to the alleviation of serious problems which can arise from certain forms of human behaviour.

The Evolution of Human Reproductive Behaviour

Warfare is behaviourally highly complex, and its relation to inclusive fitness not immediately clear. Reproductive behaviour is much more obviously linked to individual fitness, and, therefore the imprint of selection should be readily apparent on it. Yet it is well known that there is considerable variation within and among cultures with respect to many reproductive practices. Moreover, there are actions, including mate sharing, adoption of another's child, and the practice of birth control, which, on the face of it, appear to reduce fitness. In this section, we shall look carefully at some aspect of cultural diversity in reproduction and cases of appearent genetic altruism to see whether they refute the evolutionary prediction that humans will behave in ways that tend to maximize their inclusive fitness.

The family unit is universally accepted by anthropologists and others as the basic components of human societies. Human males and females have the capacity, shared with many other animals, to form a family in which male and female live together and cooperate in the rearing of the progeny they produce. The selection of a mate should have a great deal to say about the genetic success of individuals. What evidence in there that human males and females behave in ways that promote their genetic success in this phase of the reproductive process?

Mate Selection

One significant feature of mate selection is the culturally reinforced prohibition against copulation between some sexually mature individuals. Insect taboos appear to be a universal feature of human social behaviour. Thus parent and their children are forbidden to copulate, and people are usually not permitted to marry their siblings (and often first cousins). Incest taboos are often culturally extended to forbid marriage between individuals belonging to particular categories, such as the same clan. The strength of the insect taboo has been demonstrated by the refusal of children reared together in Israeli kibbutzim to marry one another. When the kibbutz system was devised, the social scientists in charge envisioned that communally raised children would eventually marry within the kibbutz. Such marriages were actually encouraged, and yet not a single one has occured; the children of one kibbutz always fall in love with members of another kibbutz. It is as if the brain had a program that reads: "Do not become sexually attracted to humans with who you have been reared." (When unrelated cactus mice were taken from different litters and experimentally reared together, they too were reluctant to copulate with nestmates. Over evolutionary time juvenile mice, and humans, that grew up together were almost always siblings or very close relatives). The inability to become sexually attracted by close relatives contrasts sharply with the ability of humans to fall in love with complete strangers. This ability, the Incest taboo, and the culturally extended proscriptions against copulating with or marrying certain classes of individuals all contribute to outbreeding. The occurrence of similar mechanisms in other animal species suggests that a major advantage of these traits is the avoidance of inbreeding. This prevents the accumulation of deleterious recessive alleles in double close in one's progeny. Humans from incestuous matings exhibit a substantially higher rate of genetic defects and congenital abnormalities than members of outbred populations. Thus humans should benefit genetically to the extent that they avoid inbreeding. An added benefit to outbreeding among humans is the formation of alliances between different genetic lineages. The progeny produced in a marriage between two separate family groups creates a tie between the two bands that may promote cooperate between them. Without a mutual interest in the children of the marriage, the two groups might have no other shared concerns and no motivation to help one another.

Female Choice

The incest taboo applies to both males and females, but there is some indication that male parents break the taboo (by copulating with their daughter) more often than females. This may reflect the lower threshold for sexual excitation among males that females that is characteristic of animal species in which males may pass on their genes at little expense relative to the female, Female humans, like all female mammals, must make a large investment in each progeny while it is in the embryonic stage and during the early years of its life when it is nourished at the breast. Because a male's investment in an offspring may be as little as the sperm used to fertilize the egg, evolutionary theory predicts that females should be more selective than males in the choice of copulatory partners. There is ample anecdotal evidence as well as scientific data collected by Masters and Johnson that female of our culture are less easily sexually aroused than males. Females should be more cautious because they have much more to lose than a male. If a woman becomes pregnant at an inopportune time (e.g., when she is unmarried and her partner has no intention of helping her rear her child), her large physiological investment may be poorly rewarded. At least in the past, the child of a husbandless woman would have had a reduced chance of surviving to maturity; if it did reach reproductive age, it would probably have a lower fitness than average because of juvenile malnourishment or the possession of few material goods with which to attract mates. Moreover, a single mother may reduce her long-term chances to acquire a helpful mate, either because potential husbands would consider her a poor risk or because they are uneager to help care for another man's offspring.

Thus in cultures in which males may provide material resources for their children, a female should permit copulation to occur only if her partner has demonstrated both a willingness and an ability to make a large parental investment in their progeny.

Male Parental Investment

It is easy to see why females might prefer mates willing and able to provide them with food and protection. If a division of labour between the sexes has been the rule throughout much of human evolution, males have controlled highly valuable resources (such as protein-rich game in many societies), when used by a female and her progeny would greatly enhance their fitness. But, as discussed earlier, the male animal typically enjoys reduced benefits relative to the female from equal parental investments per offspring. First, the reliability of paternity is often less than the reliability of maternity, and a male therefore runs the risk of helping someone else's genes. Second, the cheapness of sperm means that males can potentially inseminate many females. This elevates the benefits of a reproductive strategy that emphasizes effort to copulate with many females rather than invest heavily in one or a few mates. Male mammals, including most primates, rarely are monogamous and rarely provide much in the way of parental care for the progeny of females with which they have copulated. Why are human males an exception to this rule?

The following explanation is speculative. We begin with the hypothesis that males have consistently been more successful hunters than females. A female protohominid in estrus might attract a consort male who would guard her during the time she was receptive and most likely to ovulate. During the consort phase a male might share prey he had recently captured with the female because it would be likely that the progeny produced by the female would bear his genes; protein gifts to the female might, therefore, have genetic gain for the male. (As usual, there is no requirement that individuals have to be aware of the genetic consequences of their acts). If gift giving became an established component of courtship and consort guarding, than a female might gain if she could extend the period of guarding and thus receive more presents. Instead of blatantly advertising the time of ovulation by becoming receptive just prior to this event, female protohominids might have partly hidden the time of ovulation by lengthening the period of receptivity. The advertisement-of-ovulation strategy is practiced by female mammals as an incitation display to encourage competition among males for access to them, thus increasing the probability of mating with a dominant male. But if parental investment is offered by a male, the quality of this investment may become more important to a female than signals of male genetic quality. By making if difficult for a male to determine when his chances of fertilizing an egg were highest, females that retained their receptivity could make it advantageous for at least some males to practice prolonged guarding. The more hidden the ovulation time and the more prolonged the period of receptivity, the longer a male must remain with a female (copulating with her at regular intervals) if he is to increase his chances of fertilizing her eggs when they do become available for fertilization. A widespread use of a strategy of permanent guarding (marriage) reduces the availability of fertilizable females and reduces the time a married male has available to find these females (because of the guarding time required to protect a mate). This being so, it may become advantageous to be devoted husband and father and to expend resources on one's presumptive offspring, which are highly likely to bear one's genes. This may increase their chances of surviving and reproducing, providing a genetic gain that may often (but not always) exceed a Casanova strategy of attempts at multiple sexual conquests.

Because human males can offer useful parental services to their offspring, one can expect females to choose husbands partly on their potential ability to provide superior investments, whereas males should make the parental investment in ways that tend to promote their own genetic success. A female should choose a husband who appears to be genuinely that he is willing to help her. Males can and do deceive some females, reducing them with the intent of deserting them at some point. Female coyness and caution about copulating have the function of reducing the probability of making mistakes about males.

Table 15.2. Characteristics of male that are of potential interest to females in selecting mates.

Degree of Relatedness
Fathers, brothers, proscribed relatives not acceptable mates
Willingness to Provide Parental Investment
quality of attention given female by suitor
Gifts given to female by suitor
Reputation of suitor
Ability to Provide Parental Investment
quality of gifts given to female by suitor
Resources known to be controlled by male
Dominance status of male within the society
Intelligence

Women should also be concerned about the economic status of males, present and potential. The fact that female in a great many human societies willingly enter into polygynous marriages is at first glance puzzling. One might predict that given the potential role of male in child rearing, females should prefer to mate with a single male. Carnivorous mammals generally show a greater tendency toward monogamy than herbivorous ones. In fact, the vast majority of human marriages have surely been monogamous. Nevertheless, why should there be any polygynous marriages freely entered into, given the male's potential contribution to the welfare of his wife and children? Perhaps human men are rather like redwinged blackbird males. Most human societies offer opportunities for truly exceptional performance in various fields, with commensurate material rewards. Outstanding leaders, hunters, weapon makers, and healers have probably always secured large amounts of food directly or indirectly.

Because the difference between successful and unsuccessful men is fairly obvious, marriageable women have probably had no difficulty discriminating between the two. Given a choice between an outstanding provider and an inferior one, it may be to the women's advantage to choose the former, even though this means entering into a polygamous relationship. In general, the economic status of a potential husband should be at least as important as his physical attractiveness (and those male traits that are generally considered attractive—a well-proportioned body, broad shoulders, tallness, strength, and intelligence—are the kinds of attributes useful in hunting and in competing with other males both within the outside the group).

Male Choice

As should be apparent from the preceding section, the reproductive interests of males and females are not necessarily convergent and may often be at odds. I believe this possibility is related to the low threshold for suspicion and fear of exploitation that characterizes many interactions between the sexes (and which is well developed in the feminist movement and in the hostile response of many males to militant feminism). Males, like females, appear to behave in ways that promote their own special reproductive goals. Because of the costliness of male parental care, we can expect human males to be selective about helping children. Because the reliability of paternity can never equal the reliability of maternity in our species, husbands should be somewhat more cautious than their wives about providing material resources for their children. We predict the willingness of a male to remain with his wife and help her children should be proportional to his certainty of paternity.

Suspicion of adultery and fear of cuckoldry are extremely widespread phenomena in cultures in which males take care of their children. These emotions often play a major role in the social relations of husbands and wives. The ultimate cause for male concern about these matters is the severity of the genetic consequences of female adultery for a providing husband. If the child he helps lacks his genes, he is not only failing to promote his reproductive success but actually elevating the fitness of a competitor. From a female's genetic viewpoint, a child, whether the product of an adulterous liason or not, will have 50 percent of her genes. In many societies, male and female adulterers receive harsh treatment at the hands of aggrieved husbands. Among the relatively peaceful Bushmen of the Kalahari Desert, most murders are committed by men in disputes over women, especially if adultery is involved. Even if the reduction in certainty of paternity is not the female's fault, as in the case of rape, a husband may still divorce his wife, a response accepted or even encouraged by a variety of religious groups and cultures. Raped women often do not report the incident to their husbands for fear of the reaction.

The protectiveness of husbands about their wives and their reluctance to continue to care for adulterous (or raped) wives and their progeny seem to confirm evolutionary predictions about how males *should* behave to maximize their inclusive fitness. But what about wife sharing, a practice not unknown in our society and fully sanctioned in certain other cultures? A speculative interpretation of wife sharing is that it may be an extreme example of reciprocal altruism. If the sharing is truly equal, each husband stands the same chance as the other of eventually providing support for offspring fathered by the other individual. Therefore, neither male necessarily reduces his fitness if he sees to it that his generosity is fully reciprocated. Moreover, if the exchange of wives helps cement an alliance between the males that enables them to cooperate more effectively in the acquisition of resources, the fitness of both males might be elevated in the long run. Males that exchange the right to copulate with their mates do not lack the emotion of jealousy or the sense of possessiveness about their wives, as is shown by the violent treatment of death that adulterous (non-reciprocating) males receive in some Eskimo groups and other tribes in which wife sharing occurs.

Any number of cultures are far more sexually permissive than our own, and in them *adulterous* sexual relationships may be commonplace. The sexually open South Sea Island societies of the Pacific are a classic case. In human groups of this sort, the reliability of paternity may be very low. The standard practice within such cultures is for males to withhold parental care from the children of their wives and instead to adopt the father role for the children of their sister. In our society we consider it natural for mothers and fathers to love and care for their *own* offspring. There are many proximate mechanisms that encourage individuals to do this. For example, it is now recognized that a mother usually forms a strong, loving attachment to her baby in the first few hours after it is born. Incidentally, the recent hospital practice of separating the neonatal infant from the other immediately after it is born and keeping contact to a minimum for the first few days of its life almost certainly interferes with the development of the bond. If a strong bond is not established at this time, this may have permanent negative effects on the relationship of the mother to her child. More generally, parents, male and female, are disposed to find their children truly and charming creatures, even when more dispassionate observers are hard pressed to see why. This is especially true of helpless newborns and young children, who possess certain facial features (e.g., chubby cheeks and large eyes) that are exceptionally attractive to adults.

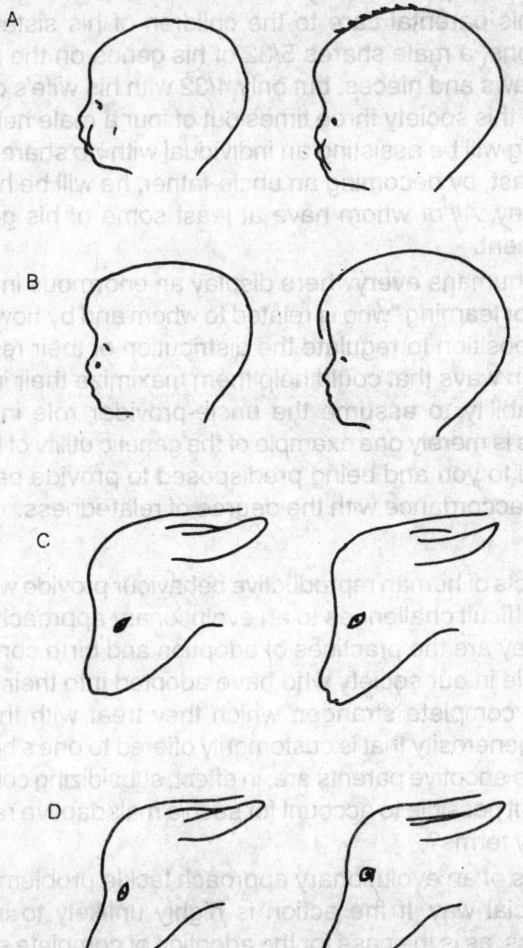

Fig. 15.5. Humans are Sensitive to Specific Cues.

From an evolutionary perspective, parents may offer their offspring love and care for many years because these individuals usually have a high proportion of their genes. But a male whose wife's children were fathered by other males does *not* share genes in common with these offspring. If most or all of his wife's children are genetic strangers, it may become adaptive to share resources with his relatives instead of his wife and her progeny. *Richard Alexander* has shown that if, on the average, a male fathers only one of four children produced by his wife, he would gain genetically by diverting his parental care to the children of his sister. Under these conditions, a male shares 5/32 of his genes on the average with his nephews and nieces, but only 4/32 with his wife's children. Importantly in this society three times out of four a male helping his wife's offspring will be assisting an individual with *no* shared genes at all. In contrast, by becoming an uncle-father, he will be helping a sister's progeny. *All* of whom have at least some of his genes by common descent.

Because humans everywhere display an enormous interest in and aptitude for learning "who is related to whom and by how much", they are in a position to regulate the distribution of their resources and services in ways that could help them maximize their inclusive fitness. The ability to assume the uncle-provider role in certain circumstances is merely one example of the genetic utility of knowing who is related to you and being predisposed to provide parentlike assistance in accordance with the degree of relatedness.

Adoption

Two aspects of human reproductive behaviour provide what may be the most difficult challenges to an evolutionary approach to over behaviour. They are the practices of adoption and birth control. We all know people in our society who have adopted into their families the child of a complete stranger, which they treat with the same affection and generosity that is customarily offered to one's biological children. These adoptive parents are, in effect, subsidizing competitor genes. How is it possible to account for such a maladaptive response in evolutionary terms?

Advocates of an evolutionary approach tackle problems of this sort in a special way. If the action is highly unlikely to increase inclusive fitness, as is the case for the adoption of complete stranger, there are two alternative evolutionary hypotheses that may help explain the behaviour.

1. The action is one of the costly effects of an ability whose benefits, on average, exceed the costs.
2. The action is *currently* disadvantageous solely because of the radically altered setting in which it is now performed; in the past, the disadvantageous response would not have occurred.

These explanation for traits that seem to harm an individual's chances for genetic success have been through this text. For example, the copulation response of a male bee to an orchid flower or the flight of a moth to a bola spider were held to be disadvantageous aspects of the generally adaptive ability of these individuals to find receptive females of their species. A hypothesis of the second type was employed to explain why robins sometimes hurl their banded offspring from the nest. The silvery band stimulates the nest-cleaning response, which is used in a way that would never have occurred in the past because human bird banders are a recent, novel addition to the bird's environment.

Similar hypotheses may help explain the modern practice of the adoption of strangers. First, adopting a nonrelative may be a genetic cost of the usually adaptive desire of adult humans to have children and raise a family. Although it is true that adults who adopt strangers are reducing their fitness, the urge to have a family and the love of children that cause them to behave this way are *on average* beneficial. Because these traits usually elevate the fitness of the humans that exhibit them, they are therefore maintained with human populations. The point is that all adaptations have costs and benefits, and one cost is that they may be misused, from a genetic perspective, especially in rare or unusual circumstances. In Joel Welty's *The Life of Birds* there si a photograph of a cardinal feeding a beakful of insects to a goldfish that has lunged out of the water, mouth gaping, to receive the food. This act certainly could not benefit the cardinal's genes. The bird, however, had recently lost its brood and was employing its food gathering and brood feeding in a rare and inappropriate manner. The ability to collect food for young and the strong desire to feed them surely elevates the fitness of most parent cardinals.

The misplaced-parental care hypothesis generates a testable prediction, which is that husbands and wives who fail to produce children themselves should be especially prone to adopt strangers. Their frustrated desire to be parents could find inappropriate expression from an evolutionary perspective in the adoption of a biological child substitute, such as an animal pet or an unrelated infant. However, the risk that parental drive will be *misused* is relatively slight, and the overall genetic benefit of the eagerness to have children is obvious.

Another prediction is that if the adopting parents proceed to have children of their own (not an uncommon occurrence), then they will be more likely to withhold care and resources from an adopted child of strangers than from their own biological children. One way to test this prediction would be examine the victims of child abuse in a family with both natural and adopted children.

The second basic alternative hypothesis is that the adoption of stranger is a novel, recent phenomenon that occurs primarily because of changes in the typical human environment over the past several hundred years. In large industrial societies, children are made available to people who do not know the parents of the adoptee. Throughout the vast majority of human evolution, in which people lived in small bands or villages, this event would be exceptionally rare. Adoption in this environment may have been practiced regularly to the genetic benefit of the adopter. There are at least three possible kinds of adoptions that could raise the inclusive fitness of the adopting individual.

1. Almost certainly acceptance of the children of relatives was the most common kind of adoption in the past. A direct way to elevate the fitness of a relative, and thereby perhaps to raise one's own inclusive fitness, would be to assist the children of that relative. In effect, males that practice the uncle-father strategy are adopting their sister's children as their own. Adoptive assistance given related persons might also occur frequently when the biological parents died or experienced some disastrous turn of fortune that greatly reduced their ability to care for their children. By stepping in to help, an individual could salvage nieces and nephews whose reproductive potential would otherwise be nil or very low.

2. In cases in which an individual adopted a stranger, the adoptee might be a captured child from an enemy group. The adopting family would probably not treat the child in an altruistic fashion, but instead exploit him or her as a servant or slave.

3. There are some well-documented cases of self-sacrificing adaptation in preindustrial societies of individuals known not to be relatives of the adopters. *Marshall Sahlins*, who believes this phenomenon cannot be explained by resources to evolutionary theory,

also points out that "adoption is, like marriage, a mode of alliance between groups", thereby providing a possible evolutionary explanation for adoption of this sort. A reciprocal altruism hypothesis predicts that for adoption of nonkin, the adopting individuals, although temporarily lowering their fitness by adoption of a non-relative, in the long run elevate their fitness by establishing a *mode of alliance* with the parents (and kin) of the adopted child. In the Pacific Island cultures where this practice was widespread, warfare was also highly developed. Adoption among non-relatives may have increased the possibility of alliance between villages for mutual benefit in defensive or offensive war maneuvers. Such alliances are often critical in determining the survival of group in culture where warfare is a major activity. Adoptive practices may also have decreased the probability that allies would turn on each other during or after the war because they would have a direct genetic stake in the survival of the other group. Finally, if one could adopt the child of an enemy tribe, one might reduce the likelihood of continuing aggression between the opponents and so end or moderate hostilities. In this context it is significant that the warlike Tahitians were willing to adopt the child of a person they had killed in a battle or a child of a close relative of the decreased person.

Birth Control

How is it possible to reconcile the contention that people love children so much so that they will sometimes voluntarily adopt the infant of a stronger with the widespread current and ancient practices of birth control, including abortion and infanticide, all of which prevent people from having children? An evolutionary approach to this problem
people from having children? An evolutionary approach to this problem
begins with the understanding that one can sometimes maximize fitness by *not* reproducing under certain conditions. The reproductive physiology of human reflects this principle. *Rose Frisch* has shown that food intake helps control both the onset of menstruation in women and their fertility during the childbearing years. In fact, a minimum of 16 kilograms of stored body fat is required for maintenance of menstrual cycling. This reserve contains 144,000 calories, an amount than sustain the development of a typical fetus and 3 months of lactation for the infant after its birth. A sharp weight loss (or emotional stress) can cause the physiological blockage of ovulation and, therefore, the prevention of pregnancy. The mechanisms, hormonal and otherwise, that make this happen are performing unconscious

birth control for the person, preventing the formation of a zygote that has a poor chance of surviving and a high probability of simply draining resources from a mother with no genetic return.

Conscious behavioural decisions about when to produce children can also have the effect of raising rather than lowering fitness by increasing the *total number of reproducing offspring* created by a parent during his or her entire lifetime. It is possible to make some predictions about when humans in a wide variety of cultures will practice birth control, if the ultimate function of this behaviour is to maximize individual gene-copying success.

Prediction 1. Unmarried women is societies in which the biological father normally assists in child rearing should be more likely to practice birth control, including abortion and infanticide, than married women, for reason already discussed.

Prediction 2. Married couples will practice birth control primarily with the goal of spacing the births of their children at intervals of several years. In many cultures, a male is forebidden to have sexual intercourse with his wife for months after the birth of a child. As you might suspect, this reduces the probability that the female will become pregnant during this time, an event that would divert metabolic resources from milk production to embryonic development. Such a diversion would often endanger the survival of the living infant. If a newborn can survive the first few years of life, it stands a better and better chance of surviving to reproduce and, therefore, becomes increasingly valuable and deserving (in genetic terms) of continued investment. In hunter-gatherer cultures, nursing continues for up to 5 years after the birth of a child. Nursing activates hormonal systems that help prevent ovulation; cultural proscriptions against copulation between new parents reinforces the ultimate effect of the antiovulation mechanisms, which is to space births widely.

There is some peripheral evidence that suggests that humans are designed to have children at intervals of 3 to 5 years. In modern cultures, some people are able to have many children with very short periods between each birth. Psychologists have found that the intelligence of children (which may influence their potential reproductive success and thus the fitness of the parents) is significantly influenced by both birth interval and family size. Later children in large families have considerably lower intelligence scores *on the average* than earlier-born progeny. Likewise, the longer the interval between births (up to 4 or 5 years), the higher the scores of children on intelligence tests. These results may indicate that human developmental systems are designed to operate in a social environment with siblings 4 years or so older or younger or that parents may be better able to assist in the intellectual development of their children if they are widely spaced.

Despite hormonal and behavioural adaptations designed to keep a mother with a very young child from becoming pregnant again, it may still happen. Under these circumstances, aborting the fetus or killing the newborn may in evolutionary terms represent ending a genetically unwise investment in reproduction earlier rather than later, thus saving energy to devote to living offspring and to promote the long-term reproductive success of the parent.

Prediction 3. In case in which infanticide does not have the effect of spacing attempts to rear offspring, it will result in the removal of progeny whose potential for genetic success is relatively low. Thus, we can predict infanticide to occur with higher-than-average frequency when the newborn has a congenital defect apparent at birth or when twins are born. (Because of their low birth weight and because their mother's milk must be shared between them, twins run a higher risk of mortality than single children).

It is possible to interpret the selective infanticide of healthy newborn females in some cultures in the light of this hypothesis. Female infanticide is practiced primarily in highly warlike cultures in which male mortality produces a skewed adult sex ratio. In the absence of female infanticide, the sex ratio would become even more strongly unbalanced in favour of females. Because such societies usually exhibit a polygynous mating system, a surviving male child in a population with many females may have a far higher potential for genetic success than a female whose reproductive success is more certain but limited by the relatively small number of children she can bear. This favour couples who produce male infants. If a mother rears a female newborn, there will be a long interval (3 to 6 years) before she can expect to conceive again. By killing her female infant, a woman can quickly become pregnant again and perhaps this time produce a genetically more valuable male child.

Prediction 4. Life-long celibacy will almost never be practiced by children that are the sole offspring of their parents. Individuals who never engage in sexual intercourse are rare, but do occur regularly in human societies. Females in this category "should" devcte themselves to their relatives' children, preferably their nieces and nephews who share a substantial proportion of genes in common with them. The tradition of the old maid aunt is well established in our society and others. Celibate males are usually found in the priesthood. We might predict them to occur more frequently in polygynous cultures than monogamous ones, with the celibate role being adopted by males who could judge (probably unconsciously) that their chances of competing with other males for mates were poor. By proclaiming their rejection of personal reproduction, they might be better tolerated by reproductively active males (perhaps sometimes giving them a chance to deceive these males). In the role of religious leaders, they might also be in the position to donate goods received for their services from non-relatives to their relatives, but only if they had relatives to help. Thus a male, if he were an only child, would generally have less to gain genetically by joining a celibate priesthood than if he were to reproduce personally. A male from a large family might have more opportunities to elevate his inclusive fitness as a celibate priest than by trying to form a family of his own.

We have argued that birth control can be and often has been used in ways that contribute to individual genetic success. It is indisputable, however, that in some modern societies birth control is used by many couples to reduce their reproductive potential. To explain this misapplication of birth control (from the gene's standpoint) have become available only in the last few years. This provides couples with the unique freedom to copulate without the subsequent burdens and demands of parenthood. It is striking, however, that almost all husbands and wives in western society, even with access to nearly perfect birth control measures, still choose to have several children. However, they tend to draw the line at two or three, probably because of the great time and energy requirements of child rearing as well as the enormous economic costs. Humans value freedom from economic constraints and freedom to enjoy the pleasurable aspects of life. In the pest, efforts to secure these freedoms have almost inevitable seen associated with producing a family with as many surviving children as possible. In the present, modern birth control technology permits us to satisfy our proximate drives without achieving the ultimate consequences of these drives, just as chocolate addicts exploit certain pleasure centers of the human brain that evolved in a much different context to serve different ultimate goals.

Intelligence and Problems in its Measurement

Before pursuing problems stemming from human behaviour, some discussion needed on the topic of intelligence. It would be impossible to arrive at a completely satisfactory definition of intelligence which takes into consideration all human populations in all their varied environmental setting. Most experts in the field would tend to agree, however, that intelligence does entail the ability to interpret symbols and to handle abstractions in a variety of ways. Many psychologists believe that the Q test reflect this ability and hence intelligence. The IQ test was actually designed by Binet in France, early in this century, as one way to predict a student's success or failure in his studies in school. One reason the IQ test has become so well recognized is that performance on it does tend to indicate the degree of both academic success and occupational achievements later in life. IQ value is determined by obtaining the score which a person achieves on a test which supposedly denotes his ability to manipulate abstractions. The raw IQ test results for an age group in the population are then weighted in various ways to produce a so-called normal distribution or curve. Some persons are found at the extremes of low score and high score. Most, however, fall somewhere in between. The peak or average for IQ has been established as 100 points for the Caucasian population. The score for any one person is determined by relating the individual score to the average for the age group.

Table 15.3. The means IQ for a given age group has been set as 100, since IQ's tend to fall into a normal distribution. The IQ for an individual is determined by dividing the mental age, as measured by performance on the IQ test, by the actual age and then multiplying by 100.

$$IQ = \frac{\text{Mental age}}{\text{Chronological age}} \times 100$$

Mental age	Chronological age	IQ
5	4	5/4 × 100 = 125
5	5	5/5 × 100 = 100
9	6	9/6 × 100 = 150
3	6	3/6 × 100 = 50

Whether or not the IQ test actually measures intelligence is a matter of some debate, and space does not permit us to enter into details of the arguments. While IQ achievement serves somewhat to indicate accomplishments later in life, it must be noted that the very factors which determine academic achievement are correlated with those for occupational achievements in the society. It must not be intelligence itself which is being measured, but some ability to deal with factors upon which a particular society places great value. The question is just what is being measured related to the very construction of the tests. A disturbing fact is that the IQ scores from he population do not just happen to fall automatically into a bell-shaped curve or normal distribution. Actually, the scores are forced to conform to this normal distribution by the very manner in which the examination is constructed.

Results have shown that boys perform better on test items dealing with mathematics, spatial relationships, and certain others, whereas girls surpass them in handling items dealing with language and memory. In order to prevent one sex continually scoring higher than the other on a test, let us say by the inclusion of more items dealing with mathematics which would favour boys, the items in the male-favouring category are balanced by a comparable number in the female-favouring category are balanced by a comparable number in the female-favouring group. The result of the selection of items

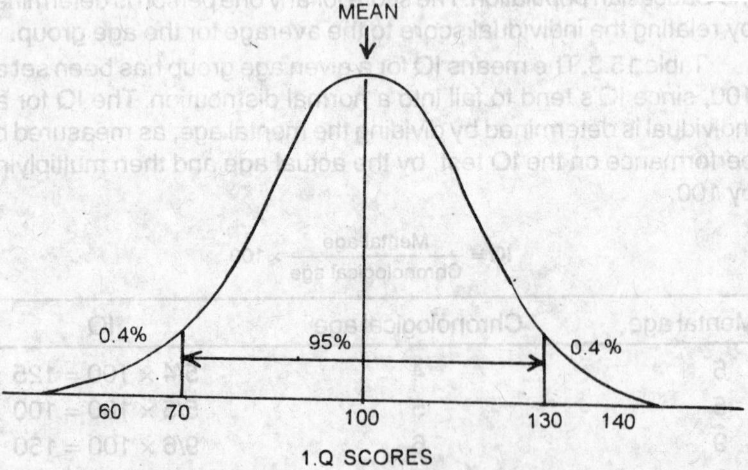

Fig. 15.6. When IQ scores are plotted for Caucasians who have been sampled, they tend to fall into a normal (bell-shaped) pattern of distribution. The mean IQ score has been set as 100 points for age groups. Approximately 95% of the population have IQ's in the range of 70 to 130. Less than 1% have IQ's lower than 60 and over 140.

eliminates high IQ performance which could be related to sex. A main point to note is that items can be selected if so desired to favour a certain group or to blur any intergroup differences which could be reflected in IQ test performance. We see that males and females of he same age group and from the same general environmental background can differ in IQ performance, depending on the items, but that this difference can be hidden by selection of the items in test construction. We have learned that different populations of humans differ in the frequency of certain genes and gene combinations. Therefore, two persons from two different population can be expected to have more genetic differences than two from the same population. Added to this would be very pronounced cultural differences. It would therefore seem that it two persons of different sex from the same population can differ on test performance, then two persons fro very different populations should differ even more in IQ performance. Genetic difference which influence the IQ scores may have nothing actually to do with the intelligence of the two groups.

Attempts have been made to eliminate as much as possible the environmental effects which can influence IQ performance. The difficulties entailed are so overwhelming that no test has been demonstrated to be entirely are so overwhelming that no test has been demonstrated to be entirely culture-free. The intangible which we call *normal intelligence* is undoubtedly the result of the interactions in a genetic component composed of a vast number of gene which influence it in many different ways. Intelligence itself, moreover, is probably not just as single entity. It must in turn be composed of many separate parts, each influenced by heredity and environment. The IQ score is just a single numerical value which tells us nothing about the separate compartments, how they interact, or the relative importance of heredity and environment in each component. Therefore, while the IQ score does have some merit in predicting academic and later occupational success in a given society. It can still be asked: "What exactly is being measured, and what is the relative importance of hereditary and environment factors in IQ performance?"

The Use of Twin Studies in Genetic Analysis

Whatever the full significance of IQ performance, there is no doubt that it does entail a hereditary component. To estimate that portion of IQ which rests solely on genetic factors, investigations have relied heavily on studies of identical twins. Identical twins (monozygotic or MZ) have all of their genes in common; their genotypes are identical. Fraternal twins (dizygotic or DZ), like any other brothers and sisters, have no the average 50% of their genes in common. Since the genotypes of identical twins are the same, monozygotic twins provide a superb opportunity to help evaluate the relative roles of environment and heredity in the expression of a trait. A truly unique opportunity is provided by those rare identical twin who have been raised a part since infancy or early childhood. Since the environmental effects would be more varied on twins reared apart than on those raised together, the contribution of environmental factors to the expression of a trait can be revealed. A pair of identical twins thus gives the opportunity to study the expression of the same genotype in a similar environment (monozygotic twins reared together) and the expression and the same genotype in different environments (MZ reared apart).

Studies of dizygotic twins (DZ) and ordinary brother and sisters reared together and reared part are also related to the data from identical twins studies. Two identical twins should always show the same trait if that trait is based entirely on hereditary factors. Such a trait would be the blood type. Monozygotic twins always agree in their ABO, Rh, MN, or other blood groupings. We say that 2 members of twin pair are *concordant* if they are alike with regard to a certain trait. They are thus concordant if they both exhibit a trait (such as the same blood type) or if they are both free of a trait (neither is a albino). Among identical twins, 100% would be concordant with regard to ABO blood type. Members of twin pair are said to be discordant if they differ with respect to a given trait. If 1 pair of identical twins in study is discordant, the operation of environmental factors in the expression of the trait under study is indicated. If MZ and DZ twins are very similar in concordance value for a trait, then the strong role of the environment is indicated, since the members of each dizygotic twin pair have only about 50% of their genes in common. If based largely on a hereditary component, the MZ twin pairs should show a much higher concordance value. We can thus use the extent to which twins, identical and fraternal, differ in their concordance values as a sort of measure of the effects of the environmental and genetic components on the expression of a characteristic or trait. Suppose 50 pairs of identical twins are studied for a trait and that 40 of the pairs (80%) show concordance. A study of 50 fraternal twins for the

	Identical Twins		Fraternal Twins	
	(pairs)	20 40 60 80 100	(pairs)	20 40 60 80 100
Beginning of sitting up	(63)	82%	(59)	76%
Beginning of walking	(136)	68%	(128)	31%
Hair color	(215)	89%	(156)	22%
Eye color	(256)	99.6%	(194)	28%
Blood pressure	(62)	63%	(80)	36%
Pulse rate	(84)	56%	(67)	34%
handedness (left or right)	(343)	79%	(319)	77%
Measles	(189)	95%	(146)	87%
Clubfoot	(40)	32%	(134)	3%
Diabetes mellitus	(63)	84%	(70)	37%
Tuberculosis	(190)	74%	(427)	28%
Epilepsy (idiopathic)	(61)	72%	(197)	15%
Paralytic polio	(14)	36%	(33)	6%
Scarlet fever	(31)	64%	(30)	47%
Rickets	(60)	88%	(74)	22%
Stomach cancer	(11)	27%	(24)	4%
Mammary cancer	(18)	6%	(37)	3%
Cancer of uterus	(16)	6%	(21)	0
Feedblemindedness	(217)	94%	(260)	47%
Schizophrenia	(395)	80%	989)	13%
Manic-depressive psychosis	(62)	77%	(165)	19%
Mongolism	(18)	89%	(60)	7%
Criminality	(143)	68%	(142)	28%
Smoking habit	(34)	91%	(43)	65%
Alcohol drinking	(34)	100%	(43)	86%
Coffee drinking	(34)	94%	(43)	79%

Fig. 15.7. Concordance and discordance. The shaded areas indicate the percentage of concordance among monozygotic (MZ) and dizygotic (DZ) twins.

same trait shows that 10 pairs (20%) are concordant. This would provide evidence for a strong hereditary component. If both groups, monozygotic and dizygotic, showed approximately the same amount of concordance for a trait (or conversely, the same amount of disordance) such as 70%, then the greater importance of the environment would be indicated.

A list of conditions along with the percent of concordance. Note from the table the eye colour shows a much higher concordance value between MZ than between DZ twins. This tells us that the genetic similarity of the MZ twins is highly significant in the expression of this condition and that eye colour has a strong hereditary component. The fraternal twins show much lower concordance, since members of a pair will not always have the same genes for a trait because only about 50% of all their genes will be common to both. Note that susceptibility to measles is not so different when concordance is compared between the 2 classes of twins. This is a reflection of the greater magnitude of environmental influence on this condition. The values for some conditions, such as stomach cancer, suggest a hereditary component significant difference between MZ and DZ twins in concordance value) but a large environmental component (73% of the identical twins are discordant).

Heritability and the Intelligence Controversy

There are certain limitation in twin studies which must be recognized. For example, the operation and impact of prenatal environmental influences on both MZ and DZ twins is hard to evaluate, and these could operate to influence the expression of a condition even when twins are separated from birth. It could thus be a parental environmental factor which causes a concordance, and this would be measured as a genetic one. Nevertheless, valuable information can be gathered from twin studies on the relative roles of environment and heredity in the expression of a trait or characteristic. The information can be sued to measure *heritability*. Heritability is a measure of that percentage of the variation shown in the expression shown in the expression of a condition of characteristic which is due to genetic factors. Any complex characteristic such as height or weight which involves many genes and environmental factors will exhibit variation. If all that variation could be shown to be due entirely to genetic factors, then the hereitability would be equal to 1. This would mean that the variation seem from one individual to the next is due completely to genetic influence. If the heritability were 0.5, then half of the variation seen is due to genetic factors the other half to environment effects. A heritability of 0 would tell us that all the variation is due to the environment.

Intelligence is a most complex characteristic, and it well known to vary greatly from one person to the next. While we do not know exactly what the IQ performance measures, it would appear that it does measure some kind of mental ability. IQ has been used as an index of intelligence in numerous studies, and he results leave no doubt that intelligence, as measured by IQ performance, has a significant hereditary component. IQ performance of identical twins, some reared apart, other together, is compared to that of fraternal twins and to that of related and unrelated persons. If heredity alone were involved, the correlation between monozygotic twins should be 1.0 and between fraternal twins and ordinary siblings, 0.5. The correlation between MZ's is significantly higher than that for the other groups. The difference between MZ's together as opposed to MZ's reared apart indicates an environmental influence. From analyses using this approach, some investigators have estimated that environment accounts for approximately 20% of the variation seen in a population in IQ performance. In other words, the heritability of intelligence as indicated by IQ is considered by some students of the subject to be in the vicinity of 80%. This would mean that 80% of the variation shown in IQ is the direct result of hereditary factors. The procedures used to arrive at a hereitability value of 0.8 are somewhat complex and entail a number of assumptions. While there is certainly a significant genetic component involved in the variability seen in IQ performance in a population, it seems highly questionable to assign a rigid heritability value of 0.8 or 80%.

Relevant to this point is the concern which has been shown in some quarter concerning the lowering of intelligence due to unequal reproductive rates among he various economic groups in society. Some have feared that the higher rate of reproduction of persons in lower economic strata, who tend to score lower on IQ than those from higher socio-economic classes, will lead to eventual decrease of intelligence in the population. There have been few tests designed to test different generations in order to determine whether or not intelligence does vary from one generation to the next. Results from the few tests scores! Besides arguing against fears of lowering intelligence, the results seem to show a very rapid change in test score performance in just one generation or less. This can hardly be due to a genetic change of monumental proportions. Changes in environmental factors are strongly indicated, and these would appear to be instrumental in influencing the test scores and causing change in a short period of time.

Observations such as these lead us to be cautions when assigning a rigid value to heritability, a value when often rests on many assumptions. Heritability values for many characteristics have been shown to vary from one population to another similar population in a different environment. Heritability data on IQ have been obtained largely from studies of white populations in various regions of the United States and Europe. Since heritability for any characteristic is a figure derived from a population in a particular environment at one particular time, we may question the application of a fixed value for heritability of intelligence to one population when it has been calculated from data obtained from studies on another. Yet this is what appears to be the case in studies where conclusions have been drawn about racial differences in intelligence based on IQ test performance.

Few would deny that scores achieved by American blacks on IQ tests tend to be lower than those for American whites. Several studies suggest that the average IQ score for the white population is 100, where that for the black is 85. Even if this difference between mean IQ values should prove to be the case, it must be understood that many blacks score higher than the average for whites and many whites score lower in test performance than the average black person. No one can say that any black must score lower than a white person. While a difference between an average score of 100 and one of 85 is mathematically significant, the interpretation of this difference has led to many heated debates. According to many, this difference reflects various disadvantages experienced by American blacks in educational and economic conditions. Little is known about the subtle effects which nutritional deficiencies, even those before birth, can have on a character as complex as intelligence. Perhaps all of the difference in IQ performance between the two groups stems from environment alone. Moreover, this performance may not be related to intelligence at all.

There are others who hold an opposing view, citing a heritability of 0.8 to which they rigidly adhere. These persons argue that 80% of the difference or variation seen in IQ achievement has a hereditary basis. Only 20% at the most is contributed by the environment. *Jensen* and other proponents of this viewpoint argue, therefore, that only 20% of this variation could eliminated if all environmental differences were eliminated. Under exactly the same environment, there would still be a great variation between the black and white groups, and this would be entirely genetic. Assuming a heritability of

80%, it can be shown mathematically that elimination of the variation due to environment would raise the average IQ score for the black population only 1.6 points. The end result would be only a slight increase in IQ performance for American blacks. Therefore, the argument goes, the mean IQ value of the whites is truly higher than that of blacks and is largely due to a difference in hereditary factors which affect intelligence.

This sort of reasoning has been used by many well-intentioned investigators who are striving to improve educational and occupational achievements for all people. It has, of course, also been seized upon eagerly by those who wish to prove that one race is more intelligent or better than another for their own unscientific reason. While the IQ controversy has led to many unfortunate debates and ill-founded statements, little can be accomplished by those persons of any persuasion who choose to avoid discussing intelligence entirely. Value for everyone can be derived from gaining insights into normal intelligence and the environmental and genetic factors which do mold it. Any differences between groups on some test such as IQ should not be construed to mean that any person or group is better in some way than any other. A test should be treated as one device which may enable the investigator to obtain a clue on some

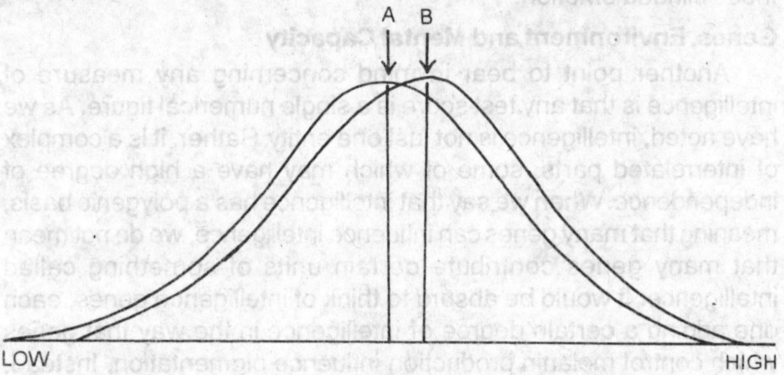

Fig. 15.8. Theoretical mean IQ differences between two races. Assume that two races, A and B, are truly found to differ significantly to mean performance on IQ tests. In such a case, there would be a very large range of overlapping between the two races. Many persons of race A, the one with the lower mean IQ, would nevertheless score much higher than the average IQ of race B, even though the latter race has the higher mean score. The differences found within any one race are greater than those between the two races.

aspect of natural ability. It should not be taken to mean that it must reflect a person's innate intelligence, a characteristic composed of so many components. The subtle effects of environment must never be overlooked in the evaluation of such a complex attribute. Strict adherence to some mathematically computed value for a component of a complex character should be avoided until it can demonstrated that such a value is constant under all conditions. A figure such as heritability can be considered truly reliable only when obtained from studies conducted on similar *and* very different populations under the widest assortment of environments. This has not been done. There is thus a complete lack of data on heritability differences seen within a group as compared to heritability differences seen within a group as compared to heritability differences which might pertain to separate groups of populations under a range of environmental backgrounds.

No true scientific investigation even sets out to prove a point, nor does it use vague or ill-defined mathematical concepts to support a chosen argument. The most effective person is the one who gathers known facts, evaluates them, and tries to apply them constructively. He or she may use them to argue successfully against those who are ill-informed, unaware of pitfalls in their reasoning, and especially those blinded emotion.

Genes, Environment and Mental Capacity

Another point to bear in mind concerning any measure of intelligence is that any test score is a single numerical figure. As we have noted, intelligence is not just one entity. Rather, it is a complex of interrelated parts, some of which may have a high degree of independence. When we say that intelligence has a polygenic basis, meaning that many genes can influence intelligence, we do not mean that many genes contribute certain units of something called intelligence. It would be absurd to think of intelligence genes, each one adding a certain degree of intelligence in the way that genes which control melanin production influence pigmentation. Instead, there are unknown numbers of genes which can influence development of the brain and other parts of the central nervous system in an unknown variety of ways. Normal intelligence must depend on a certain number of nerve cells and their spatial relationships to one another. Intelligence undoubtedly involves the physiological activities of these cells: their ability to carry on basic metabolic processes, to form messenger RNA, to carry on translation.

Any of these aspects of central nervous system development could be influenced by genetic factors. Certainly it is influenced by environment as clearly seen in environment heredity interactions in the case of certain mental disturbance. This is quite evident in phenylketonuria, a disorder we have cited frequently. Mental and behavioural disorders due to PKU clearly have a genetic component, but that is by no means all that is involved. Environmental conditions can prevent many unfortunate consequences of the untreated disorder. The recessive PKU gene is thus one which can have an effect on intelligence, but we now have some understanding of what this means from the knowledge which has accumulated on the molecular basis of the syndrome. Certainly, the normal allele of the PKU recessive is important in the development of normal intelligence, but we see that its effect stems from the control of a specific biochemical step. It would be ridiculous to say that it contributes a unit of intelligence. Many normal genes are required for expression of normal intelligence, since a defective allele many cause a derangement in cellular chemistry which in turn may lead to mental retardation. An unfortunate victim of Huntington's disease may be a person of high intelligence who deteriorates mentally as a result of central nervous system deterioration later in life. The allele of the defective dominant gene is this required for the maintenance of normal mental capacities, but again we see clearly that no unit of normal intelligence or unit of behaviour is involved. Instead, human intelligence and human behaviour are divided into many components. A gene or an environmental disturbance of some kind may influence one or more of them and operate against what is considered the normal expression of these human characteristics.

A lengthy list could be compiled of single-gene defects which are known to influence mental development and which can cause behavioural disturbances. No one of these produces more bizarre effects than the sex-linked recessive associated with Lesch-Nyhan syndrome, a kind of cerebral palsy. An early symptom of this disorder is vomiting. However, this may be associated with the aggressive behaviour so typical of the syndrome. Performance on intelligence tests indicates that victims of this disorder are retarded, but many feel that these children possess a good level of intelligence. They are certainly bright in comparison to most persons who are institutionalized for retardation. Children with the disease develop a compulsion to mutilate themselves, biting the lips, tongue and fingers.

When upset, they may vomit, covering those about them. Speech is greatly impaired in the *Lesch-Nyhan syndrome,* and spastic movements are evident. Particularly poignant is the observation that affected children apparently desire to be protected against the multilation they tend to inflict on themselves. They allow the fitting of restraints (mittens, splints) to protect them. The child tends to become relaxed with the restraints; when they are removed, he cries out, seeming to fear his own tendency to mutilate himself. There is some reason to believe that stress may trigger this compulsion. Besides harming himself, the Lesch-Nyhan child is extremely aggressive. He may kick, hit, and pinch persons caring for him. Still, he may apologize for this aggressive behaviour!

There is variation in the expression of certain of the perverse behaviour traits. Onset of self-mutilation was delayed until the age of 14 in 1 case, and 2 older children did not express this tendency. The variation may be a reflection of the amount of biochemical upset in a particular case. The molecular defect in the *Lesch-Nyhan syndrome* is known, and information on this very rare sex-linked disorder has great relevance to the study of aggressive and compulsive behaviour in general. It raises the question of the role of genetic control in the expression of excessive aggression and in the manifestation of several types of compulsiveness. The recessive gene has been identified with a deficiency of a certain enzyme (*hypoxanthine-guanine-phosphoribosyltransferase* or HGPRT). This enzyme is involved in the metabolism of purines and is either absent or occurs in an unstable or inactive form in those children with the mutant gene. Lack of the normal enzyme leads to an accumulation of a substance which, when present in excess, causes an increase in the metabolic rate. How the enzyme defect is related to the aberrant behaviour is unknown, but it apparently causes biochemical derangements which in turn are responsible for neurological damage. The biochemical upset in the Lesch-Nyhan syndrome also occurs in patients who suffer from gout, even though they exhibit no neurological damage. Like gout sufferers, *Lesch-Nyhan* symptom may be apparent in a 6-9 months old child before some of the neurological disturbances become evident. *Lesch-Nyhan* victims eventually develop gout symptoms, and they also tend to suffer kidney damage, including the development of stones. Kidney damage is usually the responsible cause of death before reproductive age. Since the trait is sex-lined, no male can transmit the gene.

Consequently, a female victim cannot arise, since such an individual must receive a defective gene from the paternal as well as the maternal parent. Attention undoubtedly will continue to be paid to this rare disease for the information it can provide on problems of arthritis, and especially for the light it can shed on behavioural disturbances and the biochemical upsets which may trigger it.

Other single genes are known (autosomal, sex-linked, dominant and recessive) which exert their primary effects in very different ways but which can bring about mental retardation as well as certain behavioural disturbances.

Heredity and Environment in Mental Illness

Mental illness is a matter of great concern, since it takes such a toll in the burden is imposes on large numbers of the population. The psychoses are the most serious of the mental illnesses, and the most frequent psychosis is recognized as *schizophrenia*, characterized by withdrawal into the victim's personal would which is pervaded by very bizarre thoughts and gives rise to most unusual and most unexpected behavioural responses. Schizophrenia exists in several forms, making recognition and accurate diagnosis difficult. This has presented a problem in the elucidation of the genetic component associated with this psychosis. Some psychiatrists have held the view that the disorder stems solely from environmental causes. However, present data show that a significant genetic component exists. Concordance values for identical twins are much higher than those for fraternal twins or for unrelated persons. Indentical twins *do* show discordance; therefore, the environmental influence must be recognized as a significant factor in the expression of the psychosis. However, the importance of the genetic component is seen not only in the higher concordance between monozygotic than between dizygotic twins, but also from other kinds of observations. For example, children of schizodphrenic parents have been separated from the mentally ill parents at birth and raised in normal families. The risk of developing schizophrenia however, in such children is only slightly less than if they had not been separated. Comparison with other adopted children has shown that the act of adoption itself is not the causative stress which brings the schizophrenia to expression.

There has been much debate concerning the responsible genetic factors which can predispose toward schizophrenia. Some investigators favour a single-locus hypothesis in which a dominant gene is implicated; others favour a recessive-gene interpretation gene is implicated; others favour a recessive-gene interpretation. According to some adherents of the latter, the predisposing recessive would produce an added risk of schizophrenia even the heterozygote. Most schizophrenics would be homozygous for the recessive gene, but according to this hypothesis, unknown environmental factors can exert a strong influence on expression of the genotype. Carriers of the gene in both the homozygous and heterozygous conditions would thus run a great risk of developing schizophrenia than would a person lacking the gene entirely. However, persons with the gene, including the homozygote, could be normal. On the other hand, even those lacking the predisposing gene could develop schizophrenia solely as a result of certain environmental stresses.

There are other students of the problem who support the concept of a polygenic rather than a single-locus hypothesis. Environmental influence is also recognized by adherents of this hypothesis. Whatever, the exact nature of many defective genetic factors in schizophrenia, no molecular or biochemical disturbances have been identified, although some derangement in biochemistry which affects nerve hormones (neurohumors is anticipated). The kind of environmental stress which can trigger the psychosis is also unknown.

Most of us have heard of the manic-depressive psychosis another fairly common mental illness characterized by extreme changes in mood which may from deep depression to great elation. A genetic component is also indicated for this condition, and it is different one from that responsible for schizophrenia. Again there si debate on the nature of the genetic component which is involved.

Studies of the genetic and environmental components in mental aberrations are very important, not only for the information they may yield to lead to successful treatment and prevention, but also for the information which may be provided in an understanding of other, more subtle aspects of behaviour. Twin studies indicate that a genetic component may exist for personality type. The data show that personality can be greatly influenced by the environment, much more so than for certain other complex characteristics such as height, weight, or IQ achievement. But the possibility that heredity may be involved in the determination of personality type is also suggested. This has a great bearing on the study of criminal behaviour, a very controversial topic which becomes very heated when discussed in relation to any hereditary predisposition. No definitive statements can be made at the moment, but the possibility cannot be dismissed that some genetic factors exists as well as environmental ones which predispose to criminal behaviour.

Ethical Consideration in Genetic Investigations

We have already discussed the possibility of an extra Y chromosome as a predisposing factor toward criminal or aggressive behaviour. This topic leads us to another controversial matter, that of the ethics and morals in genetic investigations and procedures of a certain nature. Proposed studies of children with an extra Y chromosome have focused attention on the type of heated debate which can surround genetic research. More and more, we can expect to encounter problems which pose questions with serious ethical ramifications. The relationship of an extra Y chromosome to antisocial behaviour is still unclear, and much more clarification is needed before positive statements can be made. The discovery at Boston Hospital that the XYY conditions occurs with a higher frequency than formerly suspected (perhaps 1 male birth in 1000) indicated to some a rather urgent need to accumulate more information on the condition in order to settle the matter. A study was undertaken to follow up children with XYY and those with normal XY chromosome constitutions and to observe the two groups throughout a period of years. Any behaviour differences between them would be noted. Such follow-ups, it has been argued, could yield several benefits. Any behavioural aberrations might be detected at an incipient stage when they could yield more effectively to treatment and guidance. Moreover, observations of those XYY children who happen to manifest no behaviour problems could be related to studies of aggressive XY as well as to aggressive XYY children. The information obtained could cast light on those environmental influences which may mold a child's behaviour or personality. Perhaps the extra Y may prove to contribute little or no effect.

The study has been challenged by at least one group of scientists who maintain that it is unethical and unscientific. Most important, it is argued, such a study is potentially harmful to the children involved. It immediately sets any child in the study apart by calling attention to him. Moreover, parents with an XYY child, knowing he has some kind of genetic anomaly, will inadvertantly treat him in a different fashion, which could distort has entire mental development. This in turn could set off a serious psychological disturbance which could be or could not be related to the presence of the extra Y. Many who argue against the study feel that if focuses attention on very unusual genetic conditions and thus detracts from the socio-economic problems which are known to contribute to antisocial behaviour and criminality.

Many proponents of the investigation admit that it may be better not to allow the XYZ person to know about his genetic constitution, since he might conceivably develop normally without this knowledge. If the design of the study could be altered to prevent this, it then arises the provocative question of a person's right to know and the right of parents to have such information if it is known to medical personnel. The controversy has been sufficiently heated to cause a cessation in further identification of XYY males. However, the study will continue on groups which are presently under investigation.

The XYY story illustrates the kind of moral dilemma which will continue to arise in genetic research as the procedures used in detection and treatment of genetic disturbances become more sophisticated. No one would argue that steps should avoided which could prevent human suffering. The dilemma which arises often centers around the problem of whether the procedure will cause in the long run more problems for those it tries to treat, as well as greater problems for the human population in general.

Another type of research which is becoming more publicized involves research on the human fetus. Many have attacked such measures on ethical grounds, while others appalud it for the advances it can achieve in medicine. A study made for a national commission which will advise the Department of Health, Education, and Welfare on fetal research has reported that total abstention from this kind of study would have caused delays in our advance of medical knowledge and would have cost thousands of lives. One area which profited, so the study reports, is that of the treatment of babies suffering from blood disorders due to Rh incompatibilities. Many thousands of infant deaths would have resulted with so research on fetal material, since animal research would not have been as satisfactory. Moreover, without the information obtained, many additional thousands would have become seriously incapacitated due to brain damage from Rh incompatibilities, placing burdens as well on society. As many as 12% of married couples may face some kind of risk due to Rh incompatibilities. The development of a vaccine which prevents the Rh negative mother from becoming sensitized by Rh^+ antigens has been accelerated greatly by research utilizing fetuses.

The report also pointed out that a complete ban on research involving fetuses would have prevented the development of amniocentesis, a procedure in almost routine use to diagnose fetal disorders as well as to detect genetic defects *in utero.* In some cases, the parents are presented with the option of abortion to prevent the birth of babies doomed to years of suffering.

Genetic Engineering and Problem Society

The problems which will continue to confront society in areas of genetic research underscore the need to weigh carefully the benefits and the dangers entailed in a particular approach. The best course to take may, unfortunately be unknown at the moment a decision is required. Urgency for some guidelines in certain types of genetic research is perhaps best illustrated by the decision reached at a conference of biologists assembled to debate research in genetic engineering, an area which involves the manipulation of genetic material by the insertion or elimination of DNA into a cell. An ideal solution to a genetic disorder would be the replacement of a defective gene by its normal allele in those cells in which activity of the normal gene is required. Accomplishment of such a feat is not in the immediate future and depends on research in various directions, among them the assignment of human genes to precise locations on the DNA of the chromosomes. Complete cure would require the insertion of the normal allele into germ tissue so that correct information would be transmitted to the next generation. While still a distance from effecting such complete cures, certain approaches have already been undertaken. It has been well established that some bacterial viruses can incorporate into their DNA bits of genetic information from a host cell and then transfer these to another bacterial cell which the virus later infects. Actual attempts have already been made to infect humans with viruses carrying genetic information missing from the humans' cells. Success has not been achieved, although enzyme deficient human cells *in vitro* have been shown to acquire the ability to manufacture a specific enzyme after infection with viruses carrying the required functional gene. It is still unknown whether the viral genetic material with the added gene has been incorporated into the chromosome or whether it is residing elsewhere in the cell. Another question which arises is whether infection of a person with a virus would in the long run be detrimental. Since the DNA of the virus may persist in the human cell and could very possibly undergo transcription leading to disturbances in the physiology of the cell. A safer approach might be the injection of the enzyme known to be lacking in a disorder, as has already been reported with some success in the case of *Gaucher's disease.* But this approach does not correct the genetic information and could therefore require continued treatments.

The isolation of specific genes has actually been accomplished in the laboratory, as has the storage of particular genes from certain lower animals. Isolated genes can provide the geneticist with the opportunity to decipher the actual makeup of a gene and to study the mechanisms of gene action and control. However, it is just this kind of approach to genetic engineering which has led to the conference of concerned biologists. Techniques have now been perfected which enable the transplantation of genes from one kind of animal into the DNA of another. This can permit the storage in a simple cell type of genes derived from a complex cell. This feat has been made possible by the use of some enzymes which were found to be capable of breaking DNA into segments and of other which can repair the breaks. Segments of DNA have been taken from animal cells such as those of the sea urchin or the South African toad. The segments have then been inserted into bacterial DNA through the action of enzymes which can recognize and break specific regions of the DNA. The foreign animal DNA can then be inserted into DNA of the simples cell, and all breaks can be healed. Such bacteria then reproduce and form a storehouse of specific animal genes for study as needed. Large amounts of products of the inserted genes could pile up in the altered cells. Another advantage would then be the production in the simpler cells of large amounts of a valuable substance (insulin perhaps) required by persons with serious genetic diseases.

The potential dangers of this kind of genetic surgery for the human have caused alarm within scientific circles. It is conceivable that transfer of foreign genes from higher cell types of common bacteria such as *E. coli* could impart very infectious properties to the microorganism which could then run rampant, causing novel kinds of infections. There is even the danger of transfer to the human of known cancer-inducing animal viruses from cells in which virus genetic material may be stored. The potential for harm in this field of research has promoted the scientific world to recommend guidances in the pursuit of such studies. The guidelines have no legal basis, but the moral compunction to follow them should be strong, since so many illustrious scientists have recognized the need for them. The proposal of safeguards and guidelines is very unusual in the scientific world, which has tended to avoid the imposition of restraints on investigations along certain pathways of inquiry. We see in this case complete intellectual freedom being weighed against the dangers inherent in a particular approach. To many, restraints appear preferable during the pursuit of an investigation whose capacity for overwhelming harm is at the movement unknown.

Our understanding of human genetic disorders will continue to present on end of ethical dilemmas. There are many who seriously question the morals of any kind of abortion, even when the fetus is known to be defective and the victim of an unfortunate disorder, such as *Tay-Sachs disease* or *Down's syndrome*. Perhaps even more disconcerting are problems which relate to the successful treatment of genetic disease. The life of an affected person may be spread, but the treatment required may be lifelong and experience. It may imparted as excessive burden to the victim, the family, and the medical terms which are required. There are those who ask what benefits are gained by saving from early death those children with genetic defects which require expensive treatment that permits survival for just a few additional years. Could the efforts be better utilized in another direction even though death in certain cases would be a certainty if the disorders are left untreated? Remember that many genetic disorders due to single-gene defects are very rare. As more and more continue to be understood and treated, will this divert attention to an ever-growing list of disorders, each affecting few people? Will this shift attention, thus restricting the resources and efforts needed to solve health problems which affect very large numbers of people?

Not to be overlooked is the problem presented by saving person with genetic disorders who will later reproduce and pass on the defective genes. Phenylketonuria illustrates more than one problem of this kind. No one would question the value of treatment of PKU victims. However, when a PKU woman saved from mental deterioration, becomes pregnant, additional steps must be taken to protect the child, she is carrying. This is so, since the PKU adult will very likely be on a normal diet. Procedures must be undertaken to control the diet of PKU mothers in order to avoid fetal damage as a result of high levels of phenylalanine in the maternal blood.

There are those who point to possible dangers to the population as more and more persons with defective genes reproduce and pass the defective genes into the population gene pool. Many geneticists consider it possible that a crisis will eventually arise in future generations if freedom to reproduce is left unchecked. Others answer this by saying that, while increase of a dominant gene would be quite rapid (about 6 times the original frequency in 5 generations), the frequency of the more common defective recessives would rise very slowly (only about 1% in a generations). Man's knowledge in the long run, so this argument goes, would enable him to devise ways throughout the years to cope with the increase in the frequency of affected persons. In the future, persons who are now considered defective in some way would impose no burden to the population and incur no disadvantage as a result of the perfection of procedures and treatment which are at the moment unimaginable.

In the long run, it becomes a question of human behaviour, that multisided characteristic composed of so many intagibles. The human's individual behaviour can certainly result in harm to himself. Should restrictions be imposed upon individual behaviour if it affects only the individual who practices it? The use of certain drugs is known to be harmful to the person. Even certain drugs known to be beneficial to the average person in the treatment of illness may cause violent reactions in some individuals, as seen clearly in the case of G6PD-deficient persons to primaquine and other substances. Such people are normal in all respects other than their drug response. Another example is seen in the violent reaction of barbiturates experienced by those with one or another form of porphyria a group of disorders with different genetic bases. Some of these people are essentially normal but may exhibit varying degrees of sensitivity to light. Exposure to barbiturates, however, can result in death. Many more examples of drug anomalies could be given. It should at least be appreciated that normal persons may respond differently to a given drug, in their obvious physical reactions to it and even in their ability to respond in any way at all. More and more drug anomalies are certain to appear as more therapeutic agents are designed.

Since harmful drug responses with a genetic basis are well known, what should be practiced to regulate personal use of substances whose effects are poorly defined? Should a marijuana user be punished because some evidence indicates that the immune response is lowered with habitual marijuana use? Perhaps just the dangers of usage should be publicized, as in the case of cigarette smoking, and personal choice permitted. But what about evidence that marijuana may depress the level of the male hormone in habitual male users of this substance? In mice, rats, and even in some primates there is experimental evidence which indicates that interference with male hormones during fetal development can interfere with the completion of normal male development, including behaviour. While still an unknown area, the question arises whether marijuana use by a pregnant women could adversely affect the development of the male fetus. Other reports suggest that marijuana use interferes with the assembly of RNA and even of amino acids into proteins. Such possibilities raise the point that personal use of substance may do more than harm the individual user alone. The potential for harm to future generations by any mutagenic actions of drugs in current use is also unknown but cannot be disregarded.

It must be kept in mind that the personal behaviour of each human can affect the welfare of others. The behaviour of the human species collectively can alter the existence of other species on the planet. Research into the genetic and environmental components of the human's most intricate characteristic, behaviour, may hold significant bearing on the future of *Homo sapiens* and all other forms of life. Whether the human's genetic advancement of all life forms will finally yield an answer to the tantalizing question, "Is today's human truly a higher form of life?"

Human Behaviour and Evolutionary Theory

We have at the threshold of an exciting new phase of the analysis of human behaviour, and a new generation of biologists and social scientists has an unparalleled opportunity to help develop this phase. Evolutionary theory has just begun to be used to explain the diversity of human behaviour among cultures. As we have seen, gene thinking generates a host of predictions about how humans should behave under variable ecological conditions. The extraordinary differences in the behaviour of humans of different cultures can be used to test directly where there is in fact a correlation among cultural strategy, the ecology of the society, and the inclusive fitness of members of the culture. Weighed against human cultural diversity, the variation in the social behaviour of other species is vanishingly small. It is unfortunate that much of this diversity is disappearing before it can be thoroughly described and before new information can be collected that would permit scientists to test evolutionary predictions about human behaviour. Nevertheless, there is still time for a productive union of anthropology, psychology political science, economics, sociology, and evolutionary biology, although forming such a union will be a challenging task. We have a great chance to increase our knowledge of ourselves; I hope we have the willingness to do so.

16

PARENTAL CARE AND BEHAVIOUR

Parental care behaviour is any behaviour, performed after breeding, by one or both parents, that contributes to the survival of their offspring. In some cases egg-covering behaviour is found among some salmonids, lampreys, and cyprinids. In such cases, the animal clearly protects the offspring from some environmental hazards and thus promotes their survival. Following discussion, the term parental behaviour is restricted to the *active care and protection of the offspring* by one or both parents excluding viviparous species (e.g., many cyprinodontoids) in which fertilization is internal and the females give birth to mobile young.

Parental behaviour among vertebrates ranges from little or no care in most poikilotherms to intensive and prolonged care of the young in many birds and mammals. In this chapter, we shall first describe some recent work on parental behaviour in fishes and then compare the phenomenon of parental care as it occurs among fishes and birds. It is hoped that such a comparison will contribute to a greater understanding of the origin and phylogenetic development of a type of social interaction that plays a critical role in the early life or many animals.

Parental behaviour is by no means universal among fishes. Of some 250 families described in *Breder* and *Rosen's* encyclopaedic treatise on fish reproduction, about 77% show no parental care, another 17% include species that care for the eggs only, and less than 6% contain species that are known to care for both eggs and newly hatched young. However, within this latter group are some large and diverse families, including the Cichlidae, that are a major component of the freshwater fish fauna of Central and South America and of Africa.

Two general types of variation in parental behaviour exist among cichlids. First, either both parents, or one alone cares for the offspring. Thus, there are paternal, maternal and biparental species. Second, the eggs and newly hatched young are either maintained on the substrate—that is, on plants, under stones, in excavated pits, and so on (these are usually called substrate-brooders or guarders)—or carried about in the parent's mouth (the mouth-brooders) or oral incubators). *Wickler* has suggested that "*open brooder*" and "*concealed brooders*" are more accurate labels for the two main cichlid parental strategies, because some substrate-brooding species guard their eggs and fry in caves or burrows ("*concealed broods*), where the degree of protection afforded the young is closer to that of mouth-brooders than to the substrate-brooders that guard their young in the open. In fact, as well be discussed below, some cichlid species use both strategies, first guarding their eggs on the substrate, then carrying the newly hatched offspring in the mouth.

Male-Female Parental Roles

In some biparental cichlid species there appears to be an unequal division of parental duties between male and female. Most of these species are substrate-brooders, and the differentiation between maternal and paternal roles is most pronounced when their offspring are at the egg stage. Typically, the female performs more of the direct egg-care behaviour, and the male is the more active defender of the brood against potential predators. Quantitative data related to these generalizations are, however, available for only a few species. Some examples follow.

Chien and *Salmon* found some differences in parental activities between male and female *Pterophyllum scalare*. The number of bouts of egg-fanning and the total time spent fanning were both greater in females than males on the first day postspawning, but these differences disappeared on days two and three. Egg-nipping and several activities directed at wrigglers (the developmental phase between hatching and free-swimming) and at fry (free-swimming juveniles) did not differ between parents.

In a study of parental behaviour in *Tilapia mariae,* the females performed more egg-aerating bouts than did the males, but males did more *calling-the-young,* an activity in which the pelvic, anal, caudal, and dorsal fins are repeatedly snapped open and shut, stimulating the fry to congregate below the signaling parent. On the

Fig. 16.1. King penguins form a colony to brood their eggs. Each of the birds is brooding a single egg.

other hand, parental *Etroplus maculatus* also signal their young by pelvic finflickering, and *Cole* and *Ward* found that males and females did not differ in their fin-flickering rates, nor in the time spent with the school of young up to 12 days posthatching.

Female convict cichlids (*Cichlasoma nigrofasciatum*) performed more *fin-digging* than males did while guarding 12 to 30 days old fry. In this behaviour the adult fish settles to the bottom and, with vigorous undulations of the body and fins, stirs up loose substrate materials that stimulate the fry to congregate around the parent and feed on the loosened particles. These authors also found that male and female parents were equally aggressive in defense of their broods.

Two studies of cichlid behaviour recently completed in our laboratory yielded quantitative data comparing male and female parental roles. These data are briefly reviewed here.

Male-Female Roles in Herotilapia Multispinosa

Patricia Smith-Grayton studied the Central American, biparental, substrate-brooding cichlid *Herotilapia multispinosa.* In this work she made a detailed quantitative comparison of the behaviour of males and females while guarding their young. Data were collected during 15 breeding cycles, from spawning until 11 days postspawning. The following results are pertinent to the question of male-female parental roles.

Fanning Rocking and Hovering

These are three types of behaviour performed while the adult fish remains close to the offspring. Fanning is the well-known aeration activity in which the adult maintains position 2 to 3 cm from the eggs or wrigglers and with large-amplitude, low-frequency beats of the pectoral fins moves water over the young. Simultaneous lateral undulations of the caudal fin prevent backward movement of the fish while fanning. The motor pattern is found among many egg-tending species and probably serves to ventilate and clean the eggs and young by providing oxygen rich water and removing metabolic wastes and silt. Female *H. multispinosa* performed more frequent bouts of fanning of the eggs than males did, but both sexes fanned the wrigglers equally often. When measured as total duration of times spent fanning, the mean value for females was higher than for males during the egg stage, but the difference was not statistically significant; the duration of fanning the wrigglers was almost identical in the two sexes.

Rocking is a parental activity directed only at the eggs and presumably serves ventilating and cleaning functions. In this pattern the adult fish remains close to the eggs and rhythmically pitches in the vertical plane while moving slowly about over the spawn. Brief

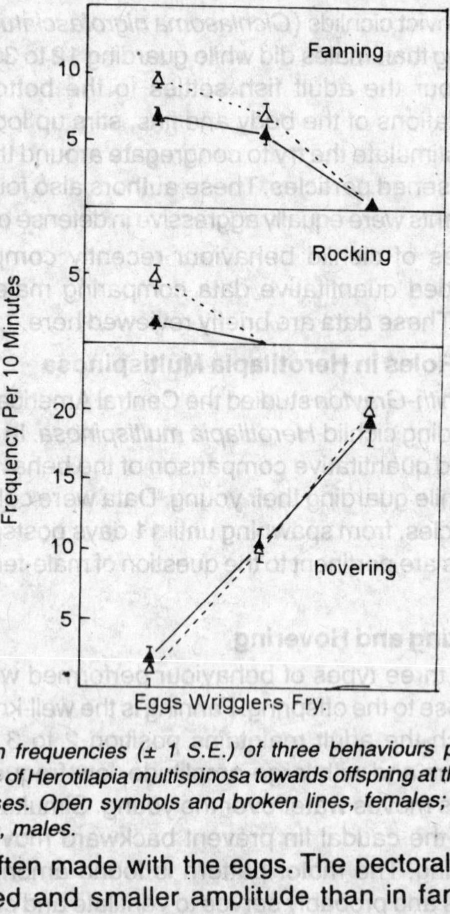

Fig. 16.2. Mean frequencies (± 1 S.E.) of three behaviours performed by adult pairs of Herotilapia multispinosa towards offspring at three developmental phases. Open symbols and broken lines, females; solid symbols and lines, males.

contact is often made with the eggs. The pectoral fins flutter with higher speed and smaller amplitude than in fanning. Females performed more bouts or rocking than males, but, as with fanning, the total time spent in rocking by the two sexes was not different.

Hovering is a pattern in which the adult fish stations itself virtually motionless 3 to 4 cm above or beside the eggs, wrigglers, or fry, and with the longitudinal body axis parallel to the aquarium substrate. Fin and body movements are slight, and they apparently serve only to maintain position. Hovering is often performed when the fish is resting between bouts of other activities. Both male and female parents showed increased hovering behaviour as their progeny developed from eggs to fry, and the only consistent difference between the sexes was at the egg stage, when males spent more total time hovering than females did.

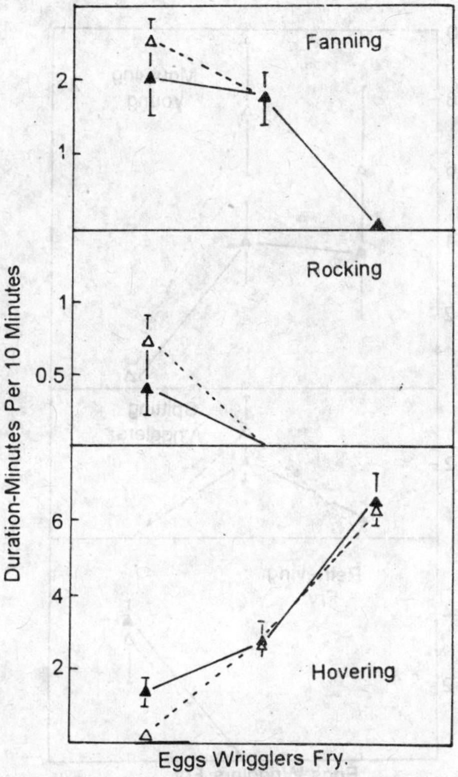

Fig. 16.3. Mean duration (± 1 S.E.) of bouts of three behaviours performed by parental H. multispinosa. Symbols as in Fig. 16.2.

Mouth-Contact Behaviour

Three activities in which the parents contact their young with the mouth were also quantified.

Mouthing the young

This includes gentle sucking at the eggs and picking up of wrigglers and immediately spitting them back onto the substrate. It appears to be primarily a cleaning action. Fungused or opaque (and presumably dead) eggs are removed from the substrate and eaten, and silt or debris is removed from eggs and wrigglers. Mouthing may also aid in the hatching process. The only consistent difference between sexes in this activity was that females mouthed the wrigglers more often than males did, although the eggs also tended to be mouthed more often by females than by males.

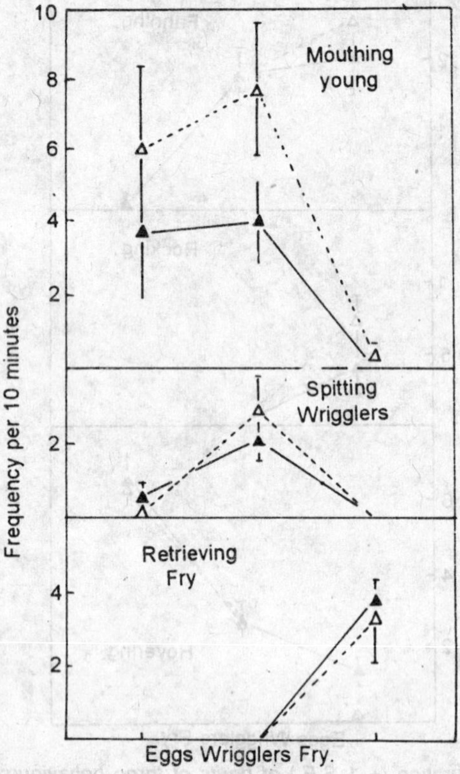

Fig. 16.4. Mean frequencies (± 1 S.E.) of three mouth-contact behaviours performed by parental H. multispinosa.

Spitting wrigglers

This occurs when one or more wrigglers becomes separated from the others in a pit or when it drops from a plant (where newly hatched wrigglers are often placed, and from which they hang suspended by mucous threads from the head). The adult fish gently picks up the stray wriggler in its mouth and spits it back with the others. Both parents performed it equally often.

Retrieving fry

This was recorded when one or more fry had moved a short distance (usually more than 10 cm) from the main school of fry, and then was snapped up in the mouth by one of the parents and spit back into the school. There was no difference between parents in the frequency of retrieving fry.

Digging

This is the behaviour by which adult *H. multispinosa* excavate pits for the retention of wrigglers. The fish thrusts its snout into the substrate, takes up a mouthful or gravel, moves up to 10 cm away and spits it out. Digging occurs frequently as the eggs approach hatching, but continues after hatching, since the wrigglers are usually moved about from one pit to another. The only sex difference in digging was that females performed it more often than males did when their progeny were at the wriggler stage.

Chasing

This was recorded whenever one parent pursued a fish other than its partner away from the vicinity of the offspring. Duration of chasing was highly variable, because the pursued fish could not escape from the aquarium and the number of hiding locations was limited. A chase of any duration was recorded as one event. Both parents chased other fish away from their brood, and the only quantitative difference sexes was at the egg stage, when males chased more often than females did.

Feeding

Feeding movements were recorded, not because these are forms of parental behaviours, but because they are a clearly functional activity that can be performed only when the adult fish is not engaged in one of the parental care actions. As shown in the figure, feeding was relatively infrequent during the egg stage and increased as the progeny developed. This coincided with declines in several of the parental behaviours, such as fanning, rocking, mouthing young, and digging, and it presumably indicated that as the progeny developed, the parents were able to spend more time with their own maintenance activities, of which feeding must be of prime importance. No clear sex difference in feeding frequency was recorded.

In summarizing the main results of this study, both male and female *H. multispinosa* performed all of the species-typical parental activities; that is, none was sex-specific. And yet quantitative differences between the sexes were found for some forms of parental behaviour. These differences support the generalization of other workers that among biparental, substrate-brooding cichlid fishes, there is some separation of parental roles, with the female being more involved in direct care-of-the-young activities and the male more concerned with defense of the brood. The differences between

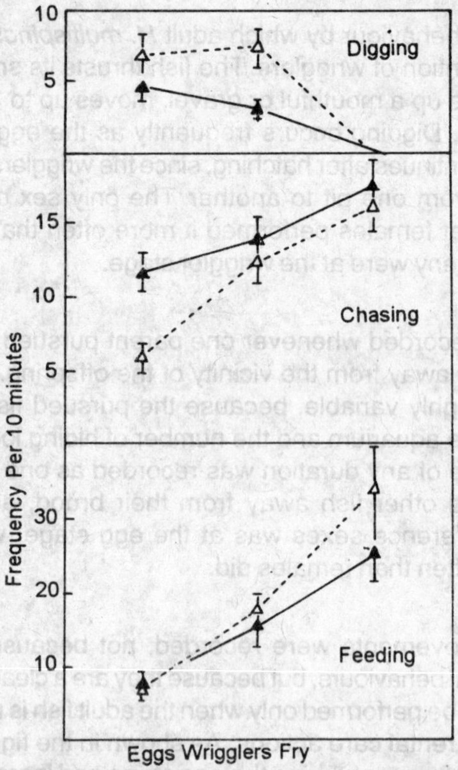

Fig. 16.5. Mean frequencies (± 1 S.E.) of three behaviours performed by parental H. multispinosa.

the parents were most pronounced when the young were at the egg stage, and they disappeared as the young became free-swimming fry.

Male-Female Roles in Aequidens Paraguayensis

A second study of parental behaviour concerned the biparental South American cichlid *Aequidens paraguayensis*. In this species, a mature male and female establish a pair bond and select as a spawning site a loose leaf lying on the substrate. During the prespawning courtship period, and also while the clutch of 200 to 400 eggs is being guarded, the pair move the leaf about by grasping one edge with the mouth and pulling or pushing it. This occurs most often when the fish have been disturbed. Just before the eggs hatch, they are picked off the leaf, and the young fish are carried orally by both parents for up to several weeks, although once the fry have become actively free-swimming, they enter the parents mouths only when disturbed and at night. Thus *A. paraguayensis* adults use both substrate-and mouth-brooding as parental care strategies. Only a few other South American cichlids are known to practice both types of brooding.

During the egg-guarding phase of a breeding cycle, female parents are more involved with direct egg-care activities than males are. For example, the mean time spent *fanning egg* per 10 minute observation period was 7.15 min for ++ and 3.34 min for OO (S.D. 0.28 and 0.40 respectively). In 9 of 10 pairs the difference between sexes was significant ($p<0.05$; Wilcoxon matched-pair, signed-ranks test).

During oral incubation of wrigglers and fry, the brooding parent regularly perform an activity called *churning,* in which the entire

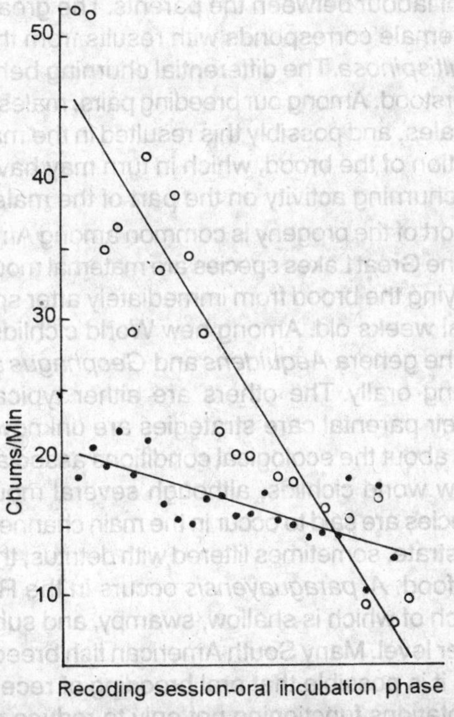

Fig. 16.6. Churning rates for male (open symbols) and female (solid symbols) Aequidens paraguayensis. Data from 10 pairs, adjusted so that each point represents pooled data for same elapsed time since eggs were picked off substrate.

buccal cavity is enlarged and then quickly decreased in size. This is performed in bouts of from one to several distinct churns, and although several functions have been ascribed to this activity in other mouth-brooding cichlids, it appears that ventilation of the young is the most likely function in *A. paraguayensis.* Our observations, based on 10 pairs, showed that incubating males churned the brood more often than females did: (a) churns/min: O mean, 26.2; + mean, 17.4; (b) churns/bout: O means, 5.58; + mean 3.57 ($p < 0.05$ in both cases, *Wilcoxon* matched-pairs, signed-ranks tests).

A more detailed examination of our churning records showed that churning activity by both sexes declined as the progeny developed, but the decline was steeper for males than for females. The data fit two linear regression lines with negative slopes that are significantly different from each other.

Thus, *A. paraguayensis* is another biparental cichlid showing some division of labour between the parents. The greater egg-care activity by the female corresponds with results from the substrate-brooding *H. multispinosa.* The differential churning behaviour is not yet clearly understood. Among our breeding pairs, males were slightly larger than females, and possibly this resulted in the males carrying a larger proportion of the brood, which in turn may have stimulated more frequent churning activity on the part of the males.

Oral transport of the progeny is common among African cichlids, where many of the Great Lakes species are maternal mouth-brooders, the female carrying the brood from immediately after spawning until they are several weeks old. Among new World cichlids, only a few species within the genera *Aequidens* and *Geophagus* are known to carry their young orally. The others are either typical substrate-brooders, or their parental care strategies are unknown. Little has been published about the ecological conditions associated with oral brooding in New world cichlids, although several mouth-brooding *Geophagus* species are said to occur in the main channels of streams over sandy substrate, sometimes littered with detritus, through which they forage for food. *A. paraguayensis* occurs in the Rio Paraguay watershed, much of which is shallow, swampy, and subject to rapid changes in water level. Many South American fish breed under such conditions, and it is possible that oral brooding of recently hatched young are adaptations functioning not only to reduce predation on the offspring, but also to reduce mortality associated with rapid rising or falling water levels.

ONE-PARENT REMOVAL STUDIES

The studies on *H. multispinosa* and *A. paraguayensis* described above showed that all of the species-typical parental behaviour patterns were performed by both members of breeding pairs. Some quantitative differences existed between the sexes, but none of the parental patterns was specific to one sex. This lack of pattern specificity suggested that the parental roles of males and females in these two species are virtually interchangeable, and led me to an investigation of the ability of a single parent to rear a brood of young successfully after the loss of the other parent.

Two questions formed the basis of this study. First, is one parent more successful than the other in raising a brood alone? Second, does the developmental stage of the progeny at which one parent is lost influence their subsequent survival? These are not likely to be trivial problems. The fish fauna of much of Latin America is rich and varied. Potential predators of smaller, biparental cichlid species are numerous, and the ability of one parent to defend its brood even if its mate has been lost would appear to be a real advantage.

H. multispinosa was chosen as the test species because it breeds readily in captivity, and its small size at maturity increases the probability that in nature an adult fish may occasionally be lost to predation while raising a brood. Adult males used in the study ranged from 7.7 to 11.5 cm total length (mean, 8.98 cm). Females ranged from 6.5 to 10.0 cm total length (mean, 7.87 cm; n for each sex, 132).

The experimental technique was to allow a pair of *H. multispinosa* to spawn, next to place potential predators in the same aquarium, and then to remove either the male or the female parent when the young had reached one of the following developmental periods: the beginning of the embryonic period as soon as spawning was complete; the beginning of the wriggler phase, when all eggs and had hatched; and the point at which all wrigglers had first become free-swimming fry. The experimental variables—that is, the sex of parent removed, and the developmental phase of young when one parent was removed—were randomized among experimental pairs. Five replications of each treatment and of control tests were run in each of four separate experiments.

Two male and two female adult *H. multispinosa* were placed in each aquarium to provide some choice of breeding partners. When a pair was clearly preparing to spawn, the other two fish were removed.

After spawning, the eggs were photographed and counted on enlarged prints. Separate tests were made of the accuracy of this method; eggs were photographed in the standard manner, then the flower pot was removed and the eggs were counted directly. The mean error, based on 10 such tests, was 3.5%; this was ignored in the later determination of fry survival. Estimated spawn size for the four experiments together ranged from 284 to 1612 eggs (mean, 787.3; n = 145).

Experiment 1

In this experiment eight treatments were applied: either the male or the female parent removed, at each of the three developmental phases; plus two controls, one with and one without predators present. Two males *Cichlasoma nigrofasciatum* were used as predators. The predators ranged in total length from 5.8 to 10.0 cm (mean, 8.20 cm; n, 64). In captivity this is an omnivorous species that feeds readily on juvenile cichlids. Its natural range overlaps with that of *H. multispinosa* (Miller, 1966), although the two species do not appear to be sympatric when breeding (Baylis, 1974).

The mean per cent survival of fry under all experimental treatment was low, and no clear differences were found among treatments. The two control treatments resulted in high survival, and the control without predators had the highest survival of all, although there was great variability in both control series.

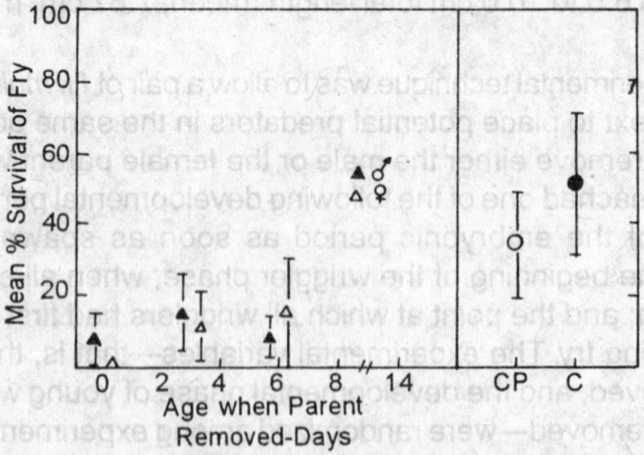

Fig. 16.7. Mean percent survival (+ 1 S.E.) of H. multispinosa fry by 15 days of age in Experiment 1. Either male or female parent was removed at beginning of egg, wriggler, or fry stage. CP, control series with predators present; C, control with no predators. Predator species, Cichlasoma nigrofasciatum.

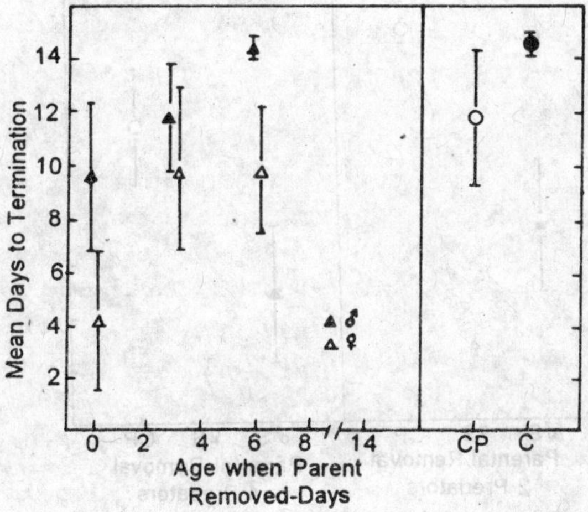

Fig. 16.8. Mean number of days (± 1 S.E.) of termination of trails of Experiment 1.

When the data are presented as mean number of days to termination, the differences between treatments were more pronounced, although the small sample sizes and high variability within treatments resulted in a lack of statistical significance in these differences. In all cases, some of the progeny survived longer when the male parent was removed than when the female parent was removed. That is, the female parent alone appeared to be somewhat better able to protect the progeny from predation that did the male parent alone. This sex difference was most pronounced when parental removal occurred at the beginning of the egg and the fry phases. Also, the termination time tended to increase as the one-parent removal occurred later in the development of the young. This increase was not simply a function of increased duration of protection of the brood by two parents. In several replicates of the removal-after-spawning treatment, all eggs were eaten within one day of parental removal. That is, predation did not generally occur at a uniform rate throughout the tests, but it was most pronounced at certain stages. This observation led to the design of Experiment 3.

Experiment 2

The object of this experiment was to measure the survival of uniparental broods in the presence of two different predator densities. The treatment consisted of removing one parent immediately after spawning was completed and then introducing either two or four predators. Controls consisted of leaving both parents with their brood

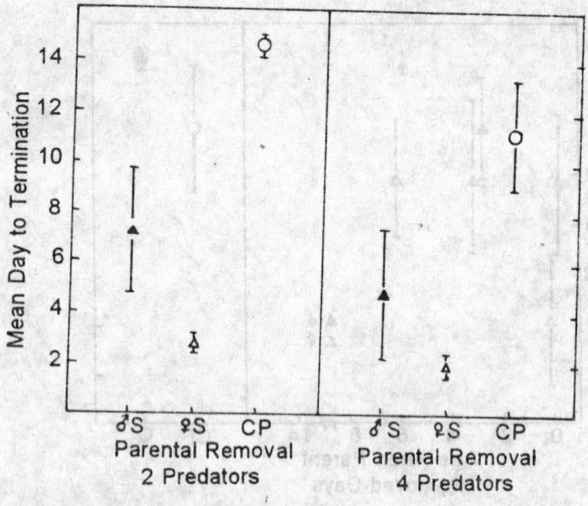

Fig. 16.9. Mean number of days (± 1 S.E.) to termination of trials in Experiment 2. Predator species C. managuense.

in the presence of two or four predators. The predator was *C. managuense,* a cichlid whose range is the Great Lakes of Nicaragua and the Atlantic slopes of Costa Rica (*Miller,* 1966); it is thus sympatric with *H. multispinosa* (*Baylis,* 1974). It is mainly carnivorous and is a likely natural predator of our study species. *C. managuense* used as predators were juveniles that ranged in total length from 2.3 to 6.8 cm (mean, 4.30; n=90).

Table 16.1. Mean percent survival of *H. multispinosa* offspring 15 days postspawning with *Juvenile C. managuense* as predators. Single parent was removed after completion of spawning.

Treatment	Mean percent Fry survival	Standard Error
2 predators		
O removal	2.40	2.40
+ removal	0	0
4 predators		
O removal	0.46	0.46
+ removal	0	0
Control		
2 predators	28.58	8.92
4 predators	10.20	6.25

The effects of this experimental treatment were clear. The percent survival of progeny on the termination data (15 days postspawning) was zero in 18 to 20 trials in which one parent was removed. Survival was somewhat higher in the controls, and highest in the control with two predators.

When the data was presented as mean days to termination, broods with the male parent removed survived longer than those with the female parent removed, especially when only two predators were present. Thus, female *H. multispinosa* again appeared to be more successful than males as single parents. Both control treatments had longer survival times, and survival was longest and least variable in the presence of two predators.

Experiment 3

Under the conditions of Experiment 1 and 2, adult *H. multispinosa* were occasionally seen eating their own progeny. This occurred primarily in two situations. First, when one parent was removed immediately after spawning, the remaining parent occasionally began eating the eggs rather than caring for them in the usual way. Usually the predators joined in when parental egg predation began. Second, as the fry developed, they became more active, and the tight school formation that is typical of newly free-swimming fry became looser. The parents typically retrieved individual fry that strayed from the school and spit them back into the group. Toward the end of the 15-day experimental period, one or both parents occasionally ate, rather

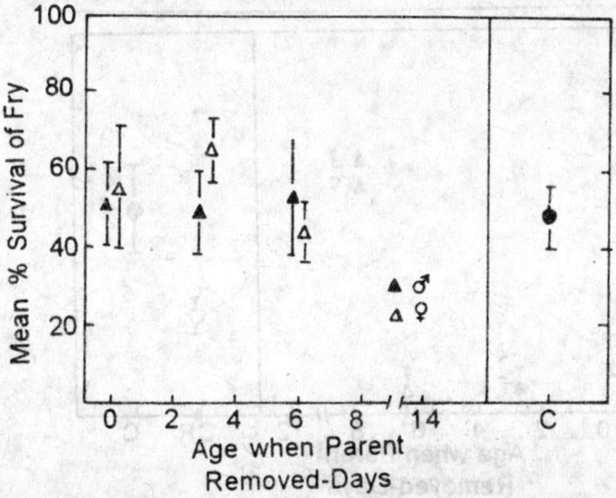

Fig. 16.10. Mean percent survival (± 1 S.E.) of H. multispinosa fry by 15 days of age in Experiment 3. No predators present in any treatments.

than retrieved, the dispersed fry. The first of the two observations suggested that removal of one parent was in some cases of disturbing to the remaining parent that its normal parental activities were seriously disrupted. In addition, the presence of potential predators of another species may have increased the level of disturbance of the single parent.

Experiment 3 was designed to test these possibilities. The treatment schedule was identical to that of Experiment 1, except that no potential predators were present. In only one of 35 trails were all progeny eaten before reaching of the age of 15 days. The mean per cent survival of fry across treatments is shown in figure. It is clear that one-parent removal, without the presence of other potential predators, did not result in lowered survival of the progeny. This strongly suggests that the cannibalism occasionally seen in the earlier experiments was related to disturbance caused by the combination of one-parent removal and the presence of predators.

Experiment 4

H. multispinosa are known to breed in shallow, weedy ponds close to the Rio Frio, Costa Rica. These ponds are flooded during the rainy season, but become separated from the river as water levels decline in the dry season. Breeding occurs during the dry season, at the end of which the ponds contain large numbers of adult and juvenile *H. multispinosa* (*Baylis*, 1974). Thus, it is possible that juveniles are important predators of the eggs, wrigglers, and fry of their own species in this habitat. Experiment 4 was designed to

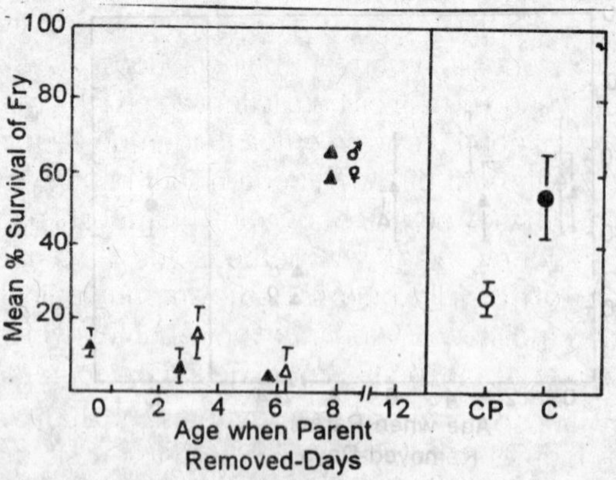

Fig. 16.11. Mean percent survival (± 1 S.E.) of H. multispinosa fry by 12 days of age in Experiment 4. Predators were juvenile H. multispinosa.

measure the ability of single-brooding *H.multispinosa* to protect their offspring against predation by juvenile conspecifics.

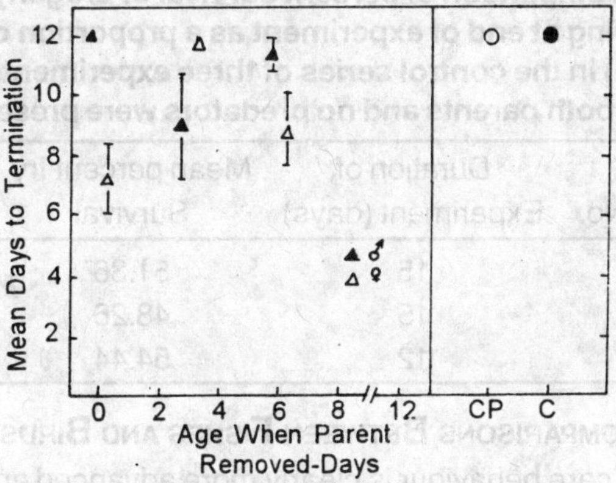

Fig. 16.12. Mean number of days (± 1 S.E.) to termination of trials in Experiment 4.

The experimental design was identical to that of Experiment 1, except that two young *H.multispinosa* were present as predators, and the maximum time allowed for each trial was reduced from 15 to 12 days, because the fry school began to disperse by age 10 to 12 days, and the aim was to measure predation before this dispersion was pronounced. Potential predators ranged in total length from 4.3 to 6.8 cm (mean, 5.56; n=80).

Under all experimental treatments, the survival of fry was below 20%; under the two control conditions survival was higher, and especially so in the control without predators. In the latter case, mean per cent survival from egg to termination was similar to that in the two comparable controls from Experiments 1 and 3. Although there was considerable variation among replicates within each of these three control treatments, it appears that in the absence of the other potential predators, approximately one half the egg laid by *H. multispinosa* pairs survive until 12 or 15 days of age under these aquarium conditions.

Survival, measured as mean number of days to termination, varied considerably among the treatments of Experiment 4. It was greater following male-parent removal than female-parent removal for the postspawning and post-free-swimming removal treatments. This corresponds to the result of Experiment 1. But unlike Experiment

1, mean termination time in this experiment was greater following female-parent removal for the posthatching removal treatment.

Table 16.2. Comparison of percent survival of progeny (number of fry surviving at end of experiment as a proportion of number of eggs laid) in the control series of three experiments. In each case both parents and no predators were present.

Experiment No.	Duration of Experiment (days)	Mean percent fry Survival	Standard Error
1	15	51.36	20.04
3	15	48.26	7.71
4	12	54.44	11.57

Comparisons Between Fishes and Birds

Parental care behaviour is clearly more advanced among birds than among fishes. In general, the association between parent and offspring among birds is of longer duration and of a more direct and intimate nature. Despite the transport by some fishes of eggs in pouches, stalks, or skin folds, and by others of eggs and young held in the mouth, the parent-offspring links in most of the fish species showing any parental behaviour at all are generally looser and briefer than those of most birds. However, comparison of some particular aspects of parental behaviour in these two large vertebrate classes indicates some interesting similarities as well as differences.

Four major functional aspects of avian parental behaviour are : nest construction, incubation of eggs, feeding of the young, and protection of the young. These features, and their counterparts in fishes, will be briefly discussed.

General Features of Parental Behaviour

Nest Construction

Most bird species build nests—that is, structures in which the eggs are laid and then incubated. Some lay their eggs directly on sections of narrow rocky ledges or on patches of bare ground that have been scraped clean of vegetation and debris. Others make excavations in the ground or in trees, posts, and the like, while still others use existing excavations. In either case, the birds may or may not build nests inside such holes. A majority of species builds true nests, which range in complexity from simple collections of sticks to those requiring varying degrees of care and precision in the formation of the completed nest cup. Among the most elaborate are the nests of the weaverbirds (Ploceidae), which carefully weave plant materials together into nests of various shapes and sizes. An extreme form is the large colonial nest of the social weaver (*Philetairus socius*), which builds first a roof of coarse straws in a large tree and under his accumulates a large mass of woven plant materials with many individual nest chambers in it.

Among fishes, the extent and variation in nest-building is much more restricted. First of all, external fertilization with simultaneous release of male and female gametes into the water, is the rule among fishes, and many species produce buoyant, pelagic eggs that are carried by the prevailing currents; these species do not make nests or provide any form of parental care. Among those with demersal eggs, some species (e.g., salmonids, lampreys, some cyprinids) dig excavation in gravel substrate, lay their eggs in the pits, cover them with gravel, and desert them. Other species (e.g., centrarchids, blennids, some cichlids and pomacentrids) spawn in excavations they have made in the substrate, often under stones, sunken logs, and debris, or inside empty mollusc-shells. At least the eggs, and sometimes also the newly hatched fish, are guarded in these locations.

Only a few fishes construct nests that are comparable in complexity to those of passerine birds. The best-known of these nest-building fish are the stickleback (Gasterosteidae). The male builds a spherical or elongate nest by collecting plant fragments, rootlets, and the like, and then binding them together with adhesive kidney secretions. The various probing, boring, sucking, and *glueing* activities of the male result in the formation of a compact nest with an internal chamber to receive the eggs. The most elaborate nest is made by *Apeltes quadracus.* With this species, a cup-shaped nest is attached to rooted plants close to the bottom. After a clutch of eggs is laid, the male builds an extension of the nest up and over the eggs, with a concave upper surface to the extension. A second clutch of eggs is laid on the new floor, and this procedure may be repeated several times, until the male has several clutches of eggs stacked vertically within a single multitiered nest.

A number of other species build less complex nests of collected materials. These are less tightly bound together than the stickleback nests and are usually placed under placed under rock or in dense vegetation. Presumably, in these locations they are protected against turbulent wave action, and the capacities of the parents to guard against predation are enhanced.

Incubation of Eggs

Incubation is defined as "the process by which the heat necessary for embryonic development is applied to an egg after it has been laid." It is universal among birds, although some species do not incubate their own eggs directly. For example, some members of the Megapodidae (such as the mallee-fowl of Australia) bury their eggs in mounds of vegetation, within which the heat from decaying plant material provides the incubation energy. Others, such as cuckoos (Cuculidae) and cowbirds (Icteridae), are brood-parasites, in that their eggs are laid in the nests of other species and are incubated by the foster parents.

Fish eggs are not maintained at temperatures above ambient, either by the parents or by surrounding substrate or nest materials, and hence on true incubation occurs. Even though some species maintain contact with their eggs by oral brooding (e.g., some cichlids), by wrapping the body around the eggs (some stichaeids and pholids, and of course by carrying them internally after copulation and internal fertilization (many cyprinodontoids), there is no evidence that these egg-tending behaviours influence the temperature of the eggs.

Feeding the Young

The chief avian methods of parental feeding of the young are: carrying food to the nest in the bill and placing it in the open mouth of the nesting; regurgitating food, which the juvenile picks up from the ground or from the parent's mouth and throat; and uncovering food items on the ground (by pecking or scratching), after which the juveniles pick it up for themselves.

Among fishes, there are no known direct counterparts to the first two of these methods, but there are two parental activities that appear to be analogous to the third. One of those is *fin-digging.* The other is a form of plunge-feeding found among *Geophagus* species, in which the fish thrusts its snout vigorously into the substrate, withdraws it and expels the mouth contents. Loose particles may then be ingested, and if an adult *Geophagus* is foraging in this manner with a school of its progeny, the young fish also gather and feed on the expelled material.

Another form of feeding the young also occurs among cichlids (e.g., *Symphysodon discus; Etroplus maculatus, Cichlasoma citrinellum*). As the young reach the free-swimming fry stage of development, they begin to graze on the mucus on their parents side. Mucus-producing cells in the adults' epidermis are most numerous at this stage of the reproductive cycle, and it is clear that mucus is being ingested by the young fish. In fact, feeding from the parent's body can be induced in a number of cichlid species by depriving the fry of other food sources. Similar parent-contacting behaviour by juveniles occurs in the coral reef pomacentrid *Acanthochromis polyacanthus,* although no direct proof of mucus ingestion has been obtained.

Protection of the Young

Here there are many similarities between birds and fishes, particularly in the use of overt aggression toward potential predators, and in the use of signals to stimulate the young to take evasive action. Visual displays, such as fin-flickering and body-jerking, performed by adults while guarding free-swimming fry, are often followed by the young fish settling and clumping more closely on the substrate. Acoustic warning signals may be produced by some species, although there is little supporting evidence for this as yet.

The specialized form of parental protective behaviour known as *injury feigning* is common among shorebirds and waterfowl, where its chief function is probably to distract predators from eggs or juvenile birds. It does not appear to have a counter-part among fishes.

Probably the most efficient protective device used by fishes while guarding their young is oral brooding. This is common among cichlids, but it also occurs in other groups. The only way a predator can capture young fish that are being carried in their parent's mouth is to injure or kill the parent, force it to jettison its brood, or wait until the parent releases its young voluntarily, then dash in and capture them. The best available evidence for this type of predation comes from several species of *Haplochromis* in Lake Victoria, which feed either exclusively or principally on the eggs and yolk-sac larvae of other cichlid fishes. The method these fish use to capture their food is uncertain because of the difficulty of observing them in the turbid waters of Lake Victoria.

17

Lekking in Fishes

Lekking is the temporary aggregation of sexually active males for reproduction. In the typical case, males gather on a lekking ground or arena. There each male occupies a territory, or court, from which he displays to females and interacts with other males. The aggregation of males on the lek is visited by females, singly or *en masse,* who select the males with whom they mate. Once mating is accomplished, the females leave the lek.

Clusters of males holding permanent, all-purpose territories, even if visited there by females for reproductive purposes, are not considered by us to constitute a lek. In a more comprehensive treatment of reproductive adaptations, their relationship to lek systems would have to considered. In the context of this paper, such consideration would be too great a digression.

Some colleagues have suggested to us that the use of the term *colonial breeding* would be more appropriate in such a view. We disagree, insofar as our aim is to make explicit comparisons between lek systems in birds and fishes. That term has tow significantly different applications in the avian literature. The first is exemplified by oceanic birds that nest as monogamous pairs in closely packed colonies. In our consideration of fishes, we are not dealing with aggregations of monogamous pairs, although colonial breeding of this sort has been reported for cichlids of the genus *Tilapia.* The second is exemplified by passerines such as weaverbirds and some blackbirds, in which a number of females mate with and nest within the territory of a single male. Such polygynous systems are closer to lekking as we define it and may even grade into it in fishes. But they suggest more appropriate comparisons with the harem societies of some mammals, such as occur in some pinnipeds and bovids, and among teleosts, of some cichlids of the genera *Lamprologus, Nanochromis, Teleogramma, Apistogramma* and *Nannacara.*

Table 17.1. Incidence of Lekking among Teleost Fishes

Order : Cypriniformes	Order : Perciforms
Characidae a	Centrarchidae
Cyprinidae	Percidae
Catostomidae	Sparidae
Order : Atherinomorpha	Embiotocidae a
	Cichlidae
Atherinidae	Pomacentridae
Melanotaenidae	Labridae
Cyprinodontidae	Scaridae
Poeciliidae	Acanthuridae a
Order : Gasterosteiformes	Callionymidae
Gasteroteidae	Belontiidae

Reproductive adaptations based on lekking were first described in birds. Because it is an extreme departure from the usual avian pattern of monogamy and joint care of the brood, lekking has long attracted the attention of behaviourally oriented ornithologists, from the pioneer studies of *Selous* to recent papers by *Robel* and *Ballard, Rippin* and *Boag,* and *Pitelka, Holmes,* and *Mac Lean.*

Concise description of what appears to be lekking in teleost fishes were fist published by *Reeves* and *Newman,* and have appeared consistently in the ichthyological literature up to the present. However, no explicit analogy was made between lekking behaviour in fishes and birds until *Fryer* and *Iles* pointed out the correspondence in the reproductive adaptations of many species of the cichlid genera *Sarotherodon* (formerly *Tilapia* and *Haplochromis*).

Lekking, or what seems to be lekking, has been described in a variety of other kinds of animals. It has been reported in a number of African antelopes living in open country, and most notably in the Uganda kob, in a variety of other ungulates, in hammerhead bats, in one reptile, in one anuran amphibian and among insects in drosophilid flies, dragonflies. One species of harvester ant also appears to engage in lekking. Although lekking may prove more prevalent among some of these groups than the literature would indicate, fruitful comparisons of lekking in different environments must draw at present upon the ornithological and ichthyological literature.

To aid the reader, we enumerate at the outset the prerequisites to lekking any species.

1. *Synchrony.* The reproductive activities of a substantial proportion of the males and females of a given population must be in phase.
2. *Lekking Ground.* The males must congregate at a given place, and the females must also proceed to that place.
3. *Mobility.* The species must be sufficiently mobile to travel to the mating ground.
4. *Parental Care.* If parental care taking exists, it requires only one parent, not a pair.
5. *Feeding.* Either no feeding occurs on the lek, or only incidental feeding that is insufficient to meet energetic needs.

Further on we provide some tables that outline the major points in our central arguments. Although they are redundant to the text, we thought they would be useful in keeping the main themes obvious. In preparing the tables, we were motivated to provide testable propositions in the hope that these would stimulate observations to refute them. The reader will discover that the propositions are a pragmatic mixture of inductive conclusions, though often based on scanty evidence, and deductions that seem to us to flow from what we have learned.

We now summarize the main features of teleost and avian reproductive biology. Next we compare lekking as practiced by members of these two groups. In closing, we speculate on the ecological factors that may have led to the evolution of lek systems in both and attempt to account for tis prevalence in teleost fishes.

A. Comparison of Teleost and Avian Reproductive Biology

Teleosts are characterized by a wide range of reproductive modalities. Oviparity with external fertilization, ovoviviparity, and true viviparity have been reported for the group. With the exception of the viviparous surfperches (Embiotocidae), for which we have only circumstantial evidence of lekking, and a single viviparous poeciliid, *Poeciliopsis occidentalis,* all lek fishes known to us are oviparous. We will therefore concentrate on the epertinent features of oviparity with external fertilization.

In most cases, the female performs a number of spawning acts, releasing some portion of the total spawn each time. In a lekking species this means that the female may move from one male to another, leaving some eggs with each one. Alternately, she may perform all the spawning acts with one male.

The numerous eggs are large and full of yolk. After spawning, the female may not spawn again for several weeks, or until the next season; because it takes time to obtain enough food to lay down such a generous energetic larder. Alternatively, as appears to be the case in cyprinodont, melanotaeniid, and some atherine fishes, the female lays just a few eggs each day over a protracted period. Males, in contrast, need to commit only a small amount of energy to produce vast numbers of tiny sperm. They can spawn repeatedly during a day and for several days. Female therefore make a large initial investment per gamete, the male a small one.

A male may fertilize the eggs of a number of females, regardless of whether the spawning pattern involves production of demersal or pelagic eggs. Thus a male often accumulates a clutch that may number up into hundreds or thousands of eggs, if he is the custodian. Such a number is huge compared to the clutches of birds.

The eggs of teleost fishes are surrounded by a permeable membrane of variable strength. Substances necessary for the development of the embryo, such as water and oxygen, cross this membrane from the external environment while metabolites move in the opposite direction. Teleost eggs are susceptible to attack by bacteria and fungi because their dependence on a steady exchange with their surroundings has in most instances precluded the evolution of effective morphological barriers against such infections. Reproductive success among those teleosts that put their eggs on the substratum therefore depends critically on the accessibility of hygienic spawning sites.

Typically, such sites are characterized by a disturbed or inherently depauperate microbial community as well as adequate dissolved oxygen to allow normal embryonic development. Many fishes exploit recurrent natural phenomena, such as the scouring effect of the spates resulting from the spring melt in the temperate zone, or the monsoons in the tropics, to deposit demersal eggs upon the favourable substrata thus produced. Most nest-building species themselves create a substratum with a disturbed microbial ecology through their nest-building activities.

A different set of adaptations accompanies the releasing of pelagic eggs into the plankton. In reef-dwelling species that do so, it becomes important to expel and fertilize the eggs at the best place to assure that the zygotes are swept into the most favourable water mass. This often means spawning at the outer edge of the reef when the tides and currents are propitious.

Teleosts are poikilothermic. The normal development of their eggs therefore depends on the ambient temperature. This imposes a marked seasonally on the reproductive cycle of temperate zone fishes. Seasonal change in water temperature does not seem to play an important role in regulating the spawning of many tropical fishes. Instead the periodicity of rainfall appears to impose seasonality on most tropical freshwater fishes, for whom the arrival of the rainy season often provides the proximal stimulus for spawning. Seasonality seems less pronounced in tropical marine habitats, but the intraseasonal cyclicity of spawning of some fishes in such habitats may be linked to lunar cycles. Taken together, these environmental factors often impose a degree of synchrony of sexual activity, from modest to great, even among coral reef fishes with prolonged breeding seasons.

Many fishes, especially freshwater species, practice parental care of their spawn. The caretaker is almost always the male. When the female is involved, the course of evolution appears to have gone from exclusively male care to joint care, then to exclusively female care. That the male is the usual caretaker in fishes stands in contrast to the situation within the Vertebrata, with the exception of the Amphibia. A further contrast is that there is no reversal of sex roles when the male fish is the caretaker, contrary to *Wilson*. The courtship behaviour of the male fish is masculine by the accepted standards.

A key factor in the evolution of such a pattern is that in species that fertilize the eggs externally, the male can be certain that the zygotes left in his care were fertilized by him. Chances of cuckolding are slight compared to species that fertilize internally, the usual case in most of the other vertebrata and in some fish groups. Thus an externally fertilizing male can significantly promote his genetic investment by protecting his offspring.

Dawkins and *Carlisle* have proposed a different explanation for why, among fishes, the male is the usual custodian. Their thinking also turns on external fertilization and is therefore applicable as well to aquatic amphibians with parental care. When fertilization is internal, "After copulation, the female is left physically in possession of the zygote, and while it is still in her body, she cannot desert it, but the male can. However fast she lays it, the male is still offered the first opportunity to desert, thereby closing the female's options and forcing her into Triver's cruel bird." The bind, as modified by *Dawkins* and *Carlisle*, is that the partner that deserts first does not necessarily condemn the progeny to death. Instead, it simply sloughs the decision off onto the parent left with the zygotes.

In fishes and many amphibians that fertilize externally, the sperm are lighter than the eggs and are hence more readily dissipated or swept away. From this, *Dawkins* and *Carlisle* reasoned that males have more to lose by spawning too quickly, on the chance that the partner delays, than do females. Thus a female can afford to go first in spawning, leaving the male with the zygotes after he has fertilized them. In our collective experience, though synchronous ejaculation of gametes seems the general case, whenever one sex spawns first, as in gobies, blennies, and damselfishes, but not in cichlids, there is usually paternal care of the spawn. This explanation, nevertheless, has some problems in wide application.

First, one would expect males to evolve sperm that would not wash away, comparable to the spermatophores of some salamanders. In fact, this solution appears to have been evolved by two maternal mouth-brooding cichlids, *Sarotherodon macrochir* and the Tanganyikan endemic *Opthalmochromis ventrailis,* although it does not appear to be inflexibly linked to prior ejaculation by the male.

Second, in many maternal mouth-brooding cichlids, the female takes the eggs into her mouth before the male fertilizes them, immediately after she has expelled them. Fertilization occurs intrabuccally. This suggests that she is eager to possess the eggs, probably in order to minimize the time they are exposed to egg predators, not that she is eager to desert them. Recall, too, that maternal mouth-brooding has probably evolved from joint male-female parental care of the spawn. Finally, it overlooks the general case in seahorses and pipefishes, in which the female *inseminates* the male by leaving her eggs in his brood pouch.

The explanation proposed by Dawkings and Carlisle provides part of the answer to the prevalence of parental care of the spawn in fishes. But we suspect that the necessity of sequetering a suitable spawning site (prerequisite to reproductive success in any species with demersal eggs), the energetic differences between the sexes in the production of gametes, and the certainty of paternity that follows external fertilization, are of equal or greater importance in its evolution. We shall return to this point.

Yet another general feature of teleost reproduction needs mentioning. When the male is exclusive guardian of the zygotes, the female is driven away from the breeding sites as soon as the spawning act has been accomplished. Females are notorious egg predators, as seen in cyprinodont fish and gouramies. Excluding all females save those immediately ready to spawn, thus increase the survivorship of the male's offspring. In addition, spawned-out females might interfere with subsequent matings by the male with other females. Chasing away the spent female is so universal among teleosts that it is remarkable that some groups have been able to evolve joint parental care of the spawn.

Parental care in fishes appears in most instances to have been derived directly from the defense of a territory for reproductive or other purposes. It provides a suitable environment for the development of the zygotes, for their defense, and, less commonly, for defense of the mobile fry from predators. The eggs and newly hatched fry are fanned and/or mouthed, while predators are driven away by the guardian's attacks. In most teleosts the period of parental care is brief, extending only to the eggs and larvae. In the few instances while parental care of the mobile fry is practiced, a behaviour whose occurrence is limited almost entirely to fresh water fishes, such a commitment rarely lasts more than six weeks.

Defense of the eggs or young differs from that seen in birds and mammals. By comparison, the eggs and fry of fishes are tiny relative to the size of the adult, and number at times into the thousands. Their predators are therefore relatively small, seldom larger than the parent and usually much smaller. Consequently, the parent can readily drive away individual spawn predators at little risk to himself, although his defense may sometimes be overwhelmed by sheer numbers of them. Adult fish are themselves subject to a different type of much larger predator. Thus, when considering the effects of predation upon teleost reproductive patterns, one must keep in mind that it operates upon two different levels, the spawn and the parent, and in ways that elicit radically different responses by the breeding adults.

An additional important factor is that much predation on the eggs, and particularly upon the fry, may be by conspecifics. This is true of cichlids and of other freshwater fishes such as sticklebacks and pupfish. Such intense predation upon the eggs and young by conspecifics is less prevalent in birds and mammals. As a consequence, there is sometimes a semantic difficulty in talking about territorial defense because driving away conspecific intruders may actually represent defense of the spawn against such predators.

Although there are too few data available at this time to permit detailed comparisons, it is worth mentioning that territorial defense in lekking fishes may be more ritualized than in non-lekking species. *Apfelback* and *Leong* compared reproductive aggressive behaviour in three species of *Tilapia.* These were *Tilapia zilli,* a monogamous substrate breeding species; *T. galilaea* (= *Sarotherodon galilaeus*), a pair-forming species in which both sexes engage in mouth-brooding; and *T. macrochir* = *S. macrochir*), a lekking species with maternal mouth-brooding. Aggression was the most damaging in *zilli* and the most ritualized in the lekking species *macrochir; galilaea* was intermediate. Future studies should be alert to the possibility that a concomitant of the evolution of lekking is a shift from damaging aggressive behaviour to ritualized threats.

There is no trophic component in the parental behaviour of the majority of teleosts practicing defense of their spawn. In some species, e.g., substratum-spawning members of the family Cichlidae, the parents may protect the young as they forage. In cichlids of the genera *Symphysodon* and *Etroplus*, however, and in two species of bagrid catfishes, the fry depend on parental mucus occurs in a number of other cichlid species. In no instances known to us does parental care include a thermoregulatory components.

All birds are oviparous, with internal fertilization of a cleidocal (shelled and self-sufficient) egg. The shell of the avian egg encloses a milieu in which all of the raw materials of embryonic development are present save oxygen, and within which provision is made for isolating nongaseous metabolites from the developing embryo. The shell is permeable only to gases and provides a virtually impregnable barrier to bacterial invasion. However, because birds are homeothermic, their eggs require a reasonably constant temperature for development. Care of the developing eggs, predominantly thermoregulatory in nature, is consequently universal among birds. With the unique exception of the Australian family Megapodidae birds accomplish this end by brooding the eggs.

During this period, the eggs and one or both parents are vulnerable to predation. Security for the clutch and the incubating parent or parents is consequently a major factor influencing the evolution of avian reproductive adaptations during the incubation period. That period may last as long as the entire interval of parental care in most teleosts. *Lack* has concisely summarized the nesting adaptations adopted by birds to maximize reproductive success. His account underlines the importance of a secure nest site and various adaptations favouring crypticity for the nest, clutch and parents. Nest site selection is therefore important in birds, but for reasons different than those dictating such behaviour in teleost fishes.

Posthatching roles vary considerably between nidifugous and nidicolous birds. A thermoregulatory component of variable intensity, nonetheless, is characteristic of virtually all birds. It is most pronounced in nidicolous species, whose young are not feathered at hatching. The precocial young of nidifugous species, in contrast, have a cost of insulating down from the time they hatch. That lessens the need for heat from the parent and is therefore less restrictive of the caretaker's activities.

The major difference between these types of birds, however, is the way the young get food. Nidifugous birds do not usually bring food to their active young, but rather guide them to food sources and provide them with some protection against predators as they forage (*Lack,* 1968). The parents or parent of nidicolous young must forage for food, which is then brought to the young at the nest. The trophic dependence of the young persists until they are fledged and, in some species, even for a short time thereafter. Regardless of how they are discharged, the trophic responsibilities of breeding birds are of central importance in their overall reproductive strategy. The seasonality so clearly evident in avian reproduction is imposed in large measure by the necessity of having sufficient food at hand both to support the parents and to feed the young during a protracted period of growth, considerations that also limit the choice of nesting site.

Parental Role of Male and Female in Lekking Birds and Teleosts

In all lek birds the male has no parental role. To many ornithologists this is a *sine qua non* for lekking. Once the female has mated and completed her clutch, she avoids the lek, incubates her eggs, and rears her hatchling alone. The ability of the female to discharge all parental functions unaided is therefore a precondition for the evolution of lekking in birds.

Among lekking teleosts, four possibilities exist with regard to parental care:

1. Noncustodial Lekking

Demersal eggs are deposited by one or more females within the male's territory on the lek with no prior preparation of a nest, as in most cyprinodonts and in the atherine families Melanotaeniidae and Atherinidae. Alternatively, pelagic eggs are shed into the plankton in a *nuptial dash* launched from the reef, as in the lekking species of the marine families Labridae and Scaridae, and possibly surgeonfishes of the genus *Naso*. There is no overt dense of the spawn, either because of the brevity of the lek's persistence of sitting the lek where there are no predators, or because of the dispersal of the eggs and larvae.

2. Lekking with Maternal Care of the Spawn

This mode of lekking is known to occur only in one poeciliid and in the maternal mouth-brooding species of the family *Cichlidae*. In the poeciliid, the female visits the male and is inseminated on his court. In the cichlids, the female visits the male and is inseminated on his court. In the cichlids, the female visits the male on his territory, where spawning occurs. The eggs are taken into the female's mouth, wherein they are fertilized if they have not already been fertilized. The egg-laden female leaves the lek while the male remains, awaiting further mates. The female swims to spatially separate nursery grounds, where she remains until the fry become independent. Or both sexes may abandon the lek and reconstitute the original school, as in several lake Tanganyika open water cichlid species. In some instances the fry are shepherded and protected by the female for a while after they have emerged from her mouth. In others, the young are simply released and abandoned by the female.

3. Protocustodial Lekking

The male constructs a nest that is the focal point in his territory. Then females approach individually and deposit their eggs there. The spawn, of one of several females, are neither cleaned or aerated by the male. Whatever protection from predation the eggs and larvae receive derives incidentally from the male's defense of his territory from conspecific intruders. The fry depart as soon as they are mobile and thus are not protected. In the examples known to us, the males abandon their territories after completing the process of reproducing. This mode of lekking occurs in many cyprinids, in the pupfishes of the genus *Cyprinodon* (Loiselle, in prep.), and in some darter perches of the subfamily *Etheostomatinae.* It may also occur in the cod *Gadus callarias.*

4. Paternal Custodial Lekking

The eggs of one or more females are deposited within the male's territory in a nest prepared for that purpose. The eggs are usually aerated and cleaned by the male, who vigorously repeal all intruders. Some defense of the mobile fry is possibly. This mode of lekking is practiced by some cyprinids, by one cyprinodont, Jordanella floridae, by many darter perches, and by all lekking centrarchids. There is some question, however, about the male parental role of one centrarchid fish, the sacramento perch, *Archoplites interruptus.*

Further examples of paternal custodial lekking are provided by sticklebacks by some lekking wrasses that produce demersal eggs, and by lekking species of the marine families *Sparidae* and pomacentridae. In Thailand, groups of male gouramies of the genus *Trichogaster* from discrete arenas under their bubble nests where they are visited by females ready to spawn.

The thermoregulatory and trophic components of avian parental care preclude the evolution of noncustodial protocustodial, and strictly paternal custodial lekking in birds. The two considerations, taken with the practice of internal fertilization, produce the sharpest difference between lekking birds and fishes. They contribute importantly to the greater diversity in forms of lekking in teleost fishes.

That so few internally fertilizing fishes have been unequivocally reported to lek may be an accident of insufficient observation or of observations unguided by hypotheses about reproductive strategies. We expect additional examples of lekking to be found among internally fertilizing fishes, as we have suggested for the Embiotocidae. An obvious group to examine is the Goodeidae. They are in the same suborder as the Cyprinodontidae, which has so many examples of lekking species; the two families also resemble one another in ecological adaptations and in morphology. We also anticipate the field studies will reveal lek systems to be more prevalent among poeciliids than present evidence would suggest.

Occurrence of Polygamy Among Lekking Birds and Teleosts

The absence of monogamous pair-bonding and the corollary occurrence of sequential polygyny are taken as defining elements of avian lekking. The extent of polygyny is precisely known only in the ruff and in various grouse species. The occurrence of polyandry is also possible in all lek birds but improbable in the galliformes because their females practice sperm storage and hence need visit the lek only once in a season to produce a clutch of the fertile eggs. However, there are no data no polyandry in lekking birds. It seems a probably corollary of the attenuation of the pair bond.

Lekking teleosts are likewise characterized by polygamy. Sequential polygyny is found in all lekking species, and sporadic instances of simultaneous polygyny have been reported among the Centrarchidae and the Labridae. External fertilization of the egg makes this departure from the typical pattern possible. Resident males, however, are normally receptive to but a single female at a time. In some instances, the male may actually repel females that attempt to enter his territory while he is engaged in the terminal phases of courtship or actual spawning.

The occurrence of polyandry among lekking teleosts is better documented than is the case among lekking birds. Polyandry has been reported among cyprinodonts and occurs among melanotaenids under quarium conditions. Females of four darter perches practice polyandry. Such behaviour has been reported as normal in one cichlid, *Sarotherodon macrochir* and in one pomacentrid, *Chromis multilineata* considered it typical of the reproductive behaviour of most centrarchids. Subsequent investigations have revealed polyandry in one sunfish not cited by Breder, the Sacramento perch.

Less information is available on the extent of which individual females of a given species indulge in polyandry. Ruwet reported female *S. macrochir* carrying a clutch of eggs fertilized by five or six males. *Breder* regarded centrarchid breeding systems as essentially nonassortative, citing an instance in which a female *Lepomis gibbosus* visited every male on a small lek of indeterminate size.

Topological Position of the Male, Female Choice, and Predictability of the Environment

The disproportionate reproductive success enjoyed by centrally located males of the ruff and many grouse species with its overtones of Darwinian sexual selection, has attracted the attention of many workers. The occurrence of such a position effect within the lek has not been documented in other lekking birds, however. This lack of information is particularly marked for forest-dwelling lek birds. Until these species have been more extensively studied, it would be premature to regard such position effects as being a universal feature of avian lekking.

In those avian species for which position effects have been determined, the classical cases of lekking among tetraonids, central males may enjoy in excess of 80% of the compulations that occur during the breeding season. Succession to central sites within the lek follows a clear protocol. If vacancies occur through mortality, they are filled by peripheral males, who are in turn replaced by marginal males. Direct competition for such sites is ritualized. The protocol of succession resembles the seniority system of the American Congress. The chairmanships of powerful committees come almost automatically to those who succeed in assuring their regular reelection and avoid antagonizing their colleagues by displays of nontraditional behaviour. To the best of our knowledge, however, the analogy breaks down in all but a few cases upon consideration of the rewards accruing to persistent males.

There is little information, in studies of fish lekking, on the mechanisms determining access to favoured territories. The evidence suggests that in some centrarchids and cichlids overt competition exists and can be intense. There is no indication of a protocol of succession to favoured sites.

There are only fragmentary indications that position effects characterize lekking in teleosts. The existence of discrete classes of central and peripheral males may be inferred for two lekking cyprinids, one cyprinodont, two darter perches, two centrarchids, gouramies of he genus Trichogaster and one cichlid. *Hunter* reported that leks developed around the first male green sunfish to spawn, and *Wright* observed that male gouramies place their nests around that of the most aggressive male. In Mediterranean wrasses of the genus *Crenilabrus,* a central male is surrounded by smaller nonterritorial satellite male who only occasionally fertilize eggs in the nest when the large male is temporarily away; however, it is not clear whether there is a position effect among the territory-holding males.

The extent to which highly dimorphic dominant males enjoy augmented success in mating has been demonstrated well in only one teleost species, the bluehead wrasse (*Thalassoma bifasciatum*) by *Warner et. al.* (These males are central in the sense that they are often surrounded by smaller nondimorphic males). The females move across the reef from the shallow to deeper water to reach the lek. There the large station-holding males are sought out. As the females approach the lek, they are solicited by the small darb males who are not territorial. Sometimes females spawn with groups of these smaller males. Still other drab males, *streakers,* join the female when she spawns with the gaudy *central* male, and yet others, *sneakers,* try to setal a spawn on the lek. Nonetheless, the *central* dominant male enjoys as enormous reproductive advantage over the small drab ones, regularly spawning about 40 and occasionally 100 times/day. In contrast, the nondimorphic peripheral makes, spawning predominantly in groups, only achieve the equivalent of about one to two pair-spawnings/day.

The situation is less complex in gouramies of the genus *Trichogaster.* Among Thai populations of *Trichogaster trichopterus,* up to 20 to 30 males nest together. The nests are most closely placed around the central male, and each nest territory there is only about 20 cm in diameter. The female swims directly to and butts the male of her choice. In all 88 spawning observed by *Wright*, the central male was chosen by the female.

The example of the bluehead wrasse draws attention to another important difference in teleost lekking made possible by external fertilization. It is the possibility of neighbouring resident males, or nonterritorial marginal males, joining the consorting couple at the moment of oviposition and participating in the fertilization of eggs. Such behaviour has been documented in one cyprinid, one darter perch, one centrarchid, and one cyprinodontid, one wrasse, and has been observed in one cichlid. It is difficult to evaluate the significance of such a breakdown in the lek system on the basis of these examples. One would wish to know, in particular, whether the relative paucity of reports is an accurate reflection of rarity of such a breakdown, or simply an indication of failure to record its occurrence in other species.

An instance of a system in which selection seems to have operated against such cheating is cited by Ruwet for *Sarotherodon macrochir.* Resident males whose territories adjoin will display frenetically to an approaching female. Once the female has entered the territory of one of the competitors, however, all display by the unsuccessful rival ceases. They turn away for the female and indulge either in nest-maintaining behaviour or interact with other resident males.

In some lekking fishes, such as wrasses and parrotfishes, a form of cheating may be a regular feature of spawning. In some parrotfishes, and in some wrasses, the situation is complicated by a combination of intra- and intersexual competition. Many individuals are sequentially hermaphroditic. Some start life as males (primary males), but most are first females, who then change into males (secondary males). Large males are gaudy, and they lek. The young but sexually mature secondary and primary males resemble the drab females and are the cheaters. They are divergent in the context of this paper, and considering the implications of their biology here would carry us away from the main theme.

The existence of position effects among some lek birds and their probable occurrence in fishes raises the issue of female choice. Reproductive success, the number of offspring surviving to reproductive age, is the measure of evolutionary fitness common to all organisms. In lekking and most other promiscuous birds, the only measure of male reproductive success available to the observer is the number of successful copulations per breeding season, since the male plays no role in the rearing of the young. Many of the behavioural and morphological features that characterize males of lek birds, such as extreme sexual dimorphism and elaborate displays, are thus adaptations serving to maximize reproductive success.

The measure of female success, on the other hand, is the number of young she brings to independence. This will be determined by her own experience, crypticity, and ability as a mother, and by the genetic endowment of the hatchlings themselves. The latter is the only respect in which the male may make a significant contribution. If therefore behaves the female to select a male whose genetic material will maximize the chances of her young attaining independence and thus, presumably, sexual maturity. In the grouse the question of choice is yet more crucial, because the sperm storage means a female probably has but one chance a year to make an optimal choice or a mistake.

The exigencies of female choice and the nature of the mechanism that determines male succession to central sites within the lek may explain the existence of a position effect in some lek birds. When male succession is largely a function of age, a central male must possess a genome well adapted to its immediate environment. Otherwise he would not have survived long enough to attain such a rank. As *Wiley* (1974) has shown in one instance, the displays of older birds appear more attractive to females, thus providing a proximal behavioural mechanisms for their selection of central males as mates.

Females may choose between different groups of communally displaying males. Females of the ruff preferentially visit arenas with a large number of satellite males. As the number of satellites present on an arena declines, so do the number of female visits and the number of copulations enjoyed by resident males. Additional evidence comes from a colonially nesting species, the village weaverbird. Colonies with fewer than 10 displaying males attract disproportionately fewer females than do the typically larger colonies. Leks may be more attractive to females, and therefore to other males, in direct proportion to the number of displaying males, a point to which we will return.

Position effects, as indicated, are not as well documented in teleosts.

One would predict their existence in the following situations:

1. Species Practicing Protocustodial or Paternal-Custodial Lekking

When defense of the spawn is practiced in conjunction with lekking, the optimal strategy for a female is to mate with a male who can provide the most protection at the best location. In a lek situation, central males incidentally benefit from the screening provided through the interaction of peripheral resident males with intruders. Because they have fewer potential predators to contend with, central residents can render more effective defense of their spawn from the few intruders that penetrate the territories of peripheral or so satellite males. At the same time, they can devote a proportionately larger amount of time and energy to actual courtship, thus providing a proximal mechanism of female choice.

2. Lekking in Relation to Predictability of the Environment

In this section, and in later ones dealing with tradition and evolution, a key concept is predictability of environment. The present digression is necessary to explain how the term is employed. We use the concept relatively loosely and at times as being synonymous with stability of environment. For a more precise treatment of the concept as applied to periodic phenomena, the readers is referred to *Colwell* (1974).

Predictability with regard to lekking sites means merely that each year during the breeding season the same set of conditions is apt to prevail on the same lekking grounds. Thus in many species of grouse, the cocks are able to use precisely the same arena year after year.

Contrast this with the situation among gouramies of the genus *Trichogaster* in Thailand. The fishes commence breeding with the onset of the rainy season. The males build their bubble nests among floating and emergent vegetation in canals, pools, and flooded fields. As the rain continues, the rising water level submerges the vegetation at the original lek. The fish then decamp to find new, better suited sites. Thus, the best place to lek is unpredictable. The seasonal pattern of rainfall, and consequent general pattern of movement of the fishes, however, are fairly predictable. Thus, while the environment may be too unpredictable for traditionality of lek sites to develop, it may be highly predictable in the sense of a given male having the best genetic endowment for coping with it.

This brings us to the issue of predictability as a factor influencing female choice. Consider the female's problem: If the environment is relatively predictable, she should mate with the male best adapted to that situation. The predictability of the environment implies that the genome that is now the most fit will continue to be so in the next generation. If, however, the females' offspring are likely to find themselves in an environment or environments that differ from the present one, the female should not invest all her gametes in the male best adapted to the present situation. Her optimum course of action is to mate with a number of different males. By thus increasing he genetic variability of her offspring, she increases the probability that some of them will be optimally endowed for whatever environment they find themselves in.

Predictability of environment is relative to the species, as is the concept of the niche. Take the case of planktonic larvae of a marine fish that are widely dispersed to coral reefs scattered about the tropical sea. A small sedentary species, such as any of several damselfishes or gobies, faces a highly unpredictable community of other species of sedentary fishes when it settles out of the plankton onto a small coral head. In contrast, a larger species such as a surgeonfish of the genus *Naso* can move about and average out local differences in community structure. Its environment is more predictable.

Similar arguments can be made for temporal predictability. If the climate is characterized by long cycles of suitable weather and water conditions, the female should pick the currently best adapted male, all else being equal. But if the onset and length of the breeding season and other features of the environment affecting the survival to maturity of the young are unpredictable, the female should be relatively polyandrous.

Annual cyprinodont fishes illustrate this point well. These fishes are able to survive in ephemeral pools by virtue of their drought-resistant eggs. The eggs are buried in the substratum of the pool; they survive the dry season and hatch with the onset of the next rainy season. Not all the eggs spawned in a given year hatch with the coming of the first rains of the next. In a proportion of each spawn, the diapause, or resting stage, of the embryo is prolonged; from several weeks up to, in some instances, several years. This is an adaptation to environments where the onset of the rainy season is characterized by one or more false starts. Even if many eggs do hatch after a light or unseasonal rainfall, and are subsequently lost, some resting eggs will survive the disappearance of the pool and will hatch with the true onset of the rainy season.

In a relatively predictable environment, selection will favour females who produce a large number of nonresting eggs. Such eggs will hatch immediately, giving the fry first access to the food resources of their environment. Such early fry can be expected to reach sexual maturity more rapidly than fry hatched later in the season and to enjoy a longer period of reproductive activity, thus producing more eggs. In an environment where the onset of the rainy season is unpredictable, the reverse should be true. Females who produce a large number of resting eggs will enjoy disproportionate genetic representation in subsequent generation.

While we have no evidence that the earliest hatched males or the largest males in a population produce more nonresting eggs than do smaller, later-hatched males, it is reasonable to assume some correlation between these characteristics. There may well be aspects of the male's behaviour that make them variously adapted to competing, depending on whether they enter the population early, mid, or late in the rainy season. In any event, assuming that males differ in this regard, we would still predict that the optimal strategy for a female attempting to hedge her bets would entail spawning with a large number of males rather than with one or a few individual males.

In predictable environments, mechanisms determining male position on the lek may arise that reflect in a direct manner the adaptive value of a particular genome. As an example, if male position is determined by aggressive interactions, older, larger males, and/or those with a superior energy balance, would be expected to dominate in such encounters. Such males could then secure the nest sites that are best for the development of the eggs, and thus attract the most females. Evidence suggesting this situation in the darter perch *Etheostoma nigrum* was presented by *Winn.* Other males should crowd around this most attractive male to maximize their own chances of attracting females, as in the green sunfish, the bluehead wrasse, or gouramies of the genus *Trichogaster.* The male attributes contributing to such victories are also correlated with adaptation that particular environment: adaptations leading to increased trophic efficiency result in larger size and/or superior energy balance. In a relatively predictable environment, the genomes of such successful males should converge upon an optimal configuration. Females would then maximize their reproductive success by spawning with such males.

It should follow that sexual dimorphism is reduced in those lekking species that are polyandrous. However, because of the brevity of mating, selection will still favour a high degree of dimorphism to enable rapid unambiguous recognition of the opposite sex. This can be based purely on colouration and shape (or possibly on sounds or chemicals), with size dimorphism being more important when polyandry does not occur.

Thus, we predict that when position effects are marked, the males will be dimorphic for size and for colour and shape. On the contrary, when females are less discriminating, the males should be dimorphic for colour and shape but not necessarily for size. Size dimorphism should still be expressed to some degree, however, because size can still be important in obtaining a position of the lek. In fact, when the breeding sites are in especially short supply as may occur where some annual killifishes breed, size dimorphism should be pronounced.

We have written as if environments were either clearly predictable or not, which is not the case. There is a continuum of situations between highly predictable and unpredictable environments. Most will be relatively predictable or unpredictable to varying degrees and with regard to different properties of the environment. Consequently, most species of lekking fishes should reflect a mixed strategy. The more predictable the environment, the more prevalent should be dimorphism, polygyny, and position effect, and vice versa.

Persistence of the Lek and of the Occupation of Territories by Individual Males

Avian leks are typically occupied for part of each day during the breeding season. The resident males spend the remainder of the day foraging. Individual males of the ruff and several lekking grouse are faithful to a particular site within the lek, as are long-tailed and white-bearded manakins. There are no data available on site attachment in other lekking species. In some lekking grouse, males will revisit the site during the fall, well outside of the breeding season.

Species inhabiting open habitats are active on the lek during the early hours of the day. One such species, the great snipe, even displays at night during the full moon. This is apparently an adaptation to minimize aerial predation faced by birds displaying in the open.

The situation is less clear in forest dwelling species. The impression conveyed in the literature is that males of those species are active on the lek during the latter part of the day. In males of the long-tailed and white-bearded manakin, peaks of activity occur at different times in different parts of their long breeding season.

Such intermittent lekking, in which the lek is occupied for only part of each day, is also practiced by some teleosts. It is characteristic of lekking cyprinodonts, melanotaenids, and atherinids, and it may possibly occur in the Sacramento perch. The period of sexual activity, as would be predicted in poikilothermous organisms, is correlated with water temperature and usually occur in the late morning and early afternoon. Occasionally, however, high temperatures interfere with lekking during the afternoon. In most marine situations, or in large lakes, the temperatures are more stable and consequently less important as phasic triggers, though thermal effects have been reported.

Intermittent lekking is feasible in noncustodial species that produce demersal eggs, such as some cyprinodontids, and that habitually breed in environments into which few or no spawn predators penetrate. This allows males to practice intermittent lekking while accumulating eggs in their territories. It is also feasible for species that shed pelagic eggs into the plankton, such as parrotfishes and wrasses. In neither case is the male's presence required to protect the spawn, nor is there an energetic investment in nest construction to be defended. Intermittent lekking may, nevertheless, be synchronous if for some reason the females have preferred times for spawning.

In contrast to some lekking birds, there is no recorded instance of sexually inactive male fish visiting the lek site outside of periods of reproductive activity. This doubtless happens, however incidentally, in cyprinodontids confined to small pools and in other fishes.

Continuous lekking, in which the males occupy the lek without interruption for foraging, relying upon stored energy reserves to sustain their activity, is known only in teleosts. (The mating systems of penguins and albatrosses, and of pinnipeds, though not examples of lek systems, provide parallel case among birds and mammals. Continuous lekking is predictable, for obvious reasons, in fish species practicing some type of defense of the spawn, be it constructing an elaborate nest or overtly repulsing predators. Sustained lekking in males of maternal mouth-brooding cichlids, however, cannot be thus explained. It may instead by correlated with the more overt competition for territories within their leks. *Coe*, for example, reported that male *Sarotherodon grahami* that left their territories to forage lost them immediately to other males and had to contest their possession, often unsuccessfully, with the new proprietors.

There are few data on how long an individual male retains a site on the lek. *Reighard* stated that sexually active male logperch, *Percina caproides*, spend 10 to 14 days on the lek, then retire to deeper water. *Neil* found that sexually active male *Sarotherodon mossambicus* hold a nest site in the aquarium from three to ten days, with a mode around five to six days. Similarly, successful males of the green sunfish have a period of occupancy of around eight to nine days. In Thailand, each group of lekking male *Trichogaster trichopterus* lasts about one week. Otherwise, it is known only that the males remain on the lek for a substantial period of the reproductive cycle, a period of time that can vary from several hours, as in the Tanganyikan maternal mouth-brooders *Xenotilapia melanogenys* and *X. ochrogenys*, whose spawning is characterized by a remarkable degree of synchrony, to several days at least.

As with intermittent lekking, activity appears correlated with water temperature and usually peaks in the late morning and early afternoon.

Environmental Factors as Determinants of Lek Sites

Traditionality of lek areas is one of the most remarkable features of avian lekking. In fact, *Wilson* gives traditionality as a criterion to distinguish lekking from the more general set called communal displaying.

Armstrong documented traditionality in five galliform species (Argus pheasant, blackcock, prairie chicken, sharp-tailed grouse, and the extinct heath hen—a race of the pairie chicken), one charadriiform species (the ruff, and two forest-dwelling passeriform species (greater bird of paradise, Gould's manakin). *Wiley* presented persuasive evidence for traditionality in an additional galliform, the sage grouse, and *Gilliard* for another passeriform, the cock-of-the-rock. Traditionality is also well developed in long-tailed manakins and in white-bearded manakins. In contrast, nontraditionality appears to be an important feature of lekking as practiced by one charadriiform species, the buff-breasted sandpiper (Pitelka, personal communication).

This dichotomy of traditionality versus nontraditionality may be explained by assuming that the location of avian lek sites is, or has been determined by environmental factors. In the case of traditionalists, the factors are presumably a predictable environment coupled with the limited number of areas from which effective displays can be presented. These factors are particularly evident in the case of forest-dwelling species. Authors who have reported on the incidence of lekking in manakins, bellbirds, cock-of-the-rock and birds of paradise; emphasize the following : (1) the performance of the displays requires open space; (2) special conditions of lighting are needed to emphasize the distinctive features of plumage; (3) the lek arenas are located in areas where environmental factors have disturbed the continuity of the predominantly closed forest canopy.

Table 17.2. Circumstances associated with traditionality.

A. Physical environment

1. Relatively predictable
2. More "best arenas exist than are generally used
3. Arenas often modified by males' behaviour

B. Animals

1. Dispersed breeding population
2. Relatively long-lived (at least more than one breeding period)
3. Delayed sexual maturity in males.
4. Central nervous system complex enough to allow of learning and memory.

The situation in marsh-and prairie-dwelling birds may not be as obvious and is open to debate. *Wiley* and *Hogan-Warburg,* however, implied that the traditional areas of the sage grouse and the ruff, respectively, were originally positioned in relatively open patches of habitat where edaphic or other environmental factors had thinned out the prevailing assemblage of forbs and grasses. Students of avian lekking regularly report the occurrence of behaviour by resident males that intentionally the occurrence of behaviour by resident males that intentionally or fortuitously preserves and perhaps enhances the suitability of the arena for lekking.

Limited lek sites alone, however, could not account for the highly developed traditionality seen in some species. It requires in addition a relatively predictable environment coupled with a reasonably long life span and the ability of the males to remember the location of the lekking grounds. Otherwise, it would be difficult to account for the fact that some sites are used year after year by long-tailed manakins, while other sites that seem to have all the necessary features are not utilized. It would also be difficult to account for the persistence of lek sites when the environments is unfavourably altered. A well-known example was provided by male ruffs who persisted in displaying on old sites that came to lie in road.

In the case of the buff-breasted sandpiper, environmental factors impose nontraditionality. The sandpipers' leks are ephemeral and transitory, males gathering and displaying for periods of a week or two in a given spot, then apparently moving elsewhere to repeat the performance. Pitalka (personal communication) suggested to us that the placement of a lek in these sandpipers is influenced by the available supply of food for the nesting female and her brood and by year-to year variations in the physical environment, such as unpredictable patterns of runoff from the melting ice and snow.

According to Petelka's model, transitory lekking permits exploiting a patchy environment by moving over a wide area and settling in to display only where the terrain is suitable and food abundant. This maximizes the probability that females impregnated by them will be able to raise their broods successfully in the short Arctic nesting season. The occurrence of suitable areas depends on a multitude of local climatic factors and is therefore relatively unpredictable. Traditionality would be maladaptive under such conditions.

Lekking may also be correlated with population density. *R.R. Warner* and *S.G. Haffman* are testing the following model, which was inspired by observations on wrasses and parrotfishes off the coast of Panama. A similar model is being developed for other vertebrates by *S.T. Emlen* and *L.W. Oring.* We present the model in abbreviated form, with due apologies.

When the population density is low, the prevailing mating system is a territorial harem society. At intermediate densities lekking develops. At high population densities dominance relationships break down and territories are forsaken; a number of females may spawn synchronously each with more than one male in attendance—the *conubium confusum* of *Breder* and *Rosen*.

The intermediate population density at which lekking occurs must be considered relative to the species. It is our impression that lekking species of birds and fishes are relatively common. In general terms, lekking is probably favoured by population densities that are relatively high, but not so high that social organization breaks down.

The dependence of many teleosts on a hygienic spawning location for their relatively vulnerable eggs means that environmental factors directly determine the location of lek sites in species practicing noncustodial, protocustodial, and paternal-custodial lekking. Traditionality in the avian sense is not well documented in such species. However, traditionality should be expected among mobile species occurring in relatively stable environments, such as the rocky littoral of the African Great Lakes or the protected coral reef. There the physical factors that dictate optimal spawning sites differ little from one year to the next. (However, the physical features of some coral reefs may at times be drastically altered in regions where violent storms occur. In variable environments, in contrast, lek sites can change in location from year to year.

The bluehead wrasse provides an example of a coral reef fish with traditional lek sites. One population of this labrid used the same area as a spawning site over a period of five years. While this represents traditionality in the broadest sense, the reproductive modality of this species introduces complicating factors not encountered on avian leks. The bluehead wrasse sheds pelagic eggs into the plankton. It thus adjusts the precise location of its spawning site within a general area from day to day, and even within a day, apparently to remain in a down-current zone that favours the fertilized eggs being swept out to sea.

Another instance of what may be lek traditionality in the avian sense is provided by the maternal mouthbrooding Tanganyikan chiclid *Cyathopharynx furcifer*. *Brichard* reported that males of this species construct sand nest on top of flat-crowned rock blocks. Sexually active males may be found using such areas continuously. Though the number of such sites is limited, intraspecific aggression is not pronounced, and resident males visit one another's territories in a manner reminiscent of such lek birds as the ruff and some grouse.

Additional instances of apparent lek site traditionality have been reported for males of several other maternal mouth-brooding cichlids.

Substratum-independent or *free-water* spawning behaviour has been described for the maternal mouth-brooding Tanganyikan cichlids *Tropheus moorii, Limmochromis microlepidotus,* and *L. leptosoma.* In most instance, however, sexually active males prepare a nest from which courtship is directed and within which spawning occur. Since their eggs do not remain in the male's nest during their development, these cichlids are not substratum-dependent in the same sense as are teleosts that produce demersal eggs. Even so their emancipation is not complete.

Regardless of their pattern of brood care, cichlids, gouramies, and other teleosts are limited in their selection of spawning sites by predation or both the breeding adults and their spawn. Ease of constructing nests will also vary among sites, as will the degree of shelter from wave action. These factors can adversely influence a male's reproductive success by obligning him to expend energy in nest preparation and maintenance that would otherwise be devoted to courting or to prolonging his time on the lek. The substratum preferences reported for the cichlid fish *Sarotherodon macrochir* by *Ruwet* were probably in response to the suitability of the substrate. And the destruction of *Haplochromis* leks by wave action, reported by *Kirchshofer* and by *Fryer* and *Iles* suggests that shelter is indeed significant in selecting a lekking ground. The place where lekking is done can be important for still other reasons as when mobile reef inhabitants seeks a favourable lauch window through which to shed their pelagic eggs.

In some instances, the siting of leks is constrained by adverse physiographic factors. An extreme case is provided by two cichlids. *Sarotherodon grahami* inhabits hot spring in Lake Magadi, *S. alcalicus* the hypersaline Lake Natron, both in the Rift Valley of Kenya and Tanzania. Thermal factors severely limit the areas available to sexually active males *S. grahami,* while salinity gradients restrict male *S. alcalicus* in their choice of nesting sites.

Evolution of Lekking

Early attempts to determine the functional significance of avian lekking are exemplified by the following passage from Armstrong (1947:225):

> The conclusion is ineluctable that the advantage of arena or lek displays must be very great. It is highly probable that not only is

sociality in itself stimulating, but that the psychological effects of pugnacious posturing have a beneficial effect on the race.

These remarks reflect the group-selectionist framework within which many previous workers approached the study of his animal reproductive adaptation and others.

In this account, we follow the lead of *Hamilton, Maynard Smith, William* in affirming that the functional basis of any reproductive adaptation is the increased reproductive success of the individual practicing it. According this principle, we should be able to explain both the functional significance of lekking and how it evolved. This requires an examination of ecological factors together with an understanding of the limitations imposed by an organisms' reproductive biology.

Avian Lekking in Relation to Feeding Adaptation

While the ability of the female to raise a brood unaided is precondition for the evolution of lekking, lek birds are not the only species in which the male is divorced from a parental role. Alternative adaptations based on uniparental care of the brood are possible, ranging from harem polygyny to simple *promiscuity*, practiced concurrently with or independent of colonial nesting. Such modes of reproduction probably arise when the trophic advantage of biparental care is outweighed by other factors, for example, by the increased risk of predation resulting from the more conspicuous activities of two parents around the next. This applies to lekking species as well. Furthermore, harem polygyny and/or simple *promiscuity* occur in such groups as grouse, pheasants, and hummingbirds—groups that contain lek species. There must be finer or overlooked differences in environmental factors that correlate with lekking. A brief digression to consider bird territoriality in its best-known form will help us bring out a salient difference.

Classical avian territoriality is feasible only if a resource, such as food, is sufficiently concentrated to make its defense economically profitable. In most territorial birds, the male's defense of his territory simultaneously assures the resources adequate to lodge and fledge a clutch, fixes his position in space, and advertises his presence to females. Conversely, the distribution of discrete advertised territories in a spatial mosaic increases the likelihood of a female encountering an unmated, territory-holding male. The system thus maximizes the fitness of both individuals in a pair.

Table 17.3. Circumstances promoting lekking in birds.

A. Properties of the physical environment

1. Natural phasic stimuli for synchronization of mating, e.g., seasonal changes in photoperiod
2. Best places for courtship and mating existing apart from feeding grounds, are characterized by:
   (a) Spatial properties that enhance signal propagation
   (b) Lack of ambush sites and/or unobstructed view of approaching predators.

B. Trophic considerations

1. Food resources either continuously or patchily dispersed.
   (a) Promote mobility, lack of site tenacity, and tendency to aggregate
   (b) Create need for effective long-range communication
2. Peaks of superabundance
   (a) Remove necessity of biparental provisioning
   (b) Promote synchronous breeding.

C. Predation

1. Promotes clustering of displaying males, which confers some protection upon them.
2. Promotes either dispersal or camouflaged nests or clustering of nests in a secure place, well away from displaying males

D. Social factors

1. Traditionality
2. Male need satisfied by small space, since territory serves for mating only
3. Relatively high population densities.

The situation has to be different in lekking birds. Without exception, they forage widely for dispersed food, ranging from fruit or nectar to seeds and insects. In some instances, not only are the foods dispersed, but they also tend to be unevently distributed in time and/or in space. And the higher the latitude, the more compressed in time is the period when food is maximally available.

Such a feeding adaptation has at least four consequences for the behaviour of its practitioners:

1. The birds must be prepared to move on to a better location when the local supply of food dwindles. They are thus less likely

to evolve the behavioural mechanisms necessary for sustaining large territories, notably high levels of aggressive responsiveness, a large individual distance, and site tenacity.

2. The nature of the food is such that when it is sufficiently abundant for breeding it is superabundant. Defense of the resource is thus economically unrewarding. Further, only one parent is then required, either as the provisioner, as in song birds, or as the caretaker of self-feeding offspring, as in gallinaceous birds.
3. Regardless of whether the food supply is uniformly dispersed or patchily distributed, there is still the risk of males and females foraging in different areas, or of individuals being widely separated. They need a communication system to bring them together.
4. We assume that there is a best place, or places, for the males to communicate to the females their presence, identity, and readiness to breed. We also assume that males, not females, congregate, because males mate repeatedly, and because the parental sex cannot afford to be conspicuous to the degree demanded by sexual competition. Males will there tend to congregate at, and complete among themselves for, the best sites. Selection will favour the larger, stronger males in such a situation, leading to the evolution of size dimorphis between the sexes. Further, the broadcast range and channel saturation of the communication system will be heightened through the summed activities of displaying males, increasing the individual fitness of each of them.

The initial stages of a trend toward lekking in birds are evident in non-lekking grouse and hummingbirds. There, male possession of discrete all-purpose territories contrasts with exclusively female brood care. The common denominator of such territorial behaviour is active defense of a suitable display site by the male. Such sites are prerequisite to reproductive success.

Birds whose foraging patterns preclude the defense of linked feeding and display territories have a serious problem. They must ensure access to essential display sites while maintaining a normal intake of food. The difficulty is apt to be acute when the physical setting or predation severely limit the number of sites available. Birds other than oceanic species like penguins are evidently unable to store sufficient energy to allow males to occupy their display sites continuously. They have to vacate the sites daily in order to feed.

Only two alternative solutions are therefore possible. The first is for the male birds to engage in physical competition for display sites after each foraging trip. Such activity would be bioenergetically wasteful. It would also increase the risk of injury to the combatants and make them more vulnerable to predators—in short, given them a pyrrhic victory. All these factors would reduce individual reproductive success.

The second alternative is to lessen overt competition by increasing the threshold of responses to stimuli eliciting aggression. That would permit males to occupy closely adjoining display sites without continual aggression. This solution also minimizes the expenditure of energy while diminishing the risk of injury and predation.

Lekking as practices by such forest species as the bearded bell bird and the cock-of-the-rock appears to illustrate this early grade of lek evolution Their simple groupings displaying males appear to lack such concomitants of classical avian leks as classes an arena, marked position effects, and clear protocols of site succession.

Classical avian lekking, as typified by the ruff, blackcock and sage grouse, is interpreted by us as aggregations of displaying males whose territories are arranged according to dominance relationships. Contrary to much widely accepted thought, the dichotomy between territoriality and dominance is not sharp. This type of lekking is characteristic of species inhabiting open country. The relatively undifferentiated topography there results in large aggregations of males, providing enhanced "artificial" land marks for females. In such large aggregations, competition for the best display sites would be increased. Selection should favour individual males who can compete for those vital positions while minimizing the risks inherent in combat. Such behaviour is apt to produce distinct dominance relationships.

Following these hypotheses, all the classic features of such avian leks may be interpreted as manifestations of a hierarchial social structure. Their existence therefore would be faciliated by long-term association of males outside of the breeding season. Those males would maintain or adjust their dominance relationships, obviating the need for high levels of aggression at the advent of the next breeding season. The delayed onset of male reproductive activity, another characteristic of birds with uniparental brood care, would also help in this respect: juvenile males are neither well equipped nor strongly motivated to contest for sites in the lek.

Lekking among birds implies the physical separation of display and nesting areas. With no further information, one cannot predict whether such birds would lek or, alternatively, form polygynous colonies, as do village weaver birds, in which display and nesting occur in the same area. The missing element we believe to be the nature of nesting sites in relation to predation. As Crook pointed out, weaver birds concentrate their nests in the few trees that afford both a large measure of protection from predators and proximity to a rich supply of food. Thus, when the predator-prey relationship favours clustered nests, given the foregoing ecological situation, a polygynous colony is predicted. But when the best antipredator adaptation is dispersed nests, as in grassland, a lekking society will evolve.

The cock-of-the-rock provide an exception, but one that illustrates the importance of an adequate display arena. The females nest in colonies in rather dark caves with restricted entrances. Visual display within them must be ineffective. The males do not engage females there. Rather, they lek in forest galleries where shafts of sun-light strike their brilliant plumage.

We hypothesize, therefore, that the evolution of avian lekking requires first that food be maximally available for a relatively brief period. This leads to synchronization of reproductive activity. Second, the food must be superabundant, making its defense as a limiting factor economically unprofitable. This also permits males to confine their activities to courtship and insemination by facilitating the evolution of exclusively female brood care. Third, the distribution of the resource in space favours the evolution of a system of communication that will bring both sexes together.

Perhaps the most critical step in this process occurs when communal displays develop at the best available sites for their performance, determined largely by characteristic physical features of such locations. This definition of "best" may include the presence of other males. Vocalizations and conspicuous movements of contrast-rich structures may thus undergo a multiplier effect that further enhances the detectability of each participating male by aiding females in finding the assemblage of displaying birds. The behavioural attributes of classical avian lekking will evolve, once this step has been taken, as selection pressures favour those modifications in behaviour that facilitate the close proximity of sexually active males. Lastly, predation pressure must favour dispersed nesting. This will result in the movement of inseminated females away from the display ground. The end result is the complete spatial separation of display and nesting ground that is characteristic of lekking birds.

We also need to emphasize that the foregoing is a general scheme, and that the factors leading to lekking may have played relatively different roles in different species. In particular, we have slighted the possible importance of predation on the lekking birds. Lekking may have been, and may still be, crucial for birds that live in open country where little cover is available, but where the males must broadcast to attract females. Then the males might congregate to reduce predation on themselves, as suggested for schools of fish or any aggregating species. In one mammal that leks in open country, the wildbeest, the males have been shown to be subject to heavy predation.

Lekking in Teleosts in Relation to Spawning-Site Dependence and Nest Predation

The evolution of lekking in teleost is linked, we believe, to dependence upon suitable spawning sites, by the localized substrata for the deposition of demersal eggs or points assuring a favourable launch window for pelagic eggs. While synchrony of reproductive activity is imposed by environmental factors, as in birds, proximity is determined by the availability of suitable spawning sites. Where these spawning are limited in either occurrence or extensiveness, their sequestration, either totally or in part, by an individual may be a positive adaptation. Males would be in a better position to implement such as adaptation because of the greater energetic demand that the maturation of eggs places on females.

As in birds, however, total sequestration of a resource is reasonable only if its defense is economically profitable. Hence, while a male fish might, theoretically, enhance his fitness at the expense of conspecific competitors by monopolizing a spawning site *in toto,* such a strategy would be practical only if the energetic cost of its practitioner were exceeded in some manner by a positive return on the investment. Here the positive return would be enhanced reproductive success.

We can find no record in the literature of any teleost whose reproductive pattern includes such behaviour. Rather the pattern that emerges is sequestration of a site only large enough to facilitate successful spawning. Because the less site itself is unimportant in the nutrition of the offspring in virtually all lekking teleosts, the defense of extensive territories is unnecessary.

Table 17.4. Circumstances promoting lekking in teleosts.

A. Properties of the physical environment

1. Natural phasic stimuli for synchronization of mating e.g., seasonal changes in temperature, rainfall, or lunar/tidal cycles
2. Best places for courtship and spawning exist apart from feeding grounds. They are characterized by :
   (a) Hygienic properties that promote the development of the zygotes after spawning
   (b) Spatial and physical properties that enhance signal propagation
   (c) Relative freedom from predation due to
      (i) Inability of predators to penetrate the arena
      (ii) Lack of ambush sites and/or unobstructed view of approaching predators

B. Trophic considerations

1. Lack of trophic component in brood care facilitates uniparental care of spawn
2. Ability to store metabolic reserves allows males to hold territories for extended periods of time.

C. Predation

1. Promotes synchrony of breeding as numbers of eggs and fry procured care "swamp" their predators.
2. Promotes clustering of displaying males
   (a) Summated brood defense
   (b) Protection against predators of adult fish by "swamping" or by "selfish herd" phenomenon.

D. Social Factors

1. Traditionality
2. Founder-male effect
3. Male needs satisfied by small space since
   (a) Territory used only for spawning or mating
   (b) Brood care strictly custodial
4. Relatively high population densities

Lekking therefore arises automatically in teleost fishes when the size of a discrete spawning territory sequestered by a male is small relative to extent of the available site. As the number of males entering the area and setting up territories increases, the size of each territory will contract to the smallest area required by each male in order to reproduce successfully. Under most circumstances, and for most animals, the resulting geometry will converge upon, but seldom achieve, an hexagonal array of territories.

We suspect that intermittent lekking is the most primitive manifestation of this reproductive strategy in teleosts. As male expend energy in obtaining and defending a site, selection will favour the evolution of site tenacity. From the standpoint of minimizing the chance of injuries sustained in intraspecific combat alone, this is a more efficient strategy than repeated contests for a suitable spawning site. The result of such selection would be the appearance of persistent lekking in the absence of any sort of parental care, such as is seen in some darter perches.

The defense of a spawning territory, even if only from conspecifics, confers serendipitously a degree of protection from predation to the eggs deposited therein. It requires little additional behavioural adjustment to broaden such defense to include heterospecific predators on the spawn. Natural selection would favour those fish two did practice such defense, or whose spawning behaviour in some other manner favoured the survival of their fry. Hence, the widespread occurrence of some type of protective behaviour among teleosts with demersal eggs.

Among teleosts that practice parental care, the primitive condition is for the male to establish a territory and for the female to visit him there. The female then has four options: (1) She can remain with the male, an advanced condition shown by few kinds of fishes, and join the care of the spawn. (2) She can assume full care of the eggs herself, while the male obtains further females elsewhere, as in some of the dwarf cichlids. (3) She can pick up the eggs in her mouth and leave his territory. As previously noted, option (3) is confined to one family of fishes; it has clearly been derived from the first option—that is, joint parental care. (4) She can leave the eggs to the care of the male—the most general case. This solution permits the male to receive a number of females, accumulating eggs from them in his territory. However, if selection favours active paternal care, such as fanning the eggs, or assisting them to hatch, or dispersing the larvae, then the male will come to accept eggs for only a brief period in order to coordinate his care with the needs of the brood. Consequently, the reproductive system will become increasingly dissimilar to avian lekking.

We believe lekking to be prevalent among fresh-water teleosts because it is compatible with the third and fourth spawning options presented in the preceding paragraph. Indeed, it may actually facilitate their implementation in certain situations. Among maternal mouth-brooding cichlids, the advantages of lekking behaviour must be similar to those that accrue to lek birds. Among custodial lekking fishes, participants in the lek birds. Among custodial lekking fishes, participants in the lek may wellbenefit from a summed territorial defense, with centrally located males enjoying a reduced burden of defending the spawn because of the activities of peripherally located males. This factor appears to have influenced the evolution of lekking in gouramies of the genus *Trichogaster.* Regardless of whether they practice parental care or not, lekking teleosts have the potential advantage of swamping spawn predators with the sheer number of eggs and fry produced within a limited area, depending on the number of eggs and fry, and on predators and their capacity to devour eggs and fry.

Thus just as predation upon nesting individuals has combined with feeding adaptations to produce lekking in birds, so has feeding biology and spawn predation, interacting with dependence upon hygienic spawning sites, led to the widespread occurrence of this type of mating system among teleosts.

An Overview of Lekking in Fishes

We would like in closing to be able to make a straightforward comparison of lekking in birds and in fishes. That is not easily done, and for two reasons. First, there are more variations on the lekking theme within teleost fishes than exist among birds. Second no one species of fish offers a pattern of lekking that is completely comparable to that seen in any bird. Yet all the cases previously cited fulfill our minimum prerequisites for lekking. These are synchrony of the breeding population, lekking grounds where males await females, sufficient mobility to move between separate areas for breeding and feeding, and performance of parental care, it is exists, by only one parent.

We do not include as a minimum prerequisite the absence of habitat constraints. *Pitelka* believes that it is necessary to show that there are available but unused alternate sites for lekking. Even when this can be demonstrated, it does not rule out the importance of suitable sites. For example, the long-tailed manakin apparently has a number of requirements for its arena, including a rare vine hat is always present. Yet traditionality is highly developed, and some seemingly suitable sites are not used. The issue is whether some attractive but limited features of the environment occur in only a few places, and thus help to concentrate the animals, or whether the animals themselves act as the focal point, given that adequate situations occur in a number of different places. We see these as interactive factors rather than opposing ones, and factors whose operation is effective to varying degrees, depending on the species.

Table 17.5. Characteristics of highly evolved lek systems in both birds and teleosts.

A. General characteristics

1. No feeding on the lek
2. The more males present, the more females attracted
3. Reduced aggression and increased intermale display
4. Clear dominance relationships
5. Well-developed sexual dimorphism
6. "Cheating" by young or subordinate males are
7. Succession to central positions determined by strict protocol

B. Characteristics of central males

1. Largest and oldest are the most dominant
2. Experience less interference with mating or spawning
3. Devote relatively more time courtship, less to status fights, territoral defense, or, in teleosts, antipredator behaviour
4. Females select central males
5. Central males mate or spawn more than do other males.

A close parallel is exemplified by parrotfishes. The males take up positions on the lekking grounds for only a few hours during the day, as in grouse. And the situation is complicated by small males who resemble females and compete with the lekking males to fertilize the females' gametes.

Another reasonably close parallel is found in the desert pupfish whose reproductive pattern is probably typical of that of most cyprinodontid fishes. The males apparently leave the lek of feed and to sleep, as do classically lekking birds. There are even peripheral males that could be called satellites. The critical difference from birds, however, is that the demersal eggs are left in the male's territory, although the male provides no overt spawn care.

Most freshwater lekking fishes seem to be of the sunfish type. The differ among themselves in the extent to which they provide parental care, from protocustodial to paternal-custodial, and therefore remain continuously on the nest site. This mode of teleost lekking diverges most strongly from the avian paradigm in terms of the persistent occupancy of courts by sexually active males and the existence of a paternal custodial role.

The last type of lekking in fishes is that of maternal mouth-brooding cichlids. Cichlids parallel birds in that the females have their gametes fertilized at the lek, and their take them to separate brooding grounds to rear and, in many cases, shepherd the offspring. Unlike birds, the males stay on the lek for periods of a week or so. And unlike other fishes, the nature of the substrate is obviously not critical for the development of the eggs. The particular site, however, can be important for nest-building and courtship, and for the avoidance of predation.

Four factors in the biology of teleost fishes stand out as being responsible for the differences from birds. One is that they regularly store metabolic reserves, a practice that allows the males to remain on station without feeding for long periods of time. The second is external fertilization. The third is the sensitivity of their eggs to the medium that surrounds them. The fourth factor is the huge clutch size of teleosts, with a correspondingly lower investment per gamete.

Give this diversity within lekking by different species of teleosts, it is difficult and perhaps premature to attempt to explore how trophic adaptations might be of overriding importance in the evolution of lekking in fishes, as we attempted to do for birds. However, some obvious generalizations can be made about ecological factors, even if they only serve as hypotheses to be disproved.

The trophic adaptations of lekking fishes must require a degree of mobility compatible with moving to breeding areas. The breeding areas must have some features that make them more suitable than those in which other activities take place, particularly feeding; these would include favourable physical conditions for the development of eggs and lessened vulnerability to predation on the fish and/or their eggs. Breeding must also be sharply phasic to bring a number or reproductively responsive animals together; phasing may be through trophic and climatic factors such as rainfall, or through tidal cycles linked to lunar periods.

The foregoing general features of lekking probably apply to any kind of animal that utilizes a lekking mode of reproduction.

18

MIGRATION IN BIRDS

Any movement between two areas is called *migration.* As a rule, it is a response of an animal population to changes in environmental conditions.

Birds are more uniformly migratory than any other group of animals. Nearly all orders of birds include species which perform migrations; in other vertebrates and in the lower groups of animals the migratory habit occurs in species scattered through a smaller proportion of orders.

Among birds there are two common kinds of migration, *daily and seasonal.* A daily migration is a movement to and from a familiar place such as a roosting area. A seasonal migration, on the other hand, involves a passage at one season from a place of hatching and a return at another season to the same general area. This section of the book is concerned entirely with seasonal migration.

For centuries the phenomenon of migration has been primarily associated with birds; indeed, judging by the references of the subject in the earliest literature, migration was first observed in birds. Since civilization developed in a temperate region of the world where migratory movements are especially pronounced, it is hardly surprising that bird migration has long received attention. Man could not help noticing the flocks of birds in the spring and fall, the seasonal disappearance of some species and reappearance of others; nor could he help being interested in what he saw and eager to investigate what he could not understand.

Today there is an enormous literature on bird migration, based on extensive studies in Europe and North America. And yet the causes and processes of migration are not fully known. The root of the problem is that bird migration, whether it occurs by day or night, is "an unseen movement." One must investigate its mechanisms indirectly through laboratory research on physiological and environmental influences; the analysis of migrant birds mist-netted or otherwise captured, of birds killed during migration by television towers and other man-made hazards, and of returns from banded birds; mathematical calculations based on kinds and numbers of birds observed through a telescope as they fly across the face of the moon; correlation of meteorological data with known migratory activity; deductions derived from direct field observations, radar surveillance, and tracking by radiotelemetry; and experiments on homing and direction-finding. The following pages present briefly some of the important facts and concepts of migrations and suggest a few studies that students may undertake in conjunction with class work.

Causes of Migration

All modern birds—even those incapable of flight—are descended from volant stock. Early in their history birds had the power of flight and presumably could migrate with facility.

Today many species in the temperate regions of the world are strongly migratory, exhibiting mass movements away from both poles as day-length, temperature, and food supply diminish, then reversing the movements at the season when there is a general augmentation of these environmental factors. Such migrations are more evident in the temperate region of the Northern, or Continental, Hemisphere since more species are involved.

Because the north-south migrations of the Northern Hemisphere include many well-known and conspicuous birds which seem to move in the same general way, there has been a tendency to conclude that migration, like bird flight, has developed along the same lien in all species, with deviations for adaptive purposes. A study of the migration phenomenon in all groups of birds soon shows the fallacy of this reasoning. One finds that :

1. Some species migrate in directions other than north-south.
2. Some species migrate irrespective of day-length, as in tropical lands.
3. Some species migrate when the temperature is mild and the food supply ample, and others when the opposite conditions are true.

4. Some species migrate as a result of seasonal alternation in rainfall and drought.
5. In certain species, some populations migrate while others do not.
6. In some populations, some individuals migrate while others do not.
7. Some individuals migrate in some years but not in others.

The only conclusion one can safely reach after considering the above peculiarities is that there is no one line along which migration in all birds developed and that there must be different causes of migration in different groups of birds.

A number of authorities have sought causes of migration in historical factors. Three of the several causes suggested bear mention. (1) Bird migration, at least in the Northern Hemisphere, was initiated by the effects of the Ice Age (Pleistocene). Prior to the coming of the great glaciers to the polar regions, birds lived the year round in the Northern Hemisphere where they originated. Though forced to retreat with the advance of the glaciers, they nevertheless continued to return to nest in the summer because of an innate attachment to their homeland. Objections to this suggestion are several. Many of the birds which are today typical migrants were in existence before the Pleistocene and there is no reason to suppose that they were not already migrating before the glaciers advanced. The retreat from the glaciers does not account for migrations in directions other than north-south, nor dies it account for migrations of birds in tropical regions that were neve glaciated. (2) Birds migrate in the fall because of the oncoming shortage of food during the winter in their breeding areas. Not that they "know" of the impending lack of food; it is the fact of the food shortage that has caused fall migration to evolve among species which, owing to the shortage, would fail to survive if they stayed for the winter. Birds return in the spring because the northern areas provide more favourable and less crowded conditions for nesting and rearing families. (3) Intraspecific and inter-specific competition for food, territory, nest sites, and so on may have been important factors initiating migration. When, as suggested by *Cox*, individuals among normally sedentary species could benefit by moving into adjacent areas where the season was favourable and competition reduced, they did so if the hazards of moving did not exceed the gains in their survival and reproduction.

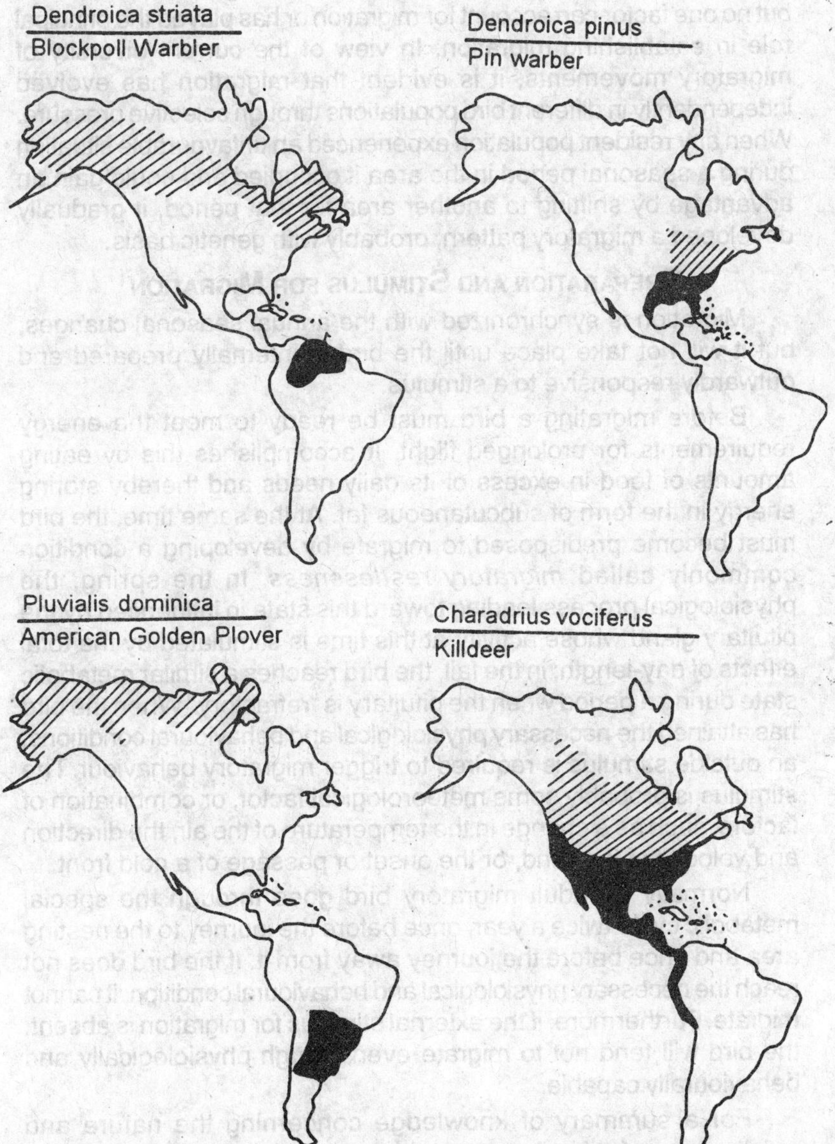

Fig. 18.1. A comparison of long-distance (left) and short-distance migrants within two groups, the warblers (Parulinae, top) and plovers (Charadriidae).

Unquestionably, certain historical factors and various factors prevailing today such as day- length, air temperature, and food supply have influenced migration indirectly by affecting the environment, but no one factor can account for migration or has played the principal role in establishing migration. In view of the current diversity of migratory movements, it is evident that migration has evolved independently in different bird populations through selective pressure. When any resident population experienced an unfavourable situation during a seasonal period in the area it occupied and could gain an advantage by shifting to another area for that period, it gradually developed a migratory pattern, probably with genetic basis.

Preparation and Stimulus for Migration

Migration is synchronized with the annual seasonal changes, but it will not take place until the bird is internally prepared and outwardly responsive to a stimulus.

Before migrating a bird must be ready to meet the energy requirements for prolonged flight. It accomplishes this by eating amounts of food in excess of its daily needs and thereby storing energy in the form of subcutaneous fat. At the same time, the bird must become predisposed to migrate by developing a condition commonly called *migratory restlessness.* In the spring, the physiological process leading toward this state in influenced by the pituitary gland whose activity at this time is stimulated by the total effects of day-length; in the fall, the bird reaches a similar metabolic state during a period when the pituitary is "refractory." When the bird has attained the necessary physiological and behavioural conditions, an outside stimulus is required to trigger migratory behaviour. The stimulus is probably some meteorological factor, or combination of factors, such as a change in the temperature of the air, the direction and velocity of the wind, or the onset or passage of a cold front.

Normally an adult migratory bird goes through the special metabolic cycle twice a year, once before the journey to the nesting area and once before the journey away from it. If the bird does not reach the necessary physiological and behavioural condition, it cannot migrate. Furthermore, if the external stimulus for migration is absent, the bird will tend not to migrate even though physiologically and behaviourally capable.

For a summary of knowledge concerning the nature and mechanisms of periodic preparation and stimulus for migration.

Diurnal and Nocturnal Migrations

Migrations proceed by day or night. Many birds—e.g., loons, geese, ducks, gulls, terns, and shore birds—travel by day or night, apparently indifferent to daylight or darkness. But this is not the case with many other birds. Herons, hawks, eagles, falcons, crows, hummingbirds, swifts, and swallows migrate only during the day, while nearly all passerine birds (excepting crows, swallows, and a few others) migrate primarily during the night from after sunset until dawn.

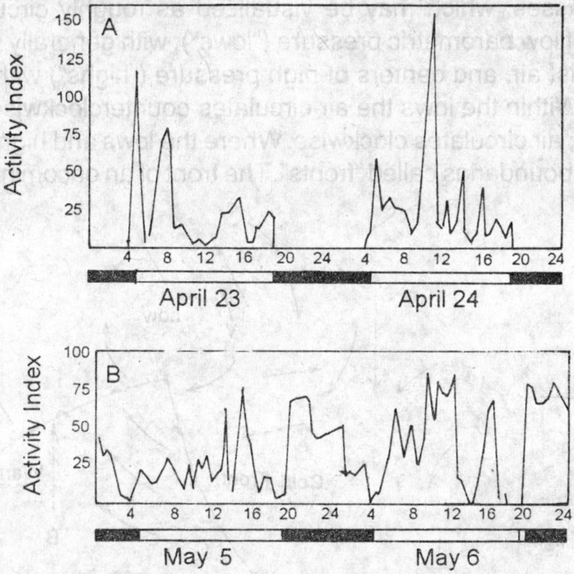

Fig. 18.2. The measurement of Zugunruhe in a White-crowned Sparrow (Zonotrichia leucophrys) before and during its normal migratory period. The black bar represents darkness and the open bar daylight. Note how much more active the bird is during early May, its normal time for migration.

At least two explanations have been advanced for the development of nocturnal migration. (1) Movement by night affords birds, which normally live in thick vegetational cover and rarely take long flights away from it, the protection of darkness against their diurnal predators. (2) Movement by night affords birds the opportunity of using all the daylight hours of feeding, thereby enabling them to build up sufficient energy resources for sustained long-distance flights.

Effects of Weather on Migration

Normal weather alternates between fair and inclement conditions. The movements of the vast majority of migrants show a close correlation with these conditions which are largely governed by barometric pressure patterns, temperature, and wind directions.

To understand how migration takes place with relation to weather, the student must first be acquainted with a few basic facts about weather elements and their sequence over a ground area.

Perpetually sweeping across the continent in an average easterly direction are air masses that vary in velocity, depending on the season and numerous other factors, from 500 to 700 miles a day. In these masses, which may be visualized as roughly circular, are centers of low barometric pressure ("lows"), with generally warmer, more moist air, and centers of high pressure ("highs") with cooler, drier air. Within the lows the air circulates counterclockwise; within the highs, air circulates clockwise. Where the lows and highs adjoin there are boundaries called "fronts". The front of an oncoming high is

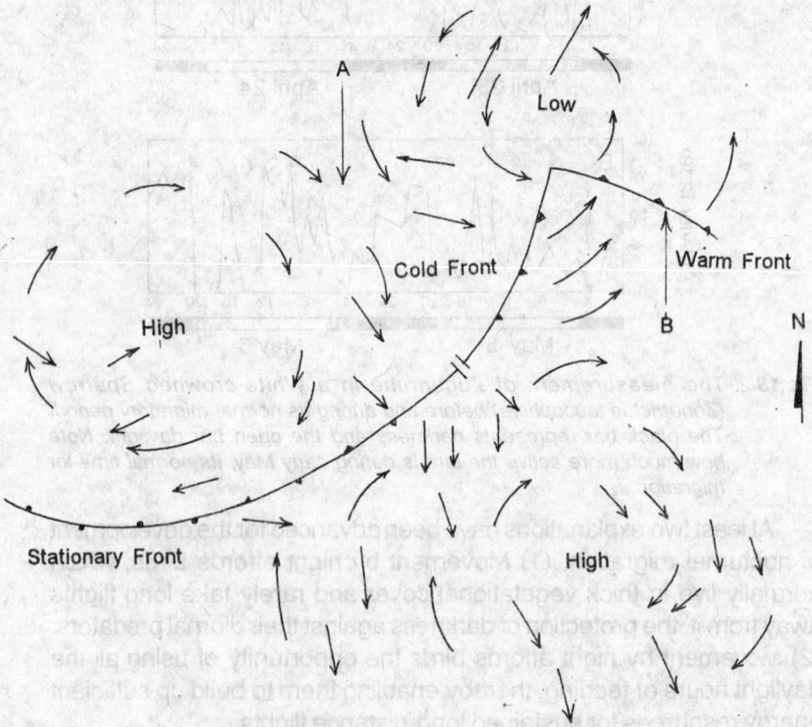

Fig. 18.3. An example of how the distribution of fronts and pressure cells can provide favourable winds for a migrating bird. Large arrow A notes a situation with favourable winds for a bird moving south, while large arrow B shows a situation favouring northward movement.

the cold front; of an oncoming low, the warm front. The area of generally low pressure between the cold and warm fronts is called the warm sector. Any weather map appearing in daily newspapers will show the lines of equal barometric pressure (isobars) as roughly concentric circles around lows and highs, and cold and warm frons as heavy lines with marks indicating the direction they face. Areas where there has been precipitation are shaded.

The table, "Relation of Weather Elements to Migration," demonstrates in a general way the sequence of spring and fall weather elements when a warm and then a cold front pass over a given area in conterminous United States. The accompanying sketch, "Pattern of Fronts and Wind Directions," shows how the sequence indicated in the table might appear on a weather map. The movements of the elements in both the table and the sketch is to the right (east). Thus the set of weather elements characterizing "Ahead of Warm Front" will be followed by the elements of "*Warm Front*" and so on. The rapidly with which one set of elements—e.g., the elements of a warm front—arrive, prevail, and finally disappear in an area is dependent on the width of the front and the speed with which the front travels. The width and speed of the front are in turn dependent on the season of the year, the temperature discrepancy between fronts, the wind direction, and continent-wide weather conditions at the time.

Migration in the spring usually takes place with warm weather. Studies of spring migratory movements in eastern United States and Canada show that movements begin at the onset of warm fronts, when barometric pressure is dropping and warm moist air from the Gulf of Mexico and Caribbean Sea is flowing in from a southerly direction. As each low (with mild temperature and southerly winds) passes, movements proceed in full force. On the approach of cold frints, the movements usually slow down—though they may occasionally be heavy, as when a cold front advances against a western flank of warm air. When cold fronts arrive, movements stop. Not until the highs have passed will movements begin again.

Migration in the fall usually takes place with cold weather. In the Chicago area migratory movements in September and October start immediately after the passage of cold fronts when there is a flow of continental polar air from a northerly direction. Movements then proceed in full force until the first part of each low passes; thereafter the movements decrease somewhat in intensity and may even cease together as the next cold front approach.

The table, "Relation of Weather Elements to Migration," relates spring and fall migration movements to the sequence of weather elements.

Many species are reluctant to initiate migration under an overcast. Air temperature is probably the principal factor in starting migration; wind direction or air flow is a decidedly critical factor in regulating migratory movement. Most movements take place when the wind is favourable—i.e., blowing in the direction of flight—and after migrants have been held back for long intervals by such weather conditions as fog, rain, and headwinds. In the spring the weather that ordinarily accompanies cold fronts is especially obstructive to movements, forcing migrants to take to the ground without delay.

The arrival of any migration movement in an area is more often controlled by the weather at the print of departure than during its course. Thus at the height of the migration season, when the weather appears suitable in his area, the student will not always observe migration movement because inclement weather at the point of departure or some intervening point has arrested the flight.

The study of weather maps in conjunction with observations on movements is very useful. By means of weather maps one can frequently anticipate a migration movement in a given area. For instance, if during the height of spring migration the map indicates a cold front moving in from the northwest, one may expect its arrival to stop and hold numerous transients in the area. Weather maps also assists in explaining the failure of a migration movement to appear. If there is no pronounced migration movement in an area for a long period, even though the season and weather are favourable, maps may reveal that a pressure area along the migration route has become quasi-stationary, either "damming up" migrants (in case the pressure area is unfavourable to movement), or (in case it is favourable) allowing migrants to move along from day to day without massing in conspicuous waves.

Exceptional weather conditions with usually high winds often deflect migrants from their usual routes and at the same time carry many non-migrating birds from their regular ranges. The hurricanes or heavy northeast storms that occasionally move north along the eastern Atlantic seaboard, by the counterclockwise direction of their winds, force many south-bound migrants and sea birds far inland.

As a result, ducks, geese, and gulls are reported in great abundance in places where they do not ordinarily occur, and such sea birds as petrels showup far inland. Two very severe northeast storms in December, 1927, and January 1966, bore spectacular numbers of Eurasian Lapwings (*Vanellus vanellus*) over the North Atlantic from their migration route in western Europe to the vicinity of Newfoundland. The student will find instructive the paper by *Bagg* showing in detail; with a series of weather maps, how the great storms brought the Lapwings to North America.

Regularity of Migration Travel

Despite the effects of weather on migration, migratory travel over a period of years is regular on the average. This is apparent to a student observing bird population from year to year in any given area of North America north of Mexico.

If one keeps records for several years of the days when common summer-resident species first arrive in the spring, and computes average dates of arrival of each species, he can eventually predict within a few days when these species will appear. Similarly, by

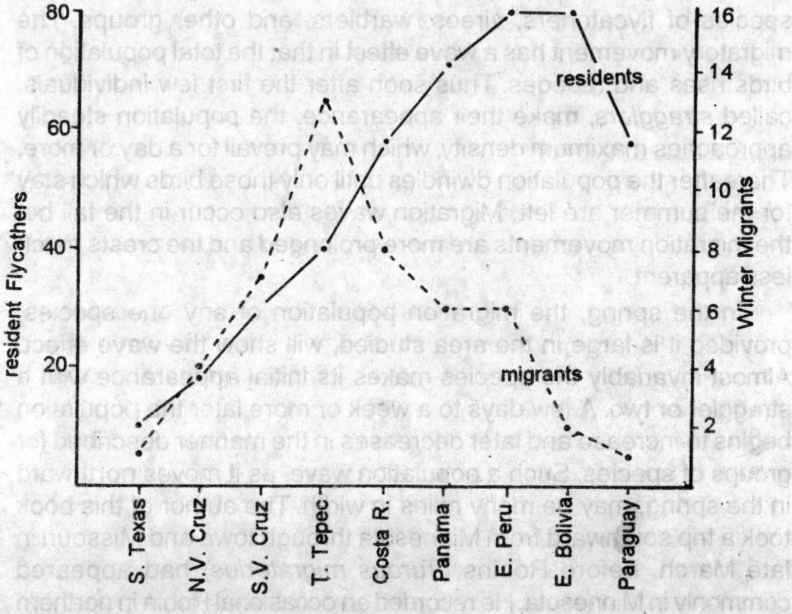

Fig. 18.4. Distribution of resident and migrant species of flycatchers (Tyrannidae) on New World wintering areas.

keeping records of departure of transient spring species—i.e., dates when species going through the area are last seen—he will known approximately when these species depart each year.

There is much less regularity in the arrival and departure of transient species in the fall, thus the securing of records is difficult. Fall migration is more prolonged. Slight variations in weather conditions have stronger effects. Cool late-summer weather, for instance, may induce species to arrive surprisingly early and warm weather may cause them to linger very long. Keeping tract of early and late individuals is complicated by the fact that many species are customarily silent in the fall and have inconspicuous plumage. While it is possible to obtain average dates of arrival and departure, fall dates are apt to be much less useful and meaningful on account of the great discrepancy in annual dates and the problems of finding early and late individuals.

In conterminous United States, north of the southern tier of states and in Canada, there are two so-called *migration waves.* The first, in the early spring, is made up of "hardy" birds—many fringillids and a species or two from various other bird groups; in the second, a month later, there is a preponderance of insect-eating birds—various species of flycatchers, vireos, warblers, and other groups. The migratory movement has a wave effect in that the total population of birds rises and recedes. Thus soon after the first few individuals, called *stragglers,* make their appearance, the population steadily approaches maximum density, which may prevail for a day or more. Thereafter the population dwindles until only those birds which stay for the summer are left. Migration waves also occur in the fall but the migration movements are more prolonged and the crests much less apparent.

In the spring, the migration population of any one species, provided it is large in the area studied, will show the wave effect. Almost invariably the species makes its initial appearance with a straggler or two. A few days to a week or more later the population begins to increase and later decreases in the manner described for groups of species. Such a population wave, as it moves northward in the spring, may be many miles in width. The author of this book took a trip southward from Minnesota through Iowa and Missouri in late March, before Robins (*Turdus migratorius*) had appeared commonly in Minnesota. He recorded an occasional Robin in northern Iowa; great numbers through central and southern Iowa; and a steady decrease in numbers through northern Missouri until in central Missouri there were only scattered individuals, presumably the birds that were to become summer residents there. He estimated the width of the migration wave to be roughly 225 miles.

In making the first spring and fall studies of birds, the student should pay special attention to the local movements of transient species. He should note the dates each species is seen, make counts or estimates of the number of individuals of each species observed on each date, and keep a record of weather conditions. This information will give him a proper conception of the length of time each species remains in the area, the way a species population rise and recedes giving the wave effect, and some of the relationships of weather conditions to population trends.

Irregular Migrations

The many cases of movements among birds populations, which either do not conform to the usual seasonal migration pattern or are not sufficiently well understood to seem a part of the pattern, are loosely classified as *irregular migrations.*

In certain permanent-resident species there may be mass movements of particular populations with some periodicity. The Blue Jay (*Cyanocitta cristata*), especially in the northern part of its range, shows some migratory movement. Each fall, small numbers usually move south past Hawk Mountain in Pennsylvania; in 1939 there was an exceptionally heavy migration (over 7,000 individuals counted), which may have been due to a shortage of beechnuts and acorns in northern forests. In northwestern Oklahoma, the Bobwhite (*Colinus virginianus*) shows a distinct seasonal population shift by moving from summer habitats in the uplands to pass the winter in bottom-lands and dunes where there is better cover. Their movements, involving distances up to 26 miles, are apparently heavier during severe winters.

Populations of species which are permanent residents may on occasion show a mass movement—*invasion or irruption* without periodicity. Now and then there is a winter when considerable numbers of Snowy Owls (*Nyctea scandiaca*) leave the Arctic and invade southern Canada and northern conterminous United States. The cause of this behaviour is sometimes attributed to a sharp reduction in the lemming on which the Snowy Owl preys to large extent. Crossbills (*Loxia* spp.) and Evening Grosbeaks (*Hesperiphona vespertina*), normally residents of the Coniferous Forest Biotic Community,

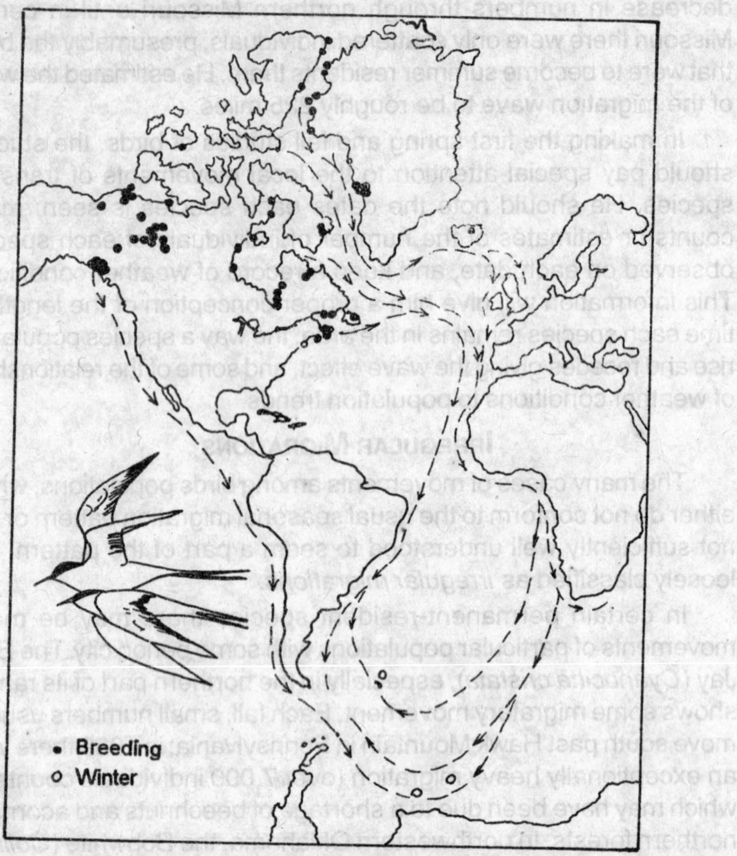

Fig. 18.5. Distribution and migration of Arctic Terns from North America. This species is distinctive for its migratory pathway, which takes it across the Atlantic and southward as far as Antarctica.

occasionally appears in flocks as far sough as Florida, presumably due to a failure of their food supply. Sometimes invasions involve mainly young birds. At Cedar Grove, Wisconsin, on the west shore of Lake Michigan, *Mueller* and *Berger* reported a southern invasion of Goshawks (*Accipiter gentilis*) in the years 1961 through 1963. Most of the birds that they trapped and examined were apparently hatched in 1961 shortly after a decrease in snowshoe hares and grouse, their principal prey. *Mueller* and *Berger* hypothesized that these young birds had come south after being displaced by adults already well established in a range that could not support large wintering populations of the species.

Young of numerous species, after attaining full growth, often wander in the late summer and fall for great distances. The movement, called *juvenile wandering,* is explosive in that the birds move in all directions from the hatching area. Among the species particularly noted for this behaviour are egrets, herons and gulls. Some young egrets and herons actually travel several hundred miles north of their place of hatching in the south. At the conclusion of the breeding season in the big colony of Herring Gulls (*Larus argentatus*) at Kent Island, New brunswick, an impressive number of immature birds go northward along the coast, although the majority seem to take a southerly direction. Many first-year Herring Gulls, reared in colonies on islands in the Georgia, Florida, and the Gulf of Mexico, and quite a few reach the coast of Mexico; second-year and older Herring Gulls tend to remain on the shores of the Great Lakes within 300 miles of their colonies. Juvenile Sooty Terns (*Sterna fuscata*) from the large nesting colony on the Dry Tortugas, islands lying directly west of Key West, Florida, move across the Atlantic to the Gulf of Guninea, West Africa, and do not straggle back to the western Atlantic until they approach breeding age; the adults, after nesting, tend to confine tneir dispersal to the Gulf of Mexico and the Caribbean. For the extensive wandering of young birds, the most plausible

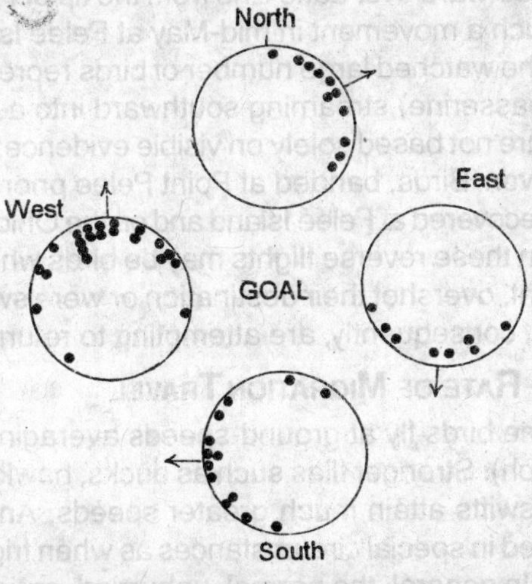

Fig. 18.6. The observed departure bearings of pigeons whose clocks had been advanced by six hours and then were released 30 km to 80 km north, east, south and west of their home lofts. Solid arrows note mean directions.

explanation is that they cannot compete sufficiently well with older birds for food and must therefore keep moving until they find an adequate supply for themselves.

Reverse Migration

Migration movements may be reversed, proceeding in a direction opposite the one expected for the season. A good example of *reverse migration* occurs in the fall of Nantucket Island, Massachusetts, and Black Island, Rhode Island, where many nocturnal migrants (mostly passerines representing well over 50 species) sometimes pass rapidly through during the daytime and leave in a north or northwestward direction for the mainland, into the wind. *Baird* and *Nisbet* interpret the movement to be the result of south-bound migrants, carried toward the Atlantic Coast by strong northwest winds, attempting to fly back and redetermine or regain their preferred lanes of passage overland.

Another example with a different interpretation comes in the spring from Point Pelee, a peninsula projecting nine miles southward into western Lake Erie from Ontario, and from Pelee Island that leis about eight and one-half miles southwest of Point Pelee and nearer the Ohio mainland. Time and again many small land birds have been seen returning southward over Lake Erie from the tip of Point Pelee. *Lewis* reported such a movement in mid-May at Pelee Island. Here for several hours he watched large number of birds representing 35 species (mostly passerine) streaming southward into a headwind. The movements are not based solely on visible evidence; they have actually been proven. Birds, banded at Point Pelee prior to starting south, were later recovered at Pelee Island and on the Ohio mainland. The participants in these reverse flights may be birds which, during the preceding night, overshot their destination or were swept past it in high winds and, consequently, are attempting to return to it.

Rate of Migration Travel

Most passerine birds fly at ground-speeds averaging 18 to 25 miles per hour (mph). Stronger flies such as ducks, hawks, falcons, shore birds, and swifts attain much greater speeds. Any bird can accelerate its speed in special circumstances as when frightened or diving earthward. In general, the normal, unhurried, cruising speed of a bird is much slower than suggested by published records, most of which, until recently, were estimated by observers in automobiles or airplanes moving parallel to the bird's line of flight.

Schnell, using Doppler radar equipment similar to that operated by law enforcement agencies in determining speed of automobiles, measured the ground flight-speeds of 17-species of birds in northern Michigan. He recorded on windless days two speeds of the Spotted Sandpiper (*Actitis macularia*) at 25 mph; four speeds of the Eastern King-bird (*Tyrannus tyrannus*) at 21 mph and one at 13 mph; three speeds of the Cedar Waxwing (*Bombycilla cedrorum*) at 21, 23 and 29 mph; and three speeds of the Red-winged Blackbird (*Agelaius phoeniceus*), one at 17 mph, and two at 23 mph. Had the wind been blowing, the speeds might well have been slower for birds flying into it and faster for birds flying with it. Strong winds can significantly affect flight-speed as Schnell proved with the 267 speeds of the Herring Gull (*Larus argentatus*) that he recorded in different wind velocities. Speeds, he found, averaged 25 mph in winds less than 6 mph, but averaged 18 mph (extremes 7 and 39 mph) into winds of 6 to 15 mph and 34 mph (extremes of 21 and 49 mph) with the same winds.

During migration, according to radar surveillance by *Bellrose*, birds appear to reduce their flight-speed somewhat proportionately to the increase in favourable wind-speed. Apparently they adjust their flight efforts in relation to the degree of wind assistance or resistance. Thus the ground-speeds of migrants tend to remain fairly constant despite variations in wind-speed whereas the ground-speeds of birds in the daily activity flights, as shown by Schnell, are definitely influenced by winds.

Many of the stronger flying birds show great ability for fast migratory travel. *Mc-Cabe*, in an airplane going at an airspeed of 90 miles per hour, was overtaken by two flocks of sandpipers flying at an estimated air-speed of 110 mph. *Speirs*, once estimated the average ground-speed of the Oldsquaw (*Clangula hyemalis*) at 61.5 mph and the air-speed at 50.5 mph.

Birds homing to their breeding sites, after being displaced at great distance away, demonstrate impressive ability for sustained speed for many hours. A female Purple Martin (*Progne subis*), taken from her colony at the University of Michigan Biological Station in northern Lower Michigan and released in Ann Arbor, Michigan, 234 miles to the south, at 10:40 pm, was back feeding her young at 7:15 am, having made the return flight in not more than 8.6 hours at an average speed of 27.2 miles per hour (Southern, 1959). A Manx Shearwater (*Puffinus puffinus*), removed from its nesting burrow on

Skokholm off the west coast of Wales and released in Boston, Massachusetts, reached its burrow after at least 3,200 miles in 12 days and some 13 hours, or an average of 250 miles a day (Mazzeo, 1953). Another sea bird, a Leach's Petrel (*Oceanodroma leucorhoa*), averaged about 300 miles a day for nine days from tis point of release at Prestwick, Scotland, back to its nesting burrow on New Brunswick's Kent Island in the Bay of Fundy. Of the 18 Laysan Albatrosses (*Diomedea immutabilis*), taken from their nests on Midway Island—one of the Hawaiian Leewards in the north-central Pacific—and released at widely scattered points in the northern Pacific, 14 returned, one from the Philippines, a distance of 4,120 miles in approximately 32 days, and one from Whidby Island off the coast of Washington, a distance of 3,200 miles in 10.1 days at an average speed of 317 miles a day. Presumably, not one of these birds homing to its breeding site had the fat reverses for energy that migrants acquire prior to their long journeys.

There is much additional evidence, obtained by other means, of the bird's ability for sustained speed during long distances of migration. *Cochran, Montgomery,* and *Graber,* using radiotelemetry, tracked migrating *Hylocichla* thrushes nearly all night from Illinois northward into Michigan, Wisconsin, and Minnesota. Although they found considerable variation, most flights were at air-speed—i.e., speeds with relation to the winds aloft—between 25 and 35 mph. These were usually less than ground-speeds—speeds with relation tot he earth—and thus suggested that the birds were aided by favourable winds. One of the most remarkable records of a long, sustained flight is that of a banded Ruddy Turnstone (*Arenaria interpres*) released by MaxC. Thompson at St. George Island, one of the Pribilofs in the Bering Sea, on August 27, 1965, and short four days later, on August 31, at French Frigate Shoals in the Hawaiian Leeward Islands. Assuming that this individual covered 2,300 miles between St. George Island and French Frigate Shoals in a steady bee-line flight, its average speed was 575 miles a day.

Recently, radar studies have provided many reliable estimates of the rate at which migratory birds travel. *W.R.P. Bourne* assessed the air-speed of Lapwings (*Vanellus vanellus*) in their flights during June over the southern North Sea to England at 35 knots (40 mph). *Lee* at the Isle of Lewis in the Hebrides, Scotland, showed that the air-speed of Wheatears (*Oenanthe oenanthe*) from Iceland approximated 20 knots (23 mph); Redwings (*Turdus iliacus*), from 30

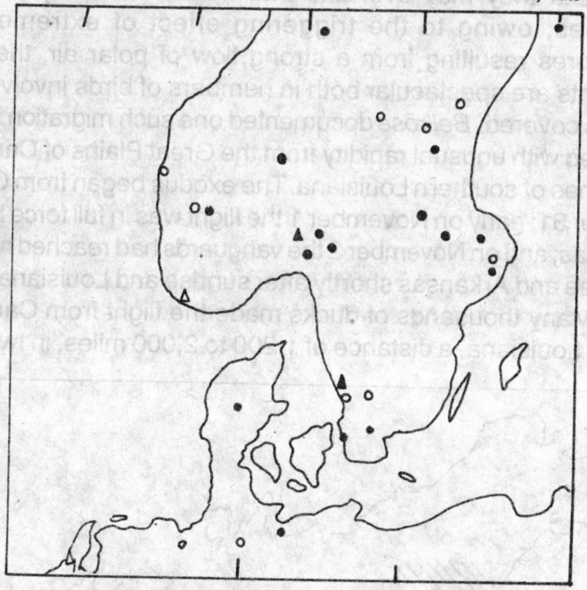

Fig. 18.7. Breeding range of chaffinches that spend the winter in the British Isles.

to 35 knots (34.5 to 40 mph); and Grey Lag Geese (*Anser anser*), from 53 to 55 knots (61 to 63 mph). *Bergman* and *Donner* demonstrated that the still-air-speed for the Oldsquaw (*Clangula hyemalis*) at low altitudes over the Gulf of Finland was 40 knots (46 mph) and for the Common Scoter (*Oidemia nigra*), 45 knots (52 mph). Once the birds reached a higher altitude inland, their speed increased by about 10 percent. After recording air-speeds of passerine birds off the coast of Norfolk, England, for a whole year, *Tedd* and *Lack,* on analyzing the results, found evidence of a seasonal difference: in the spring, the speed averaged 27 knots (31 mph), 4 knots faster than in the fall. At Cape Cod, *Massachusetts, Nisbet* and *Drury* found another seasonal difference in that the directions of migration in the spring were much less diverse than in the fall, thereby suggesting much less time lost in passage.

In the late fall, long-distance migrating ducks commonly pass from breeding area to winter quarters in a short series of mass movements, each of which carries them many hundreds of miles in one continuous flight. They start each flight immediately after the passage of a cold front when temperature has dropped and the sky is clear. But they may overtake bad weather as they proceed. Sometimes, owing to the triggering effect of extremely low temperatures resulting from a strong flow of polar air, the mass movements are spectacular both in numbers of birds involved and distances covered. *Bellrose* documented one such migration in 1955 that moved with unusual rapidity from the Great Plains of Canada to the marshes of southern Louisiana. The exodus began from Canada or October 31; early on November 1 the flight was in full force through the Dakotas; and on November 2 the vanguards had reached northern Tennessee and Arkansas shortly after sunrise and Louisiana later in the day. Many thousands of ducks made the flight from Canada to southern Louisiana, a distance of 1,200 to 2,000 miles, in two days,

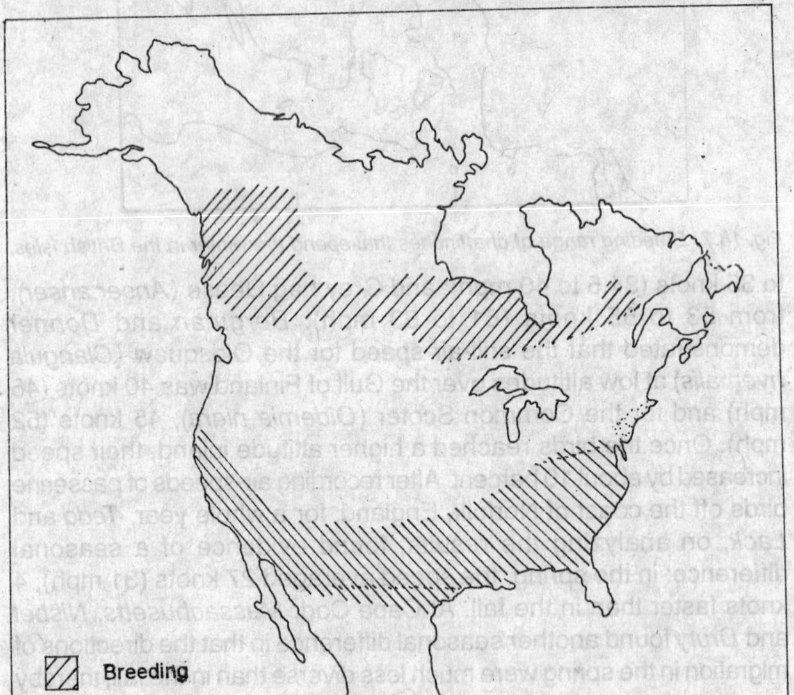

Fig. 18.8. Migration of the white-throated sparrow within North America.

or roughly 35 to 50 hours, at an average speed of 40 miles per hour. No doubt some of the birds covered the distance without stopping, accomplishing their migration in one flight.

In undertaking long-distance migrations, many small land birds tend to begin with short flights and complete them with longer flights. Indirect evidence of this procedure was reported by *Caldwell, Odum,* and *Marshall,* after comparing the fat reserves of six pieces of tropical-wintering North American passerines killed during fall migration by television towers, one near the Florida Gulf Coast and the other in the central Michigan. All the birds killed by the Florida tower showed significantly greater amounts of fat, strongly suggesting that these migrants began with low to moderate fat reserves that allowed only short flights and then increased their reserves until they had acquired a maximum amount for the long, non-stop flights such as across the Gulf of Maxico. European birds migrating south across Africa build up fat reserves of 30 to 40 per cent of their body weight by the time they set out across the Sahara.

By making longer flights as they near their destinations, birds gradually speed up their migrations. *Cooke* provided evidence for this acceleration when he analyzed migration dates of North American species, mostly passerines, approaching their northern nesting areas. "Sixteen species," he wrote "maintain a daily average of 40 miles from southern Minnesota to southern Manitoba, and from this point 12 species travel to Lake Athabasca at an average of 72 miles a day, 5 others to Great Slave Lake at 116 miles a day, and 5 more to Alaska at 150 miles a day."

From all these studies and reports, sexual generalization on the rate of migratory travel emerge. Strong winds can affect ground-speed of birds in their daily activity flights but not in migration. Birds make long, sustained flights, usually at increasingly higher speeds at higher altitudes. Spring migration proceeds at a greater rate with less time loss than fall migration. Larger birds such as ducks accomplish their migrations in a short series of a few mass movements, occasionally in one non-stop mass movement. Small land birds, however, tend to begin their migrations in many short flights, gradually building fat reserves for long, non-stop flights thereby accelerating their migrations.

Mortality in Migration

Migration is dangerous for all birds. In their long flights over land or water they are likely to meet disaster through varies of the weather.

When forced to land by cold fronts, frequently they must accept environments where, because of inadequate cover, they are easy victims of predators.

While migrations are adjusted to the normal alternation of fair and inclement weather during spring and fall, sudden and unseasonable changes in the weather occasionally have serious effects. Once, during fall migration in the vicinity of Lake Huron, untold numbers of birds crossing this huge lake were forced into the water and drowned because of a very quick drop in temperature and an exceptionally heavy snowfall. After the storm, one observer reported an estimated 5,000 dead birds washed up on a one-mile stretch of shore. In their flights north in the spring, birds are sometimes caught in severe storms and killed by becoming first exhausted and then being exposed to excessively low temperature coupled with heavy rain or wet snow. After a blinding March snowstorm in Minnesota as many as 750,000 Lapland Longspurs (*Calcarius lapponicus*) were found dead on the ice of two lakes, each of which covered only a square mile.

Man has created a wesome hazards for migrating birds by erecting lighthouse with strong light beams and by illuminating various tall structures such as the Washington Monument in the District of Columbia and the Empire State Building in New York City. During nights in the spring and fall when migration is proceeding at a low elevation because of an unsurmountable cloud layer, passing birds are attracted by the brightness and, approaching it soon become blinded and fly into its source, killing themselves.

Under certain circumstances, even street lights can be a hazard. Vast numbers of parulid warblers, driven ashore on the Texas coast by a northeast storm while migrating northward across the Gulf of Mexico during a night in early May, met their death by flying into street lights. In a part on Padre Island, *James* counted more than 900 dead birds under just one light pole plus an estimated 100 on the adjacent pavement. There were nine other light poles in the vicinity with similar tolls.

Airport ceilometers indirectly cause mortality among small, nocturnal migrants. These instruments, which are used to determine the cloud ceiling, direct a narrow, extremely brilliant beam of light straight upward. At night when the clouds are low, they produce a bright spot on the cloud ceiling that can be seen for a considerable distance. On mornings following a heavy, nocturnal migration, numbers of birds varying from three to over a thousand have been found dead near spots where ceilometers are used. Three ornithologists, *Howell, Laskey* and *Tanner*, who have investigated many of these accidents, explain the cause as follows : When there is a pronounced migration and a low ceiling, the beam attracts the migrants. After circling through the bright light, many circle back to fly in and about it. While in the beam their bodies reflect light attracting

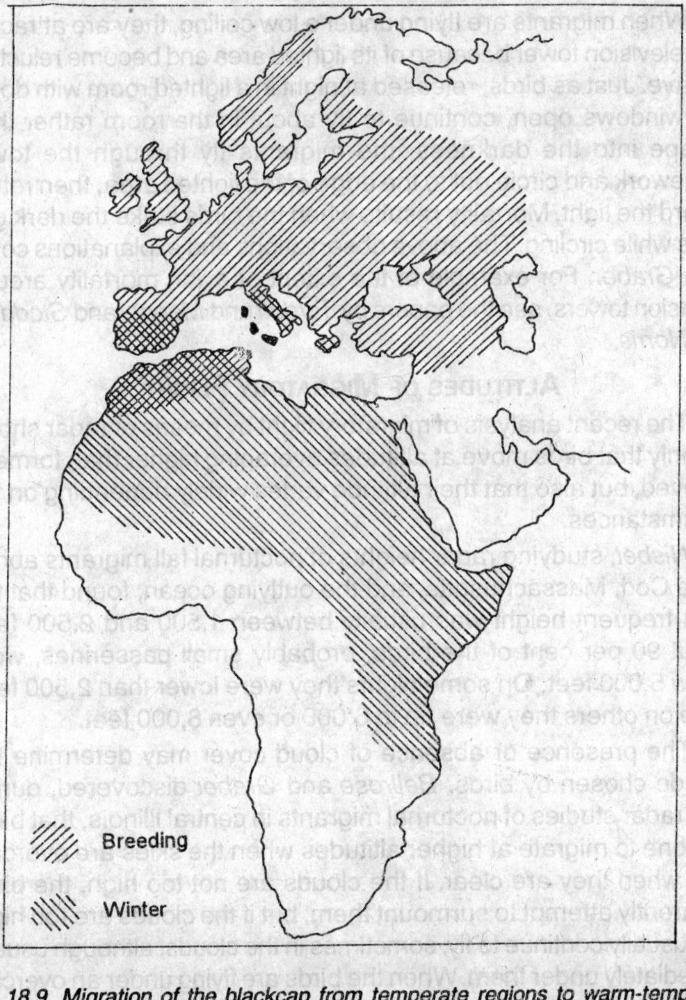

Fig. 18.9. Migration of the blackcap from temperate regions to warm-temperate and tropical Africa. There is some overlap, and a few birds winter further north than shown, for example in Southern England.

still other migrants. Blinded by the light, the birds die by collision, either with each other, with the ground, or (rarely) with a building.

A direct cause of mortality among small, nocturnal migrants are television towers erected to heights of 900 to 1,000 feet or more and, as is often the case, situated on hills or bluffs where they reach even greater heights above the local terrain. Each tower is supported by guy wires and has a system of steady, flashing red lights, mandatory on all tall structures that are potential hazards to airplanes.

When migrants are flying under a low ceiling, they are attracted to a television tower because of its lighted area and become reluctant to leave. Just as birds, released at night in a lighted room with doors and windows open, continue to fly about in the room rather than escape into the darkness, the migrants fly through the tower framework and circle out to the edge of the lighted area, then return toward the light. Mortality results when the birds strike the dark guy wires while circling. The above observations and explanations come from *Graber.* For example of the extent of avian mortality around television towers, see the papers by *Tordoff* and *Mengel* and *Stoddard* and *Norris.*

Altitudes of Migratory Flight

The recent analysis of migratory flight by means of radar shows not only that birds move at altitudes averaging higher than formerly believed, but also that their altitude varies widely depending on the circumstances.

Nisbet, studying radar heights of nocturnal fall migrants above Cape Cod, Massachusetts, and the outlying ocean, found that the most frequent height was usually between 1,500 and 2,500 feet. About 90 per cent of the birds, probably small passerines, were below 5,000 feet. On some nights they were lower than 2,500 feet, while on others they were up to 6,000 or even 8,000 feet.

The presence or absence of cloud cover may determine the altitude chosen by birds. *Bellrose* and *Graber* discovered, during their radar studies of nocturnal migrants in central Illinois, that birds re prone to migrate at higher altitudes when the skies are overcast than when they are clear. If the clouds are not too high, the birds apparently attempt to surmount them; but if the clouds are too high, they usually continue to fly, sometimes in the clouds, although usually immediately under them. When the birds are flying under an overcast and consequently, at much lower altitude, they can usually be heard from the ground.

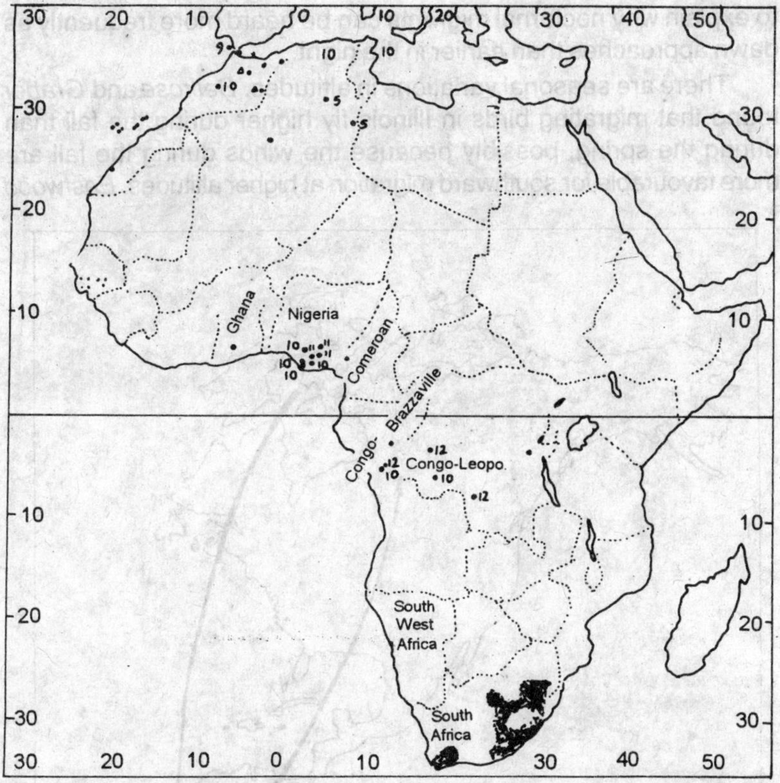

Fig. 18.10. Recoveries of British-ringed swallows in Africa. The figures indicate the month of recovery. The wintering area in S. Africa (54 recoveries) in shown shaded.

Birds fly higher by night than by day. *Lack* first noted this tendency, by radar, among migrants crossing the southern North Sa between England and the Continent. Later, *Eastwood* and *Rider* at the Bushy Hill station in England proved that the tendency is significant. From their considerable data they were able to show that 80 percent of the birds fly below 5,000 feet at night and 80 percent below 3,500 during the day. They further demonstrated that migrating birds, in a 24-hour day, have a tendency to fly at the lowest altitudes in the afternoon and the highest just before mid-night. From radar studies and tape recording to call notes in Illinois, *Garber* has concluded that migrants reduce their altitude after mid-night to 1,500 feet or less, although they continue their flight until daylight. After their descent to lower altitude, they increase their calling. This helps to explain why nocturnal migrants can be heard more frequently as dawn approaches than earlier in the night.

There are seasonal variations in altitudes. *Bellrose* and *Graber* found that migrating birds in Illinois fly higher during the fall than during the spring, possibly because the winds during the fall are more favourable for southward migration at higher altitudes. *Eastwood*

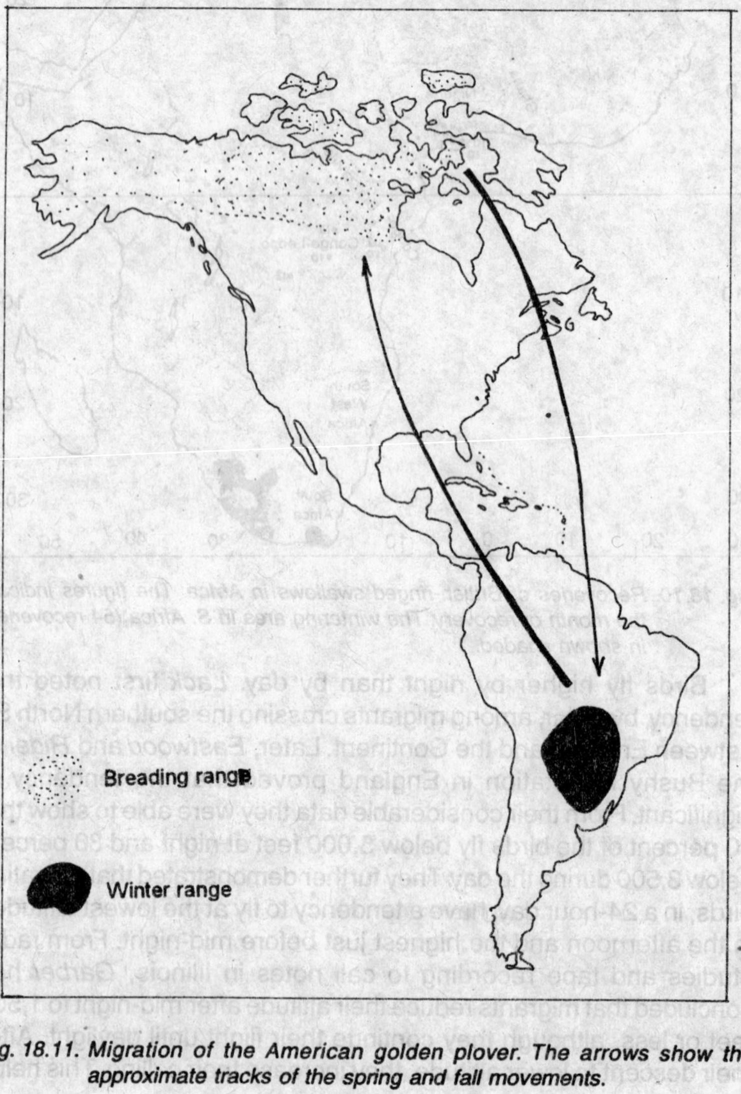

Fig. 18.11. Migration of the American golden plover. The arrows show the approximate tracks of the spring and fall movements.

and *Rider*, on the other hand, found the reverse to be the case in England. They suggest as one reason for this seasonal difference that flocks of all migrants include many young birds whose flight capabilities are inferior to those of adults and which are, consequently, unable to achieve the higher altitudes of the more mature spring migrants.

Birds migrate higher over land than sea. Common Scoters (*Oidemia nigra*) and Oldsquaws (*Clangula hyemalis*), in their spring passage over southern Finland and the Gulf of Finland, were noted by *Bergman* and *Donner* to fly at altitudes that averaged 3,400 feet over land and ranged from 300 to 1,00 feet above water. Passerine migrants, in passing from sea to land in England, were shown by *Eastwood* and *Rider* to make a similarly significant though not as great change in altitude climbing from a median height of 1,700 feet to median height of 2,200 feet.

Birds have long been known to reach very high altitudes during flight. Direct observations from aircraft proved that large birds can fly over the highest mountain ranges—for example, the Himalayas between central Russia and India. The Yellow-billed Chough (*Pyrrhocorax graculus*) was actually found on Mt. Everest at an elevation of 27,000 feet. The use of radar in determining heights of migration shows that while most birds rarely exceed 8,000 to 10,000 feet, a small proportion of migrants, particularly the stronger fliers, nonetheless attain great heights, in some cases astonishing. In his radar studies of migrants at Cape Cod, Massachusetts, *Nisbet* recorded a number of birds on several dates in September and October, usually before mid-night, or before and after sunrise, at altitudes between 8,000 and 15,000 feet and a few birds as high as 20,000 feet. These may have been sandpipers and plovers, flying over the ocean, toward the Lesser Antilles and eastern South America. Using an especially powerful and accurate height-finder at Norfolk in southeast England, *Lack* observed at sunrise on 16 dates in September a thin scattering of birds extending fairly uniformly up to at least 15,000 feet. On seven of these mornings he noted that the greatest height was 19,000 feet and on two mornings it was 21,000 feet. Probably the highest migrants were small shore birds such as Dunlin (*Erolin alpina*) which had left Scandinavia the night before.

Bearing in mind that the oxygen content of the air at 18,000 feet is 50 percent less than the air at sea level, one wonders whether high-flying birds suffer "altitude sickness". When a man sets out of climb a lofty mountain he acclimates himself gradually over a period of days, but birds take off and reach a comparable elevation in a matter of a few hours. Do birds have special adaptations that enable them to avoid altitude sickness? The answer may come someday from experimental studies of bird flight in pressure chambers.

Course of Migration and Migration Routes

Many species breeding in North America north of Mexico have their winter ranges far south and southeast; in southern Mexico, Central America, and South America. To reach their winter ranges, most of these species proceed at night from their breeding ranges to southern United States in broad fronts without notable regard to topographical features. Radar and—other observations confirm that their movements trend southeast—toward their winter ranges. This direction, coming as it does in the wake of a cold front, is with the wind and, therefore, beneficial to the migrants. In the spring, the direction trends northwest, in reverse, and again with the wind since the movements take place with the onset of a warm front. Many species, however, do not exactly retrace their course in the spring, but fly instead somewhat to the west of their fall passage. Evidence of this elliptical course—going to southern united States on way and coming back another—is borne out by kills at television towers. Certain species are well represented in the spring migration but seldom or not at all in the fall, and *vice versa.*

Owing to the narrowing of the North American continent southward and the intervention of the Gulf of Mexico and Caribbean Sea, all species moving southeastward through the United States toward their wintering ranges in southern Mexico, Central America, and South America converge on special routes. There are five altogether. Certain species use mainly one route, others two or three; no species is known to use more than three. The routes are as follows:

Route 1. From the coasts of Newfoundland, Nova Scotia, New England, and New Jersey southward over the Atlantic Ocean to the Lesser Antilles and the northeastern coast of South America. A few shore birds use this route.

Route 2. From Florida southward over the Bahamas, Hispaniola, Puerto Rico, and the Lesser Antilles to South America. The few birds which frequent this route are seldom far from land as there are many small island along the way.

Route 3. From Florida southward over Cuba and Jamaica across 400 miles of the Caribbean Sea of South America.

Route 4. From the shores of the Gulf states across the Gulf of Mexico to the Yucatan Peninsula and southern Mexico. This is the route most frequently used by the many species of birds from eastern United States and Canada.

Route 5. From Texas, New Mexico, Arizona, and California through northern Mexico. The majority of birds from western conterminous United States, western Canada and Alaska use the western side of the route.

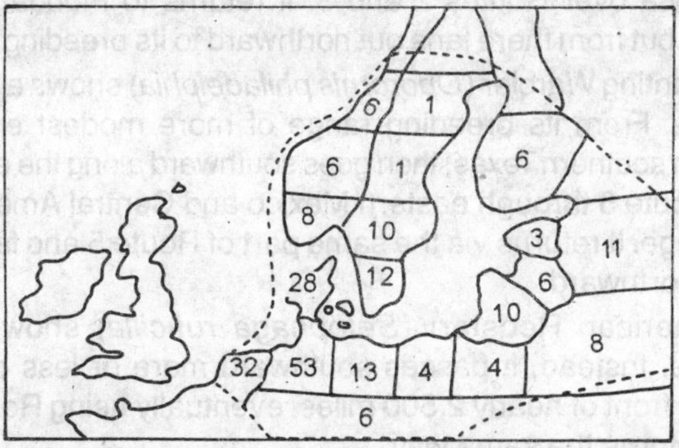

Fig. 18.12. Breeding range of starlings that spend the winter in the British Isles. Numerals indicate the breeding-season recoveries (May-Aug.) of starlings previously ringed during winter (Dec.-Feb.) in the British Isles.

Species vary greatly in their course of migration and use of routes. A few species move south over one route and return by another; a few species move eastward or westward before going south; and a few species have spectacularly long routes that take them as far south as southern South America. Among a few species which breed in western United States—e.g., the Western Kingbird (*Tyrannus verticalis*) and the Scissor-tailed Flycatcher (*Muscivora forficata*)—some individuals in the fall move eastward across the Gulf states to winter in Florida whereas most of the population moves directly south into Mexico.

To illustrate the wide variation in migratory movements and the choice of routes, the migrations of six species are described below.

The American Golden Plover (*Pluvialis dominica*) passes eastward from its breeding range to the Atlantic Coast, where it turns southward over Route 1. Once in South America, it flies directly across the continent to its winter range. It returns, however, by another route, coming up across northwestern South America and the Gulf of Mexico and reaching the United States along the coast of Texas and Louisiana. From there it continues up the Mississippi Valley and through central Canada to its breeding range.

The Blackpoll Warbler (*Dendroica striata*) shows a remarkable convergence in its southward migration. From its vast transcontinental breeding range it converges on the Atlantic coastal plain as far north as Virginia and from there proceeds to Florida where it leaves for South America over Routes 2 and 3. It returns to Florida by the same routes but from there fans out northward to its breeding range.

The Mounting Warbler (*Oporornis philadelphia*) shows a similar convergence. From its breeding range of more modest extent it converges on southern Texas, then goes southward along the eastern portion of Route 5 through eastern Mexico and Central America to its winter range. It returns via the same part of Route 5 and fans out from Texas northward.

The American Redstart (*Setophaga ruticilla*) shows little convergence. Instead, it passes southward more or less directly over a broad front of nearly 2,500 miles, eventually using Routes 2, 3 and 4. It returns the same way.

The Bobolink (*Dolichonyx oryzivorus*) migrates an exceptionally long distance, averaging farther than any other passerine species. From its transcontinental breeding range, it converges southward to leave the United States over Routes 3 and 4. Once in south America, in continues directly overland to its winter range in extreme southern Brazil, southeastern Bolivia, Paraguay, and northern Argentina. It returns the same way.

The Connecticut Warbler (*Oporornis agilis*) migrates in an eccentric manner. From its breeding range it flies directly eastward to New England, then to South America along the Atlantic coastal plain and eventually Route 3. From its winter range in South America it returns over Route 3 to Florida, but from there it passes diagonally to the Mississippi Valley and northward to its breeding range.

Altitudinal Migration

Some bird populations of high mountains in both temperate and tropical regions move down to the lower slopes and valley when winter sets in at higher elevations. In the descent of a few hundred feet they accomplish what many other populations do in their latitudinal migrations of many hundreds of miles.

Several observers have noticed that mountain birds tend to move to higher slopes after the nesting season. Probably many are young birds which, after getting their growth, always show a great tendency to move about. The reason for this may be the agonistic behaviour of their elders, but more than likely it is another examples of juvenile wandering.

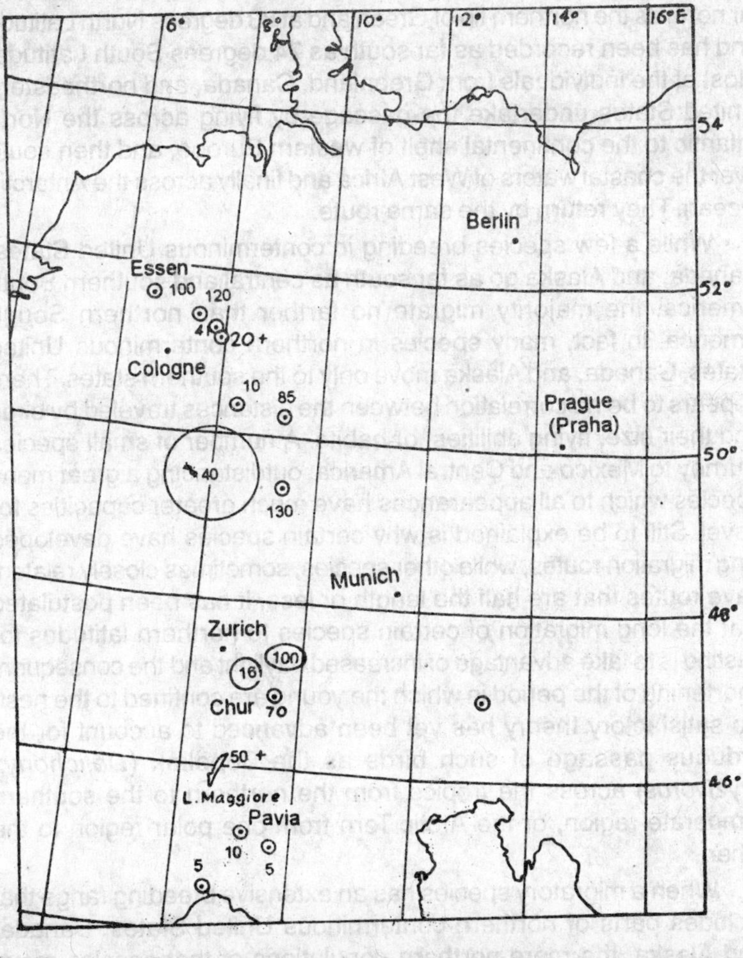

Fig. 18.13. Migration of storks. 144 birds released at Essen had been bred at Rossitten, 1000 km to the east. Figures show the numbers of birds recovered or seen. Scale : 10 mm = 100 km.

Distance in Migratory Travel

Extremes in distances traveled by migrating birds are represented on the one hand by high-mountain species, which merely pass up or down slopes of several hundred to a few thousand feet, and on the other hand by the Arctic Tern (*Sterna paradisaea*). It makes the longest flights of any bird, migrating thousands of miles from the Arctic, where a part of the population breeds, to its wintering area adjacent to the pack ice around Antarctica. This species nests as far north as the northern tip of Greenland at 83 degrees North Latitude and has been recorded as far south as 74 degrees South Latitude. Most of the individuals from Greenland, Canada, and northeastern United States undertake the passage by flying across the North Atlantic to the continental shelf of western Europe, and then south over the coastal waters of West Africa and finally across the Antarctic Ocean. They return by the same route.

While a few species breeding in conterminous United States, Canada, and Alaska go as far south as central and southern South America, the majority migrate no farther than northern Sough America. In fact, many species in northern conterminous United States, Canada, and Alaska move only to the southern states. There appears to be no correlation between the distances traveled by birds and their size, flying abilities, or habits. A number of small species journey to Mexico and Central America, outdistancing a great many species which to all appearances have much greater capacities for travel. Still to be explained is why certain species have developed long migration routes, while other species, sometimes closely related, have routes that are half the length or less. It has been postulated that the long migration of certain species to northern latitudes for nesting is to take advantage of increased daylight and the consequent shortening of the period in which the young are confined to the nest. No satisfactory theory has yet been advanced to account for the arduous passage of such birds as the Bobolink (*Dolichonyx oryzivorus*) across the tropics from the northern to the southern temperate region, or the Arctic Tern from one polar region to the other.

When a migratory species has an extensive breeding range that includes parts of northern conterminous United States, Canada, and Alaska, the more northern populations of that species move farther south for the winter than the other populations do. For example, among the several subspecies of the Fox Sparrow (*Passerella iliaca*)

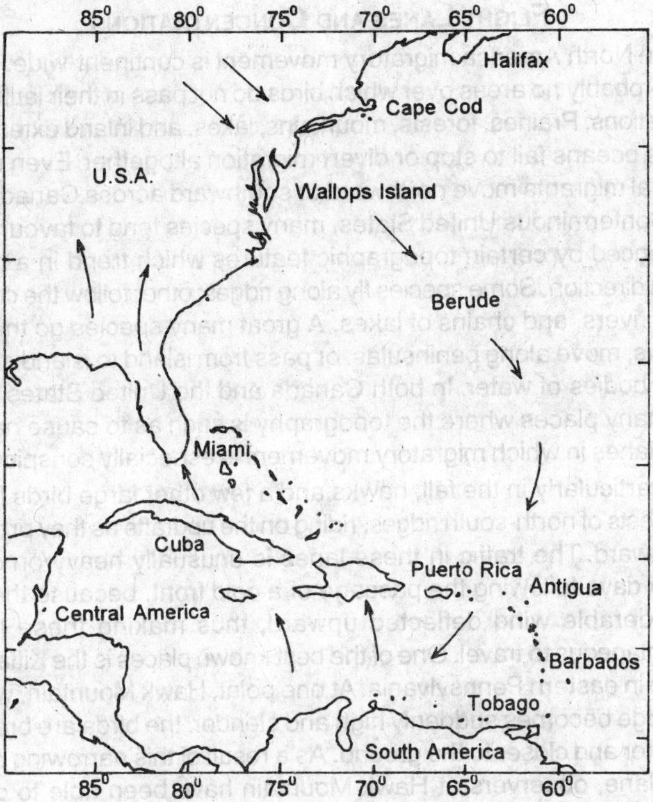

Fig. 18.14. Routes used by the Blackpoll Warbler (Dendroica striata) during its migrations.

breeding along the Pacific Coast from Alaska to Puget Sound, the subspecies nesting farthest north have been found wintering in southern California passing by other subspecies, which either do not migrate at all or move only to central or northern California. Apparently the more northern breeding populations of a widespread species acquire a stronger migratory habit than the populations breeding in more southern areas that have milder, year-round climate.

In some species there may be sexual differences in the extent of the migration. *Howell* found that females in the eastern race of the Yellow-bellied Sapsucker (*Sphyapicus varius*) outnumber the males in the southern part of the winter range by about three and one-half to one.

Flight Lanes and Concentrations

In North America migratory movement is continent-wide. There are probably no areas over which birds do not pass in their latitudinal migrations. Prairies, forests, mountains, lakes, and inland extensions of the oceans fail to stop or divert migration altogether. Even so, as diurnal migrants move northward or southward across Canada and the Conterminous United States, many species tend to favour or be influenced by certain topographic features which trend in a north-south direction. Some species fly along ridges; other follow the coasts, large rivers, and chains of lakes. A great many species go through valleys, move along peninsulas, or pass from island to island across large bodies of water. In both Canada and the United States there are many places where the topography is such as to cause narrow flight lanes in which migratory movement is especially conspicuous.

Particularly in the fall, hawks and a few other large birds follow the crests of north-south ridges, riding on the updrafts as they proceed southward. The traffic in these lanes is unusually heavy on clear, windy days following the passage of a cold front, because there is considerable wind deflected upward, thus making these lanes advantageous to travel. One of the best known places is the Kittatinny Ridge in eastern Pennsylvania. At one point, Hawk Mountain, where the ridge becomes suddenly high and slender, the birds are brought together and closer to the ground. As a result of this narrowing of the flight lane, observers at Hawk Mountain have been able to count over 22,000 hawks moving by in a single season.

Large bodies of water constitute barriers to day-migrating birds and thus cause flight lanes to curve around them. The Great Lakes are a good example. In the fall, south-bound hawks approaching their north shores from western Quebec and southern Ontario take the shortest courses around them that geography and air movements will allow. If the birds are migrating in great numbers, as on the second days after cold fronts when there are steady westerly winds and ample sunlight producing thermals, large numbers can be seen in continuous passage at such points as Port Credit on the northwest shore of Lake Ontario, Part Stanley on the north side of Lake Erie, Cedar Grove on the west side of Lake Michigan and Duluth at the westernmost extension of Lake Superior.

Peninsulas projecting into large bodies of water that lie athwart the direction of migration may become funnels for land birds in diurnal passage. Hawks moving northward in the spring through Michigan to Canada converge in large numbers on the northern tip of Lower Michigan at the Straits of Mackinac. Here, if the weather is rainy and windless, they settle on trees and other perches. With the advent of the next clear day and a favouring wind, the hawks begin to spiral higher and higher until one bird peels out and heads northward over the Straits with the others following. Small numbers of north-bound Blue Jays (*Cyanocitta cristata*), in order to get across Lake Mendota at Madison, Wisconsin, converge first at Picnic Point and then spiral upward until "barely visible to the naked eye" before crossing the 1.7 miles of open water to Fox Bluff on the north shore. Cape May, he southern tip of New Jersey between the Atlantic Ocean and Delaware Bay, is noted for its hordes of land migrants, large and small, which gather from August through November when northerly winds are strong. Some of the migrants held up here are those which regularly follow the Atlantic Coast southward, but many are birds which have been drifted by wind southeastward from their usual flight lines. Frequently, migrants at Cape May may be seen in the day flying north and northwestward into the wind as they skirt Delaware bay before continuing their journey. Similar concentrations may be observed at Cape Charles, a south-pointing peninsula separating the waters of Chesapeake Bay from the Atlantic.

Night migrants do not follow topographically determined flight lanes to any significant degree. Instead, as radar surveillance shows, birds migrate at night without regard to what lies below.

In the spring, land migrants returning to the United States from across the Gulf Mexico vary in their manner of arrival in accordance with weather conditions. If the weather is mild and the wind favourable, birds bound for more northern destinations continue inland from the Gulf for considerable distance before coming to land. The coastal area thus appears to be an ornithological "hiatus." But if a cold front with strong northerly winds moves in over the area while migration is in progress, migrants are forced to come down to the first land reached, with the result that the coastal area is flooded with birds that linger here until the weather again becomes favourable. A somewhat similar situation occurs during spring migration at Point Pelee. Here, since it is the nearest land, north-bound small land birds, on meeting a cold front from the north while they are over Lake Erie are forced to descend and remain until the cold front abates. At such times Point Pelee swarms with birds.

Flocking During Migration

During migration bird species show wide difference in flocking habit. A number of diurnal migrants, notably many hawks and other predators, are little inclined to move in groups, preferring to travel solitarily. But the majority of migrants diurnal or nocturnal, exhibit the flocking habit.

Certain species migrate in flocks strictly of their own kind. These are usually birds whose flight-speed, feeding habits, or roosting preferences are so individual as to make them incompatible traveling companions. The Common Nighthawk (*Chordeiles minor*) and Chimney Swift (*Chaetura pelagica*) are good examples of birds which migrate in their own company. Neither waxwings nor crossbills will migrate with other birds, but Cedar and Bohemian Waxwings (*Bombycilla cedrorum* and *B. garrulus*) have been seen in the same flocks and so have Red and White-winged Crossbills (*Loxia curvirostra* and *L. leucoptera*). Some of the larger birds—e.g., pelicans, cormorants, storks, swans, geese, and cranes—noted for V-shaped or linear flock formations likewise tend to travel in unmixed groups.

The majority of species traveling in flocks, whether unmixed or mixed, give call notes. This is especially true of nocturnal migrants: some species such as *Hylocichla* thrushes utter calls herd only in night migration; other species such as the Bobolink and Dickcissel (*Spiza americana*) give calls that are the same as ones heard in the day-time on their breeding grounds. If the calls are distinctive as in the case of the Bobolink and Dickcissel, it is possible to identify the species, but the calls of most species are link and Dickcissel, it is possible to identify the species, but the calls of most species are faint chips and lisps that sound more or less the same to the human ear.

Unmixed or mixed flocks of some species of smaller birds—e.g., certain sandpipers and plovers—fly in compact formations, all individuals in the flock simultaneously performing almost identical maneuvers. Many more species of smaller birds, including the majority passerines, travel in flocks that are loosely formed, though still cohesive, the individuals proceeding in the same direction. At night, the cohesion of flocks is probably maintained by call notes. At the same time, the call notes serve to space out individuals in each flock so that they will not collide with one another.

No flock of migrants appears to have a persisting leader. The direction taken by a flock, represents a compromise by each individual to the directional preference of the other individuals of the flock. At night, call notes convey directional information from one individual to the other. When a nocturnal flock shifts it direction or lowers its altitude, or when it is disoriented for one reason or another, all its members greatly increase their rate of calling.

Whether or not flocks remain intact for the duration of migration has not been determined. Nor is it known for certain whether flocks remain together during the winter. In all probability, most flocks, if they are comprised of common, widely distributed species, change from day to day during migration and in the winter split up into smaller groups, or combine with other flocks to form larger groups.

Flocking during migration is undoubtedly advantageous to the individuals concerned. Just as flocking among resident birds provides group protection against predation or increases the success in finding and exploiting food sources, flocking in migration greatly facilitates the attainment of destination. Younger birds traveling with more seasoned adults benefit from their experience. As another advantage, *Hamilton* suggests that groups of birds on an average determine their direction with greater accuracy than single individuals. Thus flocking assists any migrant that goes a long distance, or any migrant required to pinpoint its destination on a small land area in mid-ocean. Hamilton cites as examples : (1) The Broad-winged and Swainson's Hawks (*Buteo platypterus* and *B. swainsoni*). Both of these North American species travel in large flocks to winter in South America. By contrast, the Red-tailed Hawk (*B. jamaicensis*) and certain other buteos, which rarely leave the North American continent, seldom move in appreciable groups. (2) The Long-tailed Cuckoo (*Urodynamis taitensis*). After breeding in New Zealand, this species gathers in large flocks for the exceedingly long, non-stop passage to winter quarters on tiny islands in the west-central Pacific. By contrast, the Yellow-billed and Back-billed Cuckoo (*Coccyzus americanus* and *C. erythropthalmus*) of North America, which have only to reach a broad tropical region of South America for the winter, migrate solitarily.

V-shaped formations help to conserve energy by creating favourable air currents for all individuals in the flock except the leader. When fatigued, the leader drops back and is replaced by another bird in the flock. An alternate advantage hypothesized by *Hamilton* is that the V-shaped structure serves as a means of communication, enabling the individuals to profit fully from the collective direction-finding of the group. By flying in parallel alignment, each individual moves in the same direction as the leader. If, as in flocks of geese traveling in poor visibility, difficulties in establishing direction arise, the leadership changes frequently in order that collective "judgment" may prevail in maintaining the proper course. Sometimes the V-structure gives way temporarily to a crescent form; forward flight remains in the same direction but is slowed until a new leader takes over and the V-shape is resumed.

Unmixed or mixed flocks may contain only immature individuals, only adults, or only adults of one sex; or they may contain individuals of all ages and both sexes. Flocks of gene can cranes may be comprised on one family or several families.

During the fall migration the adults may precede the immature birds, the birds-of-the-year. *Hagar* reported that adult Hudsonian Godwits (*Limosa haemastica*) withdraw from their nesting grounds in central and northwestern subarctic Canada in late July; the immatures follows a month later. Among passerine species there is convincing evidence assembled by *Murray* and *Jehl* from several thousand migrants, mist-netted in the fall at Island Beach, New Jersey, an analyzed as to age, that adults and immatures travel at approximately the same time. However, in the case of the Least flycatcher (*Empidonax minimus*), *Hussell, Davis* and *Montgomerie* concluded from an analysis of 182 individuals trapped in the late summer at Long Point, Ontario, that most of the adults migrate in advance of the immatures: the adults during the second half of July and first half of August; the majority of immatures from the second half of August to the end of September.

In the spring migration many of the first flocks of certain species coming north have a preponderance of adult males which reach the breeding grounds and establish territories before the rest of the population arrives. *A.A. Allen* once carefully studied the spring migration of Red-winged Blackbirds (*Agelaius phoeniceus*) at Ithaca, New York. Since the males and females have distinct plumages and birds hatched the previous year (i.e., the immature birds) differ sufficiently in colour from the adults, he was able to analyze flocks and work out a migration schedule according to age and sex. While his findings may not hold for all Red-wing populations, it is presented below as a useful guide and basis for comparison.

Migrant adult male .... March 13-April 21
Resident adult males .... March 25-April 10
Migrant female and immature males .... March 29-April 24

Resident adult female .... April 10-May 1
Resident immature males .... May 6-June 1
Resident immature females .... May 10-June 11

In watching spring migration in his own area, the student will find that not all species follow such a schedule. In fact, males and females of quite a few species migrate together. Only rarely, however, do females precede males.

Direction-Finding

How migrating birds determine their direction when migrating or homing by day or night over areas unfamiliar to them is one of the most fascinating aspects of migration. Before considering the subject of direction-finding in migration, it is worthwhile to review some of the problems associated with homing in birds. The basic means by which birds orient themselves, whether migrating or homing, are much the same.

A great many experiments have demonstrated the remarkable ability of wild birds to return to their eggs or young after being removed great distances and released. Years ago *Watson* tested the homing ability of Sooty and Noddy Terns (*Sterna fuscata* and *Anous stolidus*), which nest on the Dry Tortugas, the islands in the Gulf of Mexico west of Florida. These species come to the Tortugas from tropical seas and are seldom seen farther north. Two nesting Sooty Terns and three Noddy Terns were captured, marked, and transported northward in a ship to point off Cape Hatteras, about 1,000 miles by sea from the Tortugas. Here they were released. Just five days later the Solly Terns were back on their nests; one of the Moddy Terns showed up after a few two more days. Some of the more recent experiments demonstrating the sustained speed of homing flight by the Purple Martin, Manx Shearwater, Leach's Petrel, and Laysan Albatross attest at the same time to their precision in direction-finding.

The inducement of birds to home is not necessarily provided by their eggs or young: it can be the breeding area or home range. Breeding Brown-headed Cowbirds (*Molothrus ater*), which are brood parasites, will home from maximum distances of 250 to 380 miles. An adult female Bobolink (*Dolichonyx oryzivorus*) escaping from captivity in September at Berkeley, California, was recaptured the following June at Kenmere, North Dakote, where it was orginally tapped as breeding bird. Six hundred and sixty Golden-crowned and

White-crowned Sparrows (*Zonotrichia atricapilla* and *Z. leucophrys*), captured while wintering in the San Jose area of California were immediately carried by plane to Laurel, Mary-land and released. Fifteen were known to have come back the following winter. In the interim they had presumably found their way in the spring to the nesting grounds in northwestern Canada and Alaska, then returned to California in normal migration.

Homing flights are more or less routine with homing pigeons which are Common Pigeons (*Columba livia*) specially bred for racing. Their precision and speed of return to the home loft is developed by training and experience. Sometimes the experience of only one flight to the home loft is sufficient to determine the proper direction for successive flights. In his work on homing pigeons, *Matthews* found that certain individuals, trained to maintain a given direction, can adhere to that direction over unfamiliar terrain and that certain other individuals can fly straight toward the home loft from unfamiliar territory regardless of the direction of the home loft. In the course of their training, pigeons must be made familiar with the area around the loft so that they will have a broad or reasonable target.

Although homing has been amply demonstrated in both wild birds and racing pigeons, the perplexing question remains: How do homing birds find their way? In seeking an answer, *Griffin* and *Hock* attempted to determine how displaced and released nesting birds find their way by following them in an airplane and watching their behaviour. For their experiment they selected the Gannet (*Morus bassanus*), a marine bird which rarely occurs inland; being large and white, it was easy to follow. Taking 17 individuals from their nests on an island off the Gaspe Peninsula, Quebec, they carried them to a point in northern Maine, 100 miles from salt water and 215 miles from the home island. There they released nine of the birds (the others were used as controls) and, from an airplane, traced their flights from 25 to 230 miles. The investigators were careful to keep the plane, 1,500 feet or more away from the birds so as not to frighten them. The experimental birds flew in all directions with no significant tendency to head directly toward their nests. Their flights paths were generally gradual curves. The first birds to reach the coast within the first few hours were the first to get back to their nests. Although 62.5 percent of these birds eventually reached their nests, in from one to four days.

The result of Griffin and Hock's work on the Gannets suggests that the ability of birds to home—that is, to return to a known or familiar site such as a nesting area—is dependent on random searching. In a strange territory birds keep circling and exploring by trail and error until they find familiar landmarks within familiar territory. However, the concept of random searching as a means by which all birds find their way cannot be reconciled with the rapid homing exemplified by Sooty Terms and other species already mentioned. Many birds, if not all, obviously have an ability to find their way by orienting themselves—determining their position with respect to their environment—and by following directional cues or navigating.

In taking up the subjects of the means by which birds orient themselves and navigate, it is well to consider first the question whether or not birds inherit at least part of their ability to find their way. *Rowan* at Edmonton, Alberta, caught young Common Crows (*Corvus brachyrhyncos*) in the late summer and kept them in captivity. In November, when winter conditions had begun to set in and all adult Common Crows had left for their winter range in Kansas and Oklahoma, he banded the captive birds and released them. Altogether 54 individuals were set free, and in the next few days he received reports of recoveries. Apparently some of the birds had not traveled very far, but those which had gone an appreciable distance were headed toward their winter range. *Schuz* at Rossitten on the Baltic Coast of eastern Prussia tried similar banding experiments with White Storks (*Ciconia ciconia*). In the middle of September, after all the local population had departed, he released 73 young banded birds. Most of them traveled eastward toward the Black Sea, paralleling the normal flight line for the local population, though some of the birds flew in a more southerly direction and three went southwestward to Italy. *Perdeck*, over a period of four years, caught and banded over 11,000 fall-migrating Starlings (*Sturnus vulgaris*) in Holland and released them in Switzerland. Recoveries totaling 354 later showed that the juveniles took their ancestral direction southwest, paralleling the normal route, to a new winter area, whereas the adults soon separated and veered westward toward their ancestral winter range. From these experiments alone, it seems clear that at least some birds must have an innate ability to follow the normal migratory route or one parallel to it. How the birds "know" when they have flown far enough is yet to be determined.

Granting that migrating birds have an innate ability to find their way does not deny that some birds acquire their ability by experience. Young Indigo Buntings (*Passerina cyanea*), hand-raised in various conditions of isolation, apparently do not attain the accuracy of orientation typical of adult buntings and thus seem to depend, at least partly, one some kind of experience. Young geese and cranes, which migrate in families or groups of families, probably learn the migration routes by following their elders. They no doubt "memorize" features of the landscape when they migrate by day as they often do. But whether or not bird inherit or acquire their ability to find their way, the fact remains that they must depend on cues other than landscape features to orient themselves and navigate when traveling at night, over the sea, or above or in a heavy overcast. What are the cues?

The first break-through came in the early 1950's when it was proved that the sun is a cue to orientation and even a guide in navigation. At Wilhelmshaven, Germany, *Kramer* placed a small, circular pavilion that was completely enclosed except for six windows, each high enough to give only a view of the sky from the cage. *Kramer* put a hand-reared Starling in the cage at the time in the spring when it would normally migrate, and from below the cage he recorded the direction in which the bird, in its migratory restlessness, showed a tendency to flutter. The bird fluttered persistently in the normal direction of spring migration if the sun was shining but not if the sky was heavily overcast. When, by an ingenious use of mirrors at each of the windows, Kramer altered the sun's apparent position, the bird deflected its direction in order to maintain the same angle as before. *Kramer* and his associates soon demonstrated in subsequent experiments that birds have some sort of "internal clock" that enables them to compensate for the sun's daily "movement" across the sky and thus can hold the same direction of flight despite the sun's steadily changing position.

Hoffmann experimented further on the internal clock as a basis for orientation by the sun. He trained caged Starlings to seek a food reward at a particular time of day and always in the same direction. Then he tested them without rewards at the time they had come to expect and at other times. The birds, he found, could allow for the daily movement of the sun. Having proved this, he subjected the trained birds to a regime of light and darkness that shifted their internal clocks six hours ahead or six hours behind the natural day outside. Later he exposed the birds to the natural day. The birds responded by shifting their directions accordingly by 90 degrees, counterclockwise if the clocks were set ahead, clockwise if they were set behind. These and later tests showed that the birds did not orient themselves by the elevation or altitude of the sun in the sky but by its azimuth or position with relation to compass direction. The birds thus determined direction by what is called "sun compass orientation."

Soon after *Kramer* reported his initial experiments with Starlings, *Matthews,* found that homing pigeons, which are non-migratory, use the sun as a guide. When he released pigeons in unfamiliar territory under clear skies, they headed in the direction of home. If the sky was overcast, they failed to do so. The concept of the sun as a compass for homing pigeons has since been supported by other experimental investigators using different techniques. For example, *Schmidt-Koenig* shifted the internal clocks of pigeons by six hours in advance and six hours behind the natural day. On releasing the birds under a clear sky at varying distances and different direction from the home loft, he found that the birds deviated appropriately to the left or right of the correct flight direction taken by the control pigeons whose clocks had not been shifted. This suggests that the sun played a role in their orientation. Experience also plays a role in the orientation. Each time a pigeon is released, the direction it takes is more accurate.

Some birds may show one-directional orientation that is not necessarily related to migration or homing. *Matthews,* experimenting with non-migratory mallards (*Anas platyrhynchos*) at Slimbridge, England, discovered that when birds of any age were displaced during a clear day for any distance in any direction at any season, they demonstrated a strong tendency to fly in one direction, namely northwest. The flights were always short and seemed to be guided by the sun, but since they had no discernible purpose other than possibly escape, Matthews called the behaviour "nonsense orientation." He believed that it might be innate or perhaps developed at an early age. Others birds may show the same phenomenon. Migratory Mallards wintering in Illinois, on being trapped and released during a clear day or night at points far east and west of their winter-home lakes, consistently flew northward regardless of the direction of their home-lakes. Common Terns (*Sterna hirundo*), displaced from their breeding colony on Penikese Island, irrespective of where they were released or the direction of Penikese Island.

Adeili Penguins (*Pygoscelis adeliae*), when displaced from their breeding colonies at Cape Crozier on the coast of Antarctica to the featureless expanse of compacted snow in the interior, oriented themselves by the sun, departed in one direction, and ultimately reached their nest sites. The investigators, *J.T. Emlen* and *Penney*, used breeding males, some of whose internal clocks had been artificially shifted in advance. Releasing them individually from three groups, each at widely separated points, they observed that the birds headed straight for the coast on courses which were essentially parallel without any convergence toward Cape Crozier. The birds seemed to have no information on their location at release with respect to their home colonies. They took their courses from the sun and, as shown by the breakdown in orientation among those birds whose clocks had been shifted, maintained their courses by referring to the sun's azimuth. If heavy clouds obscured the sun and eliminated all shadows, their orientation soon deteriorated. Emlen and Penney considered such one-directional courses to be fixed and to be maintained by an inherent time sense; they linked the procedure to nonsense orientation as described in waterfowl and other birds, but believed that it might be escape orientation with survival value by steering the penguins to off-coast feeding areas whence they would be guided, perhaps in part by familiar landmarks, to their home colonies. Later experiments by the two investigators confirmed their original findings and conclusions.

A few years following Kramer's initial discovery that migratory restlessness in caged Starlings was directed toward the sun. Sauer and his wife, undertook experiments in Germany to determine by what means nocturnal migrants find their way. The *Sauers* used three species of sylviid warblers, reared in captivity without ever having seen the natural sky. During the period when the species would normally migrate, the *Sauers* put each bird in a rotatable circular cage and exposed it to the night sky. Under a clear, starry sky, even when their cages were turned periodically, the test birds fluttered persistently in the direction that the species takes at the season of the experiments; under heavily overcast skies the birds fluttered randomly in various directions, obviously disoriented. Then the *Sauers* placed the cage in a small planetarium with a starry sky, adjusted to the local time and latitude. The birds fluttered in the direction appropriate to that season depicted. There seemed to be no doubt that these birds obtained their information for direction from the starry sky.

The *Sauers* carried their experiments further. Placing the birds under skies representing the non-migratory seasons, they noticed that some of the birds were completely disoriented. From these experiments and others, *Sauer* concluded that birds do not rely on the stars themselves for direction but on the azimuch and altitude of the starry sky. This bicoordinate system would give birds the necessary information on their location since azimuch or hour angle would denote longitude and altitude would indicate latitude.

Working with mist-netted Indigo Buntings in the United States, *S.T. Emlen* repeated many of the Sauers' experiments and made still other well-controlled tests under planetarium skies. The results confirmed some of the *Sauers'* studies but differed sharply from the others. *Emlen* found no evidence that the buntings relied on a bicoordinate celestial system for orientation but rather made use of the numerous stars, particularly "the constant, two-dimensional spatial relationships" existing between them. He believed that no single star, with the possible exception of Polaris (the North star) could give the birds sufficient information for direction. It is the configuration or patterning of the bright stars in the constellations such as Ursa Major (the Big Dipper) that gives the cues. While *Emlen* could not determine with any certainty which patterns were of special importance or essential to orientation, the evidence seemed to point to the northern astral sky, especially the area within 35 degrees of Polaris.

What prompts birds to take the appropriate seasonal direction? Seeking an answer to this question *S.T. Emlen* induced physiological readiness for spring and fall migration in two groups of Indigo Buntings and then tested them simultaneously under the spring sky of a planetarium. The group conditioned for spring migration oriented northward; the group in condition for all migration took the opposite direction. This clearly suggested to Emlen that the internal physiological changes in Indigo Buntings, rather than differences in external stimuli (e.g., in the night sky), are responsible for their direction in spring and fall.

Practically all the experiments on the responses of homing and migrating birds to the sun and stars have demonstrated that when skies are heavily overcast the birds show disorientation. The consequent implication is that birds cannot find their way when celestial cues are obscured. Very recent studies, however, indicate that this may be untrue. *Keeton* has shown that homing pigeons, while using the sun as a compass when available, can navigate accurately under total overcast and even without familiar landmarks. Radar studies reveal that birds migrate successfully when there is a heavy overcast. If the cloud layer is too high to surmount, birds will fly under it or even in it. This being the case, how do birds orient themselves without celestial cues and, if they are flying at night or in a cloud layer, without being able to see landmarks or topographical features familiar to them? The cue may be wind direction.

Nisbet once theorized that the birds might use wind—more specifically atmospheric or wind turbulence—in orientation. In central Illinois, *Bellrose, with* his associates, has gathered ample experimental evidence to show that Blue-winged Teal (*Anas discors*), a long-distance migrant that commonly winters in northwestern South America, does indeed use the turbulent structure of the wind for direction. The species migrates by day or night, and radar surveillance and visual sightings by Bellrose confirm its ability to migrate under overcast skies. In the fall and early winter. Bellrose used both hand-raised immature teal that he released at various points up to 100 miles distant from their home pens and wild immature teal, trapped during migration, that he held in pens and released from 10 to 40 days after other Blue-winged Teal had left the region. All his experimental birds were banded. The hand-reared birds had never flown over the area and none of the birds had ever viewed the landscape along the standard migration routes southward. When released under both clear and overcast skies, nearly all the birds tended to start in the same direction—with the wind. Analyzing his data from visual sightings and banding recoveries and drawing upon his knowledge of waterfowl movements form radar studies and airplane tracking, Bellrose concluded that Blue-winged Teal, and probably other species of waterfowl as well, prefer landscape features for orientation, but, if unavailable, they resort to celestial cues. Both being unavailable, they use wind direction, perhaps referring at the outset of flight to landscape cues for information on the sector of the compass from which the wind is blowing.

The present knowledge of direction-finding, the highlights of which are briefly reviewed in the foregoing paragraphs, points all too clearly tot he danger of generalizing on the means of orientation and navigation in migration. Different groups of birds in their adaptations to different modes of existence may well have developed correspondingly different means of finding their way from one place to another in accordance with the prevailing ecological conditions. It is unlikely that any species orients itself entirely by one cue. A one-directional orientation may suffice for some birds while a bicoordinate system of navigation may be necessary for others. Direction-finding, thought appearing to be basically inherent, may actually prove through eventual research to be acquired chiefly by experience. In all probability, direction-finding in strongly migratory birds is a highly complex procedure involving physiological and behavioural responses to not one but several environmental factors—celestial bodies, wind, and perhaps others presently considered as having no relationship to the problem.

Study of Nocturnal Migration

There are at least five field methods of studying nocturnal migration. (1) By observations of movements across the face of the moon. The principal equipment needed is a telescope with a power of 15 or greater. For methods and procedures, consult *Lowery* and *Nisbet* and for the application of accumulated data. (2) By radio-tracking. The equipment needed includes a transmitter and receiver. For general information on the type of equipment required, consult *Sclater, Cochran* and *Lord* and *Southern*. For information on a transmitter adapted to small migrants, as well as an example of methods and procedures in radio-tracking small migrants. (3) By radar surveillance. This requires the use of a special facility operated by skilled technicians. (4) By tape-recording. Besides a tape recorder, this requires a parabolic reflector, microphone, and amplifier. (5) By the study of migrants killed a television towers or migrants killed as a result of airport ceilometers or other man-made interferences with migration.

Of the five listed methods, the most practical for the beginning student is the study of migrants killed at television towers. It requires no special equipment. Practically every city or large community has no its outskirts one or more television towers that almost certainly take a toll of nocturnal migrants. Much as these kills are regretted, they nonetheless provide rewarding material for the study of many aspects of migration locally.

An early-morning search of the ground under a television tower and its guy wires following a night during which migration proceeded at low altitude will usually yield birds of many species. If the student chooses to make such a search as often as these accidents occur through the migration season and on successive seasons, keeping a careful record with dates of the number of individuals of each species killed, he will obtain a true sampling of the nocturnal migratory activity in the area of the tower. He may also discover the presence of species not heretofore reported, or even suspected, in the area because they have always passed through at night unseen.

Given the time necessary, the student may collect specimens for making one or more special studies such as the following:

1. Succession of sexes and age groups in different species: whether males precede females and adults precede immatures, or vice versa.
2. Geographic variation in a wide-ranging species migrating through a particular area : whether the colour and measurements of individuals reveal one or more populations or sub-species.
3. Molt with relation to migration : whether certain species are in stages of molt and, if so, which feather tracts are involved. Normally birds do not molt the remiges and their primary coverts during migration, but the rest of the alar tract and all the other tracts may be in the process of molt, depending on the species or the sex and/or stage of maturity of individuals in each species.
4. Weight with relation to the length of night-flight : whether the birds are quite heavy, having been killed soon after take-off, or quite light, having been killed after a long flight.
5. Fat condition with relation to stage of migration and duration of night-flight : whether the birds are quite obese, indicating that their migration was well under way with long night-flights, or quite lean, indicating that their migration was just beginning with short night-flights.

Before collecting any specimens, the student must obtain both a federal and a state (or provincial) collecting permit authorizing him to take dead birds "for salvage."

Collect the specimens after a kill as early in the morning as possible since the carcasses not only decompose rapidly but are soon molested by insects or eaten by house cata, crows, gulls, and other creatures that inevitably find the ground around a television tower a promising source of food. Weigh each specimen without too much delay as the body begins losing weight shortly after death. Mark the weight on a tag and attach it to a leg. Seal all specimens collected in plastic bags to avoid their dehydration, mark with the date, time of day, and place of collection, and put at once in a deep-freeze for study and analysis later.

STUDY OF DIURNAL MIGRATION

Whenever opportunity permits, the student should watch the diurnal movements of migrants in his own area. He will soon discover how greatly birds are influenced in their daytime movements by the terrain, vegetation, and waterways. He is almost certain to note that passerine migrants, although continually searching for food, follow much the same courses and in the same direction. If the birds inhabit trees or shrubs, they choose paths in which these grow and avoid, when possible, crossing wide stretches of water and open country. For an idea of how much detailed information one can obtain in a local study and some of the procedures to follow, the student is referred to the classic work by *Ball*, in which he reported on six fall migrations on a small point of land of the great Gaspe peninsula, Quebec.

19

CIRCADIAN RHYTHMS

Wallace, Craig (1918) *opened his classic paper* on appetites and aversions by stating that "the overt behaviour of adult animals occurs largely in rather definite chains and cycles......" Rhythmicity of behaviour is the dominant theme of his discussion. He culminates with the conclusion that "the active behaviour of the human being is, like that of the bird, a vast system of cycles and epicycles, the longest extending through life, the shortest ones being measured in second."

The problems of explaining the mechanisms underlying such behavioural cycles are essentially the same weather the period is a year or a fraction of a second. Annual reproductive cycles are considered by some to be 'exogenous,' times by external environmental stimuli. Others feel they are 'endogenous,' originating within the animal and thus being to some degree spontaneous (*Marshall* 1960 and *Wolfson* 1960). There are activity cycles associated with the phases of the moon and the tides (*Naylor* 1958, *Brown,* 1959, *Fingerman* 1960, *Hauenschild* 1960, *Enright* 1963, b, *Chandrashekaran* 1965). The same questions have been asked about the timing of these rhythms. More progress has been made in the analysis of daily rhythms, and this chapter will shall concentrate on them.

External Stimuli and Cyclic Behaviour

The role of external stimuli in controlling rhythmical behaviour patterns can be explored in several different ways. Does the cyclic patterning of the behaviour depend on rhythmical external stimuli? Do steady external stimulus conditions after the properties of the rhythm? What affects to rhythmical external stimuli have when they are presented and how does responsiveness to these external stimuli vary in different phases of the cycle? Do external stimuli partially or completely control the characteristics of an activity sequence—do they time its onset, continuation, and termination and do they underline differences between cycles? Are deviations from the average pattern such as bouts of activity that are longer or shorter than usual, accomplanied by changes in respectiveness? We shall try to answer these questions as they relate to activity rhythms associated with the daily light-dark cycle.

Circadians Rhythms

The behaviour of most animal is some way reflects the cycle of day and night. This influence can be seen in gross locomotor activity, in the patterns of feeding and drinking, in vocalizations and in other recurring activities. It may also affect events which occur only in lifetime, such as the emergence of an imaginal insect from the pupal

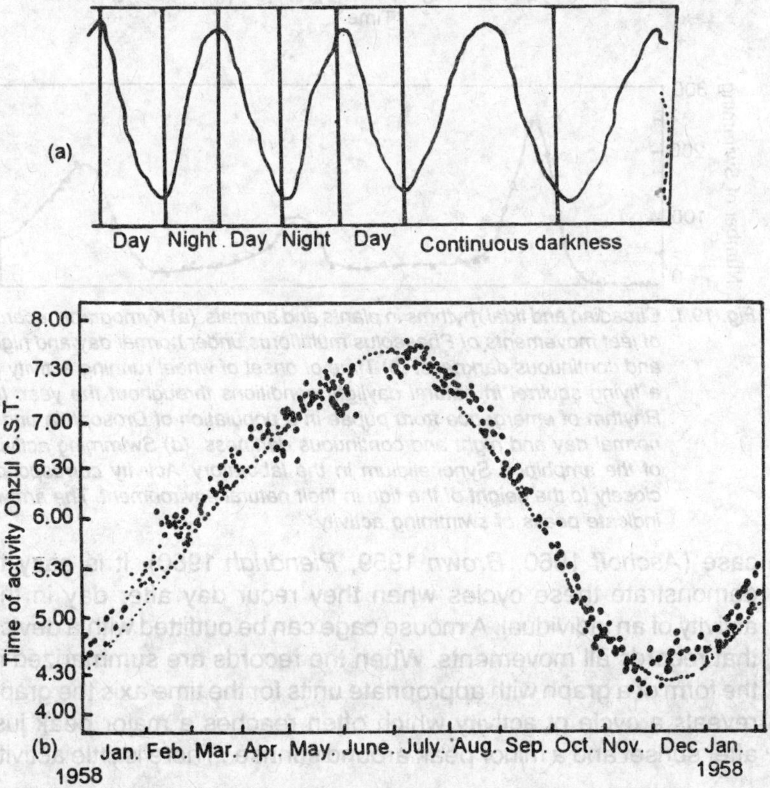

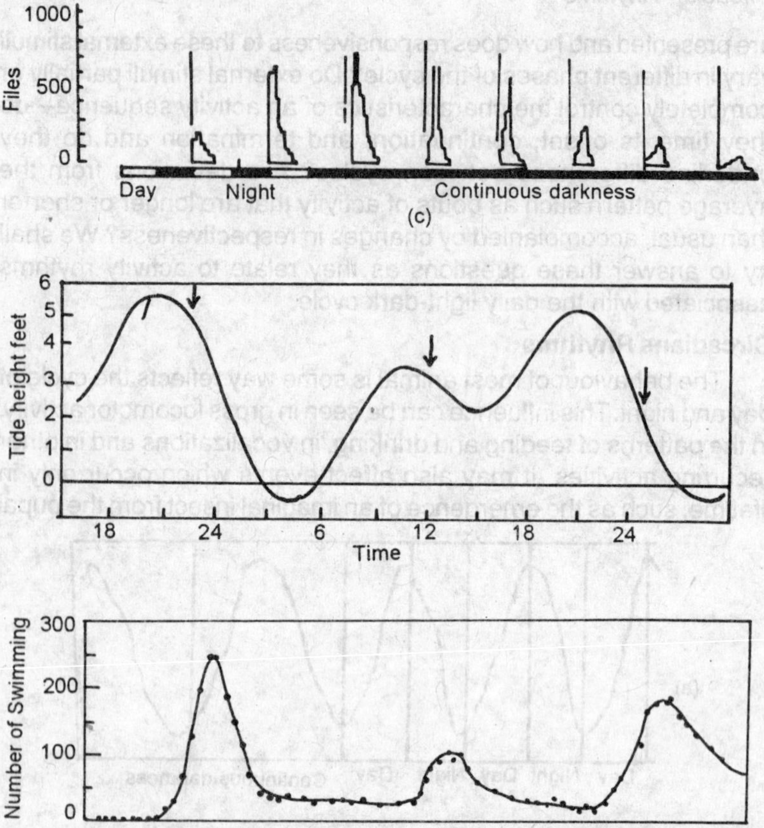

Fig. 19.1. Circadian and tidal rhythms in plants and animals. (a) Kymograph records of leaf movements of Phaseolus multiflorus under normal day and night and continuous darkness. (b) Time of onset of wheel running activity of a flying squirrel in natural daylight conditions throughout the year. (c) Rhythm of emergence from pupae in a population of Drosophila under normal day and night and continuous darkness, (d) Swimming activity of the amphipod Synchelidium in the laboratory. Activity corresponds closely to the height of the tide in their natural environment. The arrows indicate peaks of swimming activity.

case (*Aschoff* 1960, *Brown* 1959, *Piendrigh* 1960). It is easy to demonstrate these cycles when they recur day after day in the activity of an individual. A mouse cage can be outfitted with a device that records all movements. When the records are summarized in the form of a graph with appropriate units for the time axis the graph reveals a cycle of activity which often reaches a major peak just after sunset and a minor peak around sunrise.. There is little activity

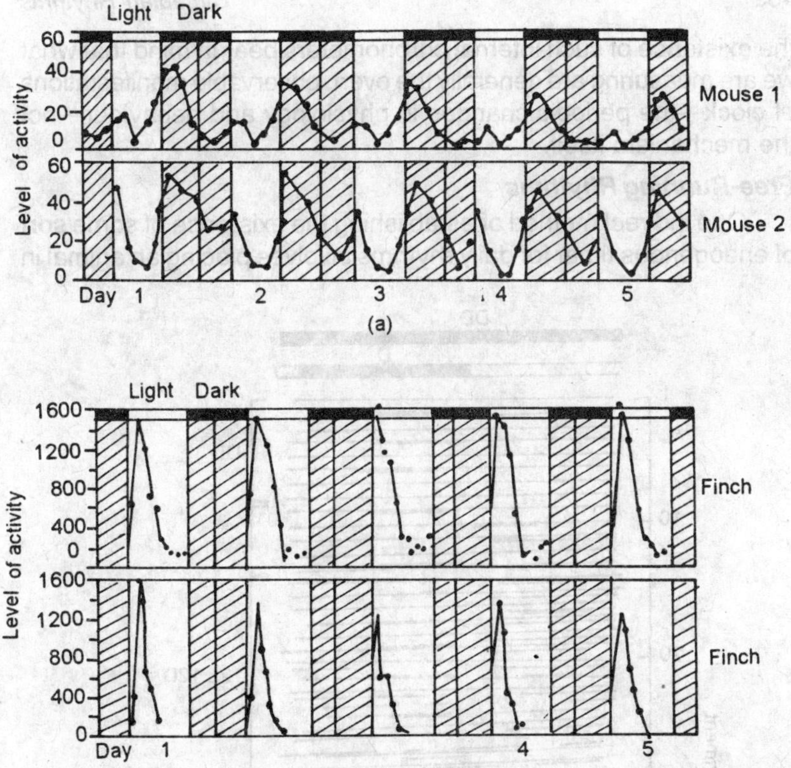

Fig. 19.2. Periodic activity of two mice, Mus musculus (a) and chaffinches, Fringilla coelebs (b) on a 24-hour light-dark schedule in the laboratory.

in the middle of the day. A diurnal animal such as a finch shows a similar pattern, but maximal activity occurs during the day. For the moment let us ignore the detailed internal structure of such a cycle and concentrate on its overall properties. These activity patterns are perfect examples of rhythmical behaviour, recurring regularly with sessions of activity separated by intervals of reduced activity or of no activity at all. What has been revealed by analysis of such daily or circadian rhythms (*circas* about *diem*-day, so called because these circles approximate 24 hours under constant conditions)?

Endogenous Pacemaker

One of the critical characteristics of biological rhythms is the existence of an internal self-sustaining pacemaker. What is the evidence for the existence of such endogenous clock mechanisms? As we examine the following sequence of information in support of the existence of such internal chronometers bear in mind that what we are measuring are generally the overt, observable manifestations of clock—the periodic changes in physiology and behaviour—not the mechanism itself.

Free-Running Rhythms

One indirect method of establishing the existence of some sort of endogenous timer for daily rhythms involves placing an animal in

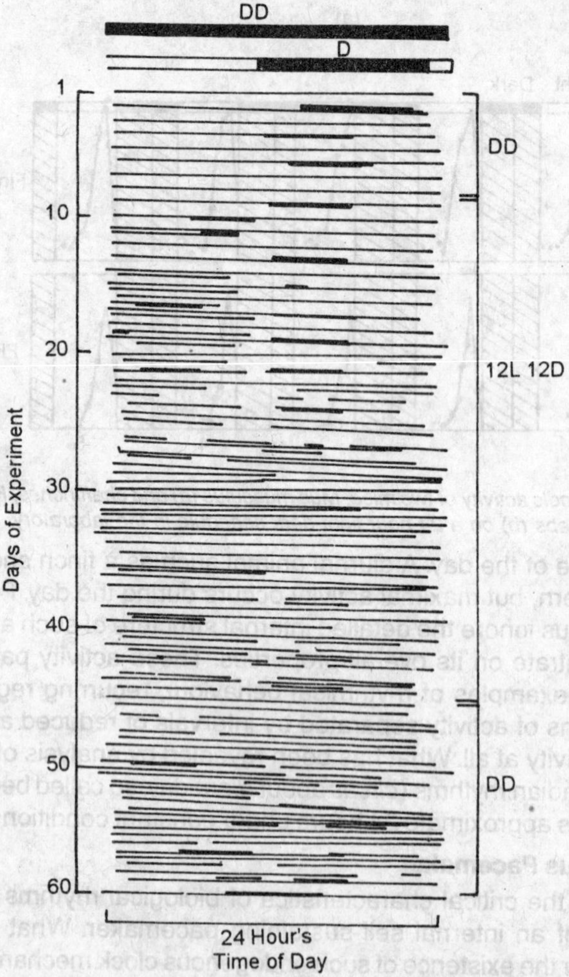

Fig. 19.3. Free running period and entertainment.

constant environmental conditions, for example, constant darkness. When this is done many organisms exhibit activity rhythms, called free-running rhythms, with a period different from that of any known cyclic environmental variable. The pattern of activity of a flying squirrel (*Glaucomys volans*) housed in constant darkness for several weeks (*DeCoursey* 1960, 1961). Since the animal's activity has a clearly rhythmic pattern in the absence of any obvious cyclic cue, there is evidence for some endogenously based system of timekeeping.

For most species studied, the free running periods follow that has become known as *Aschoff's Rule* (*Aschoff's* 1960, 1979). When animals are kept in constant darkness their activity rhythm continues with a period of nearly twenty-four hours, but it *drifts* slightly, becoming somewhat shorter (as in the flying squirrel) or somewhat longer each day. Aschoff's rule states that the direction of this drift and the rate of drift away from the twenty-four hour period are a function of the light intensity and of whether the animal is normally diurnal or nocturnal. For nocturnal animals like the flying squirrel, housing under constant dark conditions results in free running rhythm with a period of shorter than twenty-four hours; the activity begins slightly earlier each day. Conversely, for a diurnal animal housed in the dark, the free running period is slightly longer each day : the activity begins slightly later each day. There are some exceptions to this rule, but in general the data show a variety of animal species conform to the pattern.

Isolation

Some experiments have clearly ruled out "learning" or other similar influences as a mechanism for biological rhythms. Birds and some reptiles that hatch from eggs can be kept under constant conditions in an incubator from a time prior to hatching until after hatching. If the newly hatched organisms exhibit circadian rhythms, a major component of biological rhythms would appear to be inherited and endogenous control hypothesis. Hoffman maintained lizard eggs under one of three conditions, eighteen-hour days consisting of nine hours of light and nine hours of dark; twenty-four-hour days, with twelve hours of light and twelve hours of dark; and thirty-six hour days with eighteen hours of light and eighteen hours of dark. Animals from all three groups, hatched and maintained under constant conditions, exhibited free-running activity periods of 23.4 to 23.9 hours. We can therefore conclude that a component of the biological clock mechanism in these lizards appears to be inherited, remains unaffected by various rearing regimes, and is thus endogenous.

Translocation

Additional support for the hypothesis of endogenous control of biological rhythms has come from translocation experiments. Honeybees (*Apis* spp.) are known to visit particular feeding sites at the same time each day. This behaviour is functionally significant

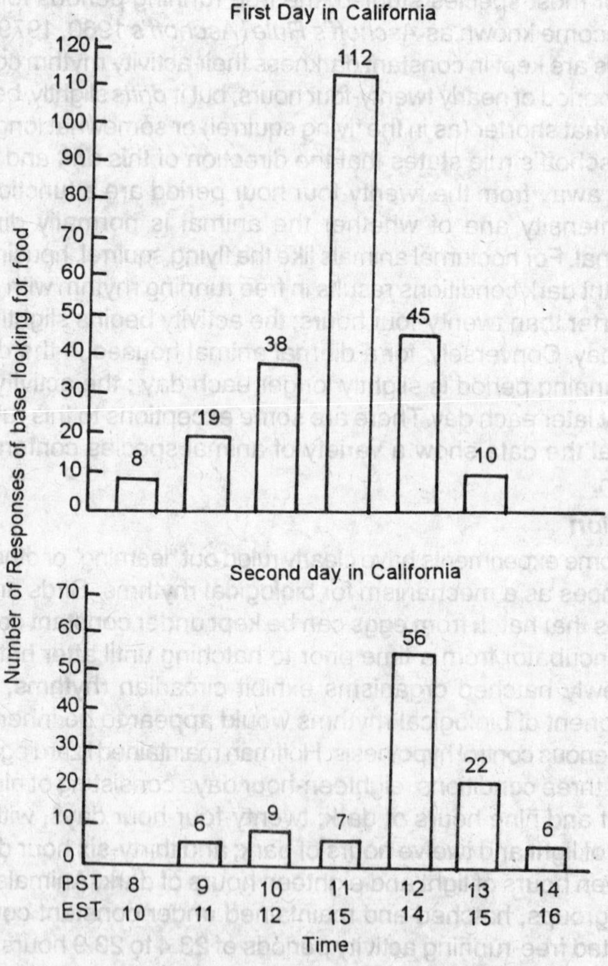

Fig. 19.4. Visiting frequency of bees.

because many flowers that provide bees with food are open only at specific times of day. A German scientist, *Renner* (1957, 1959, 1960) working in a specially designed room with constant conditions, trained bees to leave a hive to forage at a specific time each day. He trained the bees in Paris and then flew them, during the night, to New York, where he placed them in a similar enclosed room with constant conditions. On their first day in New York, the bees began to forage at a time identical to that which would have been expected if they had remained in Paris—that is, twenty-four hours after their last foraging.

In related studies, bees that inhabited outdoor hives were translocated from Long Island to California. Investigators found that, although the bees initially foraged at the same time as at the original site, they gradually adjusted to local time by foraging later and later each day.

A key aspect of current biological clock models illustrated by these experiments is that while the clocks are endogenous, external conditions can cause the biological rhythms to be reset. Similar relocation studies of circadian skin colour changes in fiddler crabs and shell valve-opening time in oysters reveal that these animals, too, react initially as if they were still in the original location, but gradually their clocks become reset to local time. The jet lag experienced by humans making long distance airline flights is another example of this phenomenon.

Variation in Period

A final source of evidence for the endogenicity of biological rhythms comes from the variation that exists in the natural activity periods of most organisms. For example, measurement of the activity periods of two frogs (*Rana pipiens*) under constant conditions reveals circadian rhythms of twenty-three hours, ten minutes and twenty-four hours, thirty-three minutes. These rhythms do not match the period of any known environmental variable. The animals must have internal chronometers that, due to individual differences, lead to deviations from a twenty-four hours rhythms.

Zeitgebers

Many organisms exhibit circadian and circannual periodicities that appear to be closely adjusted to patterns of daylength, temperature fluctuations, or other environmental patterns. The mechanisms whereby the period of a rhythm occurs repetitively and coincides approximately with the presence of some external stimulus is called *entrainment.* As we noted in the periodicity of environmental variables are called *Zeitgebers* ("time givers"). *Zeitgebers* are the entraining agents defined as those cyclic environmental cues that can entrain free-running, endogenous pacemakers. *Zeitgebers* can influence rhythms by effecting both the phase and the frequency.

Returning to the example in figure, when the flying squirrel is provided with a daily cycle of light and dark, its activity rhythm generally conforms to the light cycle—the onset of activity occurs at about the same time each evening. The squirrel is thus to said to be entrained to the daily light-dark cycle. For most endothermic vertebrates thus far tested, the light dark cycle is the critical cue for entrainment. Further, in most terrestrial organisms, the daily light-dark cycle is the *Zertgeber.* This is not surprising, since the daily light cycle is usually the most consistent environmental cue in the terrestrial habitat. For animals in the intertidal zone, the ebb and flow of the tides may be a prime *Zeitgeber.* Ectotherms such as lizards and insects that are unable fully to control their own body temperatures may use temperature or light cues as *Zeitgebers.* When entrained, they exhibit increased activity under warmer conditions and decreased activity during the cooler portions of the daily cycle.

Other environmental cues that may serve to entrain some biological rhythms in some organisms are summarized in *Moore-Ede, Sulzman,* and *Fuller* (1982). Cycles of food availability, but not water availability, can entrain activity rhythms in several species of small mammals (*Richter* 1922; *Moore* 1980). In other mammals, including humans, social cues may be involved in the entrainment of biological rhythms. Two groups of four volunteer subjects each were isolated in separate, but identical chambers (*Vernikos-Danellis* and *Winget* 1979). Each individual maintained an activity rhythm that was synchronous with the others in that room. However, the average free-running periods for the two rooms differed; 24.4 hours in one room and 24.1 hours in the other. To avoid any confounding effect from self-selected light-dark cycles, both groups of subjects were maintained in constant illumination. To further test the possible role of social cues in this synchronization of biological clocks within, but not between, the rooms, one subject was moved from one group to the other. That subject room went through a phase shift and became synchronized with the activity rhythm of the new group. The exact nature of the social cues involved in the entrainment of circadian rhythm in humans is not yet known.

Social cues may also act in correct with other environmental cues to entrain biological rhythms. When individuals who have just flow across six time zones are required to remain in their hotel rooms, they entrain more slowly to the new local time than do individuals who are permitted to leave their rooms and who thus obtain more social-environmental cues from their hew surroundings.

If we bring an animal such as a chipmunk (*Tamias striatus*) into the laboratory and with artificial lighting reverse the light and dark portions of the chipmunk's natural cycle, the animal with shift its activity phase to the light portion of the cycle within a few days. The ability to manipulate the activity phase of animal's daily rhythms has been used by zoological parks to display some nocturnally active animals. By having special quarters that are kept darkened when the zoo is open to visitors and illuminated when the zoo is closed, the public is able to see these animals as active creatures, instead of looking at them only as they sleep. As investigators we can use this type of manipulation to conduct behavioural observations under darkened conditions with nocturnal animals during our own diurnal activity phase.

The entrainment of behavioural and physiological rhythms to the cycles of various environmental cues and the phase shifting which takes place when a key environmental parameter is altered are evidence for the *Zeitgeber* components of the system of biological timekeeping.

Model

Using information thus far obtained on biological rhythms we can attempt to depict what is known in the form of a model. The model system accounts for the two major elements known to be important for biological timekeeping, an endogenous self-sustaining pacemaker and a system for entrainment to environmental *Zeitgebers*. The model utilizes light and a circadian rhythm as an example; however, the same model may be used with other types of rhythms, involving entrainment to tidal, annual or other cues.

Cyclic External Stimuli and Circadian Rhythms

The Problem of Entrainment

Under constant experimental conditions circadian rhythms do not have periods of exactly 24 hours. Usually the period length is consistently longer or shorter so that an animal originally active at night eventually becomes active during the earth's day. When an animal which has drifted out of phase with the earth's day is exposed once more to a rhythm of night and day, it rapidly becomes synchronized again. Its activity cycle becomes locked to the cycle of external conditions and maintains an accurate 24-hour periodicity. The circadian rhythm thus becomes 'entrained' to the 24-hour day.

There are several questions to be asked about this process of entrainment. What are the effective stimuli? If entrainment to a 24-hour cycle is possible, what about an 18- or 36-hour cycle? Does entrainment take place in one jump or a series of changes? If the latter, is the degree of change related to the phase relationship between the new stimulus conditions and the original circadian cycle—in other words is there a changing cycle of responsiveness during the existing rhythm?

"Entrainment of circadian rhythms is defined as the phenomenon whereby a periodic repetition of light and dark (a light cycle) or a periodic temperature cycle, or, more rarely, a periodically repeated stimulus of some other type causes an overt persistent rhythm to become periodic with the same period as the entraining cycle" (*Bruce* 1960). This definition assumes the dominant role of light and temperature as entraining stimuli in experimental situations. Homeotherms appear to be particularly responsive to 24-hour cycle of light and dark. Poikilotherms such as lizards and insects respond to a cycle of temperature change as well as to light (*Harker* 1964). Activity of the beach isopod. *Excirolana chiltoni,* seems to become entrained to the rhythm of the tides as a result of mechanical stimulation by waves on the shore.

What happens if we attempt to entrain the rhythm to a period other than 24 hours? If rarely seems possible to entrain animals to day lengths shorter than about 16 hours (*Harker* 1964). The most thorough study of this question comes from the work of *Tribukait* (1956) who subjected domestic mice to light-dark cycles ranging from a 16-hour day to a 29-hour day, with half of each day dark, the other half light. He found that the mice could be entrained to day lengths of from 21 to 27 hours. Activity rhythms failed to synchronize with days longer or shorter than this, imposed regime, however, was not totally without effect during these very long and very short days. What are normally one rhythm of activity broke into two. The evening activity peak tended to follow the endogenous 24-hour rhythm, while the morning peak followed the timing imposed by the artificial

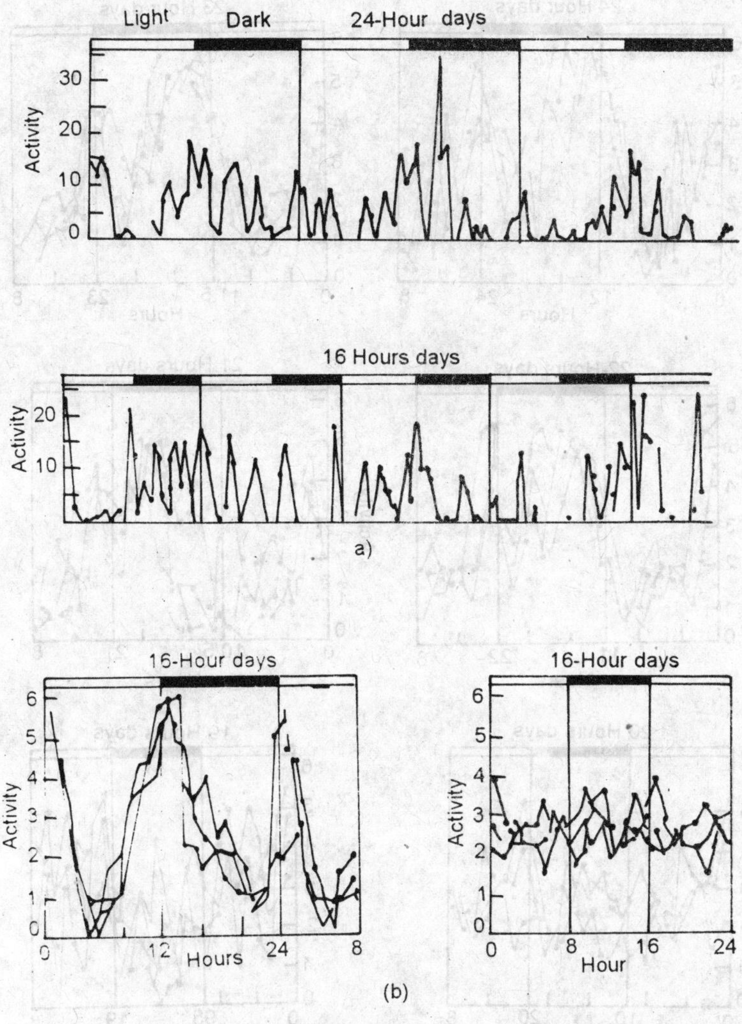

Fig. 19.5. The activity of mouse, Mus musculus, is recorded (a) on a 24-hour day with 12 hours of light and 12 hours of darkness and for a 16-hour day with equal light and dark periods. When these records are superimposed (b) it is clear that there is a rhythm with the 24-hour cycle but that it is lost on the 16-hour schedule.

conditions. Evidently the overall activity rhythm cannot be entrained by a simple one-off lighting schedule if rhythms deviate from 24 hours by more than a certain amount. Similar results have been obtained with hamsters (*Bruce* 1960).

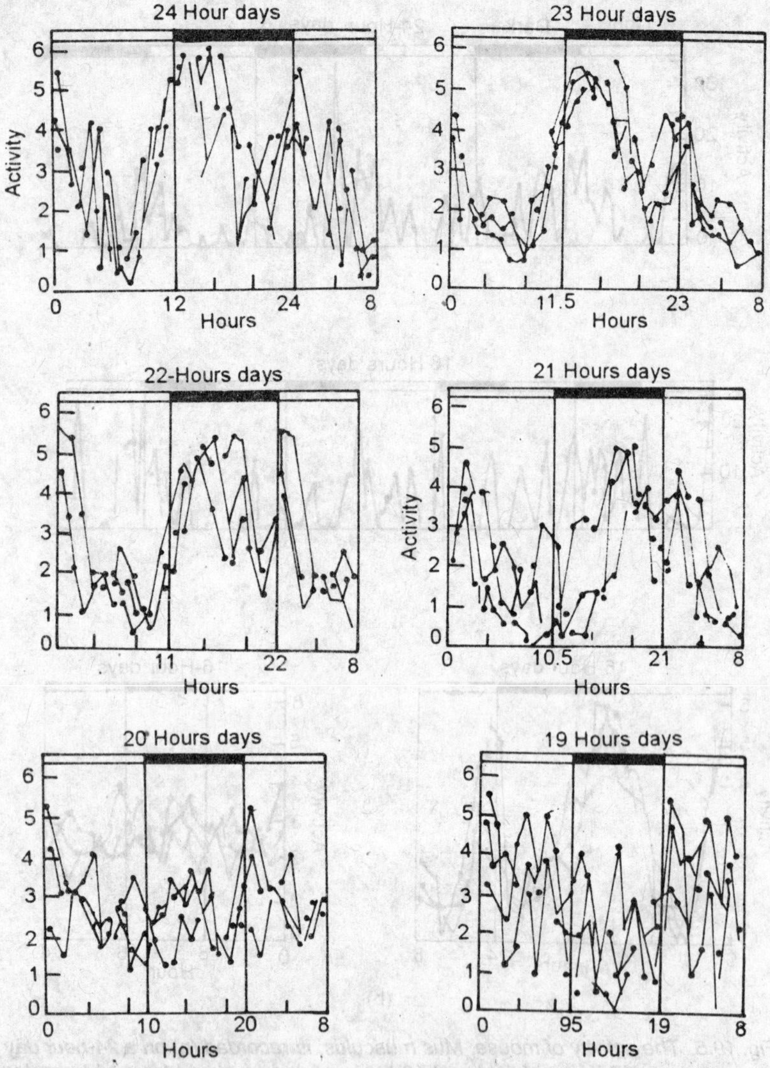

Fig. 19.6. (Continued). Tested with cycles of gradually decreasing length (c) synchrony of the activity rhythm breaks down around 21 hours.

In contrast, *Kavanau* (1962 a, b) has entrained activity, feeding and drinking cycles of deer mice to a 16-hour day (8 light, 8 dark), using artificial twilight transitions rather than simple one-off light control. More systematic analyses of activity cycles of the same species under different day-night regimes are needed, both with and without simulated twilight.

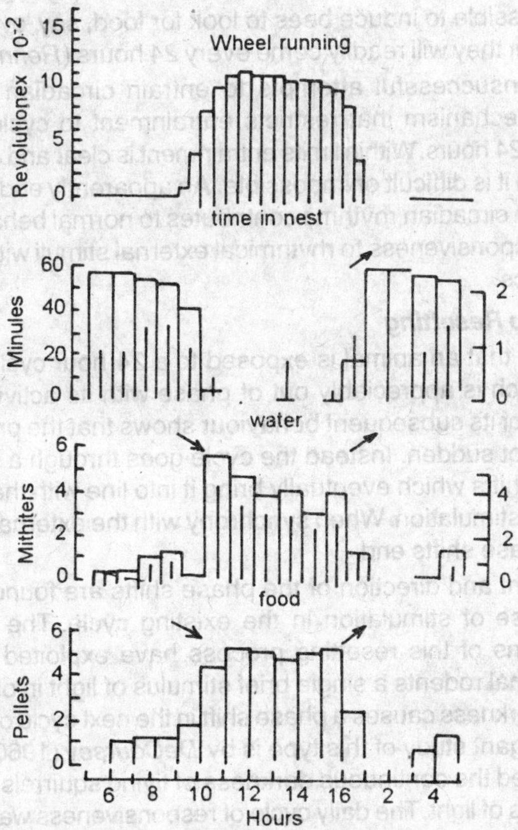

Fig. 19.7. Activity patterns of deer mice, Peromyscum, on a 16-hour artificial day. Artificial twilight transitions are indicated by the slanting arrows.

When an animal is provided with a 24-hour day composed of a series of light-dark cycles, it will often maintain a 24-hour rhythm by responding only to certain cycles. A hamster exposed to a 6-hour day, consisting of 2 hours of light and 4 of dark, extracts a 24-hour rhythm by responding only to every fourth cycle. With a 4-hour day it responds only to every sixth cycle, and so on. Deer mice and cockroaches behave similarly (*Bruce* 1960). This phenomenon, called 'frequency demultiplication', demonstrates again the predisposition to respond to stimulus cycles with a period length of about 24 hours.

If the problem is approached in a different way by trying to train an animal to perform some activity at a particular time of day, essentially the same result is obtained. "Even after weeks of training it was not possible to induce bees to look for food, say, every 19 or 48 hours", but they will readily come every 24 hours (*Renner* 1960).

These unsuccessful attempts to entrain circadian rhythms suggest a mechanism that restricts entrainment to cycles which approximate 24 hours. Within limits entrainment is clear and accurate. Beyond them it is difficult or impossible. An apparently endogenous component in circadian rhythms contributes to normal behaviour by restricting responsiveness to rhythmical external stimuli with certain characteristics.

Sensitivity to Resetting

Suppose that an animal is exposed to a 24-hour cycle of light and dark which is appreciably out of phase with its activity cycle. Observation of its subsequent behaviour shows that the process of resetting is not sudden. Instead the cycle goes through a series of daily phase shifts which eventually bring it into line with the pattern of rhythmical stimulation. When synchrony with the external cycle is achieved, phase shifts end.

The extent and direction of the phase shifts are found to vary with the phase of stimulation in the existing cycle. The clearest demonstrations of this resetting process have exploited the fact that in nocturnal rodents a single brief stimulus of light in otherwise continuous darkness causes a phase shift in the next cycle of activity. The most elegant study of this type is by *DeCoursey* (1960, 1961). She interrupted the continuous darkness of flying squirrels with 10-minute periods of light. The daily cycle of responsiveness was clearly related to the rhythm of locomotor activity. The greatest delay of the next cycle occurred if the light stimulus was presented close to the time of onset of running. The delaying effect declined gradually as the stimulus was presented later in the activity cycle, until several hours after the onset of activity. Beyond this point light stimuli caused as advance of the cycle instead of a delay. Light stimuli had no effect throughout the inactive part of the cycle, until an hour or two before the expected onset of running (*DeCoursey* 1961).

A graph reveals the rhythm of changing responsiveness in the course of the activity cycle. Comparable studies of responsiveness to brief light exposures in other species reveal curves with a rather different shape.

Maximal phase shifts occur when light stimuli are presented near the time of onset of activity, which in nature is at dusk for a flying squirrel. There is reason to consider sunset the most critical

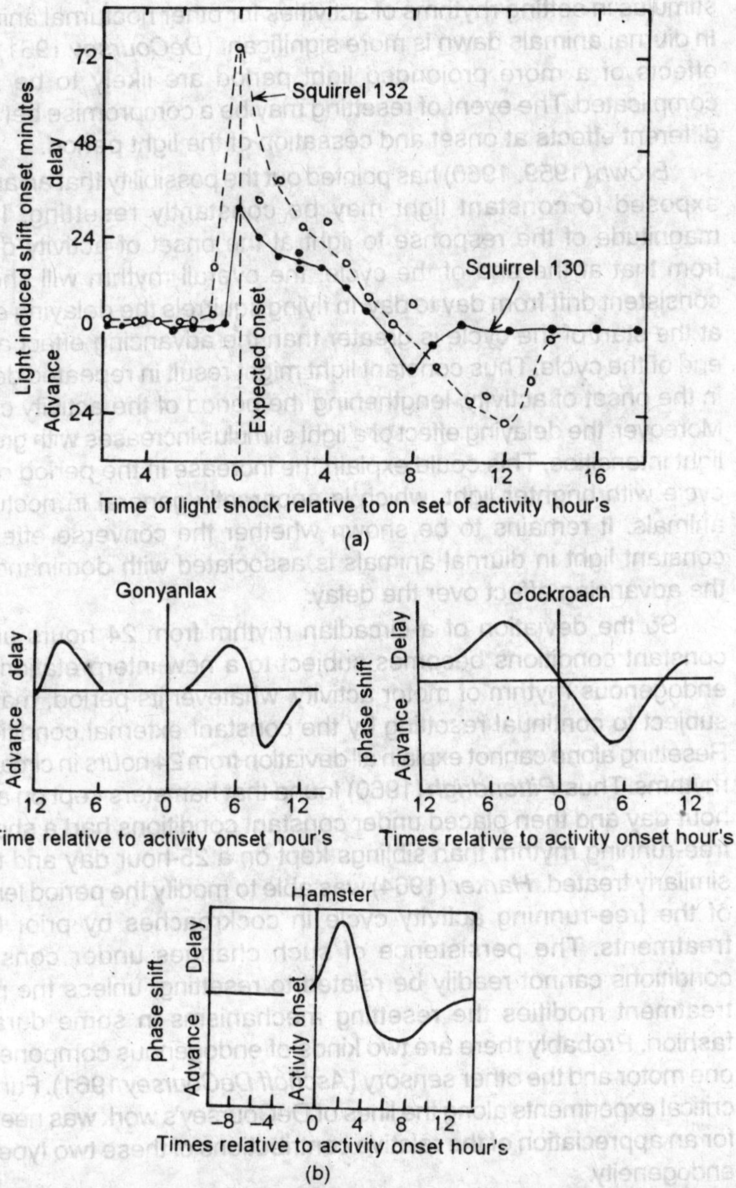

Fig. 19.8. (a) The daily rhythm of resting by 10-minute light exposures in the flying squirrel, Glaucomys. (b) Diagrammatic resetting curves for locomotor activity cycles of a hamster and a cockroach, and for the luminescent activity of Gonyaulax.

stimulus in setting rhythms of activities for other nocturnal animals. In diurnal animals dawn is more significant (*DeCoursey* 1961). The effects of a more prolonged light period are likely to be more complicated. The event of resetting may be a compromise between different effects at onset and cessation of the light period.

Brown (1959, 1960) has pointed out the possibility that an animal exposed to constant light may be constantly resetting. If the magnitude of the response to light at the onset of activity differs from that at the end of the cycle, the overall rhythm will show a consistent drift from day to day. In flying squirrels the delaying effect at the start of the cycle is greater than the advancing effect at the end of the cycle. Thus constant light might result in repeated delays in the onset of activity, lengthening the period of the activity cycle. Moreover, the delaying effect of a light stimulus increases with greater light intensities. This could explain the increase in the period of the cycle with brighter light, which is apparently general in nocturnal animals. It remains to be shown whether the converse effect of constant light in diurnal animals is associated with dominance of the advancing effect over the delay.

So the deviation of a circadian rhythm from 24 hours under constant conditions becomes subject to a new interpretation. An endogenous rhythm of motor activity, whatever its period, may be subject to continual resetting by the constant external conditions. Resetting alone cannot explain all deviation from 24 hours in circadian rhythms. Thus *Pittendrigh* (1960) found that hamsters kept on a 23-hour day and then placed under constant conditions had a shorter free-running rhythm than siblings kept on a 25-hour day and then similarly treated. *Harker* (1964) was able to modify the period length of the free-running activity cycle in cockroaches by prior light treatments. The persistence of such changes under constant conditions cannot readily be related to resetting, unless the prior treatment modifies the resetting mechanisms in some durable fashion. Probably there are two kinds of endogenous components, one motor and the other sensory (*Aschoff DeCoursey* 1961). Further critical experiments along the lines of DeCoursey's work was needed for an appreciation of the relative contributions of these two types of endogeneity.

The Internal Structure of Circadian Activity Rhythms

The manner in which circadian activity rhythms are often presented gives only a gross picture of the actual pattern of activity in time because of the summation of data over long intervals. Presentation on a finer and finer time scale gives a quite different picture (*Aschoff* 1957). There are actually many short bouts of activity in the course of a day. The interval between these bouts varies somewhat. For example, in mice they may last 106 minutes in the nocturnal phase and 170 minutes in the diurnal phase (*Aschoff* and *Meyer-Lohmann* 1954a).

The circadian cycle results from variation in the intervals between bouts, which are shortest around dusk and dawn and longest in the

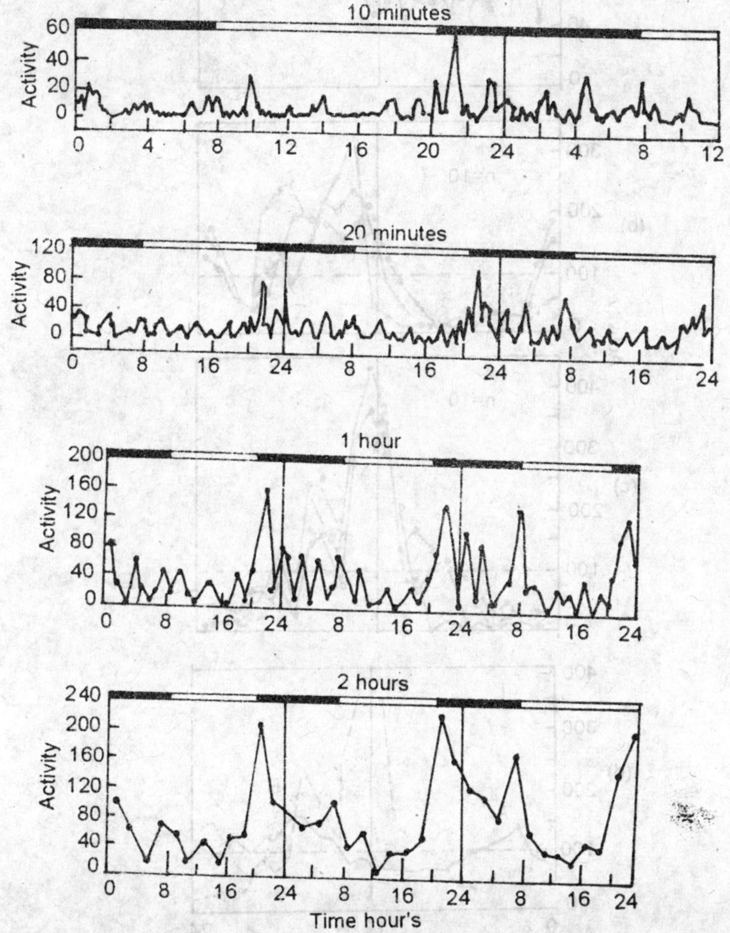

Fig. 19.9. The same activity of a mouse is plotted in increments varying from 10 minutes to 2 hours.

middle of the day. There is in fact another minor cycle much shorter in length imposed on the major one of 24 hours. In rodents, for

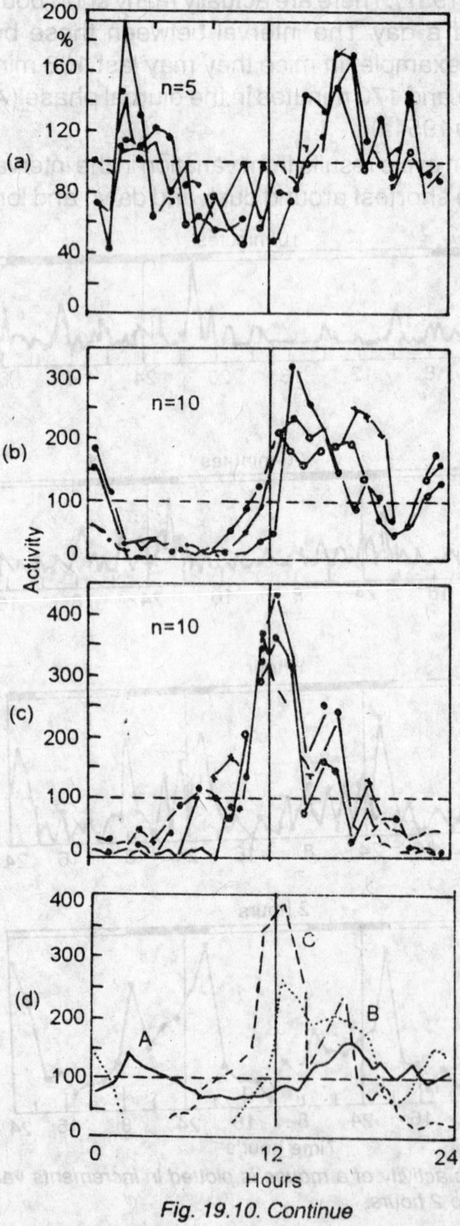

Fig. 19.10. Continue

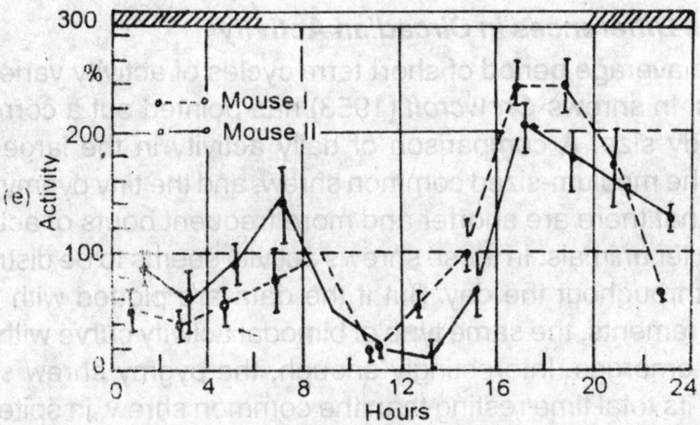

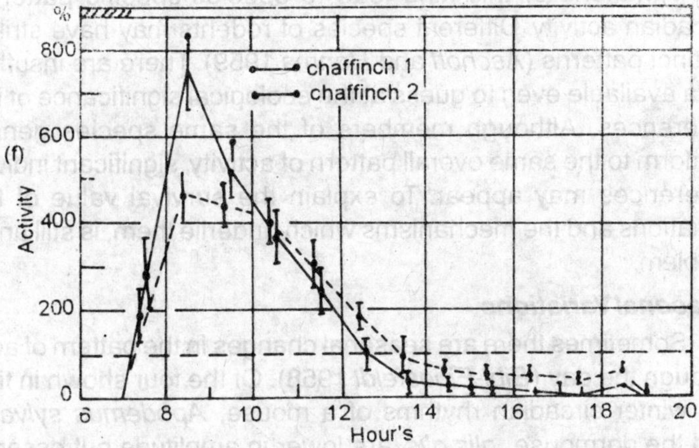

Fig. 19.10. The superimposed activity of various animals (a) during five days for the shrew, Sorex araneus, (b) ten days for the mouse, Mus musculus, and (c) ten days for the hamster, Mesocricetus auratus. Each record is a summary for three individuals. (d) These performances show species differences when summed. The individual difference between the actions, (e) of two mice, Mus musculus, and (f) two chaffinches, Fringilla coelebs.

example, it varies between about 1.5 and 4 hours. There may even be a second rhythm imposed on the first, as indicated in where (a) suggests a 2-hour periodicity, and (c) a periodicity of 4 hours. It is the summation of these sub-cycles which gives the characteristic bimodal shape to the daily activity curve of mice as it is usually presented. For this reason, the short-term patterning of locomotion, feeding, and other activities is worthy of closer attention from students of circadian rhythms.

Species Differences in Circadian Activity

The average period of short term cycles of activity varies from species. In shrews *Crowcroft* (1953) has pointed out a correlation with body size. A comparison of daily activity in the large water shrew, the medium-sized common shrew, and the tiny pygmy shrew shows that there are shorter and more frequent bouts of activity in the smaller animals. In these shrews activity seems to be distributed evenly throughout the day. But if the data are plotted with longer time increments, the same type of bimodal activity curve with occur in mice emerges. Interestingly enough, the pygmy shrew spends more of its total time resting than the common shrew, in spite of the greater frequency of activity bouts.

Variations of this kind lead to species-specific patterns of circadian activity. Different species of rodents may have strikingly distinct patterns (*Aschoff* and *Honma* 1959). There are insufficient data available even to guess at the ecological significance of these differences. Although members of the same species generally conform to the same overall pattern of activity, significant individual differences may appear. To explain the survival value of these variations and the mechanisms which underlie them, is still another problem.

Seasonal Variations

Sometimes there are seasonal changes in the pattern of activity through the day (*Eibl-Eibesfeldt* 1958). Of the four shown in figure, the winter circadian rhythms of a mouse, *Apodemus sylvaticus*, and the dormouse, *glis glis,* are lower in amplitude but essentially similar in form to those seen in summer. The mice, *Clethrionomys glareolus* and *Microtus arvalis,* become largely diurnal in winter (*Ostermann* 1955).

The role of internal and external factors in causing these individual, specific and seasonal differences in circadian rhythms is virtually unexplored. The significance of internal factors is revealed by experiments in which the long nights and short day is of inter have been simulated to compare their effects with those of shorter nights and longer day (*Aschoff* 1958, 1960). Mice are active for longer periods of time on winter nights than on summer nights. But in winter, activity starts later in relation to 'sunset' and ceases earlier in relation to 'sunrise' than in summer suggesting that the pattern throughout the year is to some degree independent of the external stimuli of sunrise and sunset. There is similar seasonal variation of the time of rising of birds (*Dunnet* and *Hinde* 1953). Nevertheless, effects of changes in the ratio of periods of light and dark upon activity should not be underestimated. There is also evidence that the intensity of the light under both constant and alternating conditions may have profound effects upon the intensity of activity and upon the ratio of activity to resting (*Aschoff* 1960). We are still far from any complete explanation of the mechanisms underlying seasonal variations of activity rhythms, however. Nor can we explain the differences between individuals and species and the variations taking place during individual development (*Aschoff* and *Honma* 1959).

Mechanisms Underlying Circadian Rhythms

There is little evidence on the physiological identity of mechanisms controlling circadian rhythms. Nevertheless, there is general consensus about the kind of thing to look for. Many investigators agree that at least two systems which behave differently in some degree must be involved. One is more strictly endogenous, and the other is more responsive to external stimuli. The overall behaviour of the animal is a result of interaction between the two.

Brown and his associates (*Brown* 1959) were led to this conclusion from their studies of colour change in fiddler crabs. Work on the timing of emergence of the adult fruit flies from the pupae led to a similar hypothesis (*Pittendrigh* 1960). If a culture of *Drosophila* eggs was raised in the dark, the adults finally emerged asynchronously. If a light was presented briefly, subsequent emergence became synchronized. It took place at some multiple of 24 hours after light onset, which would have been dawn, the time of emergence in nature. This periodicity in the behaviour of a population, involving a single event in the life of each individual, has many of the properties of a circadian rhythm and has been explored extensively from this point of view by Pittendrigh and his colleagues (*Pittendrigh* 1954, 1958, 1960, *Pittendrigh, Bruce* and *Kaus* 1958, etc.). That temperature changes can lead to a temporary resetting of the time of emergence, only to be overridden by the original cycle in the pupae that are still several days from emergence, again suggests that at lers two systems are involved.

The experiments of *Tribukait* (1956) on the effects of very long and very short days on the activity of mice imply more than one system. The evening activity peak maintains a cycle of roughly 24

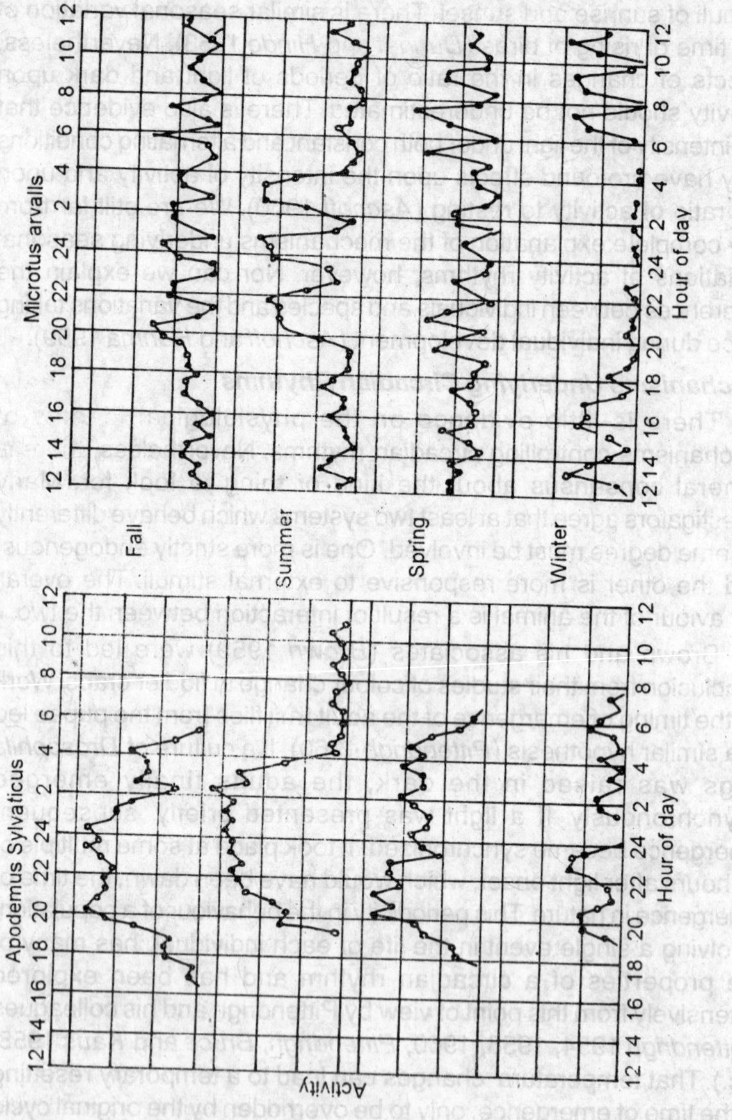

Fig. 19.11. (Continue)

hours which is presumably endogenous, whereas the morning peak keeps time with the artificial cycle. Here two systems are operating on the behaviour of the same animal at different times.

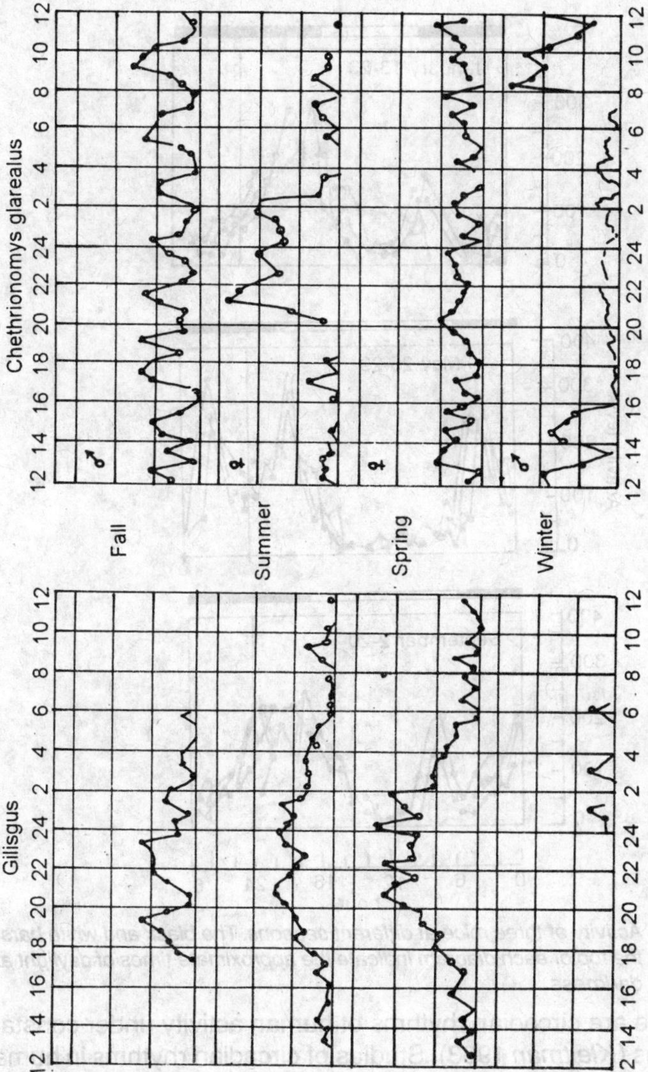

Fig. 19.11. Seasonal variation in daily activity cycles of four species of rodents.

In a general discussion of this problem, *Pittendrigh* (1960) concludes that we must "abandon the common current view that our problem is to isolate and analyze the endogenous rhythm, or the organism comprises a population of quasi-autonomous oscillatory systems."

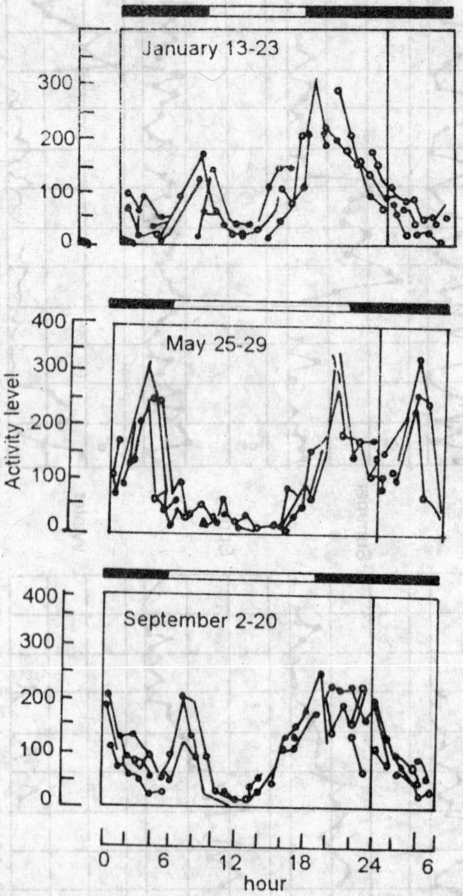

Fig. 19.12. Activity of three mice at different seasons. The black and white bars at the top of each diagram indicate the approximate times of daylight and darkness.

There are circadian rhythms of human activity under constant conditions (*Kleitman* 1963). Studies of circadian rhythms in human physiology suggest a multiplicity of mechanisms. Thus, for a man placed under constant conditions, the periodicities of maximal body temperature and production of water, calcium, and potassium in the urine are not necessarily the same as those of waking and sleeping (*Aschoff* 1965, *Richter* 1965) has reviewed physiological cycles, ranging from minutes to hundreds of days in both normal and diseased human patients. Evidently many cycles can run simultaneously.

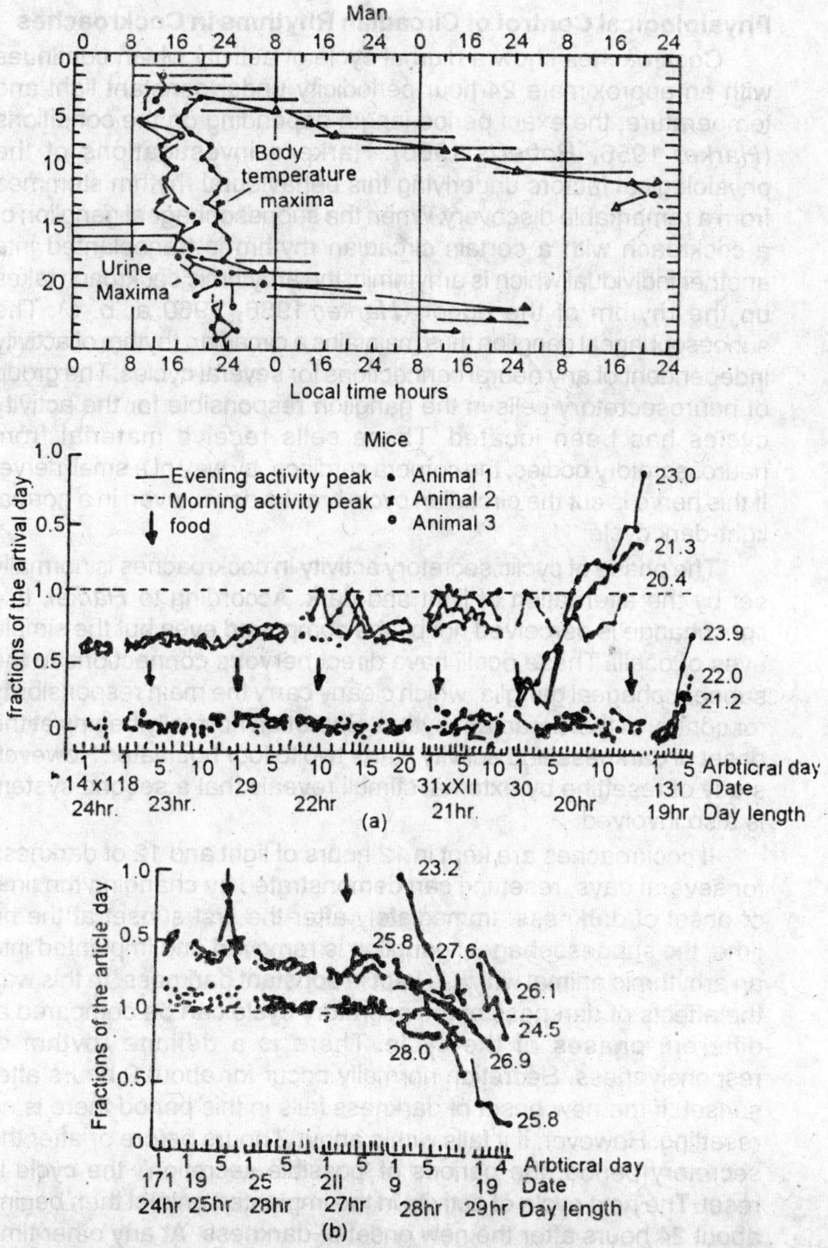

Fig. 19.13. (a) Desynchronization of circadian rhythms in a man living in isolation without time cues. (b) Graphs of the time of morning and evening activity peaks in mice kept on different day lengths.

Physiological Control of Circadian Rhythms in Cockroaches

Cockroaches show a diurnal cycle of activity which continues with an approximate 24-hour periodicity under constant light and temperature, the exact period length depending on the conditions (*Harker* 1956, *Roberts* 1960). Harker's investigations of the physiological factors underlying this behavioural rhythm stemmed from a remarkable discovery. When the suboesophageal ganglion of a cockroach with a certain circadian rhythm is transplanted into another individual which is arhythmic, the arhythmic cockroach takes up the rhythm of the donor (*Harker* 1956, 1960 a, b, c). The suboesophageal ganglion thus maintains a circadian rhythm of activity independent of any neural connections for several cycles. The group of neurosecretory cells in the ganglion responsible for the activity cycles has been located. These cells receive material from neurosecretory bodies, the corpora cardiaca, by way of a small nerve. If this nerve is cut the circadian cycle breaks down, even in a normal light-dark cycle.

The phase of cyclic secretory activity in cockroaches is normally set by the alternation of light and dark. According to *Harker,* this light change is perceived not by the compound eyes but the simple eyes of ocelli. These ocelli have direct nervous connections to the suboesophageal ganglia, which clearly carry the main responsibility for control of the circadian rhythm. Secretion normally begins at the onset of darkness and activity starts two to four hour later. However, study of resetting by external stimuli reveals that a second system is also involved.

If cockroaches are kept in 12 hours of light and 12 of darkness for several days, resetting can demonstrated by changing the time of onset of darkness. Immediately after the first sunset at the ne time, the suboesophageal ganglion is removed and implanted into an arhythmic animal which is kept in constant darkness. In this way the effects of darkness of the secretory cycle can be compared at different phases of the cycle. There is a definite rhythm of responsiveness. Secretion normally occur for about 2 hours after sunset. If the new onset of darkness falls in this period there is no resetting. However, if it falls within about 2 hours before or after the secretory period, the periods of 'possible secretion', the cycle is reset. The next cycle of activity in the implanted animal then begins about 24 hours after the new onset of darkness. At any other time during the period of 'impossible secretion', the onset of darkness has no effect on the secretory rhythm.

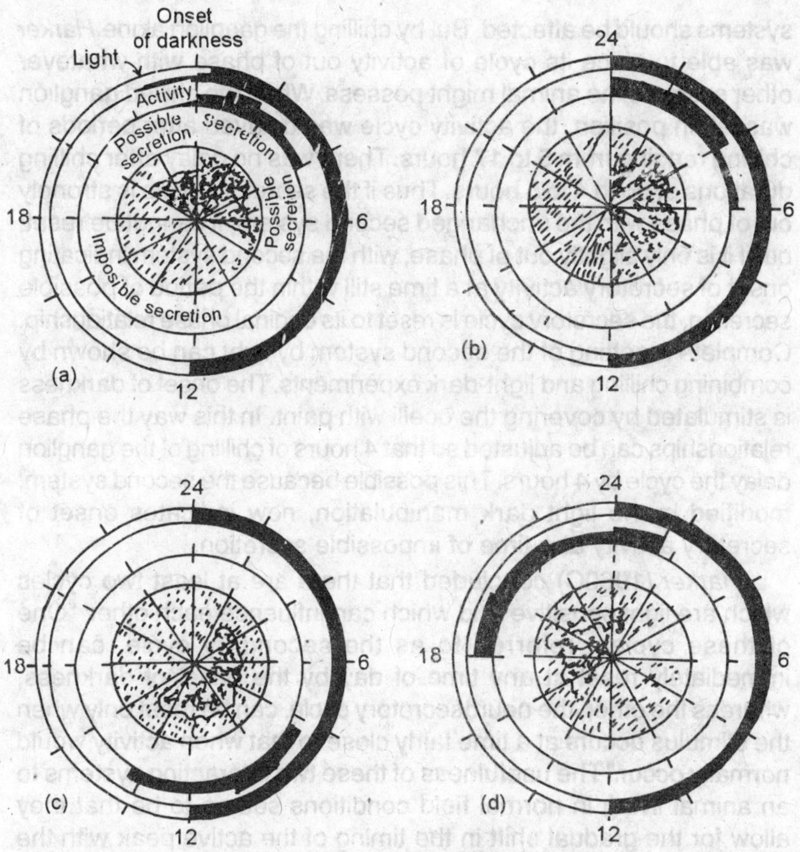

Fig. 19.14. Locomotor activity and the neurosecretory cycle of the cockroach (a) under artificial lighting of a 24-hour light-dark cycle, (b) with chilling of the suboesophageal ganglion for 4 hours, (c) 8 hours, and (d) 18 hours.

If the same experiment is done without removal of the ganglia, to observe resetting in the intact animal, there is a striking difference. Onset of darkness in what *Harker* called the period of impossible secretion now causes resetting. This outcome suggests the existence of a second mechanism, also responsive to light-dark changes, that can influence the neurosecretory cycle of the suboesophageal ganglion. *Harker* was able to confirm the existence of this second system by localized chilling of the ganglion *in situ.* The circadian cycle of cockroaches, as of other insects, can be retarded by chilling at a temperature jut above zero. If the whole animal is chilled, at systems should be affected. But by chilling the ganglion alone *Harker* was able to place its cycle of activity out of phase with whatever other systems the animal might possess. When the chilled ganglion was left in position, the activity cycle was delayed after periods of chilling ranging from 5 to 17 hours. There was no delay after chilling durations of up to 4 to 5 hours. Thus if the secretory cycle is strongly out of phase with the unchanged second system, it cannot be reset; but if it is only slightly out of phase, with the second system indicating onset of secretory activity at a time still within the period of possible secretion, the secretory cycle is reset to its original phase relationship. Complete resetting of the second system by light can be shown by combining chilling and light-dark experiments. The onset of darkness is stimulated by covering the ocelli with paint. In this way the phase relationships can be adjusted so that 4 hours of chilling of the ganglion delay the cycle by 4 hours. This possible because the second system, modified by he light dark manipulation, now indicates onset of secretory activity at a time of impossible secretion.

Harker (1960C) concluded that there are at least two cycles which are light sensitive and which can influence each other. "One of these cycles, referred to as the secondary cycle, can be immediately reset at any time of day by the onset of darkness, whereas the other, the neurosecretory cycle, can be reset only when the stimulus occurs at a time fairly close to that when activity would normally occur." The usefulness of these two interacting systems to an animal living in normal field conditions seems to be that they allow for the gradual shift in the timing of the active peak with the change in day length but prevent the resetting of phase by incidental changes of light intensity at abnormal times. A similar explanation applies to the cycles of responsiveness considered earlier, but only in the cockroach has such progress been made in locating the structures involved.

20

BEHAVIOURAL DEVELOPMENT OF ALTRICIAL MAMMALS

Altricial young provide a good opportunity to study early stages of behavioural development among mammals that might be more difficult to analyse in the seemingly more complex precocial young that possess advanced sensory and motor capacities from birth. Among altricial young, there are relatively distinct periods of thermotactile, olfactory and visual control of behavioural responses and underlying motivational processes, and the relatively simple forms of motor activity can be studied as they become organized into more complex action patterns. While it is true that limited motor abilities may hide advanced sensory and integrative processes in altricial young, it is equally true that advanced sensory and motor abilities may mask simple behavioural capacities in precocial young. The problem is, therefore, to determine, through analytical experimental procedures and theoretical synthesis of the findings, the nature of behavioural organization at each stage of development.

We are not yet prepared to develop detailed models of the behavioural organization of young at the various stages but we can sketch the major lines along which development proceeds. Among altricial young these lines are clear from many good descriptions of the development of behaviour toward the mother, siblings and nest or home site in a variety of mammalian species.

Suckling and non-suckling contact with the mother, and huddling with siblings in the nest, take an early lead in behavioural development among altricial young. At first the mother takes the initiative in suckling, but gradually the young contribute increasingly to the initiation and termination of feeding sessions, until the relative contribution of each shifts and the young play the dominant role in the feeding interaction until weaning intervenes. Similarly, huddling among siblings is a relatively simple behaviour pattern while the young are still all confined to the nest. As young become able to move about more freely, behavioural interactions among siblings become more complex: young become capable of many more responses to one another and respond increasingly to individuals and their characteristics rather than to group characteristics.

Initially the young are either passively confined to the home area or are actively oriented to it but as development proceeds they become able to move more freely in relation to it at a distance. This early form of orientation to the home area wanes after a period and the young's relationship to its companions and to various specific functions that emerge as development proceeds.

Development of Suckling

Among all species of altricial mammals suckling is initiated shortly after birth. The newborn actually begin to nuzzle the mother's nipple region during parturition, when given the opportunity in the intervals between births. Kittens, however, are only rarely successful in attaching to nipples and this is very likely the case in other altricial species. Ewer's (1961) suggestion that the delay of an hour or so *post partum* in the onset of suckling in kittens is based upon the absence of a special releasing stimulus for nipple grasping seems unnecessarily elaborate. Kittens have little opportunity to attach to nipples during parturition even when the mother is not actively giving birth. The mother is engaged in a variety of activities such as licking the kittens, eating the placentae, licking birth fluids from her fur and the delivery site, etc., which make it difficult for kittens to attach to nipple.

Not until parturition is completed does the female make herself available for feeding. The position assumed after parturition by the exhausted mother among cats, dogs and other small mammals facilities nipple searching and grasping by the newborn: she lies on her side, her nipple region facing toward the newborn, her limbs outstretched forming an oval-shaped corral in which the young are confined. Comparable positions are assumed by rat, hamster, and rabbit mothers. Most successful artificial mothers have been modelled after the immediately post-partum nursing female.

In this position the mother presents a variety of attractive thermal, tactile and olfactory stimuli. The effects of these stimuli on early suckling have been investigated through the presentation of one or several of the component stimuli under controlled conditions in artificial brooders.

Among the important features of the mother's body that stimulate approach and nuzzling appear to be the fur-textured surface, and the rounded shape with few entrapping crevices or projections that might distract newborn from locating the nipples and surrounding areola. Body surface temperature in the range of 32 to 39°C, as measured in various species, and as provided in artificial mothers, has proved attractive to newborn rats, rabbits, puppies, and kittens. The moist surface of a pulsating plastic tube was most attractive to brooder-reared rat pups.

In addition to being attracted to the mother's nipple region by these stimuli, newborn kittens are led along certain paths in their nuzzling of her body by patterns of thermal and tactile stimuli. N. Freeman found a broad but well-defined thermal gradient extending over the entire body surface of the lactating cat that could provide a basis for locating the nipples in newborn kittens. When on a floor surface at 30.2°C a thermal gradient extended from the mother's limbs, which had a surface temperature of 33.3°C, to her nipples which were at 37°C.

The pattern of fur growth offers additional directional stimulation: when the mother is lying on her side, crawling against the grain leads kittens upward from the midline to the lateral nipples. Moreover, the areola surrounding the nipples, with its bare, warm moist surface, attracts kittens. It has been suggested that the pattern of hair growth on the belly of the pig provides paths which piglets follow in finding nipples. Static features of the mother and more active ones, such as her licking that steers kittens to her ventrum, offer attractive stimuli which initially bring newborn into contact with her. More detailed and patterned features lead them to the nipples.

Suckling and Thermotactile Stimuli

Thermal and tactile stimuli provide the basis for the earliest suckling approaches to the mother in kittens, rat pups, and newly born rabbits and puppies. In the kitten, the snout and lips are supplied with low threshold thermotactile receptors and greater anterior-end sensitivity to such stimuli appears to be the general rule among newly born altricial young. The snout functions as a probe as it is moved by side-to-side head movements during crawling, a characteristic of kittens and other altricial young that move in this manner. These movements enable the newborn effectively to scan a fan-shaped region in front of it. Contact with a low-intensity tactile or thermal stimulus slows the return head swing and the newborn veers in the direction of the stimulus by repeated movements of the contralateral forelimb and cessation of ipsilateral forelimb movement.

On the other hand, contact with a strong tactile stimulus e.g., wire textured floor surface, excercise warm or cool stimulus. Interests head withdrawal followed by cessanon of forward movement and veering away. If the stimulus persists, the newborn pivots a half circle and crawls away from it. This withdrawal response, which can also be elicited by a puff of air directed at the newborn's face, has been used to establish early escape and avoidance learning in newborn of several altricial species. Okon has shown thermal and tactile stimuli with letter situation also stimulate new born.

Among puppies and kittens loss of contact with tactile stimulation from the mother and littermates stimulates an increase in activity that usually consists of pivoting and calling. In puppies, quiescence is restored when head and body contact is made with a soft tactile stimulus and a most effective stimulus for quieting kittens that have lost contact with the mother and litter is the weak tactile stimulation of the face and forehead provided by a canopy of soft bunting.

Cooling is as effective as tactile stimulation in arousing puppies and kittens to activity and in stimulating calling. In an ambient temperature of 30°C puppies remain quiescent, breathing lightly in a relaxed sleeping posture, but if the ambient temperature falls to 27°C or rises to 32.5°C, they become alter and restless, begin to call, and exhibit rapid breathing. Kittens placed on a cool surface (18°C) become active and call during the first week. Many crawl to a warmer area if this is available, but others become so active they simply walk off the table surface.

Newborns that lose contact with the mother, therefore, become active as a result of both the loss of tactile stimulation and exposure to a lowered ambient temperature. They begin to pivot and circle, emitting calls to which the mother often responds by bringing her body closer and into contact with the young or by licking them. The newborn is likely to make contact with the floor surface around the mother which is warmed by her body and with her extended fore- and hindlimbs as well as the warm furry surface of her nipple region.

These ensure that the newborn will regain contact and warmth and if it has not fed for a time, that it will initiate nipple-locating movements and finally suckle from the mother.

Specialized Responses

The specific nipple characted into which enable new bonus to grasp that have not yet been established for any at racial species and so newborns do not initially attach spontaneously to the rubber nipples of artificial brooders.

The projection of the nipple from the surrounding mound of bare skin (areola) is an important feature enabling kittens to locate and grasp the nipple. While nuzzling the mother's fur they crawl upward on her body but upon reaching the areola region their nuzzling abruptly changes to gentle nose tapping and forward crawling stops. As they contact the projecting nipple with the nose and lips, the head is withdrawn and raised and the mouth is opened; the nipple is grasped by a forward head lunge with mouth open, and often several such lungs are made before the nipple is centred in the kitten's mouth.

This is a variable response even in newborn kittens. Brooder-reared kittens develop a different pattern in adapting to the longer and more flexible rubber nipple used. They approach in a similar way but upon making contact they position the nipple at one side of the mouth, move the head sideways, thus, bending the nipple, then allow it to spring back into the mouth. If brooder nipples are placed in rounded depressions on the brooder surface, kittens find it difficult to attach to them. The reason appears to be that when probing the depression, the kittens are stimulated more strongly around the face by the edges of the depression than on the nose and mouth by the nipple. Face stimulation stimulates crawling and strong pushing into the depression but appears to inhibit a response to the nipple. The location of the nipple on a mound, as in the mother, therefore, has the opposite effects: crawling is inhibited by the nose stimulation and nipple grasping is elicited.

Each kitten rapidly develops a preference for either a single nipple or a pair of nipples. Textural differences on the surface surrounding the nipples may play a role in this discrimination, since R.J. Woll has shown, in brooder-reared kittens, that a discrimination between differently textured nipple flanges, one providing milk and the other without milk, can be learned by the second or third day.

In rat pups tactile stimulation of the upper lip appears to be the stimulus for nipple grasping Pups, whose upper lips had been desensitized, failed to grasp the nipples although they nuzzled them when placed on the nipple region of anaesthetized mothers. When they were placed on nipples, however, they suckled normally.

Once the nipple is grasped and sucking begins, young are extremely sensitive to tactile and perhaps thermal stimuli within the oral cavity. A series of postural changes during suckling has been described in rat pups, in which changing nipple stimulation and finally milk ejection stimulate first treading then stretching responses. Treading anticipates actual milk ejection and may be in response to subtle oral tactile or thermal stimuli resulting from milk let-down within the mammary gland human mothers, oxytocin released in response to infant crying raises breast temperature by 1°C as a result of milk let-down.

The likelihood of thermal stimulation playing a role in nipple grasping arises from N. Freeman's finding in the lactating cat that nipple temperature is higher (37°C) than the nearby surrounding skin (34.7°C) and G.H. Rose's observation that warming the nipple and flange of an artificial mother aids the newly born kitten to find and grasp the nipple. Stanley and his associates have shown that the newborn puppy is able to modify its sucking in accordance with the rate and pattern of milk flow, probably on the basis of oral tactile stimulation.

Suckling and Intraorganic Stimulation : Early Stage of Development

The initiation and termination of suckling cannot be explained entirely by reference to exteroceptive thermal and tactile stimuli from the mother even in the early stage of development. Young of all altricial species exhibit a periodicity of feeding that appears to depend on internal sources of stimulation generally labelled, in aggregate, as hunger motivation. The most obvious during suckling is the milk which fills the newborn's stomach. It has been assumed that filling of the stomach terminates suckling as a result of stimuli arising from muscular distention or sensory receptors in the stomach, and, conversely, that an empty stomach stimulates the young to initiate sucking and to be responsive to nursing stimuli.

James (1957) was the first to show that in young puppies (i.e., less than 26 days) suckling was not inhibited by preloading their stomachs with adequate amounts of milk. Latencies to initiate suckling and the duration of suckling were not different in preloaded and non-preloaded puppies which had been separated from their mothers for up to 3 hours, although preloaded puppies consumed less milk. Further study showed that preloading to the extent of overloading (i.e., milk regurgitated through the mouth) could inhibit sucking and reduce milk intake, but strong efforts had to be made with these fully preloaded puppies to keep them awake. Even though they attached to nipples, once they were released after being hand held and stimulated they released the nipples and fell asleep. Thus, the stomach probably does not directly regulate the responsiveness of young puppies to the exteroceptive nursing stimuli of the mother.

Hall (1975) has recently shown this even more clearly with rat pups during the first 10 days, using the anaesthetized mother to test for suckling. Hall deprived one group of pups of suckling for 22 hours and removed a second group from the mother jut before testing and anaesthetization. Both deprived and non-deprived pups were placed on the mother's exposed nipple region and their readiness to attached and suckle was measured. Their level of general activity was also measured. During the first 10 days, deprived and non-deprived pups attached with the same latencies. Latencies for attaching declined in both groups from just below 200 seconds to between 50 and 75 seconds; activity levels did not differ substantially between the two groups for most of the 10 day period. This study shows that readiness to suckle was not dependent upon any differences in stomach loading that resulted from differences in the recency of suckling.

Anaesthetized rat mothers do not release milk in response to sucking because of the inhibition of oxytocin release necessary for milk let-down. In Hall's study, therefore, pups sucked from dry nipples yet they continued to suck for the remainder of the five minute test period. In this early period, therefore, ingestion of milk is not an essential stimulus for sucking; Hall (1975) reports that pups eight days of age and younger suck for at least six hours, continuously, despite the fact that they obtain no milk throughout this period.

Similar finding have been described for kittens during the first three weeks after birth. Two groups of kittens were studied: one was presented with a normal lactating mother and the other with an anaesthetized non-lactating female in whom sucking did not, of course, cause milk release. Both groups were fed adequate amounts of milk by tube directly into the stomach at four hour intervals and were exposed to the 'mothers' for six hours each day, the exposure starting two hours before a scheduled tube-feeding.

During the first three weeks the dry-sucking group sucked as frequently and for as long as the group that obtained milk during sucking. Both groups sucked (i.e., were attached to nipples) for between 31 and 42 minutes of the first hour, indicating that latencies to attach were rather short and did not differ between groups. At the end of two hours, both groups were tube fed according to schedule and the effect was to reduce suckling during the third hour, equally in both groups, but not to eliminate it.

Thus we see that the milk intake is not essential either for the initiation of suckling during the first three weeks for maintaining sucking. However, in puppies, rat pups, and kittens, the situation changes as the young become older and milk intake plays so important role in sucking.

Among kittens and rat pups the young sleep during sucking through they may remain attached to the nipples, and when the mother rises to leave, some are so strongly attached that they are dragged from the nest.

The ingestion of milk plays a role in sucking but its role is indirect. Following stomach filling with milk the young may be lulled to sleep, as a result of a sharp reduction in general arousal perhaps through inhibitory effects of stimuli from the distended stomach. Sucking may be terminated, therefore, or fail to be initiated or only weakly initiated, in young preloaded with excessive amounts of milk.

During the first 10 to 15 minutes after preloading with milk, kittens exhibited an overall lethargy that made them unresponsive to all forms of stimulation and precluded any responsive to a nursing mother. There was then gradual recovery and the kittens became more responsive to stimuli and often began to initiate suckling approaches with nipple grasping and sucking.

Suckling has many intraorganic effects some of which derive from milk ingestion and others from exteroceptive stimuli received during contact with the mother. Hofer and his associates have uncovered an effect of milk intake on cardiac function. Two week old rat pups, separated from their mother overnight, show a decline in heart rate that cannot be attributed to a decline in body temperature. These investigators have shown that only tube feeding at regular intervals (i.e., one or four hour intervals) prevents this decline: Koch and Arnold (1976) have extended these findings to pups four to 14 days of age. In addition they have found in four to 10 days old pups that even non-suckling contact with the mother by tube-fed pups accelerates cardiac functioning more than tube feeding alone. Since rat pups in this age range are unable to thermoregulate, the effect may be based upon the thermal insulation provided by contact with the non-suckling female as well as contact with a source of warmth well above the pups' body temperature. Rabbit young suffer a loss in body temperature when isolated in 20-25°C ambient temperature. This suggests that suckling exerts an important influence on the body temperature of young partly through body contact and partly through the warming effects of milk; full body contact with the mother has similar or even greater effects.

Nevertheless a number of investigators have shown that altricial newborn of a number of species can learn discriminations of various sorts on the basis of milk reinforcement of sucking. Tactile, taste, and olfactory discriminations have been established and sucking has been 'shaped' by the presentation of milk. If the effects of reinforcement are related to the underlying motivational conditions, then in these young, reinforcement must act in a different way than in adult animals with well-defined motivational systems. The nature of this difference, however, is not yet known.

Suckling and Olfaction

Olfaction enters early into the suckling pattern of altricial newborns but it may be several days before it plays a significant role in the approach to the mother and in nipple searching. As early as three days of age, rat pups show evidence of responding to the odour of a lactating female or of nest deposits, and on the fifth day pups show behavioural arrest and EEG synchronization in response to maternal odour. By the ninth day the evidence of response to maternal odour is clear.

Bilateral bulbectomy on the second day has a serve and irreversible effect on suckling in rat pups, resulting in a gradual weight loss beginning one or two days after surgery and 80% mortality by the sixth day. Pup approaches to the mother when she enters the nest and interactions with littermates are only mildly affected by bulbectomy. At later ages also, bulbectomy produces a severe reduction in suckling and consequent weight loss, with high mortality but mild effects on social interactions with the mother and the littermates. The effect of bulbectomy appears to be specific to nipple grasping since approaches tot he mother and nuzzling in her fur are not different in bulbectomized and normal or sham-operated control pups. This suggests an early role for olfaction in nipple grasping.

Kovach and Kling (1967) similarly found that bulbectomized kittens could not suckle with the mother but were able to suckle when placed on an artificial nipple.

The rabbit pup, kitten and puppy show evidence of use of olfaction in their response to the mother during the first week. Artificial odours have been applied to the mother's nipple region and later tested alone to determine whether, in the course of suckling, young rabbit pups and puppies are influenced in their behaviour by the presence of the odour. Rabbit young thus exposed to odours during the first five days of suckling began to respond positively to the odour alone on the second or third day and puppies exposed similarly responded immediately when tested for the first time on the sixth day.

Among kittens any of the three functional pairs of nipples may be suckled by each of the kittens of a litter during the first day *post partum.* By the second and third day, however, each kitten confines its suckling to either a single nipple or a pair of nipples. Once nipple position preferences are established nose contact with non-preferred nipples does not elicit nipple grasping as it did on the first day. To determine whether olfactory cues could be the basis for these preferences, brooder-reared kittens were presented with two nipples and flanges, one of which provided milk while the other was blind. Two-day old kittens were tested for their ability to establish an olfactory discrimination. Within a day or two the kittens were able to discriminate between two artificial odours: they approached and suckled almost exclusively the nipple-flange combination with the positive odour.

The ability of rat pups to approach the mother on the basis of an olfactory stimulus has been clearly established by Moltz and Leon and Leon. From 14 days pups respond to the odour of lactating mothers in presence to that of non-lactating females. The odour emanates from a volatile component of the mother's faeces which the young ingest, and if mothers are fed different diets the offspring can distinguish the odour of their own mother. Under these conditions, the odour of a strange mother has no special attraction for them, indicating that maternal odour acquires its attractiveness through experience.

Olfactory stimuli appear gradually to become a component of the young's approach and suckling response to the mother is these altricial species. From the beginning, olfactory stimuli are able to influence the newborn's behaviour. Ammonia fumes (which may in fact have irritating tactile effects) stimulate pivoting and vigorous crawling in newborn rats. There are, undoubtedly, odours to which newborn respond positively from the birth onward, particularly those related to birth fluids etc., but, in the main, odours appear to acquire their significance through association with thermotactile stimuli that accompany them.

With the development of olfactory responses to the mother's body and particularly to those regions involved in nursing, certain changes occur in the suckling pattern :

1. Suckling becomes more specifically a response to the newborn's own mother and to specific aspects of the mother's nipple region (e.g., development of nipple position preferences).
2. Since olfactory stimuli from the mother may reach the young before actual contact is made with her, the young may initiate the suckling approach while the mother is still at a short distance and thus they may anticipate her approach.
3. Since olfactory stimuli may be spread outside the nest or home site, familiarity with maternal odour may enable young to widen the range of their activity to these areas.
4. Since olfactory stimuli may also be deposited on littermates, young may also begin to respond to their siblings on this basis.

Suckling and Vision

At the time vision becomes functional in altricial young, the suckling pattern is well established on the basis of olfactory and thermotactile stimuli. These stimuli, however, can only be received by young when the mother is close by or in actual contact with them. Use of vision enables young to perceive the mother at a distance and approach her for suckling.

The onset of visually stimulated suckling behaviour is usually taken as the beginning of distant approaches to the mother. In the rat this begins between the fourteenth and sixteenth days, in the hamster around the second week, in the puppy around the fifteenth day, in the rabbit around the tenth to fourteenth days, and in the kitten around the seventeenth day. It must be kept in mind, however, that vision may begin to function earlier in suckling but at short distances where it would be difficult without specific examination to distinguish between a thermotactile, olfactory, or visual basis for the young's approach.

While there is considerable descriptive evidence of visually guided approaches to the mother beginning sometime after eye-opening in the species we have been discussing, experimental evidence is often lacking. In the kitten, eye-opening occurs around the seventh to ninth day and there is evidence of the onset of visual functioning between the fourteenth and seventeenth days. Visually guided approaches to the mother from a distance start around the seventeenth day and are fully established by the twenty-first. In the 10 day old rabbit pup an odour associated with the mother initiates approach to the mother which is then guided by visual stimulation. The implication of the studies on maternal pheromonal stimulation in the rat by Moltz and Leon (1973) and Leon (1975) is that, form around the fourteenth day, pups identify a lactating female by odour but their approach to her is guided by visual stimulation. Altman and his colleagues have shown that rats orient to the mother and littermates in the home cage, presumably on the basis of visual supplemented by olfactory stimuli, during the third week. The beginning of visual functioning in the rat pup has been found on the fourteenth day by Turner (1935) who studied pups' ability to find an exit door, marked by a vertical black stripe, in a circular field. Pups improved rapidly from the fourteenth day onward in their ability to go from the centre of the field directly toward the door, a distance of 13 inches (33 cm.).

Visually guided approaches to the mother in puppies described by Rhein-gold (1963b) as starting around the fifteenth day and increasing thereafter, are correlated with visually evoked responses in the cortex which assume the adult form at around the same age. By four weeks of age, puppies will approach a two-dimensional drawing of a female, paying particular attention to the head and flank, and at a slightly older age to the mammary region, when they are hungry.

The onset of visually stimulated approaches to the mother marks an important stage in the young's development with implications for the organization of suckling behaviour. It is no doubt significant that the ages at which young increasingly initiate suckling approaches to the mother from a distance in the rat pup, puppy, and kitten correspond roughly with the ages at which intraorganic stimuli become increasingly important in the regulation of feeding.

If the young approach the mother when she is unresponsive or not easily available it is likely to result in either failure to suckle or in rejection by the mother. It is during this stage that maternal rejection or evasion of the young become prominent. Sucking approaches made when the mother is receptive result, on the other hand, in her cooperating with them by assuming a nursing position. Young gradually become sensitive to subtle feature of the mother's behaviour indicating her readiness or unwillingness to nurse and adjust their suckling approaches accordingly. Moreover, each young approaches the mother to suckle individually, whereas earlier the mother's approach had stimulated the entire litter simultaneously and all young suckled together. Once nursing is initiated with one offspring other young often join it, but each young acts on the basis of its own perception of the situation rather than to the stimuli that initiated suckling in its sibling.

Suckling and Intraorganic Stimulation : Late Stage

Between the tenth and fourteenth day an important change takes place in the intraorganic stimulation of suckling among rat pups. Around this age the latency to initiate suckling (with an anaesthetized mother) begins to differ between pups that have been deprived of suckling for 22 hours and those that have only recently been removed from their mothers. Latencies are very short for the deprived pups but range above 200 seconds for recently fed pups. Moreover, failure to receive milk from the anaesthetized mother, which had little effect on maintenance of suckling earlier, now begins to affect suckling: pups shift from one nipple to another, within the five minute suckling test, presumably in response to the dry nipple.

Similarly, around the third week of age, kittens suckling from a dry mother initiate nuzzling of the mother's nipples with the same frequency as kittens suckling from a lactating mother, but the frequency of attaching and sucking declines rapidly from the earlier period and the duration of sucking also declines to 20% of the previous period. Meanwhile kittens sucking from a lactating mother continue to suckle for extended periods.

After 30 days of age, preloading the stomach of puppies with a milk-chow mixture has the same effect of reducing or eliminating further eating from a dish as the same amount of food consumed voluntarily by control puppies. The contrasts with the earlier finding that preloading the stomach with either milk alone or a similar mixture has no effect on suckling.

Indications are, therefore, that arousal produced by intraorganic stimulation has become more specific than earlier so that exteroceptive stimulation received before or during suckling is not sufficient to initiate or maintain suckling. Milk intake and its intraorganic effects have become an essential part of the suckling pattern. At this stage of development, stomach filling or its absence has become a more focal source of arousal than earlier and it determines to a larger extent the young's response to the exteroceptive stimuli associated with suckling.

The kinds of exteroceptive stimuli that arouse suckling approaches undergo a change during this period. Among rat pups this is the period during which suckling approaches are increasingly initiated by the pups from a distance, in response to olfactory and visual stimuli from the mother. Among kittens there is the beginning of kitten-initiated suckling approaches to the mother, starting around three weeks of age. At this stage sucking approaches are increasingly determined by the past experience which the young has had with the mother and less by immediate stimuli for suckling.

This was shown in a study by Kovach and Kling (1967) in which they found that kittens reared in isolation from an early age until about three weeks and fed by stomach tube to prevent suckling, were unable to initiate suckling when returned to their mothers. While they attributed this to the waning of the sucking reflex, our analysis indicates that at three weeks of age the suckling reflex is only a small part of the suckling pattern. Moreover, as part of an organized suckling pattern it is initiated mainly on the basis of intraorganic stimuli rather than the previously effective proximal exteroceptive stimuii.

In a study comparable to that of Kovach and Kling (1967) kittens were reared in isolation from one to three weeks of age with feeding by suckling from an artificial rubber nipple mounted on a brooder (Rosenblatt *et. al.*, 1961). When these kittens, in whom suckling had been maintained as an active component of the feeding pattern, were returned to their mothers there was a delay of about 20 hours in initiating suckling. The previously isolated kittens had no difficulty in establishing sustained contact with the mother, which in all cases took less than three hours and in most kittens was accomplished within the first hour. Their difficulty lay in perceiving the mother as an object to be suckled. They crawled over her body, slept in close contact with her nipple region and often nuzzled, but not until they had been without milk for many hours did they finally exhibit nipple grasping and sucking. In these brooder-reared kittens the suckling pattern had been organized in relation to the brooder and artificial nipple, which differed in many ways from the mother. Under comparable conditions in the brooder situation, these kittens would initiate suckling within seconds after being returned to it.

As young develop, therefore, the relative influence of exteroceptive and intraorganic factors in the regulation of suckling shifts in favour of intraorganic factors. Moreover arousal undergoes changes : it can no longer be described simply in terms of general arousal. Arousal has become more specific as a result of experience in the litter situation and of neural maturation which gives rise to new physiological relationships between visceral and somatic sensorimotor processes. When aroused, young are channelled into particular patterns of behaviour not only by the prevailing exteroceptive stimulation, as earlier, but by intraorganic factors among which previous suckling experience plays a crucial role.

Development of Huddling and Related Behaviour

Newborn huddle from birth onward: after nursing has been completed and the mother leaves the nest or home site the young crawl into contact with one another and form a closely knit huddle. Since heat loss is dependent upon body surface exposure to the environment in young with poor thermoregulation, the huddle maintains body temperature and metabolic activity. Huddling is an early form of social behaviour among littermates that may be closely related in development to later forms of group activity (e.g. play).

Thermotactile Basis of Early Huddling

Cosnier (1965) and Alberts (1974) have shown that early huddling among rat pups is based upon thermal and tactile stimuli provided by littermates. Pups up to 10 days of age are more attracted to a warm liver pup (i.e. 37°C) than a cool dead one (25°C) but the warm pup can be replaced by warm tubing or a warm fur-lined nylon sleeve. Pups are particularly stimulated to huddle when the surrounding floor is cool.

James (1952a, b), Welker (1959) and Crighton and Pownall (1974) analysed the behaviour of individual puppies, when separated from their littermates in warm and cool environments, in an attempt to understand the thermotactile basis of huddling. Separation stimulates pivoting and vigorous crawling until the puppy makes head contact with its littermates. The essential feature of the contact is thermal, however, since contact with cooled puppies or contact with littermates in an overly warm environment elicits withdrawal rather than huddling. Within the huddle temperature is maintained at 30°C and puppies remain quiescent, breathing quietly and in a realized sleeping posture. Warming the huddle of 32.5°C causes dispersal as each puppy responds to contact with its neighbour by withdrawing; cooling the huddle intensifies huddling as each puppy responds to the warmth of its neighbour by approaching it. The greater thermotactile sensitivity of the face and snout over the remainder of the body assures that puppies face into the huddle. The few observations of kittens indicate that similar factors operate.

The role of tactile stimuli in huddling receives some confirmation from a study of Scott et. al. (1974) in which puppies that were isolated in a warm environment but nevertheless vocalized were induced to become quiet by confining them in a section of stove pipe the floor of which was lined with soft material. The tactile stimulation, and perhaps the insulation provided, are similar to that provided by neighbouring puppies during huddling.

Huddling and Olfaction

Singh and Tobach (1974) found that bulbectomized rat pups, from the second day onward, tended to be apart from the litter for longer periods than their normal littermates, although in general their social behaviour was not particularly deviant. The suggestion that olfaction plays a role in huddling was more clearly established by Alberts (1974). Using intranasal infusion of zinc sulphate to make pups anosmic he found that huddling was completely eliminated in pups 10 days of age and older, but was retained by five day old pups.

Fox (1970) has shown that olfaction may begin to play a role in the puppy's response to its littermates as early as the first week after birth. An odour applied to the mother's body, and therefore, spread to the littermates, elicited strong approach response in puppies exposed to the odour during the first five days of life, but strong avoidance responses from puppies presented with odour for the first time on the sixth day. Similarly, among rabbits, maternal odour, which is shared by the littermates, elicits a strong searching response when held close to the young's nose without tactile contact as early as the second to fifth day after birth. Evidence suggests also that hamster pups may begin to respond to litter odours at the end of the first week.

Orientation of Littermates and Vision

Huddles become somewhat less compact as young develop thermoregulation and are less in need of insulation from heat loss. Nevertheless they continue to aggregate in groups which may consist of several young instead of the entire litter as earlier. Of greater interest is the fact that orient to littermates from a distance suggesting that vision is beginning to play a role.

As late as 17 to 20 days of age, about a week after eye-opening, puppies still do not orient to littermates in the home cage from a distance. This does not mean that they do not react to them but they continue to show circular movements similar to those seen earlier. Starting around 25 days of age, however, they begin to approach littermates from a distance directly. Rheingold and Eckerman (1971) showed that kittens respond to lettermates at a distance even when they do not orient directly to them. They studied the earliest age at which kitten would be comforted by the presence of a littermate in a strange environment. At two weeks of age, the earliest age at which testing was done, kittens emitted 50% fewer vocalization when a littermate was present compared with when it was absent.

Despite the fact that only sketchy evidence is available to trace the development of huddling and other relationships with littermates among these altricial species, the main outlines of this development are clear. Stages in the development of responses to littermates are similar to those found in the development of suckling: there is an initial stage of dependence upon thermotactile stimulation, a later stage of olfaction based responses and a final stage of visually guided responses to littermates. Moreover, early responses to littermates are dominated by exteroceptive stimuli while later stages show evidence of an increase in self-initiated approaches to siblings based upon the developing social relationships. Huddling is a relatively simple behaviour pattern compared to suckling, but it represents only the earliest form of interaction among littermates that ultimately develops into the complex patterns of play.

Development of Home Orientation

The nest or home site is the socially conditioned environment in which the young undergo the early stages of their development. The young, from an early age, develop a pattern of home orientation that plays an important role in their relationship to the mother, and the organization of their behaviour in relation to their surroundings and littermates. Since the original discovery of home orientations in the kittens orientation to the nest site has been found in the hamster, rat, and puppy, in one form or another, suggesting that further research will reveal that this is a widespread and basic phenomenon among altricial newborn.

Home Orientation of Olfaction

The nest or home site differs from the surrounding regions in the odours deposited there by the mother and litter. These too provide a basic for orientation to the home site. Olfactory orientation to the home site develops after thermotactile orientation has been established and it may be based upon this earlier form of orientation.

Towards the end of the first week, a warmed floor surface is not sufficient to quite kittens and a thermal gradient between two cage regions, lacking home cage odours, does not result in kittens adopting a path from the cooler to the warmer region. Only the presence of home cage odours quiets kittens and only if the home site has the usual odours and presumably, an olfactory gradient, can kittens find their way back to it (Rosenblatt et. al., 1969; Rosenblatt, 1971). From the seventh day onward, home orientation in kittens is based mainly upon olfaction, although at the home site, the combination of olfactory and thermal stimuli is more effective than either alone.

In the hamster thermal orientation declines as orientation based upon odours from the nest becomes established (Devor and Schneider, 1974). Starting around the seventh and eighth day, hamster pups turn toward nest shaving and away from fresh shavings when placed on a screen above the border between the shavings. They spend 80% to 100% of their time crawling in the nest shavings. Shavings from nests inhabited by a mother and litter are preferred to shavings from nests inhabited by either a male or a non-pregnant female, and nest shavings from four day old litters are as effective as shavings from older times.

Earlier, hamster pups can distinguish between odours, preferring, for example, the odour of cedar to pine shavings; this appears on the third day. Nest odours must require an additional period of four or five days to become associated with thermotactile stimuli from the mother and littermates to provide a basis for orientation to the nest (Devor and Schneider, 1974).

As indicated earlier, rabbit pups and puppies begin to respond to odours from the mother before the end of the first week and it is likely that the same odours deposited in the nest would provide a basis for home orientation. It is clear that the odour acquires its attractiveness through association with other stimuli from the mother during the first five days (Ivanitskii, n.d.; Fox, 1970).

Home orientation in puppies that appears to be based upon olfaction has recently been reported by Scott et. al., (1974). Puppies were placed either in their own nest boxes, emptied of the mother and littermates, or in a similarly constructed nest box that was lined with fresh towelling; vocalizations were recorded in each. One precaution was taken, which, however, may have affected the results: to exclued the possibility that puppies might respond to the thermal difference between the litter nest box and the unused one, both boxes were warmed to about 29°C before testing. Despite this, vocalizations were more frequent in the unused nest box than in the litter nest box from around the sixth to eight day in several litters and on the eleventh-twelfth day in all litters. Vocalizations were infrequent in both conditions at first but they rose sharply after the twelfth day in the unused nest box.

The distress caused in puppies by placing them in a strange nest box is similar to the distress caused in kittens by placing them in a strange cage: the rate and intensity of vocalizations rises during tests, and with age they become more frequent and more intense. The locomotory behaviour of the puppies in the above study was not described but, in an earlier paper (Elliot and Scott, 1961) with older puppies, vocalization was accompanied by an increase in activity in a strange pen and presumably the same occurred here.

Rat pups exhibit the first signs of orientation to the home cage from a distance of less than 30 cm on the third day and even more clearly on the fifth and sixth days (Altman et. al., 1974). Placed on a 11.4 cm diameter circular platform, located midway between the home cage and a strange cage, pups point themselves toward the home during most of a three minute test but they are not yet able to crawl to it. Between the ninth and twelfth days pups become better able to locomote over long distances by crawling. On the ninth day, pups placed in a neighbouring case reach the home cage in 50% of the tests by passing through a narrow alley, with latencies ranging from the two or three minutes (Altman et. al., 1974). Latencies to traverse the 15 to 30 cm distance between the starting chamber and the home cage become shorter on the twelfth day (i.e., one to two minutes) and 85% of the pups are successful and on the sixteenth day, around the time of eye-opening, nearly all pups orient to the home with latencies ranging from 30 to 40 seconds. At 19 and 21 days latencies are 10 seconds or shorter.

Neither the paths taken to the home cage nor the sensory cues utilized by pups were analysed in the above studies. Turkewitz (1966) reported that successful orientation from a neighbouring cage to the neat site in the home cage by 9 to 12 day old rat pups was accomplished indirectly through wall-hugging. The long latencies to reach the home cage in the above studies suggests that a similar mode of home orientation occurred at 12 days of age. Old pups, appear to adopt direct paths to the home cage and as a consequences latencies are considerably shorter.

On the basis of Turkewitz's (1966) study, it would appear that olfactory and perhaps thermal stimuli are involved in home orientation by rat pups before their eyes open on the fifteenth to seventeenth day. The long latencies suggest pups adopt an indirect path to the home and this is more characteristic of young that orient on a nonvisual basis: the shorter latencies after eye-opening certainly contrast with the longer latencies earlier. Findings reported earlier, that rat pups show evidence of olfactory discrimination of nest shavings as against fresh shavings around the ninth day and respond to maternal odours by the fourteenth day, would also support this interpretation of the above home orientation (Nyakas and Endroczi, 1970; Schapiro and Salas, 1970; Gregory and Pfaf, 1971; Moltz and Leon, 1973; Leon, 1975).

Home Orientation and Vision

The onset of vision marks the beginning of the decline of home orientation among altricial young. With the ability to view the home from a distance it is no longer necessary to return to it; instead young tend to remain in the vicinity of the home or nest site for a period before they disperse. Moreover, orientation to the home declines also because the young begin to follow the mother and littermates and these become centres of orientation rather than the nest or home site. There is, however, a short period after eye-opening and the beginning of vision when home orientation continues and vision is used to enable the young to return to the home from a distance.

In the rat, as we have seen, the latency for reaching the home from a nearby region is reduced to a few seconds shortly after eye-opening (Altman et. al., 1971; Bulut and Altman, 1974). This indicates that the young begin to take direct routes to the home, and this probably depends on visual stimulation (Turner, 1935). The hamster pup shows a decline in olfaction-based orientation beginning around the thirteenth day, shortly after eye-opening, and it is completely absent by the nineteenth day. This may be based upon the gradual increase in visually based orientation.

Among kittens the use of vision in home orientation begins around the fourteenth day (Rosenblatt et. al. 1969; Rosenblatt, 1971). Testing kittens in the dark results in some decrement in performance with indications that the kittens are seeking the visual stimuli to which they are accustomed. However, they are still capable of using olfaction and most of them reach home. If olfactory cues are removed, some kittens are still able to reach the home and this increases after the seventeenth day.

The use of vision in home orientation among kittens eventually leads to the decline of this behaviour at around the eighteenth to twenty-first day. Instead of returning to the home region, kittens look toward it, or they may actually enter the home region briefly then leave it. The ability to see the home often enables kittens to wander more freely around the entire cage since the home region is always in view. However, when the home was made especially prominent visually by placing a fur piece there, home orientation persisted until 45 days of age. With home odours absent, kittens began to show this visually based home orientation around 25 days of age and gradually perfected it, traversing the cage, a distance of about 76 cm, in less than 20 seconds as they moved directly toward the home and came to rest after entering it.

Stages in the Early Behavioural Development of Altricial Young

The early behavioural development of altricial young can be divided into three stages: in the following section we shall analyse these stages and the transitions from one stage to the next.

First Stage : Thermotactile Stimulation

The neonates' earliest behaviour is organized predominantly in relation to low intensity thermotactile stimulation provided by the mother, littermates and nest or home site. These elicit from the newborn either general or specialized approach responses (Sehneirla, 1959; 1965). The general response of forward crawling is elicited by centrally applied head and face stimulation over a broad area and turning is elicited by laterally applied stimulation, through the close relationship that exists between anterior end stimulation and movement of the forelimbs in an alternating paddling motion and the hindlimbs in a joint pushing action. Specialized responses are elicited by localized stimulation of the snout, mouth, and lips; the specialized responses take their character from the peripherally organized pattern of movement characteristic of the local region and the restricted locus of stimulation.

The thermal and tactile characteristics of the mother, littermates and nest or home site provide patterned thermotactile stimuli which elicit from newborn the general and special approach responses necessary for the initiation of suckling, and huddling, and for remaining within the confines of the home. These responses are only loosely patterned sequences at first, depending upon kinds of stimulation the newborn encounters in succession as it crawls forward. Stimulated initially to approach the mother by contact with the attractive thermotactile stimulation of her body, the newborn subsequently encounters the more detailed stimuli of the mother's fur around the nipple, the areola, and finally the nipple itself.

At this early of development the central control of behaviour is not yet strongly developed beyond the mediation of the admittedly complex sensori-motor integrations underlying general and specific approach and withdrawal responses to stimulation. It is doubtful whether, for at least a short time after birth, the functionally distinct behavioural responses to the mother, littermates and nest or home site are represented by equally distinct central regulatory processes. The underlying similarities between the behavioural adaptations to the mother, littermates, and nest or home site, arise from the fact that they share in common responses to similar thermotactile stimuli. The observable differences arise from the fact that in each of these responses the neonate also responds to specific stimulus properties of the mother, littermates, and the nest or home site.

While the relative contributions of central regulatory processes and peripherally elicited responses favour the latter in the neonate (as compared with older animals in which central regulation plays the dominant role) central processes are not without effect on the neonate's behaviour. Exteroceptive stimuli (i.e., thermotactile) appears to have two concurrent effects upon the neonate. They arouse the neonate to activity, as when the mother's licking activates the sleeping neonate to raise its head and begin to crawl. Once the newborn is aroused, stimuli also elicit approach responses of a general and specialized nature. Thermotactile stimuli are therefore both arousal inducing and response-eliciting. Neonates that have been aroused are more sensitive to exteroceptive stimuli that follow and exhibit more vigorous responses to these stimuli. The central component of exteroceptive stimulation therefore is an arousal that in turn potentiates the responsiveness of peripheral processes.

Early in development central arousal is relatively non-specific in its afferent and efferent relationships and this contrasts with the specificity of the peripheral responses that are elicited by thermotactile stimuli. It is this contrast between the greater specificity of peripheral responses—their direct relationship to thermotactile stimuli—and the non-specific character of arousal and its contribution to early behaviour, that has led many to characterize the neonate's behaviour with some justification as simply reflexive in nature (Scott, 1958; Kovach and Kling, 1967).

Early suckling among newborn rat pups and puppies exemplifies the relationship between central and peripheral processes in the regulation of feeding. The compelling effect of thermotactile nipple stimuli in eliciting suckling is shown by the fact that prior feeding or stomach loading cannot prevent the young from grasping the nipple and sucking. In the rat pup, if milk is not forthcoming sucking may continue indefinitely (Hall, 1975); suckling does not continue indefinitely in the kitten but its duration is not shorter than that of kittens that obtain milk (Koepke and Pribram, 1971). This difference may reflect the fact that the rat pups were tested with anaesthetized mothers while the kittens were tested with awake, non-lactating females.

Yet normally rat pups and kittens do not suckle interminably once they attach to nipple. After a period they began to loosen their grip on the nipples and slide off, or they are detached from them when the mother rises to leave. The intake of milk has an effect on the peripherally organized response of sucking, but the effect appears to be mediated by a lowered arousal, to the point of sleep. All overt activity ceases, not only suckling, and the newborn's behaviour alternates between sleep and feeding for a considerable period after birth.

Preloading of newborn with milk has an effect on suckling when the amount preloaded is sufficient to induce sleep (Satinoff and Stanley, 1963; Stanley, 1970). With effort, even under these conditions, sucking can be elicited by bringing the newborn's mouth to the nipple but the central effect of intraorganic stimulation arising from the distended stomach or sensory receptors in the stomach; usually prevails and he newborn releases the nipple soon after grasping it. Preloading also reduces milk intake but the basis for this is not clear.

Varying levels of central arousal appear to influence the responsiveness of newborns to exteroceptive stimulation: low-level arousal, that characteristic of newborn shortly after feeding, dampens the effect of exteroceptive stimulation, while very high levels of arousal, following long period without food or upon exposure to low ambient temperature, produces an excessive response to exteroceptive stimulation. There is, therefore, an optimal level of central arousal at which the newborn exhibits its typical responses to exteroceptive stimuli.

The sources of central arousal in the newborn are interoceptive as well as exteroceptive, as studies on stomach preloading indicate. These sources have not been studied to any great extent, but interoceptive stimuli have a history in prenatal ontogeny that predates exteroceptive stimuli (Schneirla, 1965). Exteroceptive stimuli may in fact exert their influence, in part, through their effect upon interoceptive processes, as for example in responses to ambient thermal stimulation.

During the earliest postnatal stage there is a close relationship between those exteroceptive stimuli (i.e., thermotactile) to which newborn are most responsive in their behaviour toward the mother, littermates, and nest or home site, and the conditions which stimulate it to activity. Thus, newborns are stimulated by loss of contact with the mother or littermates and by displacement from the nest or home site; in the latter cases it is probably exposure to lower ambient temperatures than are normally present in these situations which stimulates the newborns as studies on kittens have shown. At this early age many different objects having attractive thermotactile properties may be equivalent in their arousal and calming effect upon newborn because the special properties of the mother, littermates and nest or home site do not yet play a role. This fact has allowed investigators to ignore the social nature of the newborn's responses to species mates and to the social conditioned home and to introduce special criteria for when the young's responses are to be considered social (Rheingold and Eckerman, 1971; Scott et. al., 1974). Since the usual criteria are based upon visual responses to species mates, responses that appear during the third stage of early development are more likely to be called social responses and social behaviour is therefore viewed as a later development. Such a view ignores the ontogenetic basis of social responses and their relationship at each stage to the behavioural capacities and behavioural organization of the young (Schneirla and Rosenblatt, 1963).

The earliest changes in the newborn's behaviour are the formation of extended action patterns, incorporating the earlier hesitating and variable crawling approach response and the specialized feeding and huddling responses into more smoothly coordinated patterns. In brooder-reared kittens, after a day or two of feeding, the kitten crawls the correctly textured flange (i.e. that associated with a nipple that give milk) in a smoother more patterned movement, with short pauses at crucial points such as the edge between the brooder and the floor and between the brooder and cover an the textured flange. At these points the kitten sniffs the brooder cover and rubs its snout against its surface as it does when it reaches the flange. A short period of flange contact is followed by nipple localization and nipple grasping and sucking.

Among puppies approaches to the nipple can be associated with either soft-or hard-textured path with gradual improvement in turning toward one or the other and heading rapidly for the nipple, grasping it and sucking (Stanley, 1970). Repeated experience with either warm or cool thermal stimuli along leading to a nipple results in gradually more rapid crawling approach to either (Bacon, 1973b).

The formation of these specialized action patterns based upon thermotactile stimuli indicate progress along several lines; approach responses become specialized in relation to the young's discrimination among thermotactile stimuli. In this respect approach responses are formed more rapidly when a low intensity, approach-eliciting stimulus is the positive one than when a stimulus that is initially either weakly approach-eliciting, or is actually withdrawal-eliciting, is used as the positive stimulus (Stanley, 1970; Bacon, 1973b). In addition, components of the approach response are integrated into a pattern that appears to be less dependent upon continuous guidance by exteroceptive stimulation and more dependent upon central regulation. In suckling approaches, striving toward the nipple indicates an early appearance of anticipatory responses in advance of actual contact with the nipple region.

These changes channel the earlier general arousal along specific lines. Thermotactile stimuli are no longer equally effective in eliciting approach responses: the newborn turns away from certain thermal stimuli and becomes highly excited when it makes contact with others. There are indictions in the study of Bacon (1974) that negative stimuli do not acquire inhibitory effects at this stage but rather that positive stimuli acquire heightened arousal effects. These may have their origin in the heightened arousal that occurs when the young are stimulated by milk during sucking (Stanley, 1970).

At this stage, new sources of disturbance may arise when the usual incentive is absent or is altered; this is in fact evidence that action patterns of a broader nature already exist in the young and that they are highly specific to the stimulus conditions in which they were formed (Papousek and Papousek, 1975).

Second Stage : Olfactory Stimulation

Although in altricial young the second stage of early behavioural development is characterized by the growing influence of olfactory stimulation, its main feature is the increasing specificity of behavioural responses in relation to the familiar objects in the environment. The young use olfactory stimuli to differentiate between familiar and unfamiliar objects.

The initial exposure to olfactory stimuli occurs during responses to thermotactile stimuli, but in the beginning olfactory stimuli can provide little guidance to the newborn. Except when it encounters strong aversive olfactory stimuli, which induce withdrawal responses (Fox, 1970), olfactory stimuli cannot elicit either general approach or specialized responses in the newborn. Not until responses have begun to become organized into specific action patterns based upon thermotactile discrimination, and the central arousal processes are channelled, can olfaction begin play a role. The suggestion is that olfaction advances further the process of discrimination among objects, along lines that are necessarily specific to the mother, littermates and nest or home site. More important, however, is the ability of olfactory stimuli to arouse the initiation of action patterns that are then guided by combined olfactory and thermotactile stimuli. Olfactory stimuli may therefore determine whether or not a response will occur.

There is associated with the onset of olfaction the appearance of olfactory orientation behaviour by means of which young explore their olfactory environment. Welker (1964) has described the development of sniffing as an olfactory exploration pattern in the young rat, and Komisaruk (1970) has added important details to the analysis of olfactory exploratory behaviour in this species. When placed in a new environment young initiate non-specific olfactory exploratory behaviour which soon gives way either to specific behaviour patterns upon identification of the odour or to further tactile exploration with the vibrissa. These exploratory patterns range from air sniffing to sniffing of the floor surface and approach and sniffing of objects in great detail. Thus, for example, by sniffing the floor surface kittens find their way to the home region and rat pups by sniffing the air are able to locate the mother at a distance of more than 30 cm (Nyakas and Endroczi, 1970; Moltz and Leon, 1973; Leon, 1975). Depending upon the distribution of odours, therefore, olfaction enables young to develop distance perception of objects and to initiate movements toward objects before actual contact is made with their thermotactile properties. In this sense, therefore, olfaction plays a large role in developing central control of action patterns begun with respect to thermotactile stimuli. Anticipation of forthcoming stimulation and the associated action pattern, exhibited during the latter phase of the first stage of development, is gradually transformed into self-initiated approaches to objects with anticipatory action patterns almost entirely centrally aroused, ready to be performed when the object is reached. Rat pups deprived of olfaction during this phase appear much less oriented to significant social stimuli and to the nest site (Singh and Tobach, 1974). The stage is set during this second stage of early behavioural development for the incorporation of vision into the central control of action patterns during the third stage of early development.

As olfaction begins to contribute to central arousal processes and the action patterns to which they give rise, it begins to play an increasing role as a motivating condition and as a possible source of distress. Rat and hamster pups, kittens and puppies begin to show distress when removed from their familiar olfactory environment. In kittens and puppies this is evident at the end of the first week and it appears at a slightly older age in rat and hamster pups; it provides the basis for these young to initiate movements toward littermates, resulting in huddling, and toward the home or nest site during the development of home orientation. Moreover, olfaction serves as a gold for these behaviours in the sense that upon reaching the familiar olfactory situation in the huddle or in the home region, young are calmed and soon come to rest. It is apparent, therefore, that not only perceptual and motor processes become more complex as development proceeds but motivational processes share in this growing complexity.

Third Stage : Visual Stimulation

Vision greatly enlarges the young's capacity in differentiate perceptually between the significant objects in its environment. Neither thermotactile nor olfactory stimuli are able to convey to young the specific actions of their mother or littermates and the specific location of their nest or home site. While olfaction may indicate that the mother is nearby it does not indicate what she is doing, while vision conveys both. This stage is marked by an acceleration in the young's development of social interactions based upon vision. Initially social interactions consist of visual approaches to species mates at a distance, at which point action patterns based upon olfactory and thermotactile stimuli take over: vision therefore contributes little to the character of the interactions but does contribute to their occurrence. Thus, for example, kittens at three weeks of age, approach the mother at a distance for suckling but when they reach her, they adopt their earlier mode of nipple searching with their eyes closed. Gradually, however, vision comes to trigger not only the approach to the mother, but also those action patterns that were formerly triggered by non-visual sensory systems. Kittens at this stage, approach the mother and reach up from beneath her as she remains standing, locating and grasping a nipple without any significant preliminary nuzzling. Brooder-reared kittens walk directly to the nipple and grasp it instead of following a path of nuzzling on the brooder surface, and cage-reared kittens walk directly across the cage from the diagonal corner to the home region instead of crawling along a path that passes through the adjacent corner (Rosenblatt et. al., 1969).

Vision accelerates the process of increasing central regulation of behaviour and with this furthers self-initiated behaviour with well defined goals. Perceptual differentiation and the multiplication of action patterns, and their growing complexity, imply highly specific central states of arousal and complex interactions, both facilitating and inhibitory, among these different states.

During the period of transition from olfactory to visual control of behaviour in the young, interoceptive stimulation appears also to become more influential and specific in its effect upon the central state of arousal. At this age, Hall (1975) found that food-deprived and non-food-deprived rat pups began to differ in their suckling behaviour, the former being faster to initiate suckling with an anesthetized mother and abandoning suckling if no milk was forthcoming. Similar findings have been reported in three to four week old puppies and kittens whose suckling approach to the mother was influenced by stomach preloading with relatively small amounts of milk (James and Gilbert, 1957).

The goal-directed character of behaviour at this stage is exemplified by the persistence kittens show in following the mother around the cage, while they look for an opportunity to suckle. At the slightest pause, they immediately reach up to her nipples and if they are detached by her movement they resume following her. At other time, when tired, they walk across the cage to join another kitten that has settled in a corner to sleep and huddle against it, falling asleep. While they remain highly responsive to environmental stimulation, their behaviour is not directly elicited by this stimulation: rather, they are capable of making perceptual discriminations among the various stimuli and which stimuli they respond to depends upon the central state operative at the time.

During this third stage of development kittens and puppies that are placed in novel environments exhibit signs of distress (i.e., vocalization and agitated movement) that are relived when littermates or the mother are placed in the same environment (Rheingold and Eckerman, 1971; Scott et. al. 1974). Rat and hamster pups orient to the mother and littermates when displaced to a neighbouring, strange cage (Altman et. al., 1974; Devor and Schneider, 1974). Thus, the absence of familiar visual stimuli becomes a source of disturbance that motivates young to regain visual contact with social companions. Often the experimenters serves a similar role if he has become familiar to the young. Since visual stimuli are likely to be a principal source of social stimuli from this age on into adulthood, and social responses of a clearly recognizable nature are particularly evident during this third state, this stage has been singled out by Scott (1958) as the beginning of true socialization. This analysis has shown, however, that social responsiveness arises almost immediately after birth and continues to grow in depth and complexity during each stage of ontogeny. Visually based social responses have their ontogenetic origin in earlier non-visual stages and they retain their close relationship to these earlier stages throughout life.

Functional Aspects of Each stage of Behavioural Development

The neonate's dependence upon thermal and tactile stimulation in its behavioural responses to tis surroundings is based upon its need to maintain an optimal thermal environment for adequate physiological functioning in the face of tis inability to regulate fully its own body temperature (Jeddi, 1970, and unpublished data; Crighton and Pownall, 1974; Leonard, 1974). Contact with the mother's body causes a rise in body temperature and huddling with the littermates, shown by Albert (1974) to result in minimal exposure of the group to ambient thermal conditions, maintains body temperature, and reduces metabolic activity (Cosnier, 1965). Leonard (1974) has suggested that overall rate of heat loss is the stimulus to which newborn hamsters respond in the thermotaxic orientation, and the effect of prolonged exposure to low ambient temperature on ultrasonic and audible sound emission by rat and hamster pups (Okon, 1971) supports this view. However, the short latency with which neonates respond to altered thermal conditions, particularly during nipple searching, indicated that thermal sensory receptors located in the snout, lips and face also play an important role (Welker, 1959; Leonard, 1974).

Neonates whose behaviour is organized in relation to thermotactile stimulation are necessarily confined to the close proximity of heat sources and their tactile representatives, and they must possess means to reaching these heat sources from short distances if they are displaced. The behaviour of neonates is, therefore, limited in spatial scope and in the range of thermal conditions under which they can function adequately.

As these thermal limitations are relaxed with the development of thermoregulatory mechanisms, neonates capable of extending the scope of their functioning, both spatially and with respect to environment conditions. Olfactory stimuli play an important role in this process. They arise from substances deposited by the mother in the vicinity of the home or nest and have special significance for the newborn, since they have been experienced in conjunction with thermotactile stimulation from the mother, and therefore they can provide means for the newborn to extend its scope of activity. The mother creates an olfactory zone around the nest or home (as well as in the home itself when she is absent) in which the newborn can function because of its growing responsiveness to olfactory stimulation. Moreover, odours are species and individual-specific they provide a basis for the specialization of responses to particular kinds of animals and particularly to the individual mother and littermates. Evidence of olfaction-based individual attachments by neonates have been reported for rats and kittens and will, very likely, be found in many different altricial species.

Home or nest orientation develops in relation to the differential distribution of odours in the vicinity of the nest or home site. It requires a responsiveness to gradients of olfactory stimulation or to orientated deposits which are the products of the mother's own position in relation to the home or nest as the centre of her maternal activity.

The maturation of vision increases even further the scope of the young animal's activity and the period of combined olfactory and visual functioning ensures that early visual functioning will be in relation to the most significant features of the young's social environment and within the zone of its social conditioned physical environment. The specialization of the young's response to qualitative features of stimuli (rather than simply than simply to quantitative features of the earlier phases) advances further with the introduction of vision. A wider range and greater variety of social signals are displayed visually than through odours or thermotactile stimuli. Play arises during the period of visual functioning and is based largely upon the young's response to the variety of visually perceived actions on the part of siblings and other familiar contemporaries.

Earlier sensory systems do not, of course, fall out of use when newer ones mature and become functional. They, do however, play a different role in behaviour than during the phase of their predominance.